D1071308

Methods for the Study
of Sedimentary Structures

Methods for the Study of Sedimentary Structures

ARNOLD H. BOUMA

Associate Professor
Geological Oceanography
Texas A&M University

WILEY-INTERSCIENCE
John Wiley & Sons
New York • London • Sydney • Toronto

Library of Congress Catalog Card Number: 69-19097
SBN 471 09120 0
Printed in the United States of America

FOR LIENEKE
AND MARK, NILS, AND LARS

FOREWORD

The study of the properties and processes of ancient and modern sediments, together with their inter-relationships, has expanded tremendously during the last few decades. This is understandable, since answers to a wide variety of fundamental questions in both pure and applied historical and structural geology, tectonics, engineering geology, soil mechanics, and geological oceanography depend on the results of these investigations. Quite recently the dramatic data obtained from deep drilling in the Gulf of Mexico and Atlantic Ocean, including the discovery of silicified turbidites at appreciable depths within the ocean sediments, has given added impetus to a more precise understanding of these inter-relationships. Achievement of this goal requires the use of diversified field and laboratory techniques of ever-increasing sophistication. Many already are available and more need to be developed or refined. Among those in existence many are not readily available, since they are either buried in the voluminous literature in several languages or have not been described or illustrated in sufficient detail to permit their effective use by others.

Hence it is the primary objective of this book to bring together and describe various techniques, processes, materials, and instruments which some of the foremost investigators in this field have developed and applied in their field and laboratory research. Emphasis is given to a number of the more recent and significant advances embodied in the use of radiography, sedimentary peels, and impregnation methods. These advances are described as well as evaluated in sufficient detail so that an investigator can, after judicious study, apply them to his research. Radiography has the unique advantage in analyzing cores and sediments of being a nondestructive method but still revealing certain diagnostic properties of a sediment sample that otherwise could not be determined.

The author brings to this book the unusual qualifications of not only being thoroughly familiar with the material discussed but of having a broad understanding of its specific use to help solve regional geologic problems. These qualities have been demonstrated in his publications in professional journals and in his textbook on the sedimentology of the flysch deposits, wherein an excellent graphic approach to facies interpretation is described. His wealth of experience gained aboard oceanographic research vessels and in the laboratories of leading oceanographic institutions in this country and abroad eminently qualifies him to evaluate as well as describe this wide variety of techniques, processes, materials, and instruments.

The extreme attention to detail is exemplified not only in the individual descriptions and illustrations of the various methods and equipment but in the extensive material contained in the five appendices. This treatment permits the reader to become proficient in their use and to learn exactly where they can be obtained, a matter of particular importance in a subject so highly specialized that it requires a wide variety of materials of high quality. The problem of language barriers presented by the many different journals in which much of the source material appears includes the use of different measuring systems and their units. This is solved by the presentation of the necessary conversion factors in Appendix IV.

It is not often that an investigator can find within a single volumn the precise information needed to conduct research in both pure and applied phases of such a specialized field of geology. Therefore this compendium will prove to be of tremendous value to a broad spectrum of earth scientists and engineers of varying degrees of professional competence and maturity. It will be valuable also to persons wishing to evaluate the validity of the research techniques used by those formulating recent hypotheses on the relationships between ancient and modern sediments and as these apply to solving regional geologic problems.

RICHARD A. GEYER, Head
Department of Oceanography
Texas A&M University

Vice Chairman of the President's
Commission on Marine Science
Engineering and Resources

January 1969

PREFACE

Friends asked me to compile a practical reference of the many techniques available and suggested that it be presented as an illustrated handbook outlining the methods, materials, variations, limitations, and applications that would be useful to anyone dealing with sediments.

Most of the techniques described in this book have been published in different languages, often in nongeologic literature. I have applied the majority myself to learn their limitations and to facilitate my writing. Because of the extent of the subject this compendium does not pretend to cover all techniques and sampling instruments.

Nearly all colleagues and manufacturers consulted agreed to review parts of the manuscript, and their reviews and comments served to improve the text considerably. However, I must take complete responsibility for the contents of the book.

Since no product is known that allows us to cement together grains of all sizes, under all humidity conditions, many products and methods have been devised to deal with various combinations of these factors. The introductory figure to each chapter is meant as a key to the selection of the proper techniques for a particular problem. Extensive author and subject indices are given for the same reason.

I should like to point out that there are some basic rules for obtaining good results and avoiding accidents.

1. Do not expect the best results on first application.
2. Work carefully. Pay attention to the required waiting times.
3. Nearly all techniques and apparatus were developed for a particular problem. Therefore methods should not be applied to other problems without testing.
4. Manufacturers should be consulted to determine whether a product is still available and bears the same name. The manufacturer can also supply the name and address of the local distributor.
5. Before you decide what method to apply read the introduction to each chapter (section) and the discussions of a group of related techniques.
6. Inaccurate sampling or preparation in the field never produces a satisfactory result in the laboratory.
7. Pay close attention to the protection rules described in this book, by local and regional safety officials and in brochures by manufacturers. Be sure that sufficient suction and ventilation are available in the laboratory space, since many products release vapors that are harmful to health.

To express my gratitude to all colleagues, manufacturers, and others who helped in the preparation of this book would require too many pages. I therefore must direct my thanks to groups rather than to individuals, even though I must acknowledge the inadequacy of this expression.

Many colleagues helped with details and illustrations concerning their methods. Some of them not only reviewed the part of the manuscript in which their work is discussed but also reviewed other parts or even chapters. Their names, except for W. R. Bryant, Miss J. E. A. M. Dielwart, A. A. Manten, R. E. Miller, L. Roskott, H. Schinkel, and B. Van Raadshoven, are mentioned in the text and in my opinion it is as impersonal to list them all as not to present them here.

H. Bakker, B. Dekker, F. Henzen, and Mrs. G. Merkel made the drawings and Z. van Druuten, J. H. Elsendoorn, F. Henzen, and H. Nyburg assisted with the photography and finishing of the figures. W. A. Burns, J. H. Elsendoorn, N. F. Marshall, and G. Ouwerkerk helped with field and laboratory experiments presented in Chapters 2, 3, and 4. Typing and correction of different parts of the manuscript was done by Mrs. M. H. Bouma, Mrs. C. Harper, Miss G. M. M. Klerk, Mrs. S. Lane, Mrs. L. Nixon, and Miss E. C. Sleeswijk; and Mrs. S. Lane read the complete manuscript for consistency of spelling. A word of special thanks only scratches the surface of my gratitude.

I am much indebted to Drs. F. P. Shepard, D. J. Doeglas, J. F. M. de Raaf, and R. A. Geyer for their encouragement and for permission to work on the book while employed at the Scripps Institution of Oceanography, the Geological Institute in Utrecht, and the Department of Oceanography of Texas A&M University. I want to thank Dr. Geyer for adding a foreword to this book.

I also wish to acknowledge the financial help of Hugh Courtright & Co., Dynamite Nobel, G. M. Mfg. & Instrument Corporation, Chemische Fabrik Stockhausen & Co., and Jean Wirtz. The work at Texas A&M University was carried out mainly under contract No. 14-08-011-10866 of the U.S. Geological Survey and contract NONR 2119 (04) of the Office of Naval Research.

Some text and several illustrations, collected from technical leaflets, were deleted to comply with the publisher's request to shorten the manuscript. I want to thank the staff of John Wiley & Sons for their assistance.

A special word of thanks goes to my wife for her untiring help.

If sedimentologists, geologists, pedologists, and coastal engineers and occasionally paleontologists, petrographers, flume experimentologists, and soil engineers can facilitate their work by applying some of the described methods, my efforts and time devoted to the manuscript have been worthwhile.

I would be grateful if colleagues would keep me informed of improvements and of new techniques.

ARNOLD H. BOUMA

College Station, Texas
January 1969

CONTENTS

Methods for the Study
of Sedimentary Structures

CHAPTER ONE

SEDIMENTARY PEELS

A sedimentary peel consists of a thin layer of material removed from the surface of an unconsolidated or consolidated sediment face by applying an adhesive, which, when dry, is peeled off. This thin peel of original material is collected without disturbing the arrangement of the sedimentary properties. It often has many advantages over a photograph, which may miss proper perspective not only with regard to scale but also with regard to grain characteristics such as size, shape, and color.

When studying outcrops of unconsolidated deposits often some relief is observed since grains are differentially blown out by wind. By applying the peel method we are able to obtain a similar relief in which the depositional and secondary structures come out clearly. Even if no relief is visible in the field, it may appear in a peel.

Lacquer peels are made for several purposes. They can be studied at home with no risk of the original sediment surface disappearing. Generally we may observe more details on the peel than in the outcrop by applying artificial illumination to obtain the most favorable result; a regular study reveals more data than an occasional visit at the outcrop.

Peels can be used as demonstration material for training students as soon as a collection is made in which different sedimentation conditions are presented. Empty walls in a geological institute can be covered by peels to form a suitable and fine decoration and, at the same time, a permanent exhibition.

In geology, prehistory, and archeology it is often necessary to conserve profiles of loose material and cultural strata for museum exhibitions or for instruction purposes (Hähnel, 1961, 1962).

Special techniques such as Araldite peels exaggerate the natural relief to obtain from an unconsolidated sediment a hard film in which laminae stand out in millimeters or centimeters, thus allowing the investigator to study a three-dimensional result. These peels with high relief can only be made of unconsolidated sandy or coarse silty material.

Lacquer peels are normally made from gravelly, sandy, or silty material which is in an unconsolidated or poorly consolidated state and which is not excessively wet. From shaly and other rocky material, which is broken into pieces by schistocity or other influences, lacquer peels can be made as long as rock pieces can be chiseled off. Wet sediments and clays can be treated only after some laboratory preparations have been made on collected samples. Also, peat and clayey soils should be sampled and brought to the laboratory before the peel technique is applied. Only using the methods of Maarse and Terwindt (1964), McMullen and Allen (1964) and McKee (1957a,b) can moist or dry clays and wet sands be treated in the field.

Peels can also be obtained from hard rock. These nitrocellulose and acetate peels do not have any relief, but on the other hand they often reveal much more than can be seen in a polished surface.

To date, no single product is known which enables the investigator to consolidate all types

1

of unconsolidated material, e.g., wet or dry, coarse or fine, or to obtain a peel with or without relief. This is the reason why so many different techniques have been developed.

Different types of peels will be described successively (Fig. 1.1). Since it may be difficult to find a suitable technique for a certain peeling problem, the reader should first consult Table I.1.

1.1 LACQUER PEELS

Lacquer peels can be made in the field from unconsolidated or very poorly consolidated (not too fine-grained) sediments as long as they are not too wet. A certain moisture content is favorable. Very fine-grained material and wet deposits should be sampled and treated in the laboratory (see Section 1.1E). In using wet deposits one is limited by their size. It should be remembered that peel surfaces above 1 or 2 m^2 are difficult to handle in the field, and many hands are necessary to recover the final product.

The method of making lacquer peels was essentially developed by Voigt in 1932 to solve the problem of salvaging vertebrates from German Eocene lignite deposits. Application of lacquer made it possible to obtain not only a better preservation of the material but also many non-skeletal animal parts that could be studied, such as the skin of a frog, the hair of vertebrates, red blood corpuscles of a lizard, worms, soft parts of insects and vertebrates (Voigt, 1933, 1934, 1935a, 1936a, b, 1937a, b, c, 1938a, c, and 1939; Fenton, 1935). These peels were well suited for study under the microscope, using the oil immersion for high enlargements (Voigt, 1935a, 1936a, b, 1937a, b, c, 1938a, c, and 1939).

Besides the application within the field of paleontology, the peel technique has been applied to unconsolidated sediments (Voigt, 1935b, 1936c, d, e, 1938b, 1947, 1949a, b). In criminal investigations the lacquer peels proved to be a useful tool (Voigt, 1938d).

Many studies describing peel techniques have been published; only a few will be mentioned. Many of these papers are introductions of a known method in journals that reach differently

oriented groups of readers, or they contain descriptions on different applications of the technique. The majority of these contributions come from the field of soil sciences.[1] Since the lacquer peels form extremely good demonstration material, contributions also come from museums (Jessen, 1938b; Pozaryska, 1958; Meyer, 1960).

Colorful peels have been made from interglacial deposits. The variation in the aspects of the structures after the periglacial deformations, together with color contrasts, make this group of sediments very suitable for application of this technique (Voigt, 1936c, d, e, 1938b, 1947, 1949a, b; Foucar, 1938; Hospers, 1950; Hähnel, 1961, 1962). Häntzschel (1938) introduced the method into the field of marine geology, while Jessen (1938a), Seidel (1951), and Kuenen (1961) used the lacquer peel in the study of fossil sediments.

From Table I.1 it can be seen that the type of sediment, the weather, and outcrop conditions all have their influence on the preparation of the lacquer peel. The variations in these types and conditions are so numerous that they cannot all be discussed. Training in different types of deposits and under various types of weather will help to build up the experience necessary to obtain optimum results.

1.1A Spraying Method

The spraying method may have some advantages over the pouring and brushing methods, especially when some routine is lacking. Basically there is no difference in the technical approach, and it is possible to obtain identical results. The spraying method requires a little more equipment, but less lacquer is used per surface unit.

Basic Procedure for Making Lacquer Peels by the Spraying Method

1. Equipment. Check the equipment before going into the field. Failures may occur if something does not operate during the preparation of a film.

2. Preparation of the Section. Cut a smooth, flat surface with a shovel or a spade. Smooth it carefully with a trowel. The moisture content of

[1]Harper, 1932; Grosskopf, 1937; Troedsson, 1938; Lyford, 1940; McClure and Converse, 1940; Orviku, 1940; Berger and Muckenhirn, 1945; Römer, 1951; Tanis, 1952, 1954a, b; Kullmann, 1953, 1954; Hulshof, 1955; Schuurmann and Goedewagen, 1955; Tüxen, 1957; Jager and Schellekens, 1963; Franzisket, 1962a, b; Van der Plas and Slager, 1964; Slager, 1964, 1966.

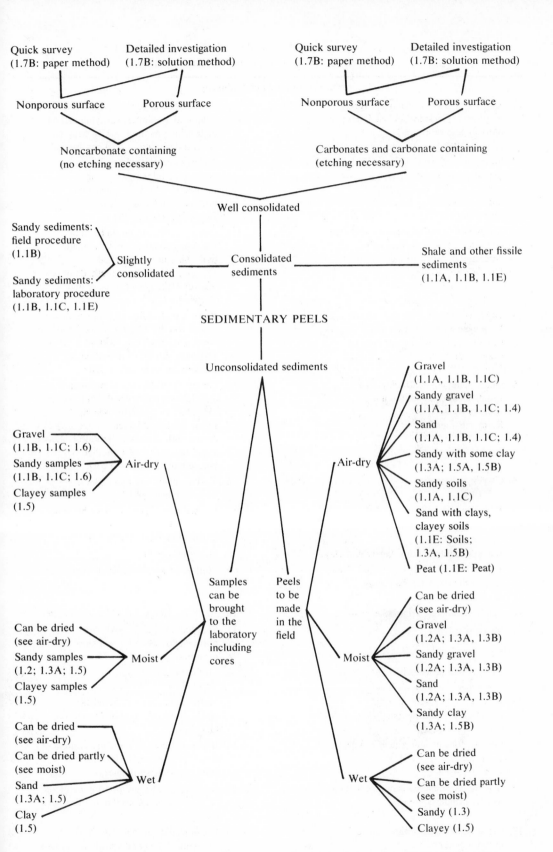

Fig. 1.1. General outline for the different techniques on sedimentary peels as described in Chapter One.

Table I.1. Key to the Different Peel Methods Applicable for
Certain Types of Sediment under Certain Conditions

A. Unconsolidated Sediments

1. Sandy and gravelly deposits. Relief of large surfaces desired. Sediment more or less dry. Dry weather.

   1. Nitrocellulose lacquer with slow evaporating thinner: spraying (1.1A) or brushing method (1.1B). May take several days before peel can be collected.

2. Sandy and gravelly deposits. Lithological composition of large surface wanted. Sediments more or less dry. Dry weather.

   2. Nitrocellulose lacquer with acetone as thinner: spraying method (1.1A). Can be carried out in less than half a day. Instant peel method (1.4)

3. Sandy and gravelly deposits. Lithology and relief of large surface to be determined. Moist to wet sediments. Dry or rainy weather.

   3. Drying with acetone (lightening). For rainy weather: build a shelter over the outcrop. Methods: see under 1 and 2, or methods of McKee (1.2B) or of Maarse and Terwindt (1.3A).

4. Sandy sediments with clay layers. Lithological composition and some relief of rather large surface to be determined. Dry weather.

   4. Be sure that sediment is not too wet or too dry (drying with acetone or wetting with water). Method of Maarse and Terwindt (1.3A); method of Heezen and Johnson (1.5B).

5. Sandy sediments with clay layers. Lithological composition and some relief wanted. Rainy weather.

   5. Sample with the box from Jager and Schellekens (very difficult) (soils). Use plastic or glue for peeling. In the field: build shelter after drying with acetone; method of Maarse and Terwindt (1.3A); method of Heezen and Johnson (1.5B).

6. Clayey sediments.

   6. Take samples and make glue peels at home (1.5).

7. Sandy and gravelly cores.

   7. Let sediment dry to correct humidity content and use pouring (1.1C) or brushing method (1.1B).

8. Wet sediments in core. Sandy and clayey.

   8. Glue peels (1.5).

9. Moist and wet sandy sediments. Short time available. Must be carried out in the field.

   9. Method of McMullen and Allen (1.3C).

10. Sandy sediments. High relief wanted.

    10. Araldite-, Epocasts- and paraffin peels (1.6).

11. Sandy dry soils.

    11. Spraying method (1.1A) with nitrocellulose lacquer and slow evaporating thinner.

12. Clayey- and peaty soils. Wet soils.

    12. Method of Jager and Schellekens (1.1E, soils).

13. Peat. Very fine detrital peat in wet clayey sediments.

    13. Method of Jager and Schellekens (1.1E, soils). Method of Franzisket (1.1E, peat).

14. Slightly consolidated sandy sediments. Large surface required.

    14. Be sure of correct moisture content. Brushing method (1.1B) with slow evaporating thinner.

15. Slightly consolidated sandy sediments. Mainly samples.

    15. Brushing (1.1B) and pouring methods (1.1C). See half-consolidated deposits (1.1F).

16. Consolidated shales and other fissile sediments.

    16. See consolidated deposits (1.1E). Spraying (1.1A) and brushing methods (1.1B).

Table 1.1. (Continued)

B. Consolidated Sediments. Samples

1. Noncarbonate containing. Quick survey. Polished surfaces do not have small holes.	1. No etching (if necessary with HF). Kodatrace or similar acetate sheet material or celluloid (1.7B: paper methods).
2. Carbonate containing. Quick survey. Polished surfaces do not have small holes.	2. Etching with hydrochloric, formic, or acetic acid. Kodatrace or similar acetate sheet material or celluloid (1.7B: paper methods).
3. Noncarbonate containing. Detailed investigation. Polished surface may have holes.	3. No etching (if necessary, use HF). Acetate or nitrocellulose solution (1.7B: solution methods). Surfaces with holes need a double pouring.
4. Carbonate containing. Detailed investigation. Polished surface may have holes.	4. Etching with hydrochloric, formic, or acetic acid. Acetate or nitrocellulose solution (1.7B: solution methods). Surfaces with holes need a double pouring.

Remarks: The investigator should bear in mind that all peels are mirror images of the original structures from which they were abstracted; consequently it should be taken into consideration that left and right are exchanged, especially if one wants to interpret former transport directions.

the sediment should not be excessive; spraying the surface with a thinner and igniting it proves to be successful in removing moisture. Deposits that are too dry must be moistened.

3. Lacquering. Spray against the section a mixture of lacquer and thinner with a sprayable maximum viscosity. Repeat this two to four times.

Mark the margins of the peel by four corner nails and a thin, strong rope stretched around them. Push U-shaped pieces of wire into the sediment to press the rope completely against the sampled surface.

Cover the peel section with patches of lacquer, using a flat, soft paintbrush and a mixture of lacquer and thinner (1:2 or 1:3 parts). When the patching is completed brush the same mixture on the entire surface.

4. Strengthening of the Back of the Peel. Paste strips of bandage or cheesecloth on the peel section with slightly diluted lacquer. The strips should overlap one another. Next apply a coating of undiluted lacquer.

5. Collecting of the Peel. Let the peel dry for ½ to 2 hours or overnight. Write the location, orientation, and other data on the back of the peel. Cut the sides of the peel loose with a knife and remove the peel carefully.

Place the peel horizontally on its back away from direct sunlight and let it evaporate.

6. Transport. The peel should be transported horizontally or widely folded.

7. Drying and Mounting. After approximately 2 days the peel is ready to be mounted on masonite or fiber board. Use relatively slow-drying adhesives and place sandbags upon the peel as weights, or apply fast-drying adhesives, which do not require any weights. A frame can be made, or strips of wood can be mounted at the back of the peel to avoid bending.

8. Restoration. Brush small areas with diluted lacquer and sprinkle the corresponding sediment on it. The finer the grain size, the greater the dilution of the lacquer should be. Larger areas to which a dye has been added can be painted with lacquer or plastic.

9. Finishing. A protective coating should be sprayed or brushed on the peel. It should not be allowed to dissolve the lacquer of the film and damage the relief.

10. Storage. To keep the peel in good condition, it should be hung on the wall or placed in a special rack.

Comments

These points give the main process of the spraying method based on normal, unconsolidated, sandy sediments. In Sections 1.1D and 1.1E, weather conditions and types of sediment are discussed.

Details Concerning Equipment and Procedure

Equipment. Spade, shovel, trowel, spray can, lacquer, thinner, mixing beaker or tins, spoon, brush, nails, pins, wire, pincer, rope, scale, gauze or cheesecloth, scissors, marking ink, compass, normal knife, spatula, long knife, and wind-screen (Fig. 1.2).

The spade and the shovel are necessary for digging the sediment away to obtain a fresh surface and to flatten it roughly afterward. For sandy material a shovel is preferable; the smaller spade is useful in more resistant material.

Trowels vary considerably in shape. In Utrecht, the Netherlands, two types are used: a longer one, which is a manufactured type, and a smaller one, which is cut into an egg-shape (Fig. 1.2). The longer one has the advantage of making a nice, flat surface and does not cut easily into the wall; the egg-shaped one is very handy for touching up small parts around small obstacles such as shells and pieces of wood. To make the work easier and to avoid rubbing the surface off, the sides of the trowels are sharpened toward their base.

Small compression sprayers, as used in yards and gardens to control insects, diseases, and weeds, are best for this technique. The only disadvantage of modern equipment is that the thinners often dissolve some of the nonmetallic parts. Spare parts should be included in the equipment, and hoses of synthetic material should be replaced immediately by metallic pipes. Two types of spray gun have been tested and proven to be very successful after the aforementioned changes were made. The apparatus must be easy to handle and clean, not too heavy and of a small size because of the small amounts of mixture used.

The galvanized steel Clipper sprayer of the Hudson Company (Hudson Clipper Compression Sprayer No. 6215) with a multispray nozzle weighs 6.5 lb (2.95 kg) and has a capacity of 2 gal (7.57 liters) (Hudson, 1961).

The spraygun used at the sedimentological department at Utrecht is a Gloria plant and flower disinfectant sprayer from Kunde and Cie

Fig. 1.2. Equipment used for the spraying method: (A) spade; (B) shovel; (C) spray gun; (D) mixing and measuring beaker with stirring rod; (E) lacquer; (F) lacquer; (G) thinner; (H) acetone; (I) long knife. Front row from left to right: two trowels, thin rope, bandage, scissors, spoon, spatula, paintbrush, scale, knife, nails, marker, pieces of bent wire, compass.

(No. 70; height 36 cm, width 21 cm, diameter 14 cm, weight 2.2 kg). The capacity is 2 liters (Kunde and Cie, 1964). The reservoir should never be completely filled since some air space is necessary.

The lacquer must form a firm, nonthermoplastic film that is not brittle and fragile, but flexible and ductile and rather quick drying (Hähnel, 1961, 1962). The lacquer must be colorless or lightly colored in order to prevent unnatural tints. Numerous lacquers are on the market, but only a few of them are specially developed for this purpose and therefore suitable for this work. Most of these products belong to the group of nitrocellulose lacquers. Only those known to the author will be discussed. Note that a spray gun filled with lacquer and thinner should not be left under pressure when placed in the sun.

The 1202 Insulating Varnish of the General Electric Company, Insulating Materials, is a clear, fast air-drying, synthetic-resin type of lacquer, which produces a hard coating. Its viscosity, about 30 cps (centipoises) at 25°C, is rather low for this group of products. As a solvent the manufacturer mentions either xylene or thinner no. 1500, but acetone or a mixture of acetone and toluene (3:1) can be used when necessary. The color of this lacquer is rather dark, but it does not harm the natural tint of the sediment (General Electric, 1958a).

General Electric 1557 Glyptal Clear Lacquer has an alkyd-nitrocellulose compound with a clear, very slight yellowish color and a higher viscosity than no. 1202. It is soluble in G-E D5 B19 lacquer thinner or in alcohols, esters, and hydrocarbons (General Electric, 1958b).

Tornol Profile Lacquer A from the Fil Company is a clear nitrocellulose lacquer, which has been specially developed for the preparation of lacquer peels. The viscosity is rather high. The official thinner is V 105, but a fairly good substitute can be made from acetone and toluene (1:1). This lacquer with its thinners is a slow drying mixture that takes 2 or 3 days to make a peel, but the relief obtained is often superior to those made by other lacquers.

The Fil Company developed a new lacquer (Tornol Profile Lacquer B), which is a clear, fast-drying, rather flexible lacquer. Thinned with the acetone-toluene (1:1) mixture or with pure acetone, it will yield results that are nearly similar to those obtained by applying the German Z4/924 lacquer.

At the request of Professor Voigt, a special lacquer (Spezial-Präparationslack Z4/924) has been developed by the Gustav Ruth Temperol-Werke. The geological institute at Hamburg, Germany, uses this for a variety of sediments and even for shales. It is a medium viscous, clear nitrocellulose lacquer, which gives a firm and flexible peel, which can be rolled up for transport.

The original German combination is not always favorable in regard to relief, since the thinner evaporates too quickly. If more relief is desired, the manufacturer advises the user to add a slower thinner such as toluene, or to make one which is rather similar to their slow-working thinner Z4/948 from acetone and butylglycol (30% butylglycol).

Since the right kind of lacquer may not be available everywhere, a lacquer solution can be prepared by dissolving celluloid film, or other such material, in acetone (Voigt, 1949b). Additional material must be dissolved in the acetone until the desired viscosity has been obtained. This inexpensive lacquer is in use at the geological institute at Kiel, Germany.

Several "Mowilith" products from the Hoechst Company are very suitable for making lacquer peels. They are all based on vinyl acetate. The polymerized product is colorless and thermoplastic.

Mowilith 20 and Mowilith 30 are both dissolved in ethyl acetate (60% in solution). Mixed with nitrocellulose adhesives, even in small quantities, the adhesive strength of the nitrocellulose products can be improved considerably (Hoechst, 1959, 1960a, b, 1965a).

Mowilith 35/73 can be obtained in the form of blocks with a slight yellow color. A 20% solution in ethyl acetate at 20°C has a viscosity of 5 to 20 cps while a 20% solution in ethanol has a viscosity of 15 to 35 cps. This product is soluble in many other organic solvents, such as methyl and butyl acetate, acetone, methylglycol, ethylglycol, and butylglycol, benzol, toluol, and xylol (Hoechst, 1964, 1965b).

These Mowilith products are also very successfully applied in finishing the peels.

In the early contributions, Voigt (1947, 1949a, b) also mentioned the Geiseltal-Sprimoloid-Lack from the Springer and Moeller Co., Leipzig-Leutzsch, Germany. Franzisket (1962a) and his group apply Kleblack Hooco W. of the Chemische Fabrik Kossack, Düsseldorf, Germany. Römer (1951) prefers a lacquer with stronger binding properties. Herrnbrodt (1954) applies Capaplex.

To strengthen the back of the peel and to prevent breaking, the operator should paste cheese-cloth, gauze, or bandage to the peel. Rolls of 10 to 20 cm (4 to 8 in.) wide are easy to handle.

The peel may become detached from the wall because of shrinkage or may have a tendency to slide down, or wind action may undermine the film. Nails, pins, and rope are used to mark the margins of the future peel and to hold it in place. This should be done especially for pure sandy deposits. The nails must be long enough to stay in place. It is useful to have a variety of lengths at hand, ranging from 5 to 15 cm (2 to 6 in.). The rope must be thin but strong. The pins are hair-pins or pieces of thin, strong wire bent in a U-shape. Normally they are shorter than the nails. It is practical to add to the equipment a roll of wire and a small pincer.

To make the lacquer mixture one can use mix-ing cans or normal empty tins. For the brushing operations a flat paintbrush (about 5 cm or 2 in.) is the only suitable brush. Marking ink and com-pass are used to indicate position, orientation, and other data relevant to the peel.

An ordinary knife, well sharpened, is neces-sary to cut the peel loose before removing it from the wall. A long knife, made of a strip of metal to which a piece of pipe is welded as a grip, and sharpened on both sides, serves as an extension of the operator's arm where the spat-ula is too short to loosen parts of the peel that are sticking during removal.

Preparation of the Section. The preparation needed depends completely on the lithology and position of the section of which a peel is to be made. In general, vertical sections are desired; sometimes horizontal ones are desired.

It is necessary to dig away enough of the sedi-ment so that the new fresh surface does not con-tain completely dry, sandy layers, desiccation cracks, insect holes, young roots, or other prop-erties not related to the environmental condition of the deposit.

The wall must be scraped to get an even sur-face, which is on all sides 10 to 20 cm (4 to 8 in.) larger than the future peel. The wall should not be vertical, or have overhang, but should have an inclination of about 75 to 80°. If there is loose sandy material above the section, it has been proven successful to dig out a step above the upper side of the peel, at about 10 to 20 cm, to prevent loose grains or clusters of the sediment above the section from falling down and damag-ing the peel surface during the preparatory stages. The flatter and the smoother the surface

is made, the better the resultant peel becomes (Fig. 1.3).

Gravel, stones, concretions, shells, roots, and faults are difficult obstacles and tend to cause holes in the surface. Such sections can be smoothed roughly with the shovel and then with the trowel. If holes are made accidentally, they have to be enlarged in order to form a platelike depression, so that lacquer can easily be sprayed or brushed on each part of it. The obstacles that stick out can either be left in place or they can be removed and cemented later into the corre-sponding holes of the peel.

The degree of humidity of the smoothed sur-face is a very important factor. The lacquer rarely penetrates into the pores between the grains without destroying the surface if the sedi-ment is too dry, but when the material is too wet, no pore space remains for the lacquer to fill up. Only through experience will the operator know what choice to make.

Dry sediment must be sprayed with water to moisten the material slightly to approximately 2 cm (¾ in.) deep. This must be done slowly and carefully to prevent the grains from falling due to the spraying action. The sediment should be moist enough to prevent grains from falling down during the time that preparations for the lacquering are made. The humidity is right if the surface can be scraped off without causing disturbance.

Wet sediment must be dried. The easiest method is to spray a thick haze of acetone or thinner onto the surface and to ignite it. Care should be taken that no flammable liquids are close to the section. This spray-igniting opera-tion can be repeated three or four times and has to be stopped when parts of the wall start to burn themselves or when some baking effect is visible. The operator can stop earlier with the application of pure acetone or thinner and can continue with a mixture of one part lacquer and five parts of acetone or thinner, which is also set on fire. The lacquer penetrates and forms a bar-rier against the water in the pores. Caution must be exercised with these pyrotechniques; the vapors are heavier than air and may still burn in small holes.

It is important to make a section at the place where the operator can reach each spot. Highly placed or long (in a vertical sense) peels can be made if a ladder is available. If possible, never make a peel at a place where too much wind or direct sun (during hot days) can influence the

Fig. 1.3. The sedimentary section is scraped smooth and ready for making a lacquer peel. Scale is 1.20 m. Estuarine deposits. Excavation in the Haringvliet, The Netherlands. Subrecent. (Photo by M. Vanossi, Geol. Inst., Pavia, Italy.)

first stages of the work. Direct sun may cause shrinkage and consequently breaking of the film. During colder periods the sun is needed for drying the lacquer.

Lacquering. A mixture of lacquer and thinner is sprayed as a fine haze onto the flattened surface from a distance of about 1 to 2 m (3 to 6 ft). Do not direct the gun on the section first, but direct it next to the section in order to check that it is working properly before touching the section. It is important not to concentrate the spray in any one spot too long. A method which works well is to first apply a very superficial spray (Fig. 1.4), wait for 5 min, and then apply a more thorough treatment. This can be repeated two more times, each time allowing the previous spraying to dry before starting again. Hähnel (1961) uses a mixture of one part lacquer with three to four parts acetone to obtain his spray mixture (lacquer Z4/924 from Gustav Ruth). The author prefers to work with the maximum viscosity that can be sprayed well.

The first coat fills the holes and penetrates into the sediment. Depth of penetration varies from sediment to sediment, since it depends on several factors (size distribution of the sediment, sorting, porosity and permeability variations, humidity, etc.). The differences in porosity cause differences in depth of penetration, which later comes out as relief on the peel.

When it is necessary to dry a slightly wet section by igniting the sprayed solvent, it is best to spray the lacquer mixture as soon as the fire is dead, to avoid progress of pore water toward the surface.

The next step is to mark the margins of the peel. Four nails are pushed partly into the four corners of the future peel. Use a measuring stick to equalize the lengths of corresponding sides. Tie a thin, strong rope around the four nails, and press or hammer the nails in completely, being certain not to disturb the sample surface. Generally the wall is not absolutely flat and the rope does not always touch the sediment. Hairpins or U-bent pieces of wire are pushed into the surface around the rope in order to get the thread at

Fig. 1.4. The first coat of lacquer mixture is sprayed onto the section. Same location as Fig. 1.3. (Photo by M. Vanossi, Geol. Inst., Pavia, Italy.)

any point to the section. Even if the rope is close to the sediment over a long distance it is good to push in U-pieces at regular distances (about 15 cm or 6 in.) (Fig. 1.5). This frame limits the size of the peel during following operations and thus prevents unnecessary use of lacquer. Furthermore, the layer of lacquer shrinks as a result of the evaporation of the solvent and therefore has a tendency to become detached from the wall, which may result in an unexpected peeling off of the unfinished peel.

The peel area is then covered with patches of lacquer (Fig. 1.6), using the flat, soft paintbrush and a diluted lacquer (1 to 3 parts of thinner against 1 part of lacquer until a viscosity of about 75 poises has been obtained). It is most convenient to begin at the top. Be sure that no drops run down along the surface, since they can later be seen on the peels as vertical, unnatural rims (Fig. 1.7). Once the bottom is reached, the patches at the top are sufficiently dry to start again. It often takes two or three tries to cover the whole surface.

When this first patched lacquer coat is dry,

brush a new, identical mixture on the surface, the rope, and a little beyond. Since the thinner will dissolve the first lacquer coat on contact, be careful not to touch each area more than one or two times with the brush in order to prevent damage.

By using only acetone as a thinner, the evaporation may be so fast that the dew point is reached and vapor precipitates against the lacquer, giving it a permanent white color. Also, the lacquer coat breaks easily because of shrinking. These splits must be repaired quickly with lacquer. Often this may not be possible since pieces of lacquer break off. It is best to paint pieces of gauze or cheesecloth over the cracks. For larger restoration, it is necessary to spray the surface before a brush can be applied.

Strengthening of the Back of the Peel. When the surface of the first complete lacquer coat is dry, the back of the peel must be reinforced by painting gauze, bandage, or cheesecloth strips over it. The strips are pasted onto the section with slightly diluted lacquer. Care should be

Fig. 1.5. (*Upper left*). Part of the sprayed section is used for a lacquer peel. Nails (A) are pushed in the four corners. Thin rope (B) is tied around the nails. U-bent pieces of wire (C) are pressed onto the sediment around the rope to get the thread to the section at any point. Same location as Fig. 1.3. (Photo by M. Vanossi, Geol. Inst., Pavia, Italy.)

Fig. 1.6. (*Upper right*). The marked area of Fig. 1.5 is covered with patches of lacquer, using a flat, soft paintbrush. (Photo by M. Vanossi, Geol. Inst., Pavia, Italy.)

Fig. 1.7. (*Lower left*). Lacquer peel from estuarine deposits with unnatural vertical rims formed by the excess of lacquer from running down the sediment face (Lacquer Tornol A with thinner V105). Excavation in the Haringvliet, The Netherlands. Subrecent. (Photo by H. Nyburg and J. H. Elsendoorn, Geol. Inst., Utrecht.)

Fig. 1.8. (*Lower right*). Strengthening of the back of the peel with vertical bands of gauze. Lacquer mixture is now pasted to the upper left side for hanging on a new strip. Excavation in the Haringvliet, The Netherlands. Subrecent.

taken to press the gauze into depressions. Protruding parts of the wall are first covered with small pieces of cloth. The strips of gauze should overlap one another and the rope should also be completely covered. It is advisable to place the gauze in vertical rather than horizontal strips (Fig. 1.8). Hähnel (1961, 1962) recommends first pasting a 10-cm-wide bandage around the edges of the peel and then filling up the rest with strips of 20-cm width. When this new lacquer coat is dry an additional coat of undiluted lacquer can be applied to the peel to secure a strong support.

It is difficult to give the exact quantities of lacquer required. This depends primarily on the coarseness of the sediment and the viscosity of the lacquer. The spraying operation generally requires about 1 to 1.5 liters of lacquer per square meter. For spraying, the author prefers a mixture with maximum viscosity, which will give a fine haze, whereas the brushing mixture should be thicker.

Collecting of the Peel. The peel must be absolutely dry before it can be removed. The drying time of the last coat may be ½ to 2 hours during dry weather when acetone is used as thinner. One should wait longer, possibly overnight, before collecting the result, especially when the sediment is still moist. However, if it is left overnight, there is a risk of the peel falling because the nails are not long enough, or the danger of some unqualified person destroying it.

The location, orientation, inclination, and other data can be written on the back of the peel with marking ink. Along the framing thread the peel can be cut off with the knife. The nails and pins along the bottom and the sides are loosened or removed, leaving the top untouched. If there are enough hands available, it is best to keep the top in place and remove the peel from bottom to top, especially when there are clay layers present in the section. Starting from the top, loose sediment often forms an undesirable piston on top of the part that still hangs at the wall. Loosening of the peel from the wall of gravelly or nonpure sandy peels must be done very carefully. A long knife (Fig. 1.2) or a long spatula is used to cut resistant clay bands.

When the entire film is loose it is placed horizontally on its back away from the direct sunlight. The inner side of the peel is now allowed to dry. Next, the peel is lifted to remove loose grains, after which missing pieces can be cemented in place either immediately or later in the laboratory. Samples should be collected from the different levels to be used for later restoration.

Transport. Voigt (1949b), as Hähnel (1961, 1962), rolls the lacquer peel with the sediment side of the peel on the outside around a carton roll of at least 10-cm (4-in.) diameter for easy transport.

Since this method of transport often disturbs much of the relief, the author prefers transporting the peels in a horizontal position. If the peels are too large, they can be folded in large folds with many old papers and cloth in the folds.

Transport panels, such as used at the institute in Utrecht, are made of 2-cm (¾ in.)-thick fibrous wood and are 100 x 80 cm (39.5 x 31.5 in.) in size. Underneath, two 5 x 2½ cm (2 x 1 in.) wooden strips are fixed along two sides.

If the peels are transported in a car without a separate section for the operators, it is important to ventilate the car well.

Drying and Mounting. As soon as the peels have arrived at the laboratory they must be laid down, sediment side up. To prevent the sides from curling during drying, heavy weights such as sandbags, pieces of wood, or metal can be placed on them. A flat film of glue is required when the peels are to be mounted without using a peel softener. In spite of the adhesive, a flat peel is much easier to mount.

The peels should be mounted on masonite or fibrous wood within 1 or 2 days since the film, due to irregular shrinkage, has an unnatural relief. The masonite plate should be a little larger than the size of the peel. The lower side of the peel can be cut with a strong knife (Stanley knife) in order to get a straight side, underneath which the legend can be written. It is also possible to cut all the sides of the peel before mounting; this is done especially when the peel is to be framed. When slow-drying adhesives are used such framing is necessary.

Hähnel (1961, 1962) covers the fibrous board and the back of the peel with a Mowilith-30 solution or a Vinnapas solution. Since these products contain solvents that slightly dissolve the film, this procedure should be carried out quickly. Small peel sizes (about 0.5 m²) are pasted all over and simply placed on the wood plate. Longer films are first placed in position on the fibrous board, and while one end is anchored with sandbags, the other end is flipped over and placed over a roll. Piece by piece the wood and the peel are pasted, and due to the rolling motion, the film is attached to the board. The edges are weighted and the entire surface covered with paper or cloth and then left for ¼ hour. The film becomes soft and possible air bubbles underneath the peel can be squeezed out by hand.

The application of these and other slow-drying adhesives require sandbags on the peel in order to get good results. Before placing the weights, a frame can be placed and nailed to the board. When working with masonite the peel should be turned over and put on a piece of foam plastic to prevent damage. It is also possible to put the frame in place without nailing it by putting weights on it. It is important to have the edges mounted flat if the frame is fixed to the peel afterward in order to avoid cracking.

By using a fast-drying product an easier method is possible. The peel is turned over carefully, dried, and cleared of grease and dirt, and then a somewhat diluted adhesive is painted on its back. The viscosity should be so low that it is not necessary to force the brush. The solvent weakens the film without damaging the relief. The dry, grease- and dirt-free back of the masonite is covered with undiluted or slightly diluted glue. When both of the glued sides have dried until tacky (after 10 to 40 min), two men turn the peel over and hold it in place above the masonite. Care should be taken in positioning the film above the masonite, since it cannot be moved once stuck together. The trimmed end is mounted first, and while one man keeps the peel from the masonite, the other presses the film onto the hard board with a piece of foam plastic, squeezing out as much air as possible. Then the entire peel is pressed gently, using a piece of foam plastic to prevent damage to the relief. The mounted film can now be placed flat on the floor. It takes several hours before all the solvent has evaporated.

Since mounted peels tend to bend, an additional reinforcement with wooden strips on the front side as a frame, or on the back side, might be useful. If the peel can be hung vertically on the wall or in a rack, this reinforcement is not necessary. However, a very slight bending may still occur. The mounted peel can be trimmed to size on a circular saw.

If the peel is to be hung in a special place, as a decoration or a display, the results can be improved by sawing all four sides of the peel into shape and gluing it on a piece of wood of sufficient thickness (16 to 18 mm), the same size as the peel. Along all the sides and underneath the film, strips of veneer should be glued, polished

with fine-grained sandpaper, and oiled (Fig. 1.9).

When mounting a peel be careful in using the same type of lacquer that is used in the preparation of the film. This will dissolve the old lacquer and damage the relief. Very good, relatively slow-drying glues are the polyvinyl-acetate solutions, such as Mowilith, Vinnapas (Hähnel, 1961), and Cetaflex.

Vinnapas B 17 from the Wacker Company has a low viscosity and is soluble in such solvents as ethyl acetate, acetone, butanol, benzene, toluene, and turpentine (Wacker, not dated a, b, c). From the extensive series of the thermoplastic synthetic resin polyvinyl acetate, which are colorless, odorless, tasteless, and nontoxic, the manufacturer recommends B 17.

The Hoechst Company includes in their Mowilith series such products as Mowilith DMC 2, Mowilith DV, Mowilith 25, and Mowilith 30 as good adhesives (Hoechst, 1959, 1960a, b, 1965a).

The Ceta Bever Company produces a glue based on polyvinyl acetate dispersions under the name Cetaflex. From this series the Cetaflex "snel" (fast) and Cetaflex "extra snel" (extra fast) types are best suited for this work (Ceta Bever, not dated a, b, c).

The Glyptal no. 1276 cement from the General Electric Company, Insulating Materials, is a rather fast-drying adhesive which does not weaken the peel but allows the two sides to be glued directly together (General Electric, 1963). Weights must be placed on the peel for about 20 to 30 min.

At the sedimentological department in Utrecht, fast-drying glues such as A-33, Bison Kit, and Snelfix are preferred. A-33 is a synthetic resin of the polyvinyl type from Ram Chemicals; it is a very viscous fluid and has a tack-life of about 2 to 5 min, depending on temperature. For best results the clear adhesive should not be diluted. Aromatic, alcohol, and ketones can be used for cleanup purposes. Allow the adhesive to dry briefly until tacky. If it becomes too dry, lacquer thinner can be applied to the surface (Ram, 1960, 1965). Bison Kit from the Perfecta Company is a glue made of synthetic rubber (polychloroprene), synthetic resins, metal oxides, anti-oxidant, and solvents. A special Bison Kit Thinner is used to lower the viscosity. The application is similar to that of the A-33 (Perfecta, 1964). Snelfix from the Ceta Bever Company is a glue made on the base of neoprene (a synthetic rubber). Snelfix is similar to Bison Kit (Ceta Bever, not dated d).

Fig. 1.9.   Lacquer peel of fluvioglacial deposits, mounted on masonite and then on fibrous wood. The sides (see separate photograph) are covered with strips of veneer. Beneath the peel, veneer is pasted for text. The upper and lower parts of this peel are part of a large block which has been tilted in frozen condition. Later a nearly horizontal solution hole is formed and filled by glacial rivers. Abandoned quarry at Tappendorf near Hohenwestedt, Schleswig-Holstein, W. Germany. Spezial-Präparationslack Z4/924 and thinner Z4/948. (Photo by H. Nyburg and J. H. Elsendoorn, Geol. Inst., Utrecht.)

Restoration. Defects occasionally occur during the preparation and mounting of the peel. Fine-grained sediment such as clay may not stick to the gauze, whereas coarse parts may be absent. The first step must be a good stock-taking. The peel is brushed with a soft brush and then blown with compressed air. Next, the mounted peel is turned upside down and the last few loose particles are tapped off gently.

Large bare spots are coated with diluted lacquer; the smaller ones, where only one piece of gravel or a shell is missing, are coated with undiluted lacquer. Large particles are pressed into the peel as sand and coarse silt are sprinkled on top of a very diluted lacquer, which is brushed on the corresponding areas. The strong dilution of the lacquer prevents an unnatural darkening afterward (Hähnel, 1961, 1962). Layers where clay is missing are difficult to restore correctly and often the gauze is visible. The best method of restoring these layers is to paint the bands with a lacquer or plastic to which a dye is added (Fig. 1.22). (For types of dyes and their application, see 2.1B.) For small areas, watercolor will sometimes restore the peel.

Finishing. The peel can be finished by mounting a frame around it and covering the film with glass. The advantage is that no dust can stick to the peel, but the disadvantages are greater, since some fine material will always fall off and cover the inner side of the glass, while larger particles fill the base of the frame. Ordinary glass may also be very inconvenient because it reflects light.

It is much more practical, inexpensive, and easier to finish the peel itself with a protective coating or to impregnate its surface (see Fig. 1.9). A superficial spraying is not sufficient since pores must also be filled in order to restore the appearance of natural moisture content. All surface particles must be solidified completely, which later makes easy cleaning possible (Hähnel, 1962). This treatment causes the film to obtain a darker color, which more closely resembles the natural situation, as the moisture darkens the sediment and makes structures more distinct. For investigations of soils by means of binoculars the front of the peel should not be coated (Slager, 1964; Van der Plas and Slager, 1964).

The easiest finishing method is the application of acrylic clear coatings such as manufactured by the Krylon Corporation (a Borden Chemical product; Krylon, 1956, not dated). These coatings dry in a few minutes and remain clear. Number 1301 is for general use and no. 1303 for artwork, paintings, drawings, etc. They are available as aerosols. Number 150 is only supplied in bulk for use in commercial spray guns.

The Flügger Company manufactures a long-oil-alkyd varnish (Impredur Luftlack 784) which is clear, does not turn yellow, has a good body, fills all pores, is weatherproof and waterproof, and does not dissolve the lacquer of the peel. It dries within 3 to 4 hours and is dry-hard within 8 to 10 hours (Flügger, 1964a). One part no. 784 must be diluted in 10 parts mineral spirits or in Solvesso 100 (Esso). This mixture is brushed over the peel, thoroughly wetting each part. The mineral spirits evaporate and only a small part of the whole mixture remains, yielding a good protective coating (Hähnel, 1961, 1962; Flügger, 1964b).

Two products of the Hoechst Corporation produce good finishings. Mowilith 35/73 is soluble in many chemicals, such as methanol, ethanol, methylacetate, ethylacetate, butylacetate, acetone, methylglycol, ethylglycol, butylglycol, benzol, toluol, and xylol (Hoechst, 1964, 1965b). If it is dissolved in ethylacetate it does not damage the peel, and a good protective coat is obtained. Similar results are obtained with Mowilith DMC 2 (Hoechst, 1959, 1960a, 1960b, 1965a).

Jager and Van der Voort pour a mixture of 25 cc vinylite synthetic resin, 50 cc acetone, and 25 cc methyl isobutyl ketone on the peel. Smith and Moody (1947) use another concentration of the same chemicals. They spray the mixture several times on the peel.

Storage. In most cases the position of the peel should be such that light falls upon it at right angles. Direct light is seldom desirable. Care should be taken not to place the delicate peel surfaces against each other. A rack made of wood and masonite, containing many small slots, will serve for storage purposes. Unmounted or poorly mounted films should be hung in order to preserve their shape. An example is a large rack, as used in Utrecht, in which many peels can be hung like maps. Each peel is framed with U-shaped steel strips. At the top several rings are attached; through this a metal rod can be inserted. The rod is placed in the rack in such a way that the peel and its rod can be moved easily for display or to make room for additional peels. This allows storage for large numbers of peels and makes them readily available for demonstration and for instructional purposes.

1.1B Brushing Method

The Shell Development Company, Exploration and Production Research Division, Houston, Texas, applies a slight variation to the method just described. The same equipment is used as for the spraying method, except that a spray gun is not required.

Summary of the Method

The basic procedure is summarized in 10 steps:

1. Cut a smooth, flat face, using a spade and a straight edge of steel, aluminum, or wood. Mois-

ture content in deposits should not be excessive. Allow sufficient time for drying. A blow torch is very useful in drying thin layers and local pockets, which tend to retain excessive amounts of water.

2. Slightly stretch a double-ply cheesecloth over the prepared surface. Pin edges of cheesecloth on the surface with nails or short pieces of bent wire. The cheesecloth must be in contact with the shaved surface.

3. Brush several coats of a relatively thin mixture of lacquer and thinner, approximately 3 or 4:1, through the cloth and into the sediment. Allow each coat to penetrate the cloth and to partly dry before the following coat is applied. Continue to apply additional coats of lacquer, until the first 10 to 15 mm of sediment are soaked.

4. Apply one or two coats of very thick lacquer and thinner, approximately 5 or 6:1.

5. Allow an ample amount of time, approximately 24 hours, for the lacquer to dry thoroughly.

6. Glue a thin plyboard to the lacquered surface of the cheesecloth.

7. Staple folded edges of cheesecloth to the back of the board.

8. Remove peel.

9. Brush or air blast poorly consolidated sediment from the surface of the peel.

10. Spray several thin coats of clear Krylon on the surface of the peel to bind poorly cemented deposits.

This is the basic procedure for making lacquer peels of poorly consolidated sand, and the results are very satisfactory (Fig. 1.10).

The preparation of the section is similar to that described in the spraying method. The Shell Company as well as Voigt (1936c, d, e, 1938b, 1949b) and Hähnel (1961, 1962) use a blow torch to dry the section or parts of it. This should be done very carefully so that no grains are blasted off the face. The basic difference between these methods is that cheesecloth or gauze is placed prior to the application of lacquer. The lacquer should be brushed carefully and evenly in order that unnatural relief of the peel will result (see Fig. 1.7). The thickness of the peel obtained makes it necessary to mount the peel on masonite in the field to prevent any breaking during removal and transport. Restoration, finishing, and storage are similar to the spraying method.

Glyptal 1276 from the General Electric Company, Insulating Materials, is used as lacquer (General Electric, 1963). Since mixtures of lacquer and thinner (xylene) largely depend upon the porosity and permeability of the sediment and also on the desired thickness of the future peel, Dr. H. A. Bernard (written communication, 1962) suggests mixtures of approximately 2:1 or 3:1 (lacquer:thinner) for the first coat and approximately 4:1 or 4:0 for the final coat, when making peels of extremely fine-grained sands and silts. For making peels of very porous and permeable sands, the first coat is generally 4:1. However, experience will help decide the proper mixture to be used for both the first and the second coats.

This lacquer was tried with different thinners: acetone toluene (4:1) and pure acetone. In both cases good results were obtained.

1.1C Pouring Method

At the Koninklijke/Shell Exploratie en Produktie Laboratorium (Royal/Shell Exploration and Production Laboratory), Rijswijk, Z. H., The Netherlands, the pouring method is in use for making lacquer peels in the field as well as for making peels from cores.

This pouring method can also be used successfully for soils. The pouring must be carried out quickly in order to avoid the forming of rims (see Fig. 1.7) (S. Slager, personal communication, 1965).

Preparation of a Lacquer Peel in the Field.

The equipment necessary is the same as used for the spraying and brushing methods, except that the spray gun is not utilized. All the actions mentioned under "preparation of the section" have to be carried out first. Next, fasten a piece of gauze, same size as the peel (if possible), to the section with nails and bent pieces of wire (see Fig. 1.5, gauze instead of the rope). Next place a flat brush (5 cm wide) at the base of the gauze, pour the lacquer mixture on the gauze just above the brush, and move the brush and mixture upward simultaneously. In this way the lacquer is spread in a vertical band, reducing the risk of forming rims on the peel (see Fig. 1.7).

After pouring, application of an additional coat of lacquer by brush may be necessary. The peel is then collected with the aid of a wooden board, which is pressed against the back of the peel. The upper side of the peel is clamped by

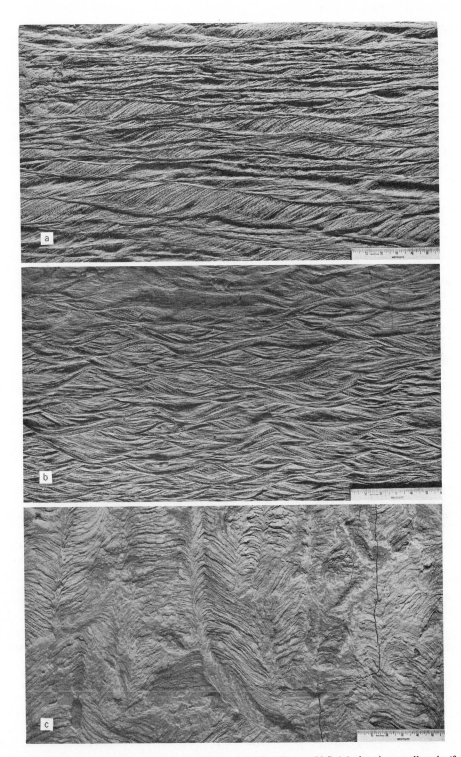

Fig. 1.10. Lacquer peels made from the Brazos River deposits (Texas, U.S.A.) showing small-scale (festoon) cross-bedding resulting from the downstream migration of small sand ripples. Two vertical sections, one in the dip (*a*) and one in the strike (*b*) direction, and one peel for a horizontal section (*c*). Scale in inches. Lacquer: General Electric Glyptal 1276. (Courtesy of the Shell Development Company, Exploration and Production Research Division, Houston, Texas, U.S.A.)

hand to the board, and by withdrawing the board, the peel is taken off. In this way the operator avoids bending the peel.

Immediately after the peel is collected it is mounted to the surface of the masonite with the same lacquer that is used in making the peel.

Preparation of Lacquer Peels from Cores.

The pouring method is the most successful method for making lacquer peels from cores (and also from box samples). The work can be divided into a number of steps.

Equipment (Fig. 1.11). Two half-tubes of zinc (the author often uses plexiglass); thin stainless steel wire; lacquer, thinner, mixing beaker or beakerglass, stirring rod; knife or spatula, trimming knife (Stanley knife); gauze strips, scissors, flat brush (2 in., or 5 cm); adhesive, board, brush. At the Shell laboratory the surface is not given a protective coating; however, the author uses a spray coating (Krylon) or special finishing products (Impredur Luftlack 784, Mowilith 35/73 or Mowilith DMC 2) with their thinners (mineral spirits or ethyl acetate).

Pretreatment of the Core. The core barrel is cut open in two halves or the core is pressed out of its holder into a semicylindrical tube. This tube can be of zinc or plexiglass (Fig. 1.12). A similar half-tube is placed over the core. The gap

between the two half-tubes makes it possible to draw a fine stainless steel wire through the core and to divide it into two halves. The cutting should be carried out by two persons to prevent disturbance of the core (Fig. 1.13). It is often useful to place a massive cylindrical object at the end of the core toward which the cutting is directed to avoid breaking the core. The core can now be opened (Fig. 1.14). Since the cut surface is somewhat disrupted by the wire, it should be smoothed with a spatula or a sharp knife (Fig. 1.15). The core should be scraped parallel to the bedding, thus preventing breakage or smearing of the structural picture.

Care should be taken that the core does not dry too quickly or too much, or shrinkage cracks may appear. Sandy cores should be lacquered as soon as possible, whereas with damp clays wait 2 or 3 hours after opening the core. Very wet clays may be left open up to 15 or 20 hours to let them dry in the air.

Pouring Lacquer over the Core. The core can be covered with a sheet of plastic, which allows the sediment to lose moisture slowly. Good peels can be made and cracking seldom appears when a core is left to dry overnight under such a plastic cover (J. M. Coleman, written communication, December 1965).

Fig. 1.11. Equipment and materials used for the making of a lacquer peel from cores: (A) sealed plastic core liner, (B) half-tube of plexiglas; (C) rubber stop; (D) bandage; (E) soft, flat paintbrush; (F) spatula; (G) scissors; (H) Stanley knife; (I) thin wire; (J) ethyl acetate; (K) masonite; (L) Mowilith 35/73; (N) Krylon Crystal Clear no. 1303; (O) nitrocellulose lacquer Tornol A; (P) beaker glass with stirring rod; (Q) adhesive Bison Kit; (R) thinner V105; (S) Impredur Luftlack 784; (T) mineral spirits.

Fig. 1.12. (*Upper left*). The core is pressed out of its barrel in to a semicylindrical tube. (Courtesy of the Koninklijke/Shell Exploratie en Produktie Laboratorium, Rijswijk, Z. H., The Netherlands.)

Fig. 1.13. (*Upper right*). Another semicylindrical tube is placed on top of the core. With a thin wire the core is cut into two halves. (Courtesy of the Koninklijke/Shell Exploratie en Produktie Laboratorium, Rijswijk, Z. H., The Netherlands.)

Fig. 1.14. (*Lower left*). The two core halves are separated. (Courtesy of the Koninklijke/Shell Exploratie en Produktie Laboratorium, Rijswijk, Z. H., The Netherlands.)

Fig. 1.15. (*Lower right*). The surface of the half-core is smoothed with a sharp knife. Core should be scraped parallel to bedding. (Courtesy of the Koninklijke/Shell Exploratie en Produktie Laboratorium, Rijswijk, Z. H., The Netherlands.)

Fig. 1.16. A strip of bandage is placed over the lacquered surface and thick lacquer is pasted on it. (Courtesy of the Koninklijke/Shell Exploratie en Produktie Laboratorium, Rijswijk, Z. H., The Netherlands.)

19

The lacquer mixture will be poured or brushed only once on sandy cores and two times on clayey cores with an interval of 4 to 24 hours between them. Fine sandy clays are treated like the clay. Very dry and coarse sands do not need this priming. A strip of gauze is spread over the sample. The gauze should have the same width as the core; it should be 10 cm (4 in.) longer than the core. When the gauze is placed in position, the mixture is poured or brushed over the core as evenly as possible, while the half-barrel is held a little inclined (Fig. 1.16). Depending on the material, the following mixtures (parts by volume) are recommended:

for clay	1 part lacquer and 2 parts thinner
for sand	1 part lacquer and 1 part thinner
for coarse, or very dry sand	2 parts lacquer and 1 part thinner

Collecting and Mounting. The last lacquer coat should be allowed to dry for 1 day if the cores are sand, and for 2 or 3 days if the material is clayey. After this drying period, the edges of the peel are freed from the sides of the tube with a trimming knife (e.g. a Stanley knife) (Fig. 1.17). The peel can now be detached from the core (Fig. 1.18) and mounted on a strip of plywood board, fibrous board, or thick masonite with the same lacquer. The author uses quick-drying glue such as A 33, Bison Kit, or Snelfix (Ram, 1960; Perfecta, 1964; Ceta Bever, not dated d). Depending on the type of glue — the adhesive should not be hardened completely — excess sediment particles can be blown off by means of compressed air, thus allowing the sedimentary structures to stand out clearly. When the mounted peel is completely dry, it can be given a finishing coat with an acrylic "crystal-clear" spray or other finishing agent (Krylon, 1956, not dated; Hoechst, 1959, 1960a, b, 1964, 1965a, b; Flügger, 1964a, b).

These monoliths can reveal a great deal of important data and are very useful for instruction and reference (Fig. 1.19).

1.1D Weather and Section Conditions

The weather as well as the conditions of the sediment section are of primary influence on the preparation of a lacquer peel. Both influence the penetration depth of the lacquer and the time it takes to evaporate the thinner. Variations in the conditions make it necessary to vary the amounts of the components of the lacquer-thinner mixture and also the time involved between the successive steps.

Normal Dry Weather

This is the most favorable weather for this work. The only variations are the moisture con-

Fig. 1.17. The edges of the peel are freed from the sides of the tube with a Stanley knife after the lacquer is dried. (Courtesy of the Koninklijke/Shell Exploratie en Produktie Laboratorium, Rijswijk, Z. H., The Netherlands.)

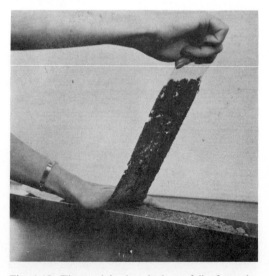

Fig. 1.18. The peel is detached carefully from the core. (Courtesy of the Koninklijke/Shell Exploratie end Produktie Laboratorium, Rijswijk, Z. H., The Netherlands.)

Labels within the image:

A — Granaatzand op het strand bij Hallum, Ameland. Laminatie zichtbaar door mineralconcentraties. Profiel evenwijdig aan de kust. N-Z N8 74b

B — Slibvangende mosselbank op de Kikkertplaat, Mytilus, cardium, Hydrobia 2dm onder halftij. Z-N N8 72

C — Oude plaat met Lamellibranchiaten waarop 30cm zand met verspoelde schelpen uit riviermond. Oost van Katwijk. N-Z N8 42

D — Getijgeul opvulling met zwakke megaribbel-structuren en uitgespoelde schelpbrokken. Oostvoorne. O-W N81

E — Ongeconsolideerde strandafzettingen met schuimige structuur. Oostpunt van Ameland. N-Z N871

Scale: 0 — cm — 10

Fig. 1.19. Lacquer peel monoliths from cores collected from various subrecent environments in The Netherlands (collected and peeled by the Koninklijke/Shell Exploratie en Produktie Laboratorium, Rijswijk, Z. H., the Netherlands): (A) beach sand with concentrations of garnet in laminae. Section parallel to the coast (N-S), Hallum, Ameland; (B) mud-catching mussel-bank with *Mytilus, Cardium* and *Hydrobia*. 20 cm under mean sea-level (S-N), Kikkert shoal, Ameland; (C) old shoal with lamellibranchs. Tidal gully of a river mouth, E of Katwijk (N-S); (D) tidal gully deposit with mega ripple lamination, reworked shells and shell remains (E-W), Oostvoorne; (E) unconsolidated beach deposits with holes of air bubbles. East point of Ameland (N-S); (F) Pleistocene river-dune deposits near a braided river. Upper part distorted by roots. Gennep (S-N); (G) top of an old dune. Oosterduinen, N of Noordwijkerhout (S-N); (H) underwater delta of the river Ijssel. Wavy lamination and indistinct

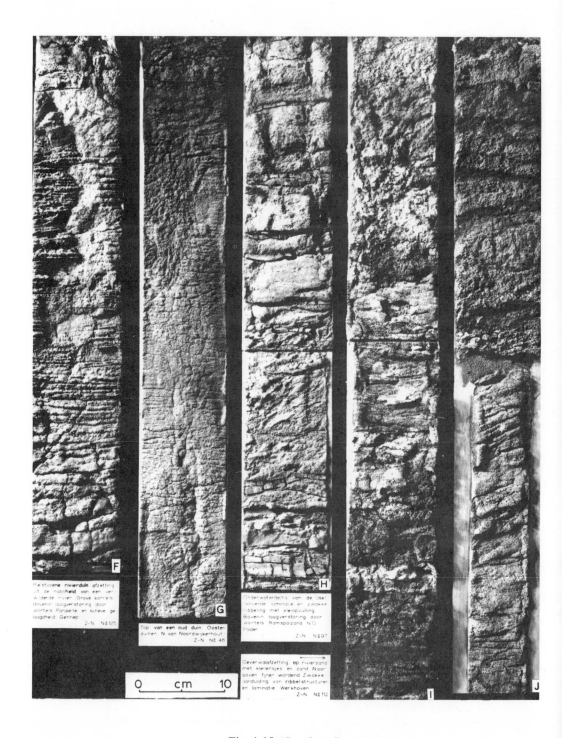

Fig. 1.19 (Continued)

ripples with clay filling. Upper part disturbed by roots. Ramspolzand, NE Polder (S-N); (I) levee deposits on top of river sand. Indistinct ripples and parallel lamination. Werkhoven (S-N); (J) bank of a braided river. Top 50 cm under surface. Mega ripple lamination. Iron oxide bands are present in the sand. Between Gennep and Ottersum (N-S); (K) levee deposit from tidal area. Bedding parallel to gully bottom. Biesbos "de Dood" (S-N); (L) sloef

deposits. Top 46 cm under the surface. Wavy and inclined lamination. Upper part with slumping. Lot D8, NE Polder (N-S); (M) muscovite containing fine sand with reworked peat near an old beach. Top 17 cm under surface. Flaser structures. Lot K10, NE Polder (N-S); (N) "katteklei," a peat containing clay without carbonate. Basic ferro-sulfate is present in fresh condition. Dessication cracks and root holes. Vijfhuizen (N-S). (Photo by H. Nyburg and J. H. Elsendoorn, Geol. Inst., Utrecht.)

tent of the section and the type of sediment. The methods so far described are based on this type of weather. For making peels of soils in the field, this is the only suitable weather condition.

Warm Weather

High temperature and direct sun are not too favorable for making lacquer peels since the sections are usually too dry. The thinner, especially acetone, evaporates much too rapidly, and cracks in the lacquer coat occur due to shrinkage. This may cause wrong relief shapes on the peel. It is imperative to select the sections in such a way that direct sunlight does not come in contact with the section until the peel is strengthened. One can also make use of a windscreen.

During such weather conditions it is better not to use pure acetone as thinner, but to add 20 to 50% toluene to it to slow down the evaporation.

Rain

There is no objection to making a lacquer peel during rainy weather. However, the result depends on how wet the section is. If the section is not wet throughout it can be treated as a wet section. After drying the surface by spraying and igniting the acetone, a screen is mounted above the section to prevent raindrops from touching the lacquer coats since water causes thin spots in the film. If the section is too wet, or becomes wet during the first steps of the procedure, the lacquer will not penetrate effectively, and the relief differences, based on sedimentary structures, are less distinguishable (Fig. 1.20).

Frost

Low temperatures or even temperatures below the freezing point do not prevent the making of lacquer peels. However, drying takes a little more time. It is necessary therefore to defrost the section, since pores that are filled with ice cannot be filled with lacquer. In general the section can be treated as a wet section. During the preparation of the peel, direct sun can be favorable since it speeds up the drying. Careful use of a blow torch will also quicken the drying.

Wet Sections

Sections that contain too much water do not allow the lacquer to fill all the pores. Water that runs out will loosen the film coat at the surface of the section.

If the moisture content is not excessive, thinner can be sprayed onto the section and then ignited; in doing this, be sure that no inflammable products are nearby. This process can be repeated three or four times, but it must be stopped as soon as surface elements start burning. If the section is still too wet, it is useless to continue the process since we are certain that under such wet conditions a good peel can never be made. In such cases, a sample of the section can be made (see 1.1E, soils and peat).

For investigations on soils, when the humus should also be observed, one is not allowed to "burn" a section.

Dry Sections

It is impossible to obtain a good lacquer peel from an absolute dry sediment since the lacquer will not cling to dry material (sand or silt). The

Fig. 1.20. Lacquer peel made from scour and fill section in subrecent estuarine deposits during rainy weather conditions. The overall picture is good, but structural details do not stand out too well. Excavation in the Haringvliet, the Netherlands. Special Präparationslack Z4/924 with acetone as thinner. (Photo by H. Nyburg and J. H. Elsendoorn, Geol. Inst., Utrecht.)

wall should be cut back until a moist surface is exposed. If this is not possible the section should be moistened by spraying water (to which a wetting agent can be added) on it.

The humidity can be checked by shaving off a vertical part of the section. Sediment particles should not run down; neither should the vertical surface break down into parts. In obtaining enough depth, it is possible that the surface may become too wet. It then becomes necessary to wait until the water is evaporated or to spray thinner onto the surface and ignite to speed up the process of evaporation.

1.1E Types of Sediment

Since the section conditions are not only based on the weather and the height above the ground water table but also on the types of sediment, a number of sediments will be discussed.

Sand

Sediment of sand size is the most suitable material for making a lacquer peel, especially when the grains are fine or medium sized. Normally sedimentary structures will stand out nicely. If the sand contains a small amount of finer material it will bind the grains and make the preparation of the section easier, often improving the result. This can be seen in Fig. 1.9, a lacquer peel made from fluvioglacial deposits.

The moisture content must not be too high or too low. Dune sands may produce many difficulties since the loosely packed material does not allow the operator to cut a very steep wall.

A large amount of water sprayed on the section may solve this problem. Another possibility is given by Voigt (1949b, p. 123). Instead of using pure water he sprays a thin solution of water and a cellulose-base glue (Glutofix, Elmer's Glue-All, Cetaflex, Tylose MH 20, -H 20 P, -C10; see 1.5) onto the section. (It is better to use a second spray gun for this work.) This operation must be carried out 1 day before the section is lacquered. According to Voigt about 1 liter of glue solution is necessary per square meter.

Completely wet sand can only be sampled and treated in the field according to the method of McMullen and Allen (see 1.3C), or in the laboratory by the pouring method, glue peel or araldite peel concept.

Gravel, Shells, and Other Inclusions

Sections which contain larger particles can never be made as smooth as sandy outcrops (Fig.

1.21). Grains of sand size and smaller often stick to the gravel and boulders. According to Hähnel (1961, 1962) it is necessary to free those large pieces from adhesing sand and/or loam, because the lacquer will not stick when the rocks are covered. A small paintbrush is a suitable tool.

There is no general rule that covers all the problems one may encounter when making a lacquer peel from sections containing very coarse particles. It can be a gravel section with sand or finer sediment as matrix, or a sand section with some occasional pebbles. The size of the gravel particles can also vary considerably.

If the gravel is not larger than about 5 cm (2 in.), the exposed gravel layer(s) must be cleaned as much as possible of sand and other material. Then the section can be sprayed, next brushed with lacquer, and then strengthened with gauze. The application of the lacquer must be carried out very carefully so that the holes and all sides of the gravel pieces are well coated.

Since the surface is not smooth, it is important to paste, with lacquer, small strips of gauze or cheesecloth over small groups of pebbles which lay next to each other. The strips must follow the irregular surface. As a result, all pebbles are connected by an unsystematic pattern of strips. The whole peel can then be covered with a complete gauze-back as is carried out with normal peels.

Lacquer peels in which rather large particles (> 5 cm) are present are difficult to collect from the section. It must be done very carefully and many of the large particles must be loosened from the wall.

Gravels, belonging to the peel, which are left in the section during collecting, must be numbered in order to mount them later in the corresponding places. Also, these types of peels cannot be glued to masonite or similar material; a special trellis frame should be constructed. The "horizontal" strips of the trellis should be placed in such a way that the largest pieces are supported. The other particles should be hung on the trellis by means of strips of nylon or perlon which have been glued to the peel with lacquer. Strips should be pasted at regular intervals of 20 to 30 cm (8 to 12 in.) to the peel in order to hang it in the right position in the frame. These strips should be secured to the wooden lattices with nails (Hähnel, 1961, 1962) and should be very sturdy.

Shells and other inclusions present in the section can be treated in the same manner as gravels. Those that stick out far, and are not too

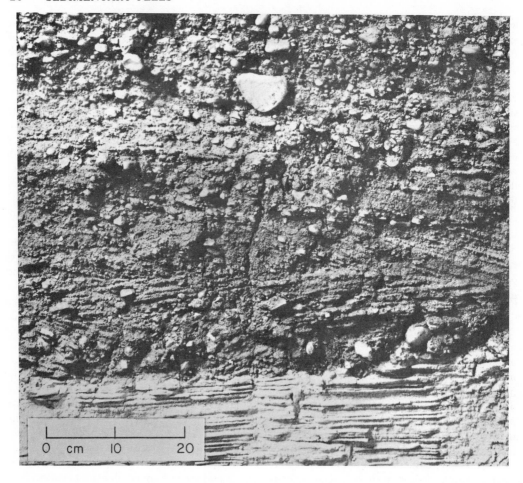

Fig. 1.21. Lacquer peel made from coarse Meuse deposits (high terrace river deposits) on Pliocene sand. De Kling near Brunssum, The Netherlands. (Peel made by J. H. M. van Dijk, Geol. Inst., Utrecht, The Netherlands.)

thick, can be broken or taken out. Collecting these peels may be rather difficult if not enough hands are available. A bag with reserve shells or inclusions may be useful for restoration purposes. If the coarse particles are only part of the peel, it can be mounted normally on masonite, even if parts of the peel do not stick to the wood. It may sometimes be useful to cut holes in the masonite to give the peel a more natural flatness.

Clay-Sized Sediments

Often it is nearly impossible to make a good lacquer peel in the field from a clay section according to the methods discussed. The moisture content makes the cohesion between the clay particles stronger than the binding lacquer-clay. These sections must be treated in different ways, as discussed in the sections on soils and on the method of Maarse and Terwindt (1.3A).

Streaks or layers of clay in a sandy or gravelly section rarely adhere to the peel backing. A white or yellowish color on the gauze or cheese-cloth will indicate the locations where the clay should have been (Fig. 1.22).

This missing material seldom can be mounted afterward during restoration activities, since the natural layering cannot be regained. Cutting pieces of clay to shape is more theoretical than practical. The best way is to paint the empty bands with lacquer or a plastic, to which a dye is added (see also Section 2.1B). By comparing the left side (not dyed) of Fig. 1.22 with its right side (dyed) one can decide if the restoration is an improvement or not.

Soils

Lacquer peels of soils can be made according to the previously mentioned methods, if the main object is to preserve the distinct colors (Hähnel, 1961, 1962). One can carry out fast and accurate investigations without difficulties on good peels using enlargements up to 150 X (Slager, 1964, 1966; Van der Plas and Slager, 1964).

If the crumbly aspect of the soil mast be preserved, a lacquer peel can only be made in the field when the soil is completely dry. The peel, made in the field, cannot be collected in a normal way such as a sand section but must be loosened from the wall at a distance of about 10 cm (4 in.) from the back of the peel. The surface of the peel must be freed from surplus material, as given by Jager and Schellekens (1963).

Many investigators have followed Voigt (1936c, d) and have contributed to the art of making lacquer peels of soils. A few of them are mentioned in the introduction to Chapter One. Some of the methods contributed will be discussed (see also J. Bouma, 1965).

In this work, Römer (1951) uses "Profillack," which produces good soil monoliths. The soil section must have the proper moisture content. A water-soluble lacquer is sprayed on the section. A second coat is painted with undiluted lacquer. If the soil has a coarse texture, the painting must be repeated. Equipment must be cleaned with water directly after use, since this lacquer becomes insoluble when dry. The dry

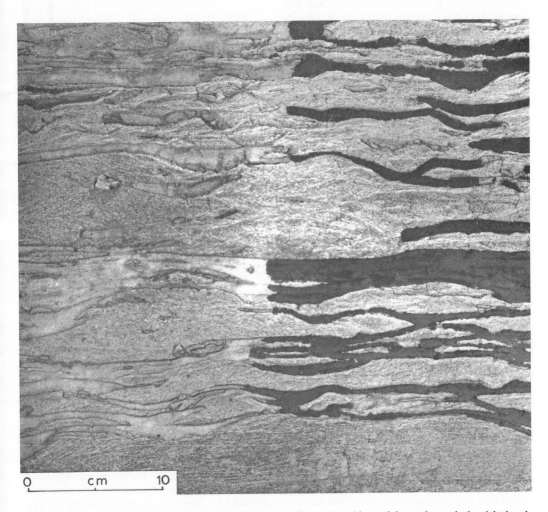

Fig. 1.22. Estuarine deposits containing many clay layers. Since clay seldom sticks to the peel, the right-hand part was painted in order to obtain more contrast. The upper right half is painted with thinned lacquer (Tornol B with acetone: sprayable viscosity) to which a number of dyes are added. The lower right half is painted with thinned plastic (synolite 333 with styrene monomer = 3:2) to which the same dyes are added. Excavation in the Haringvliet, The Netherlands.

lacquer coat produces an impression like latex. It can break easily, and therefore must be strengthened with gauze.

Collecting of the peel must be carried out carefully, loosening the section piece by piece with a knife. If the collected peel becomes too long to handle, it can be rolled up around a rod with the sediment side outside. If coarse material is present in the section, this rolling is not possible. In the laboratory the peel is first restored. If the soil becomes too dry, it can be moistened carefully with water. All traces of the knife, obtained during collecting, can be brushed away. After 2 days the peel can be mounted and finished.

Römer (1951) indicates that for some purposes peel strips of 5 cm (2 in.) width are sufficient. For these peels it is not necessary to smooth a wall in an excavation or to dig a pit. He uses a soil drill (Fig. 1.23a), which is hammered into the soil. The cutting part is a little smaller in diameter than the rest of the grooved tube, and consequently the core is influenced little by the drill-wall itself (Römer, 1934, 1951). Rods can be placed in the head of the drill and the apparatus can be turned while pulling. This facilitates extracting the samples out of the soil. These peels can only be made from soils that do not contain stones.

The core is then cut along the groove. Römer collects the peel in parts by using strong cardboard strips (about 1.5 mm thick) of the same width as the peel, and about ⅓ m (13 in.) long. These strips are painted with Profillack and when tacky the strip is pressed onto the peel and torn off carefully again. The short strips are mounted behind each other and framed (Fig. 1.23b). Instead of using Profillack one can use different lacquers. For the collecting of these short strips the fast-drying glues can be applied. Römer did not succeed in making peels from wet peat and clays and from heavily textured clays.

Jager (1959) described a method of sampling wet and heavy soils with the help of a nail-board. It is constructed of heavy wooden planks, reinforced by three strong wooden clamps. The size of the board is about 120 × 40 cm. At its front side long nails (ϕ 1.5mm) are hammered in at regular intervals of 5 cm. The heads of the nails are cut off, leaving a length of 5 cm, and they are then sharpened. With a mallet the board is driven into a vertical wall of an excavated pit. Grooves of about 15 cm depth and 8 cm width are cut along the vertical sides and along the

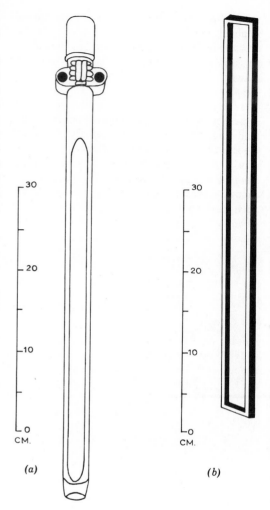

Fig. 1.23. (a) Soil drill after Römer (1951). The cutting part is a little smaller in diameter than the rest of the grooved tube. (b) Frame for monoliths (redrawn from Römer, 1951).

lower side of the board. The sample is cut loose with a steel cable, about 3 m long and 3 mm thick, with two handgrips, sawing at the end of the grooves. When several centimeters have been sawed, the lower parts of the vertical grooves and the lower groove are refilled to support the board.

At the laboratory the surface is smoothed and left to dry. The sample can be sprayed with a formalin solution to prevent mildew from forming. As soon as little cracks start forming the sample can be lacquered, using the pouring method. The peel must be strengthened with cheesecloth. When the peel is completely dry, it is turned over, the nailboard is taken off, and all nonfixed soil elements are removed carefully with a knife. The peel is then mounted and left to dry. Some soil elements may come loose and have to be reglued in their places.

Jager and Schellekens (1963, a revision of Jager, 1959) constructed a strong sampling box of galvanized sheet-iron (Fig. 1.24), with sharpened edge. A very strong iron frame was used as a reinforcement, welded to the bottom in order to drive the sampler into the wall. A spade with a heavy ball head (Fig. 1.25) and a knife for leveling the soil in the box (Fig. 1.26) were added to the equipment.

First a large work pit is dug. The wall to be sampled is cut vertically. The sample box is then placed with its upper side at the surface and next hammered into the wall with a sledge of 2000 g (4.41 lb) until the bottom touches the section (Fig. 1.27).

Grooves are then dug along the box, about 10 cm deeper than the sides, in order to reach into the sediment. The special spade is pushed and/or hammered behind the box at a distance of about 10 cm from the open side. The sampler

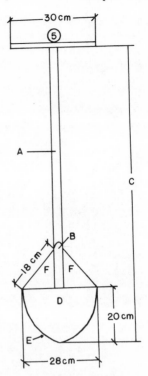

Fig. 1.25. Spade with a heavy ball-head for driving in: (A) massive iron shaft (2.5-cm diameter); (B) flattened joint with spade blade; (C) length of sampling box, + 10 cm; (D) steel spade blade (3 mm thick); (E) sharpened edge; (F) sheet-iron reinforcements. (Redrawn from Jager and Schellekens, 1963.)

must be kept in position to avoid breaking the sample in the wrong place.

When the spade is at the specified depth, the box is turned back carefully by means of the spade (Fig. 1.28). The sample box is placed at the surface next to the pit and all protruding material is cut off with the special knife (Fig. 1.29). The sample is now ready to be transported to the laboratory.

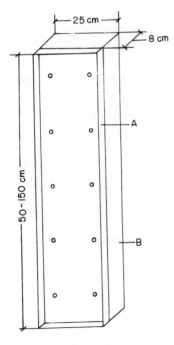

Fig. 1.24. Sampling box of galvanized sheet-iron: (A) reinforced iron frame (2.5 cm deep, 0.5 cm thick) for driving the box into the sediment face; (B) sharpened edges. (Redrawn from Jager and Schellekens, 1963.)

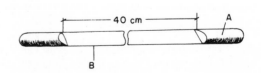

Fig. 1.26. Knife for leveling soil in sampling box: (A) handle; (B) sharpened knife edge. (Redrawn from Jager and Schellekens, 1963.)

Fig. 1.27. (*Upper left*). The sampling box is placed against the vertical section and then hammered into the wall. (Courtesy of the Soil Survey Institute, Wageningen, The Netherlands; Jager and Schellekens, 1963.)

Fig. 1.28. (*Upper right*). After deep grooves are dug along the box sides, the spade with heavy ball-head is pushed behind the box at a distance of 10 cm. The sample can now be removed. (Courtesy of the Soil Survey Institute, Wageningen, The Netherlands; Jager and Schellekens, 1963.)

Fig. 1.29. (*Lower left*). Excess of soil is cut off with the special knife (Fig. 1.26). (Courtesy of the Soil Survey Institute, Wageningen, The Netherlands; Jager and Schellekens, 1963.)

Fig. 1.30. (*Lower right*). After the sample is allowed to dry, the box is stood up at a slight angle and diluted lacquer is poured over it. Cheesecloth is then placed in position and undiluted lacquer is poured. (Courtesy of the Soil Survey Institute, Wageningen, The Netherlands; Jager and Schellekens, 1963.)

After several days the sample will be dry enough to be further treated. The occurrence of small shrinkage cracks in a heavy clay indicates that the soil is dry enough. The box is then placed against the wall at a 75° angle, and strongly diluted lacquer is poured over the sample. Next a lacquer coat is pasted and the peel is strengthened with cheesecloth. A few nails will keep the gauze in position (Fig. 1.30). Sometimes the authors first put the cheesecloth in position and then pour a thick lacquer mixture. Later another thick mixture is applied with a brush. It takes approximately 2 days before the lacquer is dry both superficially and internally. (For sandy soils it is not necessary to apply as much lacquer on the section, since less weight will hang on the peel. The authors do not even strengthen the peel with cheesecloth.)

After the lacquer treatment, the sample box is placed horizontally and the peel cut loose by pushing the knife along the sides of the box. A strong piece of wood or thick masonite is placed on top and the entire box is turned over, after which the box can be lifted off carefully. Resistance can be overcome by tapping against the outside. A large block sample, with only the lower side reinforced, is now placed on the table. Surplus soil is removed cautiously, first by hand and later with a knife. It is better to allow the soil to dry until structural elements can be easily isolated (Fig. 1.31). The half-cleaned peel is then mounted on masonite. Additional drying will allow the soil to shrink further. By tapping against the sides and the back with a mallet, all nonfixed soil elements will fall off (Fig. 1.32), and the obtained result can be framed (Fig. 1.33).

The peel should not be finished with a lacquer coat, since it will fill the small pores, and it is necessary that these properties remain visible for stereoscopic research. Soils with large amounts of humus, and also peats, have large shrinkage cracks after drying. This cannot be prevented. It gives an impression of what occurs in nature after a long period of dehydration (Figs. 1.34 and 1.35).

Jager and Schellekens (1963) indicate that this method is inexpensive since only one field trip, by two men, is necessary to collect the sample. The field time involved is therefore short, and the operators are less dependent on the weather.

At the Soil Survey Institute, Wageningen, The Netherlands, large-sized monoliths (120 x 80 cm) of soils were successfully made using a large steel box (Jager and Van der Voort, 1965). The bottom of this box is reinforced with two diago-

Fig. 1.31. When the peel is dry the edges are freed from the box walls and the box is turned over and lifted off. All surplus of soil is now removed cautiously along the faces of the structural elements. (Courtesy of the Soil Survey Institute, Wageningen, The Netherlands; Jager and Schellekens, 1963.)

Fig. 1.32. The near-finished monolith is glued to a sheet of masonite. After drying, loose soil particles and structural elements are removed by tapping the sheet gently with a mallet. (Courtesy of the Soil Survey Institute, Wageningen, The Netherlands; Jager and Schellekens, 1963.)

nal steel strips to keep the bottom flat. Two strong iron eyes are welded to the boxtop to facilitate removing the box from the pit. The box has to be pressed into the section by means of three jack-screws. Since the weight of the filled box can be as heavy as 200 to 300 kg (444 to 666 lb), it must be removed by means of a pulley-block. Two U-shaped iron bars are placed in the pit over which the back of the box can slide.

Reijnders (personal communication, May 1966) made lacquer peels from soils in Syria. These attapulgite clays contain a high percentage of salts. The content of organic matter is low, and the stability of the soil is small and consequently it falls apart easily. A sample was col-

Fig. 1.33. (*Left*). The monolith is framed. (Courtesy of the Soil Survey Institute, Wageningen, The Netherlands; Jager and Schellekens, 1963.)

Fig. 1.34. (*Upper right*). Monolith of a clayey soil with high humus content. Note the small structural elements of the soil. Height 80 cm. (Courtesy of A. Jager, Soil Survey Institute, Wageningen, The Netherlands.)

Fig. 1.35. (*Lower right*). Monolith of a peat-moor. Length 1.70 m. Emmercompascum, The Netherlands. (Courtesy of Z. van Druuten, Geol. Inst., Wageningen, The Netherlands.)

lected with the help of a box, and the soil face treated with a mixture of sapon lacquer, alcohol, and water prior to the application of lacquer. The mixture of Tornol lacquer and thinner was made at a ratio 1:1.

Peat

Peat sections can be sampled and treated according to the method of Jager and Schellekens (1963) already described. Voigt (1949b) discusses a preparation in the field. Thinner is sprayed on the smoothed surface and ignited about two to three times, followed by four to eight sprayings of thinned lacquer, which is also burned. Enough pore water is now removed to lacquer the peel. The peel must be removed carefully and the use of a knife or spatula is necessary. Hähnel (1961) mentions that only slightly weathered peat sections can be peeled, using the method of Voigt (1949b).

Franzisket (1962a) describes the preparation of a large peel (2 x 3 m) from living peat. The digging of an excavation and the smoothening of a section offered no problems, but continuous discharge of water made it useless to try drying the section by burning sprayed thinner or lacquer mixtures. Discharge of water at the base could be decreased by digging a deep pit underneath the front of the section. The smoothed surface was then plastered with gypsum (plaster of Paris). Parts of this gypsum coat became loose because of running water. However, continuous application of the gypsum to the section finally sealed the face. Next, this face was reinforced with jute and plaster of Paris, and with horizontally placed laths, about 3 m in length, spaced approximately 20 cm apart. The laths were then cemented to the section with more gypsum. After hardening, the entire section was cut in horizontal bands of about 40-cm width, using a compass saw. Each band normally contained two laths. (The gypsum sheet hangs so well to the peat that no special securing is necessary.) A foxtail of about 1 m in length was used to cut the peat at a distance of about 10 cm behind the plastered surface. The samples obtained were 40 cm high and 200 cm long and had a weight of 100 to 150 kg (220 to 330 lbs), gypsum and laths included.

At the laboratory the sample parts were laid in order with the peat side up, and the surface was leveled and smoothed. Methylated spirit was sprayed on the sample and ignited, after which a 1 mm thick layer with very fine cracks was formed to which a lacquer mixture was sprayed. The water, still present in the section, gives this lacquer a white color, and a black coloring powder was therefore added to the lacquer.

The "Kleblack" was diluted with acetone (1:1) and about 5 liters were sprayed on a 6 m² surface. Five times this mixture was sprayed and ignited. The pore water was then replaced by lacquer, and as soon as a closed lacquer coat was made, the water boiling underneath was observed. After this burning treatment, undiluted lacquer (about 10 liters) was directly applied to the peel, followed by pressing jute by hand into this unhardened coat. (The peel must be collected when the lacquer is completely dry, but not yet stiff, to prevent the formation of false shrinkage cracks.) Protruding jute was therefore nailed to a lath to facilitate the peel removal. Parts that stuck were loosened with a knife. The peel was mounted directly on a frame, restored, and placed in a glass showcase. During drying, cracks formed; this cannot be prevented. (Franzisket indicates that the peel can also be finished directly after mounting by wetting the surface thoroughly with a solution of shellac in alcohol. It may produce a milky appearance in some spots by the emulsion of water and shellac, but after drying, alcohol can be applied to the spots to obtain the right color again.) It might be useful to add some dye to the lacquer to prevent the formation of the milky appearance or lighter colors.

Poorly Consolidated Deposits

Sediments in which grains can be rubbed off by hand will be referred to as a half-consolidated deposit. Lacquer peels can be made without large difficulties. However, the collecting may be hard, and therefore several experiences will be discussed.

In the Torrey Pines State Park at La Jolla, California, two peels of the Pleistocene sandy deposits, whose grains were cemented by iron oxides, were made with little success, even though grains could be rubbed off by hand. The dry material was too hard to make a smooth face with a spade or a trowel. The spraying method was applied without wetting the section with water. The binding by the iron oxides proved to be stronger than the lacquer (Z 4/924 from Gustav Ruth) binding. Empty spots on the bandage resulted when removing the peel. A third, small peel was made in the same way, but was collected by spraying water above and later behind the peel. This loosened the particles, and a few millimeters of the peel could be collected. The procedure gave good results, but was rather time

consuming and required a large amount of water.

From the Ventura Basin, California, a number of Pliocene samples from the turbidite formations were collected. Stratigraphically vertical sides were flattened by wetting them and using the cement floor as an abrasive plate. The samples were left to dry and the brushing method was applied next. Even the rather small pieces were strengthened with bandage. Collecting these peels produced the same difficulties as encountered in the State Park. The samples were therefore placed in a large tray with water. Each day loose material was brushed and washed away. Depending on size and sand/clay ratio of a sample, it took 2 to 6 days to remove all noncemented grains. The peels obtained in this manner were excellent and showed an abundance of details (Fig. 1.36).

Samples were collected in an abandoned diat-

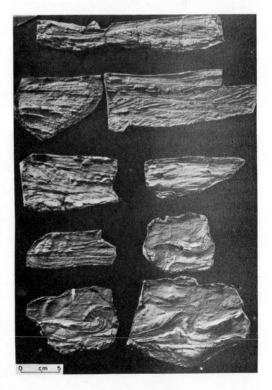

Fig. 1.36. Lacquer peels made from semiconsolidated samples from Pliocene turbidite formations from the Ventura Basin, north of Los Angeles (California, U.S.A.). The peels show parallel lamination, current ripple lamination, and convolute lamination. Made with Tornol A and thinner V 105 (Fil Co.).

omite quarry near Hollerup (Jutland, Denmark). The air-dried specimens were smoothed on one side and thinned lacquer was brushed, followed by an undiluted coat, and then strengthened with bandage. The result obtained with nitrocellulose lacquer proved to be better than with other products (Fig. 1.37).

Kuenen (1961) produced a lacquer peel, made by the Koninklijke/Shell Exploratie and Produktie Laboratorium in the Netherlands, from a fine-grained Lower Barremian sandstone. This peel was also made from a sample. The peel was collected by tearing it off carefully.

Consolidated Deposits.

Voigt (1949b) and Hähnel (1961, 1962) both give descriptions of lacquer peels made from consolidated sediments and other solid rocks. It is necessary that loose material crumble from the section if the operators are to be able to separate the peel from the subject. Hähnel (1962) states that "hardly any difficulty will be encoun- tered with sandstone or friable or weathered granite, but should it prove difficult to remove the film, it must be chiseled off."

Both of these authors discuss the preparation of a lacquer peel from a strongly folded and faulted schist (Stinkschiefer) of the Middle Zechstein from Lieth, near Elmshorn, Germany. This peel was prepared with the spraying method. The upper and lower sides of the rein- forced back were directly connected to two strong laths, of which the upper one was an- chored with a strong rope. The dry film had to be chopped loose. (It is important for the prepara- tion of such a peel that all the protruding parts are free of dirt and that these parts, as well as the holes, are well covered with thinned and undi- luted lacquer. The peel can be glued to thick masonite or fibrous board when the back is flat; otherwise it must be hung in a trellis.)

At the Geological Institute at Hamburg, Germany, there are peels made from Cambrium, Ordovician, Old Red, Carboniferous (coal seam), Keuper, and Turonian (limestone). At

Fig. 1.37. Peels from diatomites (Riss-Würm interglacial). The right peel is made with Tornol profile lacquer B (Fil Co.) and acetone; the left one with Mowilith 20 (Hoechst) in ethyl acetate. The bedding is vaguely visible in the right peel; the left one cracked by peeling off. Hollerup, Jutland, Denmark.

the Greifswalder Institute, Germany, there is a peel collected from Cretaceous deposits with chert concretions.

1.1F Three-Dimensional Peels

For display purposes it is advantageous to make peels of several related faces. One can make several types of three-dimensional peels. The simplest method is to make two or three single lacquer peels, which represent two (three) sections at right angles to each other (see Fig. 1.10).

The peel can also be made in one piece, of which parts are collected from two or three faces. These are the easiest ones to interpret, since the observer must realize that the peels are mirror images of the sediment sections, and that consequently all directions are turned 180°.

Voigt (1947) described the procedure of making a block peel, which is a cube or rectangular body of which five or six sides are represented. Of course it is not necessary to make a spatial peel with all sides at right angles to each other. A peel can even be made of completely irregular bodies, like concretions. The observer must realize that he is dealing with real sediment and not with a copy of a model; they only are mirror images. For this reason the present author prefers a two-or three-faced peel to a complete block, especially when sedimentary structures must be examined.

Examples of spatial peels were previously given by Voigt (1936d; and later discussed in 1949b). Also Hähnel (1961, 1962) briefly describes the procedure. A spatial peel can be obtained from all sediments of which a two-dimensional peel can be made. This work requires patience, careful handling, and experience in the method of making lacquer peels. The main problem is that absolutely vertical faces must be made when preparing a block, and it is very possible that when the sections are nearly smoothed and ready to be sprayed, a piece of sediment may fall out. Sometimes faces can be brought back a little and good results still obtained; however, it is often smaller in size than was originally intended.

Two-Faced Peels.

A two-faced peel can be made easily from two vertical sides or even slightly inclined sides, or from a horizontal and a vertical section. Instead of making two vertical peels at right angles to each other, one can make three faces which

have angles of 135°. For example, one peel at right angles to the current direction, one parallel, and one oblique.

In every instance it is less risky if the sections are first cut to size, after which one face is smoothed, sprayed, and painted once. Then the following face is treated in the same way. If one section must be horizontal and the other vertical, it is best to start with the one that has the highest position.

During spraying, the nonsmoothed face also will receive some lacquer. This can be avoided by holding a piece of cardboard against it. Another solution is to smooth just a few centimeters of the other section along the edge or to cut the sprayed surface along the edge, when dry, with a razor blade to prevent damaging the first section by pulling off small pieces when levelling the second face. If the sediment is loamy or slightly clayey, both sections can be smoothed before starting to spray.

When both faces are painted they can be immediately strengthened with gauze. The edge must be cut and smoothed very accurately to obtain a sharp peel edge which enables the investigator to continue lines and beds from one side to another.

Both peel sides are mounted on masonite. The sides of the masonite toward the edge must be beveled to allow the peel faces to make good contact after being turned to an angle of 90°. The inner side of the spatial peel can be fixed in position with a piece of wood glued along the edge. The peel can be demonstrated as two dimensional, but this creates some difficulties in determining the orientation (Fig. 1.38).

Three-Faced Peels.

These normally consist of one horizontal and two vertical faces at right angles to each other. The horizontal plane can be a peel from a face that lays either above or below the vertical ones (Fig. 1.39). The mounted spatial peel just represents the contrary (Fig. 1.40).

If in the quarry the horizontal plane lies above the vertical ones, the operator can start smoothing, spraying, and painting the horizontal face, after which he starts the two vertical sides in succession. The gauze should be laid on one at a time, taking care that the bandage strips also overlap the edges. Before removing the peel, orientation, etc., should be indicated. The outer peel edges are first cut loose from the surroundings. Then the peel can be torn off carefully. If

Fig. 1.38. Lacquer peel from rippled sands of Pleistocene age (Drenthe-Warthe glaciation). This is a two-faced peel: the upper part represents the horizontal plane and the lower third part a vertical plane through the deposits. Scale in centimeters. Hamburg, Rothenbaumchaussee, Baugrube NDR. (Courtesy of W. Hähnel, Geol. Inst., Hamburg, Germany.)

the total peel is not too big, one can lift it with one or two shovels and move it away. Sediment parts will fall out and soon most of the material is lost.

Since it would be a pity to cut open one side and turn all the faces, the whole peel should be mounted on masonite and connected with laths, maintaining the real situation. When starting from the situation given in Fig. 1.39b one obtains a peel which represents Fig. 1.39a. The peel should be looked at obliquely from below.

If the situation of Fig. 1.39a exists in the field, it is better to immediately prepare all three faces before starting to spray. The operator can also smooth all three planes roughly, followed by the peeling procedure beginning with the bottom. During spraying, pieces of cardboard can be held along the vertical planes to avoid cementing of grains. When the horizontal plane has reached this point, it can be protected with a piece of cardboard and the vertical faces can be treated successively. Before spraying a vertical face the operator should remove the cardboard and brush and blow loose grains off the bottom part and along the sides since the cardboard will never fit the sides exactly.

Many-Faced Peel: Block Peels.

For the preparation of a cubic, tetragonal, or polygonal form often an abundance of sediment must be dug away in order to obtain enough space around the pillar to enable the operator to

reach each side sufficiently (Voigt, 1947, 1949a; Hähnel, 1962).

If a rather small cube has to be made, and if the sediment is not too loose, one can cut and smooth the upper and the four vertical sides at once. With looser deposits the operator starts with the top side and treats the vertical sides successively. When all the faces are sprayed and covered with a painted coat, the whole cube can be strengthened with gauze (Fig. 1.41a). It may be practical to first strengthen the top before starting with the vertical sides. When all the sides are dry the vertical edges are cut open (Fig. 1.41b), and the peel can be lifted (Fig. 1.41c). A corresponding form is made of fibrous wood on which the film can be mounted (Fig. 1.41d).

If it is desired to obtain a cube of which all sides are present, the cube (which should not be large) is cut from its base when all sides, except the bottom, are covered with gauze, turned upside down, smoothed, lacquered, etc. To collect this peel, seven sides must be cut open (Voigt, 1947). The whole can be mounted on a preformed wooden rack or on a lightweight block.

Large blocks are more difficult to make. They can be shaped roughly beforehand, but it is better not to smooth all sides at once. First the top is smoothed, after which the operator checks the horizontal position of this section with a level. The sides of the top are cut straight with a razor blade to facilitate the preparation of sharp 90°

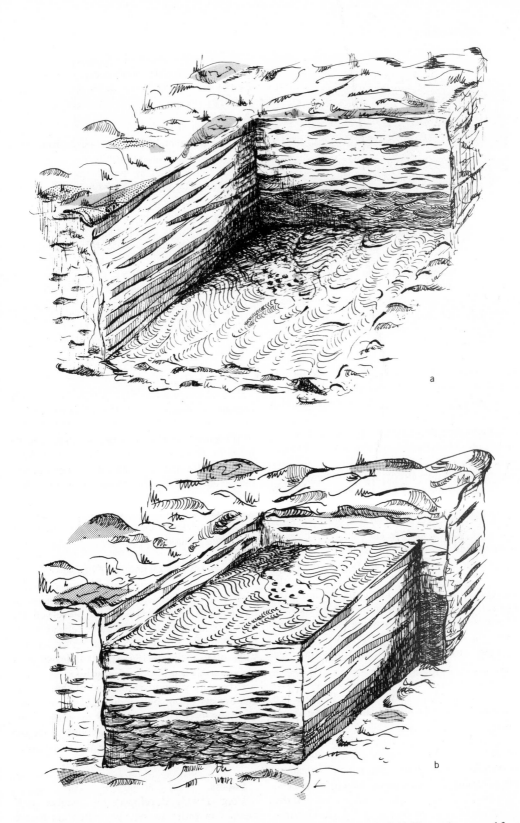

Fig. 1.39. Schemes presenting the two ways a three-faced peel can be made in the field. The peels removed from the sediment faces are mirror images of the original.

Fig. 1.40. Three-faced peel from subrecent Yselmeer deposits. (Compare with Fig. 1.10.) Made with Tornol profile lacquer B and acetone (Fil Co.). Spraying method. S. E. Polder, The Netherlands. (Peel made by J. R. Boersma, Geol. Inst., Utrecht.)

edges (Fig. 1.42a) and then the top face is sprayed, brushed, and strengthened. The first vertical plane is made absolutely vertical and is further treated.

If the layers are dipping (Fig. 1.42a), it is possible that the block will slide, and therefore the up-dip side is taken first. In such cases the first vertical side is also strengthened with gauze and connected to the top. The two faces next to it are then treated correspondingly. Each horizontal or vertical edge, between two finished sides, is also bandaged, and the last face is then ready to be treated.

All edges are cut open when the block feels absolutely dry, after orientation, etc., is indicated at sides and top. The parts are loosened carefully from the pillar with or without the help of a knife or spatula, depending on the type of sediment, and a pillar is left behind (Fig. 1.42b). The parts are laid down and left to dry for some time. Most of the loose sediment can be tapped off. The peels are placed in position and then examined carefully (Fig. 1.42c). If parts are missing, they can be sampled from the remaining pillar.

All peel parts are mounted on masonite with beveled sides. These panels are mounted on a preformed frame. The "sharpened" masonite edges enable the operator to match the panels. The final restoration is now carried out and the block peel can be finished (Fig. 1.42d). The only disadvantage is that the block is a mirror image of the former sediment pillar.

Voigt (1947) glues black paper to the back side of a block when so little clay, coal-containing deposits, or other sediments stick to the peel that they become partly transparent.

Irregular Bodies.

Even irregular objects can be reproduced if they consist of material of which a lacquer peel can be made such as lignite, marl clods, and loam lumps (Voigt, 1947, 1949a; Hähnel, 1961, 1962).

After drying, the peel is cut through the center and turned inside out. The two halves are filled with cotton saturated with lacquer and then glued together.

1.1G Other Applications

The application potential of lacquer peels appears to be enormous. Some possibilities are given by Voigt and Hähnel and will be discussed briefly.

Archeological Purposes.

The lacquer peel method is very valuable for prehistoric studies (Voigt, 1935b). Horizontal as well as vertical sections of the culture strata may be preserved. These show discolorations caused by wood posts and skeletal remains or "corpse shadows" formed by chemical reactions during the deterioration of corpses (Hähnel, 1961, 1962).

The preservation of prehistoric or archeological subjects can be carried out according to the discussed methods.

Biological and Zoological Purposes.

By studying plant associations with regard to their soil characteristics or the succession of associations, examples can be preserved by means of lacquer peels. A similar subject is discussed under the heading Peat based on the work of Franzisket (1962a); examples of soils are given in Figs. 1.33, 1.34, and 1.35.

Especially when investigations of shape, size, and density of burrows of different types of animals, the three-faced peels have excellent application. A slow-drying mixture should be used to obtain a maximum of relief (see also under Araldite Peels).

Fossils.

Voigt (1933) developed the lacquer peel method in order to salvage vertebrate fossils from the Eocene lignites of the Geiseltal in Germany (Fig. 1.43).

Fig. 1.41. (*a*) The preparation of a many-faced small-sized lacquer peel. The sediment block is covered completely with bandage. (*b*) The vertical edges of the block are cut open to facilitate removal of the peel. (*c*) The five-faced lacquer peel after having been peeled off. (*d*) The resulting peel after mounting and finishing. All block faces are mirror images from the sediment faces from which they are removed. Hamburg-Barenfeld. Spezial Präparationslack Z4/924 and thinner Z4/948 (Gustav Ruth). (Courtesy of W. Hähnel, Geol. Inst., Hamburg, Germany; Hähnel, 1961.)

Fig. 1.42. (*a*) Large-sized block peel. The material was not strong enough to cut a pillar directly. The upper surface is prepared for making a peel. Fluvioglacial deposits. Locality same as Fig. 1.11. (*b*) The sediment pillar left behind after the lacquer peel was removed per side. Same locality. (*c*) The lacquer peel strips are put down in position for a careful examination. Same locality. (*d*) A finished large-sized block peel. Height 80 cm. Estuarine deposits (Holstein Interglacial). Hamburg-Lohbrügge. (Photos *b, c,* and *d* courtesy of W. Hähnel, Geol. Inst., Hamburg, Germany.)

Fig. 1.42D (Continued)

Fossils from other deposits such as sandstone, clay marl, and chalk were preserved by Voigt. They included fish, moles, bats, frogs, lizards, reptiles, leaves, fruits, twigs, and fossil impressions. (It is also possible to preserve badly weathered and fractured bones as well as the original position of vertebrates by applying the peel method.)

Lacquer peels made from vertebrate remains even enabled the detection of soft parts such as striated muscles, nuclei, bacteria, hair and fly larvae after examination of the peel under the microscope (Voigt, 1933, 1934, 1935a, 1936a, b, 1937a, b, c, 1938a, c, 1939; Hähnel, 1961, 1962). It could even be proven that frogs, fossilized in the Geiseltal lignites, had died of suffocation since the melanophores of the frog skin were not contracted (Voigt, 1935a).

Micropeels.

Small-sized lacquer peels can be made of sediments and may replace thin sections in certain cases (Voigt, 1935a, 1936a, b, 1937a, b, c, 1938a, c, 1939). These films can even be polished. To obtain a better microscopical examination the peel should be mounted on glass with Canada balsam or polystyrene. The films themselves must be free of air bubbles. To prevent this bubble formation, Hähnel (1961, 1962) indicates that rather than acetone, a less volatile solvent like amyl acetate should be used as thinner. The surface should first be saturated with amyl acetate before lacquer is applied.

The first peel should not be kept since the surface of the sample may be somewhat disturbed by the treatment. After this first film is pulled off, a clear, undisturbed surface is obtained. This is essential for the investigation of small-sized structures in the second peel.

Microlacquer peels also have application in the field of criminology. Small films have been made for the detection of blood traces, textile fragments, hair, and fingerprints. These are then examined under the microscope (Voigt, 1938d).

Fig. 1.43. Lacquer peel of a big snake (*Paleryx ceciliensis* BARNES) in Eocene lignite. Length of the snake is 2.5 m, its number of vertebrae is 243. Peel prepared by E. Voigt (Geol. Inst., Hamburg, Germany), Geiseltal Museum, Halle/Saale (Germany). (Courtesy of E. Voigt, Geol. Inst., Hamburg, Germany.)

1.2 LATEX PEELS

For damp or wet sandy sediments, McKee (1957a) applied a latex cement instead of lacquer. This product had a special advantage for making peels of the deposits he obtained with his flume experiments since a film could be made when the sand had a rather high moisture content (Fig. 1.44).

McKee (written communication, June 2, 1964) prefers a latex to a lacquer peel because

"(1) it fills in irregularities on the surface and does not require a perfectly flat face to sample, (2) it works with wet or damp sand, and (3) it is flexible and not brittle when finished."

The procedure for making a latex peel is similar to preparing a lacquer peel. This procedure will be given as a summary after the written instructions of McKee (1959). It has been described briefly by McKee and Sterret (1961), and examples can be found in papers by McKee (1957b) and McKee et al. (1962a, b).

Fig. 1.44. Latex peels. (*a*) Colorado River flood plain deposits with ripple laminae in rhythm with overlying irregular climbing ripples. Lake Mead, Arizona. (*b*) Transverse dune deposits, main trench. Windward side of dune. Section cut in direction of wind movement, right to left. White Sands, New Mexico. (*c*) Laboratory experiment: delta deposits showing top sets and fore sets with rising water level. String indicates the last water level. The strata are marked with magnetic grains. (*d*) Laboratory experiment: foreshore beach and shoreface terrace deposits. String indicates the last water level. The strata are marked with magnetic grains. (Courtesy of E. D. McKee, U. S. Geol. Survey, Denver, Colorado, U.S.A.)

1.2A McKee Method

Equipment

Shovel, spade, trowel, knife or machete for cutting and smoothing a surface; Krylon Crystal Clear #1301 or 1303 as an aerosol; soft flat paint brush, tin with water for cleaning the brush; Cementex #600 latex cement from the Cementex Company; cheesecloth, nails, pins and scissors; masonite and an adhesive.

Cementex #600 is based on prevulcanized natural rubber latex. Postheat curing is not necessary, although a slight postheat drying will shorten the hardening process (Cementex, 1964). The rubber content is approximately 60%. This and other Cementex products are primarily manufactured for mold-making in commercial casting plaster, cement, and other water-hardening articles (F. M. Nishio, President, Cementex Company, Inc., written communication, March 22, 1965).

The manufacturer recommends keeping the brush in slightly soapy water when not in use. Squeeze the brush until dry before dipping it into the latex. When the brush is too thickly coated, place it in naphtha or benzol until the rubber is loose enough to be combed out.

Procedure

1. Cut a smooth even surface on the sediment. A very smooth face is not necessary; small irregularities will do no harm. Too much moisture will inhibit drying of the latex; but, in general, if the sand is dry enough in a vertical or nearly vertical face, no difficulties will be encountered.

2. Firm up the surface by spraying with three or four coats of Krylon crystal clear. Allow time for each sprayed coat to dry; only a few minutes outdoors in warm weather is sufficient.

3. When the sprayed surface is firm to the touch, start brushing on liquid latex, using a wide, soft brush. Avoid excessive brushing on the first layer. When the section is larger than a few square inches, the latex coat should be reinforced with cheesecloth. The cheesecloth also aids in attaching the peel to the face until it is ready for removal.

The cheesecloth can be applied in either of two ways. One large sheet of cloth can be pinned to the face with small nails before the latex is applied and brushed through the cloth. The other method is to pin a narrow strip of cheesecloth (4 to 6 in.) to the top and side of the vertical face, which should be covered with the first application of latex. With the second application of latex, additional strips of cloth can be "painted in," especially around the edges, where the reinforcement helps prevent tearing.

4. When the first coat of latex is dry (approximately ½ hour in direct sunlight on a warm day, and longer in the shade or in very damp sand) a second coat should be applied. It is advisable to apply a third and a fourth coat, especially with large sections. When the latex dries the color changes from white to yellow-brown.

5. The peel should dry several hours, preferably overnight. After orientation and other data are written on the back, the "pad" of rubber can then be peeled off the face. It is best not to disturb the surface of the peel for an hour or two after removal to ensure thorough drying of that side.

6. When the peel is prepared for permanent storage, it should be finished with a thin coat of Krylon crystal clear, after loose particles are brushed, tapped, or blown off cautiously. The peel can also be glued to masonite, but it is very difficult to eliminate all wrinkles (Fig. 1.44). Peels can easily be rolled up for transport, and they will not be damaged in this position. They can also be trimmed, if desired, with an ordinary pair of scissors.

1.3 POLYESTER RESIN PEELS

Investigators were forced to develop techniques and to find chemicals to avoid the restrictions of nitrocellulose lacquer, which does not penetrate far enough into nondry clay layers and is not able to overcome the cohesion of nondry clay particles.

Maarse and Terwindt (1964) used a polyester resin for their work in the estuarine deposits, exposed in an excavation in the Haringvliet, the Netherlands, which mainly consists of sand, clay, shells, and peat remains (Oomkens and Terwindt, 1960). This type of resin is very valuable when dealing with material rich in clay. The sections can be peeled completely and fine sandy laminae in thin clay layers can be recovered.

This procedure requires less field time than the lacquer methods. The obtained peels are hard and can be finished under water, which saves time and produces better results. (The resins are discussed in Chapter Two.)

McMullen and Allen (1964) studied recent sandy sediments in tidal flats and beaches, where only a short period of time was available during low tide. The sediments were wet and could not be dried sufficiently on the spot, and vertical sections could not be dug because of the high water content.

McMullen and Allen use low viscosity, quick-setting polyester resins based on the liquid-liquid system, where the impregnating liquid penetrates the sample by displacement of water. Depth and rate of penetration depend on gravity, difference in densities of both liquids, viscosity and gelling time of the impregnating liquid, immiscibility of both liquids, and pore and water content of the sediment. A wet sample can be treated only if the impregnating liquid is heavier than water.

The products used by McMullen and Allen also belong to the group of unsaturated polyester resins. Since the application is part of this chapter some characteristics will be given here.

1.3A Method of Maarse and Terwindt

Maarse and Terwindt (1964) carried out ex-periments with several types of unsaturated polyester resins such as Vestopal H from the Chemische Werke Hüls and Synolite from Synres, and with two catalysts, butanox (50% methyl ethyl ketone peroxide in dimethylphtalate) and cyclonox E.B 50 pasta. However, butanox is easier to work with since it can be measured in the field with a pipette.

There proved to be no difference between the types of resins. To decrease the viscosity of the plastic, an amount of styrene monomer was added. However, this increased the gelling and hardening times of the plastic. Cobalt octoate (1% Co) was used as accelerator. The combination *V*estopal H, *Sty*rene, buta*nox*, and cobalt octoate is called VESTYNOX (Maarse et al., 1962; Maarse and Terwindt, 1964).

Two types of Vestynox proved to be useful to cover their problems. Vestynox A has a gelling time of 1 to 2 hours and a hardening time of 5 to 6 days. It should be applied to sediments containing layers of clay and loam. Vestynox B is a useful combination for deposits consisting of sand, shells, and gravels. It has a gelling time of ½ to 1 hour and a hardening time of 2 to 3 days (Table I.2).

Table I.2.   Composition of Vestynox A and B by Volume

Vestynox	Vestopal H	Monostyrene	Butanox	Cobalt octoate
Vestynox A	60	40	2	1
Vestynox B	85	15	4	2

If it is known which type of Vestynox is to be used, the Vestopal H and the monostyrene can be mixed in the laboratory. Should the mixture be applied within a few hours, the catalyst (butanox) can be added too. All quantities must be measured very accurately. The accelerator cannot be added ahead of time since the gelling time is rather short. Increase of gelling time implies an increase of hardening time.

Catalyst and accelerator should never be added together or be mixed with each other since they can react with explosive violence.

About 1500 cc Vestynox is required per square meter.

Equipment.

Shovel, spade, trowel, and a wooden lath to prepare the section; spray gun with water to moisten the section if necessary; resin, styrene monomer, catalyst, accelerator, measure-mixing beaker, 2 pipettes (up to 10 and 20 cc), stirring rod; acetone for cleaning purposes; bandage or cheesecloth in one sheet, 2 wooden clamps; paintbrush, paint roller or sheet of foam plastic (0.8 x 0.5 m), with a similar-sized tray; strong rope and some heavy pins to secure the upper wooden clamp; nitrocellulose lacquer and thinner (acetone); rubber pad of the size as the peel; knife or spatula; transport units for transporting the peel. The beaker in which the mixture is to be prepared must be of sufficient size to hold the total quantity required.

The system is based on the size of the peel. Each of the wooden clamps consists of two laths which are about 30 cm longer than the width of the peel. At each end a hole is drilled through both laths, through which a long bolt (at least 10 cm) fits; this bolt clips the laths by means of a nut. The holes can also be provided with a screw-thread, making the nut unnecessary. A piece of cheesecloth must be clipped between both clamps (Fig. 1.45A, B). The lower end of each bolt is pressed into the sediment to keep the clamp in position so that the gauze is in contact with the section.

The use of a paintbrush to apply the mixture to the section will produce difficulties as long as the deposit consists of coherent sediments such as clay layers. If coarse sand, shell debris, or complete shells are present, sediment will roll down and the section will be damaged (Fig. 1.45C). Loose sediment will fill the space between the section and gauze, and, as a result, the resin mixture will not penetrate deep enough into the section. The application of a paint roller produces better results since the pressure is spread more evenly (Fig. 1.45D). For noncoherent sediments a piece of foam plastic is saturated with plastic and then pressed carefully against the gauze (Fig. 1.45E). All instruments should be cleaned in acetone directly after use.

Preparation of the Peel

The section must be prepared very accurately and must be very smooth, since the gauze must completely touch the surface of the sediment. The face should be vertical. Maarse and Terwindt use a lath to make the surface completely smooth. The sediment should not be dry; if necessary, it can be sprayed with water.

Above and below the section, a step must be cut for placing the clamps. First the gauze is clipped in the upper clamp, which is then placed in the step (Fig. 1.45). If the sediment is rather loose, rope and long pins can be used to secure the laths. The gauze is now placed against the face, and the lower clamp is mounted and put in place. The gauze should be stretched slightly over the section.

The resin mixture can be made and applied to the surface. This operation need be carried out only once. It takes about 4 to 5 days before the Vestynox is hardened.

Recovery of the Peel

Maarse and Terwindt remark that it is advisable to apply one coat of nitrocellulose lacquer, slightly thinned with acetone, after the Vestynox is dry in order to strengthen the gauze. This coat should dry within an hour.

Above the section, especially when it is clayey, sediment is dug away to obtain a stand where the peel can be loosened with a spade.

If the section is about 1 m² or larger, three persons are required to collect the peel. All sides are loosened with a knife or spatula. A rather strong rubber pad (about 0.85 cm thick and of the same size as the peel) is pressed to the section by two men. The pad and clamp are held together and the rest of the pad is against the section as the peel is loosened by a third person, while the other two operators bend the rubber pad and loosen part of the peel from the section, thus reducing peel damage as much as possible.

Transport, Drying, Mounting, Restoration, and Finishing

Since the peel is stiff and not flexible, and therefore limited in size, it is necessary to handle and to transport it carefully. The peel should be transported on a board.

At the laboratory the peel should be treated as soon as possible. If clayey, it must be immersed in water for 24 hours, after which the nonhardened sediment can be washed off easily. This reduces the risk of damaging the sometimes very fine structures of the sediment. Nonclayey sediments can be treated with a soft brush or by blowing air over the peel. A small preparation knife or needle may be necessary to loosen some nonimpregnated parts. The clay layers must be cut away so that surrounding sand layers protrude far enough.

The peel should be mounted either on masonite or on wooden or fibrous board depending on its size. When the peel is dry, the clay layers should be painted with the Vestynox A mixture to reduce shrinkage, while the rest can be sprayed or brushed with nitrocellulose lacquer or one of the other products mentioned earlier.

Combination of Lacquer and Polyester Resin

After the section is smoothed, a nitrocellulose mixture coat can be applied to it with the paint roller, followed by application of the Vestynox mixture. The resin partly dissolves the lacquer and thus increases the hardness of the peel.

This combination has many advantages. The resulting peel is stiff and rather thick. When using only Vestynox, the structures in the sand do not stand out very well, but a combination of lacquer and plastic makes it possible to preserve both the sand structures and the clay layers (Fig. 1.46).

Comments

Many other resins produced good results, but it was necessary to gain experience with those

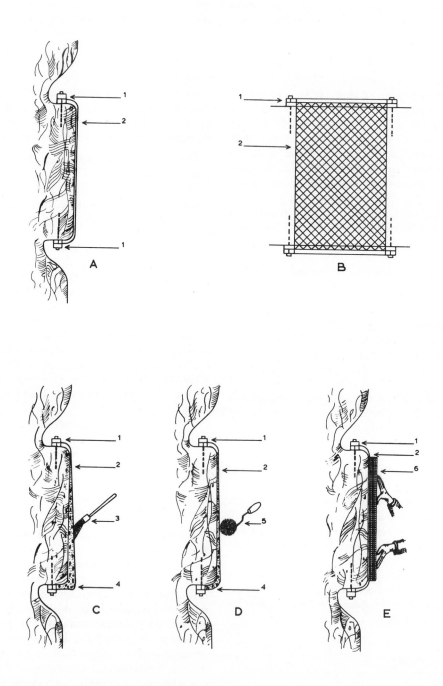

Fig. 1.45. Schematic presentation of methods used to make a Vestinox peel. (A) View from the side. (B) View from the front side. (C) Method for using a paint brush. (D) Method for using a paint roller. (E) Method for using a sheet of foam plastic: (1) wooden clamp with long bolt; (2) cheesecloth; (3) paint brush; (4) concentration of loose grains that are rolled down; (5) paint roller; (6) sheet of foam plastic saturated with plastic mixture. (Redrawn from Maarse et al., 1962.)

Fig. 1.46. Lacquer-Vestinox A peel made from estuarine deposits. Excavation in the Haringvliet, Delta project, The Netherlands. (Courtesy of J. H. J. Terwindt; Rijkswaterstaat, The Netherlands; Maarse and Terwindt, 1964.)

polyester resins which harden under wet and moist weather conditions and do not keep a very tacky nonhardened surface. Complete hardening of the surface often occurred after several days in the laboratory. The combination of nitrocellulose lacquer and unsaturated polyester resin is preferable to resin only when sand is present in the section.

Vestynox can be applied to wet sediments but not to completely saturated material. This resin mixture is also very suitable for making peels of clayey cores. The procedure is more or less similar to the one discussed later as the Pouring

method. About 50 cc Vestynox is required for a core of 60 cm.

The present author made peels using Vestopal (Chemische Werke Hüls) as well as Plaskon (Allied Chem. Corp.) without using monostyrene. Sometimes the sections were first sprayed with lacquer and next the plastic mixture (Vestopal 100 cc, butanox 2 cc, cobalt octoate–1% Co-0.8 cc; Plaskon 100 cc, butanox 1 cc, cobalt octoate-1% Co-0.6 cc) was poured over the section. This procedure required a little more resin, but less acetone was needed for cleaning purposes. Because of their small size, it was possible to collect the peels without any rubber pad. After mounting on masonite, they were finished with Krylon 1301 (Fig. 1.47).

1.3B Method of McMullen and Allen

The technique introduced by McMullen and Allen (1964) makes it possible to preserve sedimentary structures in wet unconsolidated or partly consolidated sandy deposits. The investigators use the Senckenberg sample boxes to collect a sample. The boxes must only be carried out of reach of the water for the time necessary for the resin to harden. The hardening takes about ½ to 1 hour.

McMullen and Allen apply two variations within their method. The plastic mixture can be either poured on the section or sprayed. Vertical sections as well as horizontal surfaces can be preserved. For surface structures it is not necessary to take a sample before. As long as the deposit is kept away from the water for about 1 hour, the resin can either be poured on flat surfaces or sprayed directly.

The polyester resins used have a specific gravity of about 1.1 and viscosities ranging from 100 to 2.000 centistokes at ordinary temperatures. The resins are not miscible in water, but they are freely miscible with small amounts of styrene monomer. The styrene's specific gravity of about 0.91 causes it to reduce the density and viscosity of the mixture, but it increases gelling and hardening times. Before going into the field the resin may be precatalyzed to facilitate field handling.

Temperatures of the wet sand, as well as of the air, influence the gelling time. The ideal weather condition for applications outdoors is a hot, cloudless day with no wind (>21°C, 70°F). However, McMullen and Allen have applied it satisfactorily on a breezy, overcast, cool day (< 15°C, 60°F). These polyester resins should not

Fig. 1.47. Plastic peel made of fluvioglacial deposits which show periglacial deformation. The section was sprayed first with nitrocellulose lacquer (General Electric Glyptal 1202, acetone, toluene) and then a mixture of 500 cc plastic (Plascon 800 cc, Butanox 8 cc, NL-49 5 cc) was poured onto it. Locality same as Fig. 1.9. Width of the peel is 45 cm. (Photo by H. Nyburg and J. H. Elsendoorn, Geol. Inst., Utrecht.)

be used in strong winds or with temperatures below 10°C (50°F). Suitable resin mixtures are those that gel after about 20 min. (Table I.3). For spraying application the gelling time will be longer since the mixture is somewhat diluted with monostyrene.

Crystic 189 LV (low viscosity) and Crystic 191 E of Scott Bader have viscosities of 3.5± 1 poise and a specific gravity of 1.15 and 1.11, respectively. Catalyst Paste H, and as accelerator type E (Scott Bader, 1965, pp. 82–88) are used. Gelling times, depending on the amount of accelerator added to a precatalyzed mixture and the temperature, are given in Fig. 1.48 and

Table I.3. The Scott Bader products can be used for pouring as well as for spraying applications.

From BXL Plastics Materials Group Ltd. (formerly Bakelite Limited) the resins DSR.19098 (which replaces the now obsolete SR.17497 used by McMullen and Allen) and SR.19038 with catalyst Q.17447 and accelerator Q.17448 produce good results with the pouring method, but they take too much time to set when used as spray. The viscosities are slightly lower than those of the Crystic products (Bakelite 1962, 1963; Table I.3). The accelerator is a solution of cobalt naphthenate in styrene containing 1% of cobalt metal (Bakelite, 1958a). It is used in combination with catalyst Q.17447, which is a solution of methyl ethyl ketone peroxide in dimethylphthalate, containing approximately 60% of the peroxide (Bakelite, 1958b).

The catalyst and accelerator should never be mixed together. The manufacturer states that it is essential that the catalyst is added and mixed with the resin before the accelerator is added.

Pouring Method

Equipment. Sample box with loose bottom plate (McMullen and Allen use a box with sizes 30.5 x 29 x 7.5 cm, 12 x 11½ x 3 in.), spade; polyester resin, monostyrene, catalyst, accelerator; beaker glass with stirring rod, measuring cylinder of 25 cc or a calibrated pipette; polythene squeeze bottle with acetone for cleaning purposes; gauze, knife.

Procedure. A sample is taken with the sample box. (This operation is discussed in Chapter Six). The box parts can be turned in such a way that a vertical sediment face is exposed. This surface is roughly leveled to allow pouring the resin in a uniform distribution. A piece of gauze or cheesecloth slightly larger than the exposed face is placed on top of the section. The resin-catalyst-accelerator mixture is made—taking care to measure the quantities accurately—and then poured on carefully and uniformly.

When the gelling time is about 20 min., the operator can free the gauze from the sides of the box 1 hour after the resin is poured. The sides of the box can be freed with a knife, and the peel can be removed by lifting it, starting from one edge. It is then placed on a flat surface with the reinforced side down. Another 24 to 36 hours are required before hardening is complete and before excess grains can be washed off with water to reveal the structures more clearly (Fig.

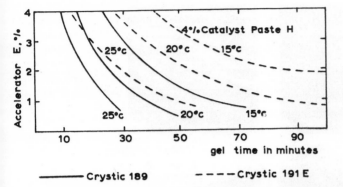

Fig. 1.48. Pot life curves of Crystic 189 and Crystic 191E. (Redrawn from Scott Bader, 1965, Figs. 21, 22.)

Table I.3   Characteristics and Composition of BXL Plastics Materials Group Ltd. and Scott Bader Products Used to Preserve Structures in Wet Sand

Manufacturer	BXL Plastics Materials Group Ltd.		Scott Bader and Co. Ltd.	
Resin	SR. 19038	DSR. 19098	Crystic 189 LV	Crystic 191 E
Appearance	pale yellow clear liquid	very pale clear liquid	straw clear liquid	water clear liquid
Viscosity at 25° C (77° F)	250 centistokes	100 centistokes	2.5–4.5 poises	2.5–4.5 poises
Specific gravity	1.11 (25° C)	1.1 (25° C)	1.15 (25° C)	1.15 (25° C)
Stability in the dark − 20° C (68° F)	>6months	>6 months	6 months	6 months
Catalyst	Q. 17447	Q. 17447	Cat. Paste H.	Cat. Paste H
Accelerator	Q. 17448	Q. 17448	Accel. E	Accel. E
	Composition for a gelling time in minutes at 25° C (77° F)			
Resin in grams	100	100	100	100
Catalyst	4 cc	4 cc	4 g	4 g
Accelerator	2 cc	2 cc	4 g	4 g
Gelling time	14	13	10	15

1.49). The peel can be mounted on masonite and then finished.

Spraying Method

Equipment. Spade, knife, trowel; polyester resin, styrene monomer, catalyst, accelerator; 1 liter polythene beaker with stirring rod, 25 cc polythene measuring cylinder; polythene squeeze bottle with acetone for cleaning purposes; air compressor or a cylinder with compressed air, spray gun. The spray gun can be an ordinary one; a nipple is added to facilitate connection to an electrically driven compressor (laboratory procedure) or to a small cylinder containing compressed air (field procedure). McMullen and Allen use a gas pressure of 10 to 15 lb/in.^{2} (0.7 to 1.05 kg/cm^2).

Procedure. This can be carried out in the field as well as in the laboratory to preserve sections or surface parts of structures and other charac-teristics. Since the polyester mist is a strong irritant to the throat and nose, it is advisable to spray, if possible, out of doors.

The surface to be preserved is first marked out with a knife, trowel, or spade to get a neat shape with sharp edges. Surfaces up to 30 x 30 cm (12 x 12 in.) can be treated without special precautions, but larger areas make it difficult to ensure a sufficient and uniform spraying with the small-sized spray gun used. Large surfaces are also difficult to lift without causing breakage.

The resin should be diluted with 10 to 20% styrene monomer to obtain a sprayable mixture. (The mixing ratios for a rather short gelling time are given in Table I.3). The amount of time involved before the preserved surface or section is hardened enough to be removed depends on the weather. Under bright, sunny conditions 1 hour will be sufficient, but more time is necessary during cool, cloudy days. At the laboratory the length of the hardening time can be decreased by using infrared lamps.

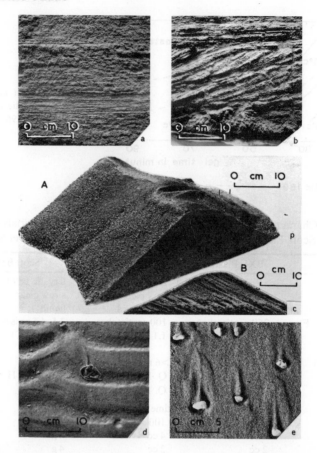

Fig. 1.49. Peels made from wet sandy sediments. (*a*) Vertical section of a channel edge on beach. Inclined lami-nae are visible near the base. The peel face is parallel to the current direction. Near Poole, Dorsetshire, England (pouring method). (*b*) Vertical section of a slipoff edge of a tidal scour pool on a tidal flat. Foreset beds are pass-ing into topset beds; near the right-center an *Arenicola* burrow can be observed. Near Poole, Dorsetshire, En-gland (pouring method). (*c*) Underwater dune made in a laboratory flume. Note the current scours on the stoss side and the avalanche tongues of sand on the lee side with grain size becoming coarser toward the periphery of each avalanche. The vertical section presents foreset beds (spraying method). (*d*) Tidal flat surface with wave-planed asymmetrical ripples and *Arenicola* castings and siphone. Near Poole, Dorsetshire, England (spraying method). (*e*) Current crescents formed in laboratory flume (spraying method). (Courtesy of J. R. L. Allen, Univ. of Reading, Great Britain; McMullen and Allen, 1964.)

The peel must be collected very carefully since the product needs much more time to be hardened completely. The peel must first be loosened from the underlying sand by applying a slight twisting motion, after which it can be slid onto a stiff metal sheet. Mounting and finishing can be carried out as in the pouring method (Fig. 1.49).

Remarks

To avoid any possible risk of damage to the peel during its removal, rather than lifting the peel off, place a piece of metal or wood on the gauze-side, after the sides are cut loose, and turn the entire peel and sample box upside down. The box can then be lifted off and the sed-iment and peel are left to dry as long as possi-ble. If it is possible to leave the peel in this posi-tion for a few hours, those grains which are only partly cemented by the resin will be bound much more strongly to the film.

The relief obtained in vertical sections cannot be compared with the one obtained with nitro-cellulose lacquer applied under favorable condi-tions. This application is especially unique for the preservation of structures with wet surfaces.

When the peeling as well as the spraying technique is done in place, a steel plate should be pushed under the hardened peel, enabling the operator to place it higher (outside the reach of tides), allowing it to harden further before lifting it up.

The results obtained by these methods are hard enough to make thin sections of.

1.4 INSTANT PEELS

Dr. Alan Thomson of Shell Development Company, Houston, Texas, successfully developed a method to prepare a sedimentary peel of sandy material onto a colorless transparent material (personal communication, January 1967). The advantage of his method is that it takes very little time to make the peel. It can be studied with normal reflecting light as well as with transmittant light (Fig. 1.50).

1.4A Products and Procedure

Chemicals

Thomson uses an epoxy resin from Shell Chemical Company, Plastics and Resins Division. It is a light colored, epichlorohydrin/ bisphenol A-type epoxy with a low molecular weight sold under the name Epon Resin 828. Its viscosity at 20°C is 100 to 160 poises (Shell, 1962a, b). The hardener used is Genamid 250 liquid epoxy coreactant from General Mills, Inc. (1961, 1966). This highly refined resinous amine adduct has a viscosity of 5 to 10 poises at 25°C. It should be stored in a tightly closed container in a dry area at about 10°C (50°F).

Equipment

Epon Resin 828 and Genamid 250; acetone for cleaning purposes; plexiglas plates, 13 x 20 cm (5 x 8 in.) and 6 to 3 mm (¼ to ⅛ in.) thick.

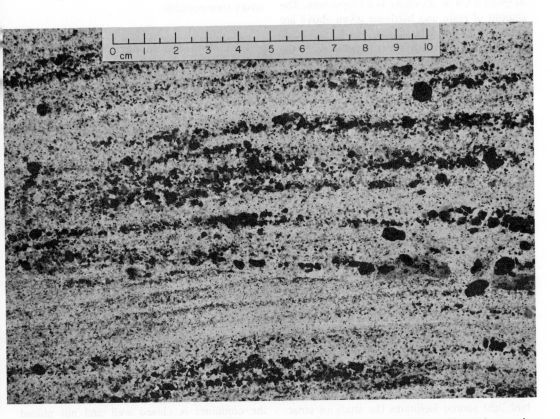

Fig. 1.50. Photograph from an instant peel using a transmittant light source (copying box). The peel was collected from a medium scale cross bedded zone. Brazos River deposits, south of Richmond, Texas, U.S.A. (Courtesy of Alan Thomson, Shell Development Co., Houston, Texas.)

These plates form the base of the peel. For measuring quantities and stirring the mixtures one can use disposable laboratory cups and tongue depressors. Also needed is a roll of thin filament tape, a knife, towels or rags, and a tungsten carbide pencil for labeling.

Procedure

1. Shave a sedimentary surface of a size slightly larger than 5 x 8 in. It should be very smooth.

2. Use the tape to make a ⅛-in. (3-mm) rim around the plexiglas plate to contain the fluid mixture. Mark the plate with number, arrow, etc.

3. Mix 60 cc Epon Resin 828 with 40 cc Genamid 250 (Thomson found this to be the most suitable mixture ratio). The curing process is exothermic and dependent on the total volume, so this ratio may not be correct for making larger peels. Since it is difficult to get larger sedimentary surfaces completely smooth, Thomson suggests 5 x 8 in. plexiglas as a proper size. The amount of resin and hardener given above are related to this plexiglas size.

4. When the mixture starts to get warm in the mixing cup spread it on the plexiglas board. This is extremely critical because once the mixture gets warm in the cup the curing reaction proceeds at an accelerated rate. When it is spread on the plate in a thin layer, however, the reaction slows down. The viscous mixture is usable for approximately 1 hour, depending on the air temperature.

5. When the mixture thickens and does not run off, it is time to make the peel.

6. Place the plexiglas tray against the shaved section for a few seconds, being careful not to move it during this time.

7. Remove the tray and place it on a flat surface. Allow it to completely harden overnight.

8. Brush off all uncemented particles. The tape is glued on by the mixture.

9. The peel can be placed on a copying box for study and black and white and/or color pictures can also be taken (Fig. 1.50).

These instant peels are ideal for collecting data of an unconsolidated sediment body. Little field time is involved, and the use of a transmittant light source facilitates the study of structures and textures.

1.5 GLUE PEELS

One method for making peels of wet sediments is to use products that are miscible with water. In this way the gluing molecules can penetrate the water-filled pores. When the molecules have enclosed the largest part of a sediment particle, the binding to the peel will be stronger than the cohesion of the particles, especially when dealing with clays. The difficulty with clays is that the pores are rather small compared to the size of a glue molecule and this prevents deep penetration of glue into such fine-grained sediments. The peels need little space for storage and do not dry out, shrink, or break into pieces as the corresponding core eventually does.

Glue peels dry only in hot, not humid climates. In dry, warm climates these peels can be made with the same techniques as applied with nitrocellulose lacquer. Glues dissolved in water can be applied for dry sediment. An advantage is that no acetone is required and that glues are easily transportable.

1.5A Method of Snodgrass

Snodgrass (1960) uses Weldwood Plastic Resin Glue of the United States Plywood Corporation. This glue is a light tan-colored, cold-setting resin adhesive of the urea-formaldehyde type, in powdered, water-soluble form (U. S. Plywood, not dated). The product is unaffected by oil, gasoline, and solvents; it resists the growth of fungi or bacteria, insects cannot harm it; and it is not harmful to skin if normal cleanliness standards are used. Since this product is hygroscopic, it should be kept free of moisture during storage, for it becomes insoluble when in contact with water.

One hundred parts of resin glue can be mixed with 60 parts of water (parts by weight), but Snodgrass suggests a more viscous mixture like an S.A.E. 30-weight oil which is used in cars. Experiments have shown that 100 g of glue powder with 50 cc of water form a suitable mixture. It does not cure below 20°C (68°F). A workable temperature lies between 23 and 30°C (74 to 86°F). The resin glue in original powder form can be stored more than a year provided the container is closed well and not placed where the temperature rises above 37°C (99°F).

Equipment

Two semicylindrical pipes of plexiglass, plastic or zinc, wire, spatula, knife and cylindrical stop to cut a core in half and to smooth its surface, gauze, scissors; glue, water, balance, mixing beaker or empty tin with stirring rod, brush; air-tight container, wet sponge or pan with water, infrared heat lamps or other temperature raising outfits; strips of masonite, strips of white cardboard, adhesive, brush, piece of foam plastic; protecting clear lacquer.

Preparation of the Peel

The core is extruded from its core barrel into a semicylindrical holder. An identical half-cylinder is placed on top and with a thin wire, the core is cut into two halves. The two halves can be separated with a spatula (for more detailed instructions, see the section on the Pouring Method).

The half-core should not be left open to avoid drying out and forming cracks. If it is left open, the glue will penetrate the cracks and form unnatural rims on the peel. According to Snodgrass, the water content is not critical, although one should be cautious with coarse, dry sands since thick peels of over half a centimeter may result. When the peel cannot be made directly after opening, the half-core should be stored in a sealed plastic tubing.

On the core surface a strip of bandage is centered; the strip is smaller than the core width to avoid cementing to the tube sides. The bandage should be longer than the sediment sample to facilitate collecting the peel. The Weldwood Plastic Resin Glue powder is now mixed with water. To avoid lumps, the water should be added to the powder slowly, in small increments, with slow, strong stirring motion. Additional water is then easily added to achieve the desired viscosity without the formation of lumps. The brownish color does not interfere with the coloration of the sediments.

With a small brush the gauze is saturated with the glue mixture. A thickness of about 2 to 3 mm is sufficient. If there are any cracks or holes left in the core, do not spread the mixture over the bandage above them. The working life of the glue at 21°C (70°F) is 4 to 5 hours, at 26.6°C (80°F) about 2½ to 3½ hours, and at 32°C (90°F) 1 to 2 hours.

The core with the peel is now placed in an air-tight container during the curing of the glue.

Drying out of the core should be prevented and the container should therefore be kept at 100% humidity by means of a wet sponge or a pan of water. Snodgrass constructed a sealed plywood box with shelves and a hinged door at one end, capable of holding 12 cores. It can also be done by sealing two semicylindrical core barrels or two molds, used for the imbedding of sediment slices in plastic or plexiglas. Two 250-watt infrared heat lamps can be used, by directing them upon the box sides to raise the inside temperature. At a temperature of 25°C (77°F) a period of about 96 hours is necessary for curing. On land cores the possibility of fungi growing may be present. Snodgrass therefore recommends the use of a translucent container or ultraviolet light.

Collecting, Mounting and Finishing

Before collecting the peel test to see if the glue is well cured after a period of 96 hours. It should be strong enough so that it is not possible to penetrate the surface by touching, scraping, or indenting it.

Peels of clays must be especially well cured. If in doubt, test a small portion of one end so as not to ruin the entire peel. If the peel is ruined, the procedure can be completely repeated after the half-core is smoothed again.

After the core with the peel is taken out of the air-tight container, the peel is collected by hand, starting at one end. Avoid only sharp bending of thick sandy peels because of damage. Excess bandage can now be cut off.

The peel should not be left to dry before it is mounted, or curling will develop. A piece of white cardboard is mounted on a similarly sized plywood or masonite strip. The strip should be 2½ cm (1 in.) wider than the peel and 5 cm or more longer, to have space for text. The cardboard can be mounted with several types of paper-wood binding glues. Care should be taken not to press the peel to the cardboard before a piece of foam plastic is placed between peel and fingers.

When the mounted peel is dry, the sides can be tapped gently to dislodge the nonglued particles. Compressed air can also be used. The text is written under or above the peel, and the peel is finished with a protective sprayable clear lacquer (see Fig. 1.51).

Brushes and other equipment should be thoroughly cleaned with hot water immediately after use. Caustic soda solution can be used for faster cleaning but must be immediately rinsed in clear water. Remove all glue from hands, clothing, etc., with cool water before it dries. Once the glue hardens, it cannot be washed off. If the operator is hypersensitive, minor skin irritation may develop. To avoid this rubber gloves should be used, and the skin should be frequently washed and smeared with certain products (see discussion of Protection and Cleaning in Chapter Two).

1.5B Method of Heezen and Johnson

Heezen and Johnson (1962) were faced with the problem of preserving over 2500 long, deep-sea piston cores in the Lamont Geological Observatory collection. The cores were 2½ in. in diameter and had an average length of 10 m. From the time they were split lengthwise, photographed, logged, sampled, and stored in enameled metal trays, they slowly dried out, formed cracks, and developed saline crusts which hid the true texture and color of the sediments.

The application of a pure white polyvinyl acetate emulsion glue, Elmer's Glue-All, from the Borden Co. (Borden, 1961, 1965), proved to be very suitable. The glue can be diluted with water to decrease its viscosity. It dries clear.

Equipment

Tools to split a core and to smooth its split surface (see discussions of Pouring Method and Method of Snodgrass); glue, small beaker glass with stirring rod, measuring cylinder, water; scissors, bandage or cheesecloth, knife, masonite, cardboard, adhesive, finishing lacquer (Krylon Crystal Clear #1301 or 1303).

Procedure

When the core is split and the freshly obtained surface smoothed with a knife or a spatula, the surface is placed horizontally. Wet clays, silts, and sands should be allowed to dry for 3 to 4 days before making a peel. A layer of diluted glue (Table I.4) is poured with a thickness of 3 mm (⅛ in.). The mixture can easily be adjusted to fit the permeability of the treated sample. It should be left to soak for about 10 to 15 min.

Next, a piece of bandage, cotton muslin, or cheesecloth is placed on top of the glue and pressed in gently. Another layer of undiluted glue is added on top of the cloth so that it becomes thoroughly impregnated. It is left to dry for 24 hours.

Fig. 1.51. Glue peels made from wet sandy-pelitic sediments according to the method of Snodgrass (1960) applying Weldwood Plastic Resin Glue. Tidal flats near Wilhelmshaven, Germany.

When collecting the peel, use a knife to free those parts that are not easily lifted up. The excess cloth is cut off. The peel can now be mounted to a cardboard-masonite combination, as in the method of Snodgrass, or to another suitable surface. The peels should be sprayed six or seven times with Krylon Crystal Clear, allowing 5 min drying time between sprayings. Any loose grains are now cemented to the peel, while at the same time it is protected against saline crust formation or the growth of fungi or bacteria. The relief obtained produces much bet-

ter information on the structural content of the sample than the split surface will (Fig. 1.52).

1.5C Comments

Using the same procedure as that of Heezen and Johnson (1962), some experiments have been made on other glue products. Only those products that produce similar or reasonable results will be mentioned here.

Cetaflex from the Ceta Bever Company (Ceta

Table I.4 Suggested Glue Mixtures (Elmer's Glue-All) Used for Some Common Sediment Types

Sample Type	Glue Mixture	Remarks
Hard dry clays	1 part water/4 parts glue	If necessary, scrape the
dry silts	1 part water/6 parts glue	surface clean of saline crust
dry sands	Undiluted glue	
Gravels and dia-	Undiluted glue	Impregnate backing cloth with
tomaceous oozes		glue and press firmly on ooze or gravel
Wet clay	1 part water/4 parts glue	Let dry 3-4 days before
Wet silts	1 part water/6 parts glue	applying glue
Wet sands and	1 part water/5 parts glue	Let dry 1-2 days before
diatomaceous oozes		applying glue

After Heezen and Johnson (1962).

Bever, not dated a) is a polyvinyl acetate dispersion whose viscosity can be lowered by adding up to 20% water to it. This lengthens the curing time, but it will not exceed more than 12 to 20 hours. This product should not be applied at temperatures below 8°C (47°F). The results are comparable to those obtained with Elmer's Glue-All (Figs. 1.53a, b, c).

Several Hoechst AG products (Kalle Company) were tried out with varying results. Glutofix 600 is a cellulose glue of which a stock solution can be made by mixing 4 g powder with 100 cc water. It should be stirred until the particles no longer sink to the bottom. It then is left for about 30 min, and then stirred again in preparation for use. It can be diluted from the 1:25 stock solution to a 1:50 mixture (Kalle, not dated a). When following the same procedure as

given by Heezen and Johnson, the curing time is about 24 hours at a temperature of 20°C (68°F). This product has already been used by Neuenhaus (1940). Good sediment penetration is obtained, especially in sandy parts. The binding to the gauze is not strong enough to collect the film, and therefore the back should be strengthened with a coating of undiluted lacquer. The obtained peel shows a good relief, but is not always sufficiently detailed (Fig. 1.53c). A slightly more viscous mixture will improve the result.

Some experiments have also been made from the Tylose C group (Carboxymethylcellulose), types C 10, C 300, and C 600, and from the Tylose MH 20, MH 200, and MH 2000. If these products are stored dry, their shelf-life is unlimited. The Tyloses are hygroscopic, comparable to paper.

Fig. 1.52. Glue peel made from marine varves according to the method of Heezen and Johnson, applying Elmer's Glue-All. The sequence consists of laminae of olive-green lutite alternating with laminae of white diatoma-
ceous sediments. North wall of the Cariaco Trench, 10°47′ N-65°07′ W, depth 392 fathoms, 550 to 560 cm below top of core V12-98. (Courtesy of B. C. Heezen, Lamont Geol. Observatory, U.S.A.; Heezen and Johnson,
1962.)

The Tylose C-types are soluble in water at
any temperature, but they are insoluble in organic solvents. However, certain amounts of
various water-miscible solvents such as ethyl
alchohol, glycerin, glycol, and acetone may be
added to the aqueous Tylose solutions for faster
drying (Kalle, not dated b, c).

The Tylose types can be poured into water at
an even rate, while the solution is stirred briskly.
Stir vigorously during the first few minutes after
pouring. Occasional stirring will be sufficient
while the mixture is allowed to stand until a
homogeneous, pastelike solution is formed. The
stirring intensity is very important, and therefore a mechanical stirrer should be used. These
products tend to form lumps when stirred into
cold water. The stock solution should not be
highly viscous, and certainly not pastelike, or
further dilution will be difficult. Ceramic, plastic,
stainless steel, and glass vessels are preferred
for dissolving and storing, rather than wooden or
metal ones. To keep the viscosity low, it is necessary to dissolve only a few percent in water

(Kalle, not dated d, e). The viscosity decreases
only slightly as the temperature rises, except
Tylose type MH, which coagulates at about
80°C.

By applying solutions of 2% of Tylose in
water at the same temperature (20°C, 68°F) and
with the same type of sample, it has been found
that the peeling results with the C types are not
very good but with the MH types are fairly
good. Curing times are about 36 to 48 hours.
The penetration is only a little less than with
Elmer's Glue-All*(Fig. 1.53). As with the Glutofix 600, the binding to the gauze is negligible,
and therefore the back of the peel should be
strengthened with a coating of undiluted lacquer. The best products of the Tylose series are
Tylose MH 20 (10 g in 90 cc water) and Tylose
H 20 p. The last forms the most clear solution
(Kalle, not dated f).

At the Geological Institute at Kiel, Germany,
synthetic resin dispersion binders, sold under a
variety of trade names in paint stores, have been
successfully used in the field and in the labora-

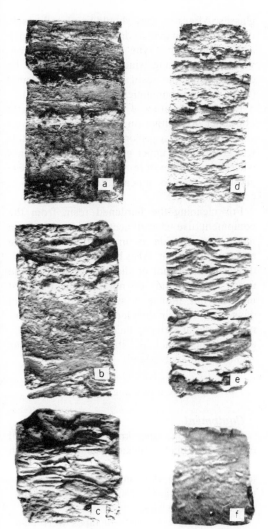

Fig. 1.53. Different types of glue peel and a nitrocellulose peel made from the same types of sediment. (*a*) Elmer's Glue-All glue peel (Borden Company). (*b*) Elmer's Glue-All glue peel (Borden Company). (*c*) Glutofix 600 glue peel (Farbwerke Hoechst AG). (*d*) Tylose MH 20 glue peel (Farbwerke Hoechst AG). (*e*) Nitrocellulose lacquer peel, Tornol A and thinner V 105 (Fil). (*f*) Cetaflex glue peel (Ceta Bever Company). Tidal flat sediments: (*a*), (*b*), and (*f*), Wilhelmshaven, Germany; (*c*), (*d*), and (*e*), outer side of the dike of the excavation pit in the Haringvliet, Delta project, The Netherlands.

tory (under wet conditions.). The resin is diluted with water in a ratio from 1:3 to 1:5. The white emulsion becomes transparant after drying, and the peels show good relief and are extremely durable (E. Werner, written communication, November 1965).

1.6 HIGH-RELIEF PEELS

Peels with an extraordinary relief can be made from sandy sediments. The sediments have to be sampled and treated in the laboratory. The large box sampler and the Senckenberg boxes are most suitable as sampling units (see Chapter Six). The relief is such that most of the two-dimensional peels show a completely three-dimensional picture.

Various people have attempted to harden soft sediments (see discussion of Impregnation). Reineck (1958b, p. 50) mentions that the impregnating liquid must have the following properties:

1. Suitable optical characteristics.
2. Good penetration possibilities.
3. Hardness of the final result must be right for grinding purposes.
4. Little solubility to the adhesives later used for mounting.

Based on results with Araldite Casting Resin B (used in soil sciences), Reineck (1958b) started experiments with Araldite Casting Resin F at the suggestion of the Ciba Company. The advantage of epoxy resins is that only two components are required. Curing can take place at room temperature or at high temperature, depending on the type of hardener used. At high curing temperatures (120 to 140°C, 248 to 284° F) only sediments that do not shrink or form cracks can be treated.

For peeling procedures only the penetration properties of the casting resin mentioned by Reineck (1958b, p. 50) are of direct importance. For the high-temperature curing resin-hardener combination, the technique as well as the results are given by Reineck (1958a, b, 1961a, b, 1962, 1963a, b). Other combinations and products suitable for room temperature hardening are also given by Bouma (1964b) and by Bouma and Marshall (1964).

The Ciba Company produces a series of resins under the name of Araldite, which, together with several hardeners, make a large variety of applications possible (Ciba, 1963, 1964a, b). The author recommends the use of Araldite Casting Resin F (CY 205), a solvent-free, syruplike synthetic resin of the epoxy ethoxyline type. (For American equivalents see Table I.7). No volatile substances are evolved during the curing process, and shrinkage is minimal. At room temperature this resin has a high viscosity, but by raising the temperature and adding a hardener the viscosity can be lowered enormously. The color of the resin is light yellow-brown, and the shelf life is at least 12 months if it is stored in a dry, cool place.

The choice of the hardener depends on various factors such as the type of sediment, the desired viscosity, pot life, application method, and possibilities. With HT 902 the mixture has a very low viscosity and a relatively short pot life (Table I.5). The curing is possible only at temperatures between 100 and 200°C. This hardener is a white powder. The mixture is made of 65 to 70 parts by weight hardener HT 902 and 100 parts by weight of resin at a temperature of 100 to 120°C (Ciba, 1953, not dated a). The mixture should not be applied at a temperature below 85° C (185°F) since the hardener will then separate from the resin. This temperature problem limits its application to sandy and coarse silty samples. Samples with coarse particles such as gravel and shells in sand or with little nonclayey material should also be treated with hardener HT 902.

Hardener HY 951 enables curing at room temperature. The viscosity of the mixture at room temperature is higher than the CY 205-HT 902 combination at 100 to 120°C, but lower than the viscosity of the resin itself since HY 951 is a low viscosity liquid (Ciba, 1964b, not dated b). The combination can be used at room temperature up to about 40°C (104°F). The mixture is made by 100 parts by weight of resin and 10 to 12 parts by weight of hardener (Ciba, 1953a, b, not dated a; Table I.5). The penetration into the sediment is less than that of the high-temperature curing combination. It also can be applied directly in the field when the air temperature is not too low.

The Furane Plastics Corporation produces a similar product under the name Epocast. The resin Epocast 530-2A as well as the hardener Epocast 530-2B are both liquids, mixed in a 10:3 parts by weight ratio. The resin has a viscosity of 750 cp, and the combination has a medium viscosity (Table I.5) and cures at room temperature (Furane, 1962). Even slightly moist samples or sample parts are hardened by this mixture. This product can also be applied under damp conditions.

Since it is difficult to clean hardened resin from the equipment, it is easier to use ordinary cans as mixing beakers. These can be thrown away after use. It is possible to use beaker-

glasses. After use, they should be placed upside down on newspaper to drain the glass as much as possible. Polyethyne beakers are very convenient for mixing, since they can easily be cleaned after the resin mixture has hardened (Ciba, written communication, January 1967).

During the heating of the Araldite resin with Hardener 902, either on an electric hotplate or supported above a gasburner, tiny white crystals will grow at the upper end of the mixing glass where the temperature is much lower. These crystals may also appear after the process is completed.

For cleaning the hardened resin from the equipment use glacial acetic acid with some methylene chloride as an additional softening and swelling agent. Also heavy chlorinated hydrocarbons, phenol or dichloro-ethane can be used. These liquids do not dissolve the hard material but initiate a swelling after several days, and the resin then loosens from the equipment. Several stripper fluids are commercially available now, such as Cital 1212 from Citosan AG (Zürich, Switzerland).

1.6A Araldite and Epocast peels

Equipment
Sampling box, tape, knife or spatula, small trowel, razor blades, tincan or beaker-glass with stirring rod, resin and hardener, measuring cylinder, acetone, marking ink, balance, oven, electric hotplate or gas burner with support, ring, and square gauze wire.

Experiments with different tapes were made (The tape is used to close the openings of the sample box to prevent sand grains from escaping during drying of the sediment and during curing of the mixture.) The different types of tape were stuck to empty boxes and placed in the oven for 5 days at a temperature of 120°C (248°F). After this period the tape should still stick to the box, but it should be possible to remove it without too much difficulty in order to open and clean the box afterwards.

Of the Scotch tapes from the 3M Company, No. 33 (Scotch Vinyl plastic electrical tape No. 33) stuck well to the box and loosened easily without tearing and leaving too much adhesive on the box. Scotch electrical tape No. 49 (aluminum backing, pressure-sensitive adhesive tape) was even better. (No. 69, glass cloth thermosetting silicone, is good but may become

Table I.5   Properties and Data of Products Used to Make Peels with High Relief

	Araldite Casting Resin F (CY 205)		Epocast 530-2A
Name of the resin	Araldite Casting Resin F (CY 205)		Epocast 530-2A
Name of the hardener	HT 902	HY 951	Epocast 530-2B
Manufacturer	Ciba Ltd.	Ciba Ltd.	Furane Plastics Inc.
Mixing ratio (parts by weight):			
resin	100	100	100
hardener	65–70	10–12	30
Mixing temperature	100–120°C (212–248°F)	room temp.	room temp.
Initial viscosity of the mixture in cp at:			
20° C (68° F)	—	4–5,000	—
24° C (75° F)	—	—	1,400
60° C (140° F)	—	400	—
90° C (194° F)	20–30	—	—
100° C (212° F)	10–20	—	—
110° C (230° F)	10–20	—	—
120° C (248° F)	5–10	—	—
Pot life of the pure mixture in hours at:			
20° C (68° F)	—	½–1	⅓–½
60° C (140° F)	—	5 min.	—
90° C (194° F)	14–18	—	—
100° C (212° F)	7–10	—	—
100° C (230° F)	5–7	—	—
120° C (248° F)	3–4	—	—
Application temperature	90–120° C (194–248° F)	room temp.	room temp.
Minimum curing time in hours at:			
20° C (68° F)	—	24–26[b]	—
24° C (75° F)	—	—	36–48[c]
60° C (140° F)	—	2–3	—
100° C (212° F)	48	⅓–½	—
120° C (248° F)	24–36[a]	—	—
140° C (284° F)	10–14[a]	—	—
160° C (320° F)	5–7	—	—

After Ciba, 1953a, b, not dated a, b; Furare, 1962.

[a] Preferred curing conditions.

[b] The combination Araldite F and Hardener 951 can be applied at a temperature of 30 to 40° C (68 to 104° F) in order to obtain a lower viscosity of the mixture.

[c] The Epocast products should be used, if possible, at a temperature between 21 and 27° C (70 to 80° F).

partly loose while in the oven.) These three tapes still have active sticking properties after 5 days. The Scotch tape No. 202 can be supplied very well (Bouma, 1964b). It sometimes has the tendency to burn lightly and to become crisp, which makes it difficult to remove from the box. None of the Scotch electrical tapes No. 4, 38, and 60, the Scotch masking tape No. 232, or the Scotch transparent tape No. 600 passed this test successfully.

Acetone can be used as a fast-evaporating measuring liquid to find to what height a can or a beaker glass is filled after a certain quantity is poured into it. This level can be marked.

The oven must be large enough to contain the sample box(es) and should have a variable heat control with thermostat ranging from about 40 to 150°C (104 to 302°F).

Choice of Hardener

The higher the curing temperature, the lower the viscosity of the mixture of resin and hardener (Tables I.5, I.7), and consequently the better the penetration of the liquid into the sediment sample, and thus the larger the peel relief stands out. The sample has to be completely dry. The Epocast 530-2A and B mixtures can also be used at "high" room temperature, and the sample need not be absolutely dry.

Preparation of the Peel

The method of collecting a sample with the Senckenberg box is described in Chapter Six. In this method, when the sample has been returned to the laboratory, the box is placed on its back and the tape is removed. When loosely packed sediments are dealt with it is advisable to keep the top end higher than the bottom end to prevent the grains from rolling if settling of the sample has occurred during transport. Next, a thin plate is placed between the sample and the top side of the cover, and the cover removed by sliding it over the base very carefully. After sliding it a quarter of the length of the box, one may assume that sediment will no longer stick to the cover, and the cover can be lifted off cautiously. It is advisable to press, with one hand, the sides of the base part against the sample during removal of the cover to prevent a sudden relaxing of the tension of the base sides and, consequently, possible disturbance of the sample. The inner side of the box cover is now cleaned of sediment grains and pushed underneath the base (see Fig. 6.6). The sample will be sustained now on all sides, leaving only one of its original vertical planes exposed. If settling has occurred, the space formed should be filled with sediment, for example, of another grain size or another color, to avoid confusion in interpretation afterward. The openings between the box parts are closed with tape.

The following step is to remove a thin layer of about ½ cm (⅕ in.) from the exposed sediment face. It is scraped away carefully, leaving low rims along the sides (see Fig. 6.6). The new surface must be flat and smooth. This can be effected with a small trowel, while difficult parts can be leveled with a knife or a spatula. If parts of the rim are destroyed, they can be restored easily by hand. Sometimes the use of a razor blade for the final touch will produce good results.

Araldite Casting Resin F (CY 205) and Hardener HT 902. The sandy sample is placed with its box in an oven for about 24 hours at a temperature of 100 to 110°C (212 to 230°F) to dry completely.

Before the drying period is finished, the mixing can is calibrated, marked, and dried (see Tables I.6 and I.7 for mixture ratios). Araldite resin F is poured into the mixing beaker and heated to 100 to 120°C (212 to 248°F). Next, the hardener is added and the mixture stirred well. It takes about 5 to 15 min before the liquid is clear.

The sample box is then carefully removed from the oven since in heating some sediments will disrupt intergrain cohesion. The hot resin mixture is now poured onto the sediment surface. Since the mixture is very fluid, care should be taken to see that it is distributed evenly over the surface, leaving the rims free of resin (Fig. 1.54). The liquid soaks in immediately, and therefore even pouring is essential. The sample is then returned to the oven to be cured for at least 36 hours at a temperature of 120°C (248°F). After this period the box can be taken out and cooled. It is advisable to wear thick gloves, or their equivalent, to handle the hot sample box and the mixing beaker.

When working with large sample boxes the method is the same as discussed above. Reineck (1961a, b) made an oven of an old filing cabinet with four drawers in which a number of electrical heating rods had been mounted. When the sample was not sustained on all sides he placed the box on a special swiveling rack, which could be turned in such a way that all noncontained sediment faces had a dip of about 60° to prevent disturbance during drying and curing.

Epocast 530-2A and 530-2B. In Table I.6 the mixture ratios are given. The sample should be as dry as possible since too much moisture prevents the resin from curing. The procedure is similar to the preceding one. The curing time is about 36 to 48 hours, depending on the temperature. The manufacturer advises the application and the curing of this Epocast combination at a temperature between 21 and 24°C (70 to 81°F) (Table I.5; Furane, 1962).

Strengthening of the Back of the Peel

When there are some weak places in the sediment sample, such as those formed by a nonimpregnated thin clay layer, it is advisable to strengthen the back of the peel. A piece of cheesecloth or gauze, the same size as the surface of the sample between the rims, is placed over the surface before the resin mixture is poured. When the Araldite F-Hardener 902 combination is applied, there is no reason to use

Fig. 1.54. (*Left*). Pouring the hot Araldite F-Hardener 902 mixture onto the dry and hot sediment surface.
Fig. 1.55. (*Right*). Removal of nonimpregnated material from an Araldite peel, using a hose with gentle spray.

Table I.6   Mixture Ratios of Casting Resin and Hardeners

Type of Sediment[a]	Total Mixture Volume for 16 in² (1 dm²)	Total Mixture Volume for 16 in² (1 dm²)	Total Mixture Volume for 16 in² (1 dm²)
Coarse sand	35 cc Araldite F : 20 cc / Hardener 902 : 17 g	35 cc Araldite F : 30.5 cc / Hardener 951 : 4.5 cc	35 cc Epocast 530-2A : 27 cc / Hardener 530-2B : 9 cc
(Medium) and fine sand	50 cc Araldite F : 29 cc / Hardener 902 : 24 g	50 cc Araldite F : 44 cc / Hardener 951 : 6 cc	50 cc Epocast 530-2A : 37.5 cc / Hardener 530-2B : 12.5 cc

After Reineck (1962) and Bouma (1964b)

[a]For coarse, well-sorted sands it is advisable to use less mixture than 35 cc/dm², since the liquid runs too easily to the bottom of the box. For these kinds of sediments the other mixtures are preferred. For fine sand 50 cc/dm² is recommended and for medium sand 35 to 50 cc/dm². The Senckenberg sample box, as used by the writer, exposes a sediment surface of 3 dm² (48 in.²).

cloth (bandage), since all liquid passes through the cloth and does not glue it to the sediment. More viscous mixtures will mount the bandage much better.

After curing, it is advisable to coat the gauze with undiluted nitrocellulose lacquer to be certain that each part of it sticks well to the sample.

Completely muddy or clayey samples and samples consisting of an alternation of sand and clay cannot be treated in this way (Reineck, 1958b).

Collecting and Finishing

Orientation, location, and other data are written on the back of the peel, after which it is loosened from the box sides by cutting with a knife. If the rims are made properly and no resin is poured on it, the peel will not stick to the metal. The tape is then removed from the box openings and both box parts are pulled apart. The peel can now be taken out.

If parts stick to the bottom, the wrong mixture, or too much mixture has been used, or the pouring has not been done evenly. The peel can be cleaned with water, spraying it gently with a hose (Fig. 1.55). Care must be taken not to wash out any clayey layers that may be present.

The peel should be handled carefully since the high relief laminae can be destroyed easily (Figs. 1.56–1.58).

It is advisable to spray the peel with an acrylic resin such as Krylon Crystal Clear #1301 or 1303 to prevent the formation of a salt crust.

Special Applications

When the investigator wants to impregnate part of a sample to make a thin or thick section rather than to make a make a peel, he should use more mixture per square unit than indicated in Table I.6. If more resin per dm² is used the relief differences of the peel may decrease considerably (see discussion of Impregnation in Chapter Two).

The peel still seems to be porous and therefore sometimes a little too fragile for sectioning or polishing purposes. Undiluted or slightly diluted unsaturated polyester resin can be poured on the back of the peel (Bouma, 1964b). This procedure will be discussed further in Section Two (see also Basumallick, 1966). A combination of a peel and resin impregnation can also be made. A peel is made and then photographed, after which the back of the peel is impregnated with plastic for polishing or sectioning purposes. This combination often takes less time than the direct impregnation with plastic, and first sedi-

mentary characteristics are obtained from the peel. A thicker peel is desired when making a slice for X-ray radiography (Bouma, 1963).

Remarks

The Ciba products available in the United States are not always the same as the European ones. In Table I.7 a comparison is given. Many experiments were carried out and the best results were obtained with the combinations Araldite F-Hardener 951, Araldite MY 740-Lancast A, Araldite D-Hardener 951, and Araldite E-Hardener 956.

1.6B Paraffin Peels

For sandy sediments paraffin can be used instead of an epoxy resin (Reineck, 1958b, 1962). The advantage is that it is very inexpensive and does not take much time after pouring. However, the disadvantage is that the obtained peel is not as finely detailed as those obtained with the Araldite F-Hardener 902 combination (compare Figs. 1.56 and 1.59). The peel must be handled very carefully since it will break very easily. It is very practical for first examinations or control tests within a series.

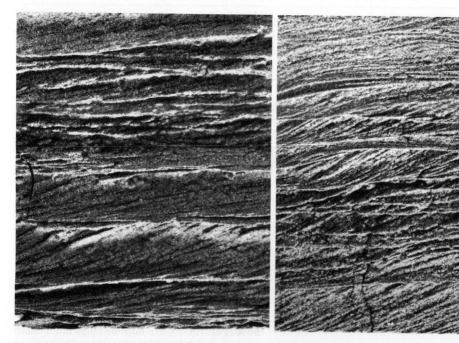

Fig. 1.56. (*Left*). Peel with "high" relief made with Araldite Casting Resin F (CY 205) and Hardener HT 902. River deposits. Excavation in the Rhine near Amerongen, The Netherlands.

Fig. 1.57. (*Right*). Peel with "high" relief made with Araldite Casting Resin F (CY 205) and Hardener HT 951 at a temperature of 18°C, using dried sediment. Locality same as Fig. 1.56.

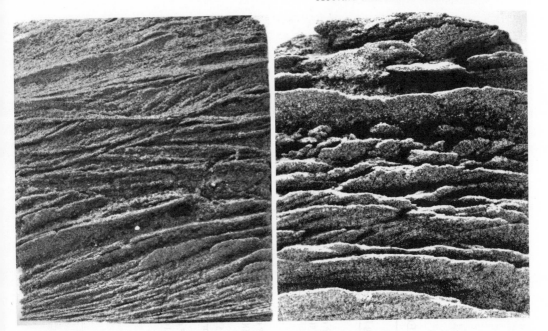

Fig. 1.58. (*Left*). Peel with "high" relief made with Epocast 530-2A and B from dried sediment. Locality same as Fig. 1.56.

Fig. 1.59. (*Right*). Peel with "high" relief made with paraffin. Locality same as Fig. 1.56.

Preparation of the Peel

The exposed vertical surface of the sample is scraped off slightly, after which the sample box is placed in the oven for about 24 hours at a temperature of 100 to 110°C (212 to 230°F), completely similar to the drying procedure prior to the preparation of an Araldite F-Hardener 902 peel.

About 35 cm³/dm² (16 in.²) of molten paraffin is required for coarse, sandy sediment, and 50 cm³ for medium to fine sandy sediment (see Table I.6). The sample box is next taken out of the oven while still at a temperature of about 100°C, and the paraffin is poured evenly to the sediment surface. The sample is now left to cool. When the temperature of the sample has reached room temperature, a similar quantity of molten paraffin is poured on the section. With a low-burning gas burner the surface is made even and pores that are still open are filled. Immediately following the heating procedure a piece of cheesecloth or gauze is pressed into the soft paraffin surface.

Collection of the Peel

When the peel is cooled completely, the sides can be cut loose with a knife. Sometimes it is

easier to use a hot knife. The peel must be taken out very cautiously. The relief can be cleaned of non-impregnated grains with water, using a hose with a gentle spray (Fig. 1.59). To protect the result as much as possible, it should be mounted directly with a fast-drying adhesive on a piece of masonite.

1.7 ACETATE AND NITROCELLULOSE PEELS

It has already been mentioned (see Section 1.1E) that peels can even be made from slightly consolidated deposits and shales using nitrocellulose lacquer and a few special techniques. This is not always easy or successful. Well-consolidated sediments cannot be treated in this manner.

The methods of peeling discussed in this section have to be carried out in the laboratory. Of the samples collected, one or more faces have to be sawed and polished. Depending on the type of sediment, the surface has to be etched, after which a mixture can be poured on it forming the back of the peel. The replica obtained is very thin and nearly lacking relief. Therefore it must

Table I.7 Comparison of Ciba Products Manufactured in Europe and in the United States

European names		American names		Mixing Ratios	Curing Data	Results and Remarks
Araldite	Hardener	Araldite	Hardener	by Weight		
Araldite F (CY 205)	HT 902	Araldite 6010	not available			See Fig. 1.56
Araldite F (CY 205)	HY 906	Araldite 6010	Hardener 906	100 : 80	2 hr at 100° C and 2–4 hr at 150–200° C	Does not harden properly. Need to add an accelerator. Resin mixture penetrates far, but laminae do not stand out separately.
Araldite F (CY 205)	HY 956	Araldite 6010	Hardener 956	100 : 25	24 hr at 40° C or 2–8 hr at 100° C	Thick peel with little relief. Good for impregnation rather than for making a peel.
Araldite F (CY 205)	Lancast A	Araldite 6010	Lancast A	70 : 30	7 days at 25° C or 2 or more hr at 120° C	Curing 4 hr at 100° C and 4 hr at 120° C resulted in a rather good peel, comparable to Fig. 1.58.
Araldite F (CY 205)	HY 951	Araldite 6010	Hardener 951	100 : 13	24 hr at 40° C or 2–8 hr at 100° C	One of the best results obtained. Relief about 6 mm (¼ in.) high. Comparable to Fig. 1.57.

Araldite MY 740	HY 951	Araldite 6005	Hardener 951	70 : 13 : 24 24 parts RD-1	16–24 hr at 40° C or 2–8 hr at 100° C	Result not bad, slightly less than Fig. 1.57. RD-1 (European DY 021) reduces viscosity of the mixture.
Araldite MY 740	HY 956	Araldite 6005	Hardener 956	100 : 25	24 hr at 40° C or 2–8 hr at 100° C	Thick peel with fair relief. On parts of the peel the relief is lacking.
Araldite MY 740	Lancast A	Araldite 6005	Lancast A	70 : 30	2–4 hr at 120° C	Very good peel. Result is in between the 6010–951 and 502–951 combinations.
Araldite D (CY 230)	HY 951	Araldite 502	Hardener 951	100 : 10	16–24 hr at 40° C or 2–8 hr at 100° C	The best combination for fine and medium grained sands. Nice, distinct relief, about 6 mm (¼ in.) high (see also Fig. 1.57).
Araldite D (CY 230)	HY 956	Araldite 502	Hardener 956	100 : 20	16–24 hr at 40° C or 2–8 hr at 100° C	Poor result.
Araldite D (CY 230)	Lancast A	Araldite 502	Lancast A	100 : 35	2–4 hr at 120° C	Very poor result.
Araldite E (CY 232)	HY 951	Araldite 506	Hardener 951	100 : 14	16–24 hr at 40° C or 2–8 hr at 100° C	Good result, better than the 506–956 combination.
Araldite E (CY 232)	HY 956	Araldite 506	Hardener 956	100 : 25	16–24 hr at 40° C or 2–8 hr at 100° C	Rather good result with fair relief. In between Figs. 1.57 and 1.58.

From Ciba (1963, 1964b, not dated, a to o).

Note: All experiments have been carried out with similar sediment, fine-medium grained sand from a point bar of the Brazos River, near College Station, Texas, U. S. A. When more than one curing temperature is mentioned, the highest temperature is always used. Also the maximum curing time is applied. All samples were dried in an oven for 24 hours at 105° C. The resin mixture is always poured on hot, dry samples. If the viscosity of the mixture was high it was placed in the oven for about 5 min prior to pouring.

not be regarded as a painting containing sedimentary material, but more as a thin, transparent print or negative from the polished surface. These "peels" are not suitable for petrological investigations, but they can be used directly, or through photographic printing techniques, to show several microscopic textural and structural data which often are not visible on the polished rock surface (Appel, 1933; see micropeels, Voigt, 1935a, 1936a, b, 1937a, b, c, 1938a, c, 1939).

The method for making peels from hard sediments was originally invented for paleontological purposes (Walton, 1928; Butler, 1935; Darrah, 1936; Easton, 1942). Several lacquer combinations were used as peel material. Later, commercially available sheets of drawing films proved to be more suitable and easier to apply.

The mixtures and films are primarily used for the making of peels of carbonate-containing sediments (Beales, 1960; Overlau, 1963). The cut and polished sections are first etched with hydrochloric acid, acetic acid, or formic acid. The silica-containing sediments must be etched with HF, a liquid many people are afraid to work with. Recently the Koninkijke/Shell Exploratie en Produktie Laboratorium in The Netherlands succeeded in making peels from sandstones and shales without any etching. This was indicated by McCrone (1963, p. 229) for limestones: they were accountable by differential abrasion during the polishing process.

The peels can be damaged very easily, and it is therefore advisable to mount them between two pieces of glass with tape or Canada balsam. The films can be studied under the microscope or they can be placed in a slide projector. For many purposes they replace the more expensive thin sections.

1.7A Pretreatment of the Sample

A flat surface of a sample is obtained by cutting the specimen with a diamond saw (see Chapter Five). Size and directions of this surface depend on the technical possibilities and the indications of the investigator.

Next the surface is polished with silicon abrasives. This can be carried out on a rotary polishing disk with carborundum powder or on wet abrasive paper. Several grades of abrasives should be used, as is done for the preparation of thin sections. However, it is not necessary to obtain a very highly polished product.

The final polishing should be carried out with no. 360 size grit (Swarbrick, 1964) or with no. 400 (Andrews, 1961; Lane, 1962). McCrone (1963) recommends an extremely fine (grade No. 1000) abrasive on a glass plate. The plate has to be replaced frequently. The author found that a final grit size of 600 is quite suitable when working with a prepared acetate sheet, and grit no. 800 when a mixture is applied. The coarser grades are used in connection with a rotary disk polisher, whereas the final polishing is carried out on a wet plate of plexiglas. The author feels that plexiglas, due to its flexibility, keeps its surface better and maintains its flatness longer.

Both the polishing disk and the polishing plate must be cleaned regularly to avoid scratches on the polished surface caused by bits of rock that are broken off. After the surface is dried (with or without forced air) one can examine it to determine whether or not the polishing has been done correctly, and if all scratches have been removed.

1.7B Preparation of the Peel

Acetate Paper Method

The application of acetate film sheets is definitely the fastest method of making an acetate peel (Germann, 1965). However, the advantages are accompanied by some restrictions. It is very difficult to apply the acetate paper to the polished surface in such a way that air bubbles are not trapped. These bubbles produce vacant spots on the final peel. As a result, the use of acetate paper is restricted to virtually nonporous sediments with polished surfaces, flat and without holes. However, for reconnaissance studies the application of acetate paper sheets is the quickest, easiest, and best method.

Walton (1928) writes that the flat surface of the sample is "immersed for a definite length of time in an etching solution of hydrochloric acid, the optimum concentration being found by trial. The acid dissolves a film of the carbonate of which the mass is largely composed, and the plant substance contained in the film is left behind, standing in relief above the surface." He uses the technique for preparing sections of fossil plants contained in coal balls or in other types of petrification. Lane (1962) and Miller and Jeffords (1962), as well as McCrone (1963), prefer a 1% HCl solution and an etching time of 10 sec when dealing with limestones. The Koninklijke/Shell Exploratie en Produktie Laboratorium uses a 5% formic acid solution and varying etching times. Lane (1962) states that "clayey or bituminous samples usually require less etching and dolomitic samples may require more." Swarbrick (1964) prefers HF for making peels of silicious samples.

Like acetate sheet, the single-matte commercial acetate film with a thickness of 0.005 in. (0.127 mm) is recommended by McCrone (1963) and 0.002 to 0.003 in. (0.508 to 0.762 mm) by Lane (1962), while Miller and Jeffords (1962) use 0.01 in. (0.0254 mm) thick clear acetate paper. The last-mentioned authors experimented with different papers.

Under the trade name Kodapak an acetate paper is sold which tears with a sharp border, dries rapidly, and can be removed after 2 to 3 min or longer. The etching time for slightly argillaceous, coral-bearing Paleozoic limestones ranges between 10 and 30 sec. The paper Vuepak (trade name) dries slowly, but it should be removed after 2 to 3 min since the film tends to become wavy and distorted when it is left longer. It tears with a jagged border. An etching time of 2 sec is sufficient since the Vue-pak adheres very tightly to rock samples. Vinyl acetate film is more limber and flexible, and shrinkage wrinkles are rare. The film is so slightly affected by acetone that etched surfaces are not reproduced adequately. Ethyl n-butyrate is a more effective solvent.

In Utrecht, as well as in many other institutions, single-matte acetate film is used, sold under the name Kodatrace. It is applied normally for drawing charts and other designs that have to be scale-fixed during time. Many thicknesses are available, but the 0.01 mm (0.0004 in.) thick film produced the best results.

Equipment. Diamond saw; grinding disk, plexiglas or glass plate and two or three grades of carborundum of which the finest size is a 600 to 800 grit; 5% (1 to 10%) solution of hydrochloric, acetic, or formic acid (or hydrofluoric acid) in a container which is large enough to place the sample in; one container with normal tap water and one container with distilled water; Kodatrace (0.01 mm, 0.004 in. thick); scissors; polythene squeeze bottle with acetone; paint roller with a soft foam plastic or a flannel roller; lantern-slide glass, tape or Canada balsam.

Procedure I.

1. The carbonate-containing sample is cut on the diamond saw and the face is smoothed and polished.

2. The acid bath is made at the desired strength. The specimen is held with the polished surface down. This face is dipped carefully into the acid. It is better to hold this surface a little inclined and to move the sample to allow the formed gas bubbles to escape. This etching is continued for 10 to 30 sec, depending on the type of rock, and mostly determined by trial and error. The etching time should be kept rather accurate in order to obtain comparable peels.

Immediately after etching, the sample surface is rinsed thoroughly and carefully in a bath with plain water, followed by rinsing in a bath of distilled water. The baths should be refreshed regularly. The etched surface should never be rinsed under running water, which could damage the delicate relief formed.

3. The sample is then left to dry at room temperature. Compressed air can also be used if applied carefully. Care should be taken not to touch the surface.

4. The sample is now placed in a box filled with clean, coarse sand to a depth of at least 5 cm (2 in.). The horizontal position can be checked with a few drops of acetone. A small level mounted to a piece of wood, the bottom covered with felt, can also be used.

5. Cut a piece of the Kodatrace slightly larger than the polished surface.

6. Squeeze acetone on the specimen face (as much as the surface will hold; Fig. 1.60). The squeezing must be carried out carefully to protect the delicate relief of the sample surface.

Take the piece of Kodatrace between thumb and forefinger of both hands with the dull side of the acetate paper toward the specimen. The Kodatrace is placed in contact with the straightest edge of the sample. This edge must not be the longest side. With the help of the paint roller, the acetate paper is rolled over the wet surface, making certain that no air bubbles become trapped. This rolling must be carried out very carefully; to prevent smearing or breaking of the delicate relief, no force should be applied to the surface (Fig. 1.61).

McCrone (1963, p. 229, point 7) writes "by holding the corners of the film between thumb and forefinger of each hand the film can be bent downward into a U-shape. Apply the dull side of the film to the rock so that the base of the U is to touch the acetone-wet surface first. In this way the film can be gently applied with a progressive rolling action that tends to prevent the trapping of air bubbles beneath the film." However, the possibility of turning the sheet slightly when part of it touches the surface and the chance of trapping air bubbles increases greatly. One can see even the tiniest air bubbles while acetone is still between film and specimen by applying incident light.

Care should be taken *not* to press the film with the fingers directly after application since impressions will result (Lane, 1962; McCrone, 1963). The film will "float" on the acetone and develop its own adhesive force, drawing gently against the rock.

7. Allow the film to dry for about 30 to 60 min. Then, starting at one corner, peel the film off slowly from the sample.

8. Cut off the excess Kodatrace and place the film between two pieces of thin glass. Both pieces are then bound together by tape, and data can be written on the tape.

Procedure II. At the Koninklijke/Shell Exploratie en Produktie Laboratorium at Rijswijk (Z. H.), The Netherlands, formic acid (5%

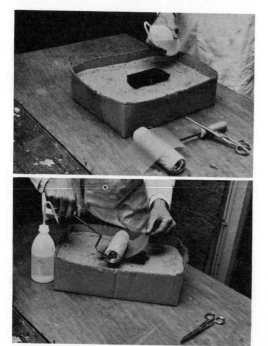

Fig. 1.60. (*Upper*). Acetone is being squeezed onto the surface of a leveled sample. In the foreground the paint roller with a piece of Kodatrace and a pair of scissors are visible.

Fig. 1.61. (*Lower*). The rolling of a piece of Kodatrace onto the sample surface wetted with acetone.

solution) is used for etching carbonate-containing sediments. The working steps to follow are similar to the ones given in Procedure I, except the last point. The collected peel is dipped in the acid bath again for a few seconds and then rinsed. It is left to dry and mounted. The advantage of this second etching is that the obtained picture is clearer. During drying the film wrinkles slightly.

Procedure III. At the Koninklijke/Shell Exploratie en Produktie Laboratorium at Rijswijk, it was found that it is unnecessary to etch a sandstone or a shale. After cutting and polishing, the specimen is allowed to dry completely. It is then placed horizontally in the sandbox, wet with acetone, and a piece of Kodatrace is rolled onto it. After several hours the peel is completely dry and can be removed (see Figs. 1.68 and 1.75).

Procedure IV. Swarbrick (1964) uses hydrochloric acid (HCl) for etching carbonate-containing samples and hydrofluoric acid (HF) for non-carbonate sediments. He writes "the specimen is placed in a plastic disk in a fume-cupboard with the polished surface approximately vertical. HF of 50 to 60% strength is then sprayed onto the surface from a plastic wash bottle. This washing is repeated three times with intervals of one minute following each spraying. This particular method of etching has been adopted to prevent, both uneven etching caused by particles freed by the etching remaining on the surface, both uneven etching caused by particles freed by the etching removing or the surface, and to avoid concentrations of acid on a very flat surface. Specimens are not etched in acid fumes because this method requires support, being in contact with the surface, which prevent etching taking place at the point of contact with the specimen." The rest of the procedure is similar to the one given before, only Swarbrick uses Clarifoil photograph of an acetate peel made from gravel.

When working with HF one should wear plastic or rubber gloves and keep the window of the running fume-cupboard as far down as possible.

Procedure V. Sternberg and Belding (1942) describe a dry-peel technique for small sections using 0.5 mm (0.02 in.) thick clear celluloid which is slightly larger than the sample. The sample is polished, etched, and dried at room temperature.

The piece of celluloid is placed on a smooth, hard surface and the polished rock surface is thoroughly wet with acetone. The sample is now pressed to the celluloid. After 2 to 3 min the peel is ready and can be removed, cut to size, and placed between glass plates. This last step can be done by cutting the celluloid to the desired size and shape, holding it firmly in place on one of the glass plates, and painting acetone around the edges with a fine brush. Some acetone gets between glass and celluloid and dissolves enough celluloid to form a cement.

This celluloid material also produces good results from nonetched sandstones and shales.

Procedure VI. Dr. W. Remy, Geological Institute Münster, Germany, applied Filmomatt from Neschen (1964) to collect coal material from sediments such as fossil plants. The Filmomatt foil is transparant and will stand acids, such as nitric acid.

Procedure. The sample is cut and ground and nitric acid is applied to remove mineral fillings from the pores in the plant cells. The depth of etching depends on type and condition of the sample.

The freed coal is next removed by pressing a piece of Filmomatt to the sample.

This procedure can be repeated until the sample is gone. The peels can be photographed. Enlarged prints reveal an abundance of inner plant structures. The peel can also be mounted on celluloid or glass for projection. The Filmomatt proved to be shrinkproof.

Remarks. These "dry-peeling" methods require experience before air-bubble-free results can be expected. The placing of the acetate paper of the celluloid sheet must be carried out rather quickly after the polished surface has been wet with acetone.

It is very important to wash the complete sample after treatment on the polishing disk before a finer grade is applied to avoid the forming of scratches by coarser grains. The sample has to be thoroughly dry before the acetate paper or celluloid can be applied since traces of water, escaping from cracks, prevent binding.

The tray with acid must be placed next to the wash basin to obtain an accurate etching time. We found that working with HF, this could be carried out fairly easily. A large plastic beaker is used as a container and placed in a running fume-cupboard. It can be closed with a thick sheet of plastic or with a large watchglass. A tall, narrow (5 cm) tin, with or without sand, is placed on the bottom of the beaker and the sample is laid on it, taking care that the polished surface is almost vertical and does not touch the tin or the beaker. A small plastic beaker with a large cover is filled with a little HF. The cover must have an eye with a thin rope running through it. When the HF container is placed in the beaker as far as possible from the specimen to prevent the acid from splattering on it, the cover is lifted off the HF box. The beaker is covered until etching time is complete. Then the acid container is closed again and the sample is taken out.

Another method is to use a small metal or plastic disk with three little holes along its side to hang it horizontally on a nylon rope. A small plastic container with a cover, partly filled with HF, is placed on the disk. The sample is placed in a plastic beaker as mentioned previously. The rope is held in one hand and a pair of tweezers in the other. The little cover is taken off, the acid container is lowered, and the beaker is covered. After etching time the beaker is opened, the acid container pulled up and covered, and the sample taken out.

Etching of the finished peel improves the result and emphasizes small textural differences. The wrinkling of the peel does not influence the picture.

If a sample contains bitumen it is impossible to make an acetate peel since the acetone dissolves the bitumen.

Acetate Solution and Nitrocellulose Solution Methods

It takes more time and extra equipment when the investigator wants to make a peel using liquids to form the film. The oldest methods known are based on liquids (Walton, 1928; Dollar, 1942; Andrews, 1961; Butler, 1935; Buehler, 1948; Bissel, 1957). These liquids are related more to cellulose than to acetate, but the application is similar.

"The method as introduced by Walton (1928) has been adopted especially to the well-indurated limestones of the Primary Era, where over and above the macrostructures, it allows a very exact investigation of the microstructures. Moreover, the organic and insoluble materials, generally more tender, which are destroyed during the preparation of thin sections, are preserved with their original microstructures by the very slow attack they underwent" (Overlau, 1963, 1965a).

Application of liquids rather than sheet material allows the entire surface to be evenly touched with no empty spots resulting. This method is especially necessary when dealing with porous sediments. The finest parts of the relief of the etched surface are not ruined when using the liquid application.

The preparation of the sample is similar to that described for the acetate-paper method. However, the following steps of the peel preparation differ considerably among the investigators.

Procedure I (Walton, 1928; Butler, 1935).

These investigators use a product known by the trade name Durofix. This cellulose compound has a considerable tensile strength after the solution has formed a film. The Durofix is dissolved in butyl acetate (1:1). Butler etches his samples during 5 to 10 sec in a 10% HCl solution.

Procedure II Darrah, 1936; Buehler, 1948; Bissel, 1957; Andrews, 1961). The mixture made is based on a recipe given by Darrah. Minor variations occur among the authors mentioned (Table I.8).

The surface of the specimen can be etched with either HCl or HF. It must be prewetted with butyl acetate before the mixture can be poured onto it.

Table I.8   Components of a Solution for Making Peels of Consolidated Sediments

	Buehler (1948)	Bissel (1957)	Andrews (1961)
Parlodion	28 g	28 g	28 g
Butyl acetate	250 cc	250 cc	250 cc
Amyl alcohol	10 cc	10 cc	30 cc
Castor oil	3 cc	3 cc	3 cc
Ethyl ether	3 cc	3 cc	3 cc
Xylene	10 cc	–	10 cc

Procedure III (Overlau, personal communication, October 1963, 1965a). This peeling method is a technical improvement of the procedure of Walton (1928), carried out by Professor F. Kaisin of the Geological Laboratory of the University of Louvain, Belgium. The systematic application to sedimentology was realized by P. Overlau in Louvain during the period 1959 to 1960 and later in the laboratory of Petrofina at Brussels, Belgium.

Equipment. Diamond sawing machine; a set of three glass plates frequently renewed, corresponding to the three grade sizes of carborundum used: Nos. 150, 300, and 600; two plastic trays, adapted to the size of the blocks to be treated, for the acid attack and for finishing (the distilled water in the rinsing tray must be refreshed frequently); acetic acid (4 cc in 1000 cc of distilled water); a fume-cupboard or a ventilated oven; a bell jar on a glass or a plexiglass plate or a glass box with cover in which the sample can be placed; a few trays with silicagel, $CaCl_2$, or another desiccator; a sandbox for placing the specimen under the bell jar or in the box; one half-liter polythene squeeze bottle with

acetone, one wide-necked half-liter bottle containing the prepared acetate solution + 10% acetone, one dropping-tube; a knife, a series of glass plates, and tape for mounting the peel.

The acetate solution is made by dissolving an acetate powder in acetone in such quantities that a viscous mixture (thick syrup) is formed. Two types of powders (cellulose acetate M and cellulose acetate K) of Fabelta (1963) have been tried out, both with good results. These powders were made at the suggestion of Professor Kaisin (Table I.9).

Preparation of the sample. The sample is sawed and then polished on a glass plate with a 150-grade carborundum powder. After careful, thorough rinsing, the sample is polished on the second glass plate with grade No. 300 carborundum, followed by grade No. 600 with a good rinsing before and after (Fig. 1.62).

The sedimentary structures normally appear very distinct after polishing with the grade No. 150 carborundum and they disappear with the use of the finer grades. After the application of the coarsest grade it is a good time to obtain preliminary information.

Table I.9. Components of an Acetate Solution for Making Peels of Consolidated Sediments According to F. Kaisin

Powder Components	27%	Solvents	73%
Cellulose acetate	83.3%	Acetone	81.3%
Triphenyl phosphate	5.6%	Methyl ethyl acetone	9.6%
Ethyl phalate	11.1%	Water	4.5%
		Methyl cellosolve (2-methoxy ethanol)	4.6%

Overlau (written communication, 1965a)

If the sample contains hydrocarbons or has been contaminated by oil or greases, it is advisable to clean the specimen, since these products might prevent the acid attack or corrupt the peel.

Acid attack. The container is filled with the 0.4% solution of acetic acid and the specimen is placed in such a way that the polished surface is upward and more or less horizontal.

The duration of the attack varies from 20 min for sandy or dolomitic and saccharoid limestones to 50 min for middle- or fine-grained, slightly argilaceous limestones. Calcilutites and fine-argilaceous or pigmented limestones become covered with totally opacified films after 40 min.

The process of attack risks local alteration by the formation of CO_2 bubbles moving along the face and may create false corrosion structures. It is therefore advisable to place the face almost horizontally and cover it with at least 3 cm of solution to avoid localized concentrations and attacks.

Fig. 1.62. The final polishing of a sample on a wet plexiglass plate using carborundum powder with a size grade of no. 600. The square bucket is filled with tap water for rinsing.

Rinsing. The sample is taken out of the acid solution, taking care that no liquid-current disrupts the treated surface, thus making false structures. Therefore the block is taken out slowly with the attacked face slightly inclined. With the same care the specimen is soaked twice in a large quantity of pure water (face slightly inclined) (Fig. 1.63). A careful rinsing will dilute and remove all the soluble residues without deforming the original structures and the delicate relief.

Drying. Complete drying is not possible in air but must be carried out in a well-ventilated fume-cupboard or in a well-ventilated oven, never exceeding 60°C (140°F). After a few hours the temperature may be raised to 100°C (212°F) maximum without risking disturbance of the argilaceous material. The full success of the later peel depends on the drying, because the humidity of a badly dried sample reacts with the acetate, giving it a milky appearance often combined with the forming of a cloud of microscopic bubbles.

Under the influence of water a considerable shrinkage of the acetate peel will take place (1 mm/15 mm). (Drying time increases with an increase in the sample thickness.) The sample should be placed on two strips instead of a flat bottom to provide better ventilation. Samples exceeding 3 cm thickness need to dry at least overnight.

After drying time is completed, the specimen should be left in the same place to cool to room temperature.

Preparation of the film. The specimen is placed on the dry sand within the sandbox. It can also be placed on the ring of the tripod of the laboratory gas burner. A small container with sand should be placed underneath (Fig. 1.64).

The treated face must be absolutely horizontal. This is tested carefully by dripping one or two drops of acetone on it with an eye-dropper.

Fig. 1.63. (*Upper*). After the acid attack (10 sec in 5% HCl) the sample is rinsed very carefully in normal water (square bucket), followed by washing in distilled water (round tray). Note the inclined face of the sample to prevent a sudden rush of water over it.

Fig. 1.64. (*Lower*). Acetate solution is poured onto a sample surface that is leveled on a tripod. Two trays are filled with desiccator and one cardboard box with sand. The sand is moistened with acetone to obtain a saturated atmosphere under the bell jar which is placed over the sample.

The acetone will spread out in a perfectly circular wet spot if the surface is horizontal. The least dip gives an egg-shaped wet spot toward the maximum slope.

When the block is leveled the face is covered abundantly with anhydrous acetone (avoid a sudden flood of liquid). To prevent the acetone from evaporating too rapidly, the surface is wetted with the diluted acetate mixture (10% volume acetate solution and 90% acetone). Next the viscous acetate solution can be poured on in a layer of about 3 mm (maximum) thickness. It is important to pour in strips so that the acetate walls can run together without trapping air. The viscosity of this mixture may prevent the forming of an even surface, and therefore acetone is poured carefully on the acetate solution.

If the specimen is placed on a tripod, the sand in the small box underneath is wetted with ace-

tone (in a sandbox enough acetone will normally flow into the sand). Two trays with desiccators are placed nearby and the entire peel is covered with the bell jar (or the special box). It is important that escape of acetone be impossible. Therefore a tripod or sandbox is placed on a glass plate and the base of the bell jar is greased with vaseline or a similar product. The air underneath the jar will now become completely saturated with acetone.

After 24 hours the glass cover is slightly raised and propped up by a small object such as a match. This assures slow ventilation and dilutes the fumes of the acetone and solvents. It should be kept in this position for an additional 24 hours. After this period the glass cover can be removed and the peel allowed to dry completely, which will take about 2 days.

Peeling. The sides of the peel are loosened with a knife (Fig. 1.65). When the operator is certain that the peel is absolutely free along its sides, the knife is inserted between peel and sample on one corner. Gentle pressure at 45° is usually sufficient to loosen the peel for a few square millimeters. This small loose corner is lifted carefully in order to extend the peeling step by step (Fig. 1.66).

Loosening the peel at any place other than this small sacrificed corner should be avoided unless the peel breaks somewhere else. The knife is used only to lift the peel when it sticks too closely to the rock and to cut the overlaps along the sample sides.

After removal of the peel from the sample face, it is exposed to the air for a few hours, then mounted between two glass plates bound together with tape (e.g., small cut strips of Scotch No. 232). Data should be written on the tape (Overlau, 1966).

Technical notes

1. When the scraping along the sides of the peel has been done carefully, the sample can be polished again very easily in preparation of a second peel. It is especially useful to make a series of successive peels when studying several fossils and/or very small-sized structures.

2. Coarse-grained material needs less etching than fine-grained sediment. Increase of the content of nonsoluble matter decreases the etching time.

3. Stormy weather or variations in pressure produce some difficulties since they cause the peel to become somewhat milky and bubbly.

Fig. 1.65. (*Upper*). The peel is cut loose from the edges of the sample surface. Of the two samples lying in the sandbox one sample still has a piece of Kodatrace mounted to the treated surface. The small sample is covered with an acetate mixture.

Fig. 1.66. (*Lower*). After the edges are loosened the film is peeled off very carefully.

This can be explained by the abrupt modification of the fume-pressure equilibrium of the acetone in solution or in fumes.

4. If the acid is prolonged for too long a period, or if there are too many insoluble materials, the obtained peel will become too dark in color. A few drops of carbon tetrachloride (tetrachloromethane, CCl_4) are put on the mat face of the peel to prevent this dark color. Until complete evaporation, the peel will remain much more transparent in this condition than if alterations are attempted.

5. Some fine-grained limestones seem to be homogeneous at outcrops or on peels. They sometimes contain many microstructures, but a uniform pigmentation hides pellets, intraclasts, calcisphaera, etc. A very simple test for this is to scrape a band with a knife or a piece of glass held in a near vertical position. This scraping must be carried in a careful and uniform motion in one direction only. In this way the microreliefs are truncated, and transparent zones will define their shape, which will contrast with the micrograined background. Sometimes it is sufficient to vary the timing of the acid attack.

6. In Utrecht this peeling method was also applied to noncarbonate-containing sandstones and shales; the polished surfaces were not etched. The results obtained were excellent.

7. The author placed a coral limestone peel in the acid bath for 10 sec. The result, compared with an unetched peel from the same sample, revealed a greater amount of details. This effect can be obtained with all coarse-grained and crystalline limestones.

8. Inspection of moderately weathered outcropping limestone surfaces, washed thoroughly by the natural flow of rainwater, often shows a large amount of sedimentary structures.

In active limestone quarries, Overlau progressed to artificial alterations. Repeated spraying (two to three times) of a solution of 20% acetic acid was accomplished with the use of a hand vaporizer, followed by washing the peel without rubbing with plain water. This kind of surface treatment, after drying for a few days, generally reveals many new details.

This exploring will determine:

(*a*) Immediate detection of structures that will produce spectacular peels.

(*b*) Evident structures that need not be studied.

(*c*) Beds without any apparent structures which should be tested from time to time.

9. Experience has demonstrated the advantage in sampling complete beds, or at least the base and top of a sequence. Ordinary reference marks must be further completed by information such as on both top and bottom, geographical north.

It is good to remember that sections which are parallel to the stratification are actually tangential to several structures, and therefore can produce complicated and confused figures.

10. The use of acetic acid on impure carbonates may produce chemical by-products, not all of which are water soluble. These newly formed chemical residues may cloud the peel (McCrone, written communication, December, 1965).

Advantages of this method (Overlau, 1965a)

1. The organic microelements are preserved with their original structures, cells of noncalcareous algae, etc. (Conil et Lys, 1964).

2. The original color and the insoluble matter of limestone samples are preserved.

3. Dimensions of about 30 x 30 cm (12 x 12 in.) are easily released without any inconveniences other than their bulkiness.

4. The handling times are very short, 15 to 30 min per peel, whereas on the other hand the manual polishing and the time spent in waiting are more tedious and time consuming.

5. The maximum contraction of the peel is less than 1%.

6. Microstructures, ranging in sizes from 5 to 10 μ, are easily observed through the microscope.

Procedure IV. At the Koninklijke/Shell Exploratie en Produktie Laboratorium at Rijswijk (Z. H.), The Netherlands, an acetate mixture is used that makes it possible to complete the entire process within 36 hours. Peels can be made from calcareous sediments as well as from sandstones and shales.

Equipment. Sawing and polishing outfit: grade 800 carborundum powder, spread on a wet glass plate, is used for the final polishing; containers with acid and with water; polythene squeeze bottle; box with sand in an exhaust room; knife, glass plates, and adhesive tape.

A standard acetate solution is made by mixing 30 g cellulose-acetate powder {tricellulose acetate: $[C_6H_7O_2 (CH_3COO)_3]_n$} in 130 cc acetone. Several hours are required for the powder to dissolve completely. Therefore the bottle should be kept in a warm place. If the bottle is well stoppered, the solution can be kept for long periods without deterioration.

Before use, it is necessary to dissolve this stock solution in an equal quantity of tetrachlorethane ($CHCL_2 \cdot CHCL_2$). This diluted mixture cannot be stored longer than 2 weeks.

Tetrachlorethane is harmful to health, and the maximum permissible concentration in air at any time should be no more than 5 ppm or 35 mg/m³ in an 8-hour working day. It is advisable to prepare the diluted solution in a running fume-cupboard. A 5% solution of formic acid is used for etching carbonate-containing samples.

Preparation of the peel. When dealing with calcareous sediment, the specimen is first cut and polished, and then etched. Sandstones and shales need not be etched. Next the sample is allowed to dry completely.

The sample is then placed in a box with sand inside a running fume-cupboard, the treated face leveled horizontally, and the surface wet with a small quantity of acetone. Before the acetone

can evaporate, the peel solution is poured on the sample (using a polyethylene washing bottle), spreading over the entire surface. Care is taken that no mixture flows over the edges of the sample. The mixture is then left to dry for several hours. A second coat of acetate solution should be applied when the polished surface forms pits or becomes pitted.

When the last coating is completely dry, the edges of the film are loosened with a knife or a razor blade, and the film peeled off carefully.

It is recommended that peels collected from calcareous sediments be etched in order to increase the clearness of the picture. The peel can be immersed in a 10% HCl solution until effluence ceases (generally 5 sec is sufficient). To avoid damage of the film it should be mounted between glass (again in a running cupboard).

Remarks. The author did not find any difference in results of peels made according to procedures III and IV. The disadvantage of the last method is the application of tetrachlorethane. Long-term damage to lungs and brain by tetrachlorethane is extremely dangerous (McCrone, written communication, December, 1965).

In Utrecht two coats of mixture are always applied to the sample because the coat thickness greatly decreases during drying, and a very thin film is difficult to peel off without tearing.

Procedure V. Peels can also be made by using nitrocellulose (Collodion cotton) as base of the solution. The mixture (see Table I.10) is very suitable for application to hard rocks, and it has the advantage of drying within a few hours.

A recipe by Gwinner (1963) is given in Table I.10. The material is not dangerous if moistened with alcohol (65:35, as delivered).

Because not all nitrocellulose dissolves directly, and many air bubbles become trapped, the mixture ordinarily becomes cloudy. It is best to let the mixture set an hour after it has been stirred well, allowing the air bubbles to escape and the nitrocellulose to dissolve. The mixture is not dangerous to health, but care should be taken since the liquid is flammable.

Experiments have shown the advantage of using a nitrocellulose type with a low molecular weight. This implies that more nitrocellulose must be added to the solvents, and thus a smaller percentage of the mixture evaporates, which results in a thicker film than would be obtained if a high molecular weighted nitrocellulose has been used.

Table I.10.   Components of a Cellulose Lacquer for Making Peels
of Hard Rock Sediments

---

90   cc butyl acetate
10   cc xylene
 0.5 cc softener, such as rincinus oil (= castor oil)
Nitrocellulose (moistened with 35% ethyl alcohol) must be added till a
   viscous solution is obtained (under continuous stirring).

After Gwinner (1963).

The nitrocellulose type A from the Springstof-fenfabrieken in the Netherlands has a nitrogen content of 12.0 to 12.4%. The viscosity of a 4% solution is more than 1500 cP (personal communication, July 1965). About 3 g in 100 cc of solvents (see Table I.9) produces a good mixture. The obtained film is excellent, but it is advisable to apply two or three thick coats of solution so that the film is thick enough to be peeled off easily without tearing (see Fig. 1.69). The viscosity of a 4% solution of type C is 25 cP; one coat is sufficient to form a good film.

The Dynamit Nobel in Germany produces a series of nitrocellulose. Experiments have been conducted with two ester-soluble types; their nitrogen content is 12% (Dynamit Nobel, not dated).

About 9 g of type HP 3000 (Norm 7E) in 100 cc of solvents, and about 30 g of type HP 50 (Norm 21E) in 100 cc solvents (see Table I.10) produce the proper viscosities. By dissolving 7 parts of HP 3000 in 93 parts of solvents, a viscosity is obtained that is similar to that of a mixture made from 21 parts of HP 50 in 79 parts of solvents. After evaporation, the obtained film made from the HP 50 mixture is three times as thick as the one with HP 3000 (Dynamit Nobel, written communication, November 13, 1964). Both types produce very good results. From the preceding remarks, it is evident that one should apply a double coat when applying type HP 3000 and a single coat when using HP 50.

Remarks. The peels made with nitrocellulose solutions produce as exact results as the ones obtained with acetate mixtures.

For samples containing $CaCO_3$ it is necessary to etch the polished surface beforehand. The nitrocellulose mixture can be applied directly for noncarbonate sediments. It is not necessary to wet the polished surface with acetone before pouring the nitrocellulose.

If the mixture is too viscous, and the surface of the poured film is not flat, a small amount of acetone may be applied. The acetone lowers the viscosity of the surface, allowing a smooth surface to form.

Care must be taken not to form or trap air bubbles in the poured mixture layer, since they will form holes in the future film. After pouring, the surface must be examined carefully and all bubbles must be eliminated.

1.7C Photography of Acetate Peels Nitrocellulose Peels

Since these types of peels are very thin, transparent, and without any real relief, they cannot be photographed in the same manner as other nontransparent peels.

The peels can be mounted between glass plates and studied directly under a binocular microscope. They can also be placed between slide glasses and used for direct projection. Staining the sample before making a peel increases the result considerably (see Chapter Four).

Oblique Illumination Method

Easton (1942) published an improved technique for photographing peel sections of corals. On one side the peel is a plane surface, whereas on the other side some faint relief is present. The finished peel is mounted between two plane-parallel glass plates and is then placed vertically in front of a camera with the plane surface of the peel toward the camera. An ordinary, but strong microscope lamp, emitting nearly parallel light, is used without its blue filter. The lamp is placed in such a way that the angle between light rays and peel does not exceed 30° (Fig. 1.67). A dull black background is placed close to the peel to prevent any false light falling upon the peel. The exposures should be taken with only the light from the microscope lamp. The diaphragm should be closed as much as possible in order that all details will be distinct and not partially overexposed.

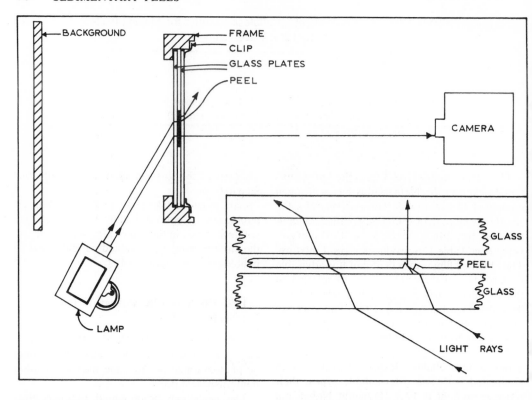

Fig. 1.67. Scheme of the oblique illumination method to photograph acetate peels. (Redrawn from Easton, 1942.)

At places where the peel has two parallel faces, a light ray is so refracted that it continues its way in the same direction. Where irregularities are present in the peel surface, the light ray will be refracted so that the outgoing one has another direction than the incidental one, and thus may be caught by the camera (Fig. 1.67).

The main difficulty is that prints become increasingly obscured when the grain decreases in size (Easton, 1942). This method can be used when the peel shows little contrast.

Slide Projector Method

Buehler (1948) places the peel between glass plates, puts it in a slide projector, and projects it on a screen. A camera is set for time exposure and the picture is taken from the screen. Printing on contrasting paper is recommended.

This definitely is the easiest method, and the results are excellent. The advantage is that the investigator can photograph either the whole picture or a detail of it (Figs. 1.68 to 1.71).

Photographic Enlarger Method

Fay (1961) as well as Gwinner (1963) place the mounted peel in a photographic enlarger, using the peel as a negative. Not applying film material as a negative implies an increase in contrast and sharpness of the print, therefore making it possible to produce enlargements up to microscopical level (Gwinner, 1963).

It is necessary to use short exposure times and small diaphragm openings. One advantage of making enlargements is that the peel can be placed between glass plates and kept completely flat, which is not possible for contact printing. Normally, the peels are always wavy and the degree of contact differs from spot to spot, resulting in blurred peels.

Overlau (1965b) places a thin section or an acetate peel into an enlarger (Meopta 6 x 9 cm equipped with a Schneider Componan lens, 150-watt lamp, diaphragm 16, 1 4/10 sec) and enlarges his object directly four times. He uses a semi-transparent film, AETr Mimosa, Kiel, Ger-

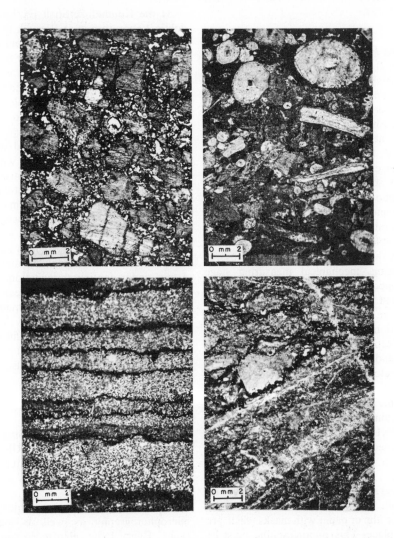

Fig. 1.68. (*Upper left*). Kodatrace peel of a cross section through a large groove cast. The kidney-shaped hole near the center is a hole in the peel formed by an air bubble. The sample has not been etched. Photograph made by the slide projector method. Eocene turbidites, section Pas de l'Escous-la Cabanette, Peira-Cava area, Alpes Maritimes, France. (Bouma, 1962.)

Fig. 1.69. (*Upper right*). Nitrocellulose (Collodion Cotton A: Koninklijke Nederlandsche Springstoffenfabrieken N. V.) peel from a crinoid limestone. Photograph made by the slide projector method. Visean, Belgium.

Fig. 1.70. (*Lower left*). Nitrocellulose HP50 (Dynamit Nobel Company) peel from algae (*Collenia*) limestone. Photograph made by the slide projector method. Visean V3b, Belgium.

Fig. 1.71. (*Lower right*). Acetate solution K (Fabelta) peel from limestone. Note the calcite veins and the irregular bands of clayey material. Photograph made by the slide projector method. Visean V2b, Chokier, Belgium.

many. The negative is developed in developer G5 of Gevaert (diluted with water 1:1) for 2½ to 3 min. The very fine-grained negative is very suitable for enlargements, contact and reduction prints (Fig. 1.72).

Measuring Projector Method

Lees (1962) describes the best, but also the most expensive method to make prints of peels by using a standard industrial measuring projector. Lees as well as scientists from the Koninklijke/Shell Exploratie en Produktie Laboratorium at Rijswijk use a Shadomaster, model VMP, from Watson, Manasty and Co. Ltd. (Watson et al., 1958).

The light source consists of a convection-cooled lamphouse with a fully adjustable, 12 V, 50 W, compact-filament, high-intensity lamp which is operated by a built-in transformer. Two interchangeable condensers (1.5 and 2.5 in.) can be placed at the lower side of the lamphouse. A large, focusing, mechanical measuring stage serves as support of the peel. The stage provides movements of 1.18 x 2.45 in. and gives a maximum coverage of 2.35 x 3.93 in. using the smallest projection lens. Interchangeable True-Gauge projection lenses (x10, x25, x50, x100) are placed underneath the stage. At the bottom of the projector a mirror is mounted which reflects the picture on to a 12 x 15 in. finely etched glass screen. The resolution of the lenses is good, and grains smaller than 10 μ can be clearly observed at x100. For such fine grain sizes the standard etched glass is replaced by a clear glass screen and a matt plastic (Astrofoil) overlay (Lees, 1962).

For photography, the image is first focused on the Astrofoil, after which this is replaced by photographic film or paper with masks which are pressed in place. This photographing must be done in a room that can be blacked out. Lees (1962) uses a density filter to increase the exposure time to a few seconds so that it can be timed without a shutter.

The mechanical stage makes it possible to measure with an accuracy of 2.5 μ. Simple grid screens can be used for crude estimates on grain size and for grain size distribution.

The Shadomaster can also be used for photographing thin sections. Polarizing filters, which fit the condensers and the projection lenses, make it possible to study grain orientation by extinction methods and to carry out routine mineralogical analysis. This last investigation can be done nicely with peels collected from stained samples (see also Chapters Two and Four).

At the Koninklijke/Shell Exploratie en Produktie Laboratorium at Rijswijk the rough side of the peel is wetted with a solution of Agfa Agepon in water (1:200) in order to obtain more contrast. However, this mixture may make the peel too light. The peel is mounted as flat as possible to an object glass by double-sticking tape (Fig. 1.72). Enlargements up to 100x can be easily obtained.

Thin sections are not covered with a cover glass. They too can be wetted with the Agfa Agepon solution to raise the contrast. The mixture should be washed off the section with water because it will dissolve the calcium carbonates.

Sheet film (13 x 18 cm) replaces the etched glas. Agfa-Gevaert IP 24 blanc is the film applied. It can be developed in a Agfa-Gevaert Rodinal solution (1 Rodinal:25 water) at 20°C for 5 min under continuous agitation. A stopbath containing 3% acetic acid is necessary. The fixing time is 10 min. Prints can be made on Kodak Bromesko WSG, using the developer D 163 (thinned down 1 to 3). A green filter is used to achieve a higher quality product (written communication, December 1966). Also a soft grade of positive paper can be applied.

Aristophot Method

Scientists of the Koninklijke/Shell Exploratie en Produktie Laboratorium at Rijswijk (Z.H.), the Netherlands, also use an Aristophot set from Leitz for making enlarged prints of a peel or a thin section.

The Aristophot is a photographic unit used for microphotography as well as macrophotography. Every Leitz microscope can be applied (Leitz, 1961, 1963).

For macrophotos, with a negative size of 9 x 12 cm (4 x 5 in.), the unit consists of a support on which a macrodia case with adjustable objectable, glass plate holders, diaphragm, and condenser is placed. The case is equipped with a 6 V, 5 A light source with frosted glass, blue glass, and green filter. A camera, 9 x 12 cm (4 x 5 in.), with bellows, lenses, and a mirror-reflex outfit is connected in an adjustable way to the upper part of the support. The Summar type objective with f = 42 mm allows enlargements from 2 to 13 times.

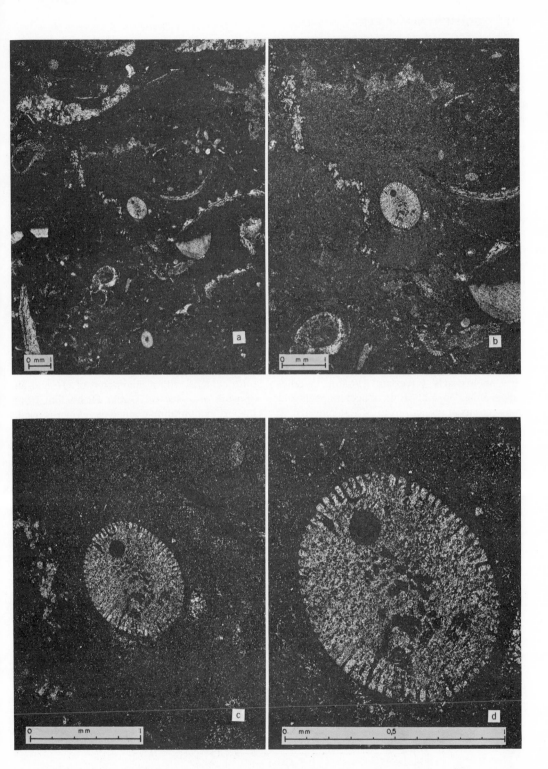

Fig. 1.72. Nitrocellulose HP 3000 (Dynamit Nobel Company) peel from coral limestone. Photograph made by the measuring projector method. (Courtesy of the Koninklijke/Shell Exploratie en Produktie Laboratorium at Rijswijk, Z. H., The Netherlands.) Original enlargements 6x, 12x, 30x, and 60x, respectively. Frasnian (Devonian) bioherm, Taillfer, Belgium.

Macrophotos can also be made with the Aristophot unit, using a Leica 24 x 36 cm camera with Leica bellows, micromirror reflex outfit, and the same Summar objective. The macrodia case and light source are similar to those preceding.

The transparent object should not exceed 90 mm in size. Its frosted surface can also be wetted with the Agfa Agepon solution to increase contrast. A blue filter produces the best light.

The objects can be photographed on Illford HP 3 film. As average exposure f4 and $\frac{1}{10}$ sec can be used. The developing at 20°C for 5 min in Agfa rodinol developer (thinned 1:30) must be carried out with continuous agitation (turning and vertical movements) of the film-spool during the first minute, followed every 30 sec by a good turn of the spool. The printing can be done on the same paper as in the previously described method.

Copying-Box Method

This method is used at the Petrofina laboratory at Brussels (Overlau, 1965b). Overlau makes either contact prints of acetate peels and thin sections or reductions to 24 x 36 mm size.

For contact printing, a copying box (Hohn and Hahn, Germany) with 4 rod-shaped lamps of 15 watts each is used. After removing the acetate peel from the mounting glass plates, an opalin and a frosted glass plate are mounted underneath a clear mirror-glass window on which the peel is placed with its mat surface face-up.

A semitransparent negative film (AETr type of Mimosa, Kiel, Germany) is placed in contact with the peel. A cover, consisting of a piece of foam plastic and wood, is put on top in order to secure good contact between peel and negative film (Fig. 1.73).

In reducing the size of the sedimentary picture of an acetate peel to a 24 x 36 mm surface, Overlau uses an ordinary copying box consisting of 4 neon tubes of 5 watts each, covered with an opalin and a clear mirror-glass plate (Fig. 1.74). The acetate peel, mounted between glass plates, is placed on top with its mat face upwards.

The photographs are made with a normal miniature camera, with a diaphragm of f16 and an exposure of 1 sec on Duplo Ortho film from Gevaert. An ultrafine-grained developer is used.

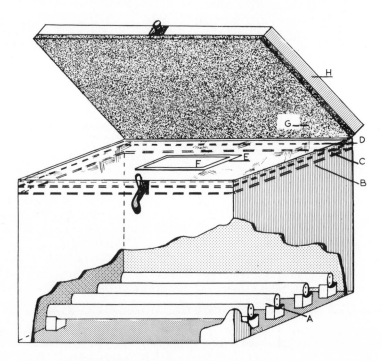

Fig. 1.73. Scheme of a copying box used for contact printing from acetate peels: (A) rod-shaped neon tube of 15 watts; (B) opalin glass plate; (C) frosted glass plate; (D) mirror glass plate; (E) acetate film with mat face upward; (F) semitransparent photographic film; (E) foam plastic; (H) wooden cover.

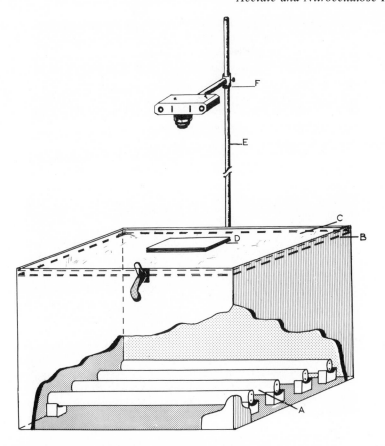

Fig. 1.74. Scheme of a copying box used for making 24 x 36 mm pictures from acetate peels: (A) rod-shaped neon tube of 20 watts; (B) opalin glass plate; (C) mirror glass plate; (D) acetate peel between protecting glass plates, mat face upward; (E) camera support; (F) miniature camera.

From these negatives normal enlargements and black and white slides can be made (Fig. 1.75).

The part of the mirror-glass window which is not covered by the peel should be covered with a frame of black paper.

1.7D Comments

Acetate and nitrocellulose peels can easily be made from consolidated sediments, where the other types of films described in this chapter do not produce good results or no results at all.

Acetate film sheets, such as Kodatrace, and celluloid sheets can be used as a quick method and/or as a first reconnaissance. However, it must be kept in mind that these sheet materials will give poor results if applied to sediment surfaces that contain small holes since air bubbles become trapped in such depressions.

Carbonate-containing samples should be etched before a peel is made. This is not necessary for noncalcareous sediments. Good results have been obtained from sandstones, shales, and tonstein. If the collected peel from a siliceous sample is not satisfactory, one can try to etch the polished surface with HF. A Kodatrace peel from a porphyry was made without any difficulties and without etching.

The acetate-nitrocellulose peels are not applicable in mineralogical investigations. They are primarily developed for the study of fossils, grain size, grain shape, and orientation, and for the structures of carbonate rocks. Katz and Friedman (1965) combine staining with acetate peels for mineralogical analyses on carbonates (see Chapter Four).

A number of general remarks are given as technical notes at the end of procedure V of the

Fig. 1.75. Kodatrace peel from a Devonian sandstone. Note the erosion relics on top of the laminated part and the large burrow through the center. Photograph made with the copying box-miniature camera method. (Courtesy of P. Overlau, Petrofina, Belgium.) Core sample obtained from P. Dollé. Coal mines near Hénin Liétard, Pas de Calais, France.

part on Acetate Solutions and therefore will not be repeated here.

Making photographs of acetate and nitrocellulose peels is a technique that differs considerably from normal photography. The investigator can make a choice of the different methods given, depending on the apparatus available. The investigator has been supplied with only a working plan, and he must rely on experience to decide if the samples should be etched, what type of acid should be used, time of etching, and type of peel. This is also true when making photographs. Certain techniques can be applied directly, such as slide projectors and enlargers, which are generally available. These simple methods restrict the size of the peel that can be photographed immediately. If the investigator wants to copy all sizes of peels and to enlarge details, a copying box should be constructed. For frequent use at

standard sizes, a Shadomaster is essential. A duplicating type of film will produce the best results.

1.8 DISCUSSION

In spite of the large variety of sediment types, one can always find a method for making a peel from them, which consists of, or contains some of, the original material. These peels always are true to scale.

At the present time there is no product known that fulfills all necessary requirements. The variation in types and conditions of the sediments necessitates the application of different techniques and different products in order to obtain replicable peels from consolidated as well as from unconsolidated deposits, which may be wet, moist, or dry, and coarse- or fine-grained.

Before adopting a certain technique, one must

consider which method is the most suitable for a particular situation. In spite of the time involved, the advice given is to read at least the introductions and remarks on all techniques described before making a decision. The key given in Table I.1 and Fig. 1.1 may be of some help.

As stated in the introduction of this chapter, an investigator cannot expect maximum results at the first treatment carried out, in spite of consulting Table I.1 and Fig. 1.1, plus the appropriate sections of the text. There is such a wide variety of sediments and conditions that some experience must be acquired. It is also recommended to start with the air-dry, unconsolidated sandy sediments, applying the spraying method, under dry weather conditions, before other types of unconsolidated deposits are treated. The method of McMullen and Allen and the prepara-tion of glue peels according to the technique of Heezen and Johnson are fairly simple. For the preparation of peels with high relief, one should start with paraffin in order to find out if it is worthwhile to buy more expensive and better products.

Concerning the acetate and nitrocellulose peels, the investigator should start by applying Kodatrace to discover whether it is useful for one's study to make better peels, and also to try out type of acid and duration of etching time.

Peels of unconsolidated deposits are very useful for studying and instructional purposes, as comparison standards, and for wall decoration. Acetate and similar peels, collected from consolidated sediment samples, often increase the amount of observable characteristics, are very suitable for projection, and are easier to handle than the actual samples.

CHAPTER TWO

IMPREGNATION

Impregnation of unconsolidated and poorly consolidated sediments involves the artificial binding of the material by the injection of a cementing agent without disturbing the arrangement of the individual grains. Once impregnated, the sample can be treated as a consolidated specimen. For many types of investigations, such as microscopical ones, it is not possible to collect data without the help of thin sections or polished surfaces. Especially in the fields of pedology and petrography, many successful attempts have been made to "silicify" unconsolidated samples.

The several known impregnation techniques differ according to the results wanted, optical properties of the impregnating material, types of sediment, and the moisture content of the samples. Like peels, sands are easy to impregnate, but deposits containing pelite are difficult to impregnate.

Ross (1924) founded the technical basis for artificial hardening, and many methods have been developed since. The majority of them follow the principle of impregnating, usually under vacuum, air-dried samples. The process should be carried out either at room temperature or slightly warmer, using a liquid with a low viscosity during the impregnation, which will afterwards become so hard that it can be sawed and ground.

The most common materials used at the present time are members of the large "plastic" group, although some previously used products are still applied.[1] The number of contributions written on impregnation are numerous and come mainly from the soil scientists. It is impossible to name all of them,[2] and therefore only a few will be mentioned in the text.

Necessary apparatus used in impregnating

---

[1]Bakelite lacquer diluted with a mixture of one part ether and one part methyl alcohol (Ross, 1924), Dioxan and paraffin (Goemann, 1937), I. G. Wax (Goemann, 1940), Kollolith (Kubiena, 1937; Redlich, 1940; Jongerius, 1957), Plasto resin no. 15 and Resinel no. 3 (Volk and Harper, 1939), Resinol (Rotter, 1941), Vernicolor (Frei, 1947; Jongerius, 1957), Canada balsam and shellac (Van Straaten, 1951), and Canada balsam only (Hickman, 1956; Kurotori and Matsumoto, 1958), Polestar (Hagn, 1953), Polylite (Kawai and Oyama, 1962).

[2]Schlossmacher, 1919; Sayles, 1921; Keyes, 1925; Ross, 1926; Ahrens and Weyland, 1928; Legette, 1928; Schwarz, 1929; Schaffer and Hirst, 1930; Krumbein, 1935; Harper and Volk, 1936; Weatherhead, 1940, 1947; Kubiena, 1942; Nuss and Whiting, 1947; Fowler and Shirley, 1947; Bourbeau and Berger, 1947; Nikiforoff et al., 1948; Day, 1949; Emery and Stevenson, 1950; Lockwood, 1950; Brison, 1951; Ritch and Cardwell, 1951; Brown and Patnode, 1953; McMillan and Mitchell, 1953, Alexander and Jackson, 1954, 1955; Ingerson and Ramisch, 1954; Altemüller, 1956, 1962; Brewer, 1956; Conkin, 1956; Exley, 1956; Hepple and Burges, 1956; Debyser, 1957; Dalrymple, 1957; Osmond and Stephan, 1957; Thissen, 1959; Amstutz, 1960; Cavanaugh and Knutson, 1960; Taylor, 1960; Borchert, 1961a, 1962; Buol and Fadness, 1961; Burges and Nicholas, 1961; Catt and Robinson, 1961; Mackenzie and Dawson, 1961; Richardson and Deane, 1961; Gadgil, 1962; Geyger, 1962; Vernet and Gautier, 1962; Wells, 1962; Werner, 1962; Haarlov and Weiss-Fogh, 1963; Jenny and Grossenbacher, 1963; J. Bouma, 1965.

sediments is discussed in Chapter Five. Since the preparation of thin sections can be considered a known technique, only the general application will be discussed, based mainly on impregnated samples and on mounting materials other than Canada balsam.

In order to understand the processes acting during the hardening of the impregnating resins, the curing of polyester resins will be discussed first, since unsaturated polyesters are the main liquids applied. Some other products will also be mentioned.

Thin sections cannot be made from most of the lacquer and other peels since the grain binding is too weak without further impregnation to withstand the forces exerted on them during grinding.

Because the moisture content influences the hardening of the impregnating materials in most cases, this chapter is divided according to the absence or presence of water. Poorly consolidated sediments are usually too soft to saw and grind directly, and since their grains are already slightly bound together, relatively fast methods should be applied. Fissile sediments consist of hard parts, but the binding forces are so weak that some artificial treatment has to be applied before the sample can be placed under the diamond saw.

Working with synthetic products may cause danger to the health of the operator. Therefore some notes deal with toxicity and precautions. These protection rules also refer to the products mentioned in the Chapters One and Four.

To give the reader an overall picture of the different types of applications, mainly related to type and moisture content of the samples, a tentative division is given in Fig. 2.1 and in Table II.1.

2.1 LIQUIDS USED FOR IMPREGNATING AIR-DRIED SEDIMENTS

To impregnate a dry sample at room temperature, a product should be used in which the viscosity can be lowered enough to obtain sufficient penetration. The liquid should be capable of hardening in this temperature environment, and the final product should be hard enough so that the impregnated sample can be sawn and ground, and color and optical properties of the binding material should not differ much from those from Canada balsam.

Great improvements in impregnation have been obtained with the application of unsaturated polyesters during the last 15 years. Hagn (1953) and Alexander and Jackson (1954) belonged to the small group who had used these materials previously, but it was Altemüller (1956, 1962) who, by introducing the unsaturated polyester resin Vestopol-H from the Chemische Werke Hüls, made an important contribution to the solution of many of the difficulties encountered previously with the impregnation of dense samples. Applying the work of Altemüller and also using Vestopal-H, Jongerius and Heintzberger (1962, 1963) developed a method for preparing thin sections of all types of soils only 15μ thick or thinner and as large as 8 x 15 cm.

2.1A The Curing Process of Unsaturated Polyester Resins

"Plastics are man-made materials which can be molded into useful articles. The first of these plastics was produced in 1862 in England by Alexander Parkes. It was known as *Parkesine* and was the forerunner of Celluloid.

Since then a large variety of plastics have been developed commercially, most of them in the last twenty-five years. They extend over a wide range of properties. Phenol formaldehyde (PF) is a hard thermoset material; polystyrene is a hard, brittle thermoplastic; polythene and plasticised polyvinyl chloride (PVC) are soft, tough thermoplastic materials, and so on.

All plastics have one important common property; they are composed of macro-molecules, i. e., large chain-like molecules consisting of many simple repeating units (*polymers*). Man-made polymers are called *synthetic resins* (or simply resins) until they have been molded in some way, when they are called *plastics*. Most synthetic resins are made from chemicals derived from oil or coal" (Scott Bader, 1965, p. 1).

In impregnation the most commonly used plastics are the unsaturated polyester resins. This means that such resins can be cured from a liquid to a solid state when subjected to the proper conditions. Most of the resins consist of a solution of an unsaturated reactive resin in an allyl or vinyl monomer, usually monostyrene. The styrene performs the vital function of enabling the polyester to cure from a liquid to a solid by cross-linking the molecular chains of

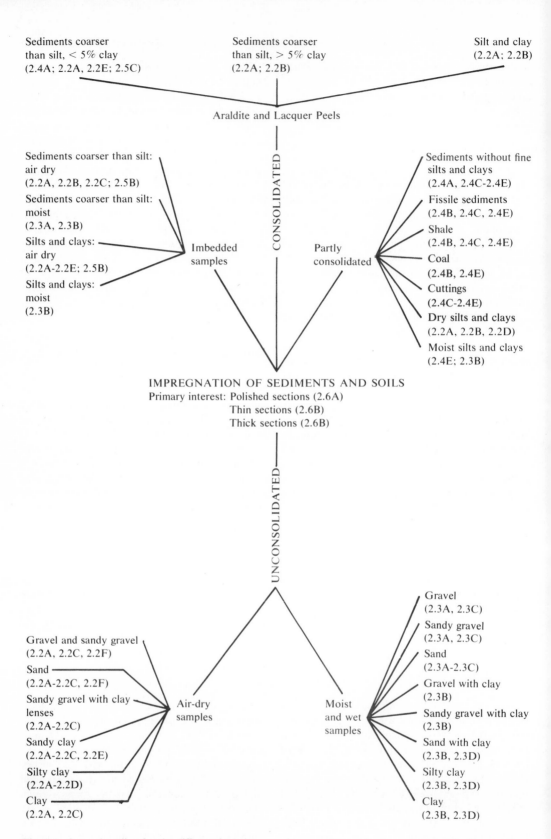

Fig. 2.1. General outline for the different techniques on impregnation methods as discussed in Chapter Two.

Table II.1. Key to the Different Impregnation Methods Applicable
for Certain Types of Sediment with Regard to Moisture Content,
Size and Shape of Sample, Results Desired, and Time Available

A. Air-Dry Unconsolidated Sediments and Soils

Thin, thick, or polished sections desired.

1. Gravel without clay and less than 5% sand. Samples as well as cores.	1. Pour undiluted unsaturated polyester resin mixture onto sample. If plastic soaks in apply vacuum. Method of Curray (2.2A) or method of Shell (2.2C). Also an epoxy resin can be used which cures at room temperature (1.6A) or at high temperatures (2.2F).
2. Sandy gravel	2. Same as (1) but vacuum may be applied.
3. Sandy gravel with clay lenses. Samples.	3. Method of Jongerius and Heintzberger (2.2A) or method applying Pleximon (2.2B).
4. Sandy gravel with clay lenses	4. Method of Shell (2.2C), using plastic mixture of Jongerius and Heintzberger (2.2A).
5. Sand. Samples < 2 cm.	5. Method of Altemüller (2.2A), method of Curray (2.2A), method applying Pleximon (2.2B). Application of room temperature curing epoxy resin (1.6A) or high temperature curing epoxy resin (2.2F).
6. Sand. Samples > 2 cm.	6. See number 2, above. Method of Curray (2.2A), method of Jongerius and Heintzberger using less styrene monomer (2.2A).
7. Sand. Cores.	7. Method of Curray (2.2A).
8. Sandy clay or sand and clay. Samples < 2 cm.	8. Method of Altemüller (2.2A), method applying Pleximon (2.2B), epoxy resin using vacuum and pressure (2.2D). When little clay is present, epoxy resin using vacuum and curing in oven (2.2E).
9. Sandy clay or sand and clay. Samples > 2 cm.	9. Method of Jongerius and Heintzberger (2.2A) (if little clay is present the mixture given by Altemüller–2.2A–can be applied). For thin samples epoxy resins may be applied (see number 8).
10. Sandy clay or sand and clay. Cores.	10. Method of Shell (2.2C). When clay is dominant the mixture given by Jongerius and Heintzberger (2.2A) can be used.
11. Silty clay. Samples < 2 cm.	11. Method of Altemüller (2.2A), method applying Pleximon (2.2B), or use an epoxy resin and apply vacuum as well as pressure (2.2D).
12. Silty clay. Samples > 2 cm.	12. Method of Jongerius and Heintzberger (2.2A). For thin (< 1 cm) samples see also point 11.
13. Silty clay. Cores.	13. Method of Shell (2.2C). The mixture given by Jongerius and Heintzberger (2.2A) can be applied.
14. Clay. Samples.	14. Method of Jongerius and Heintzberger (2.2A).
15. Clay. Cores.	15. Method of Shell (2.2C) applying the mixture given by Jongerius and Heintzberger (2.2A).

B. Moist and Wet Unconsolidated Sediments

Thin, thick and/or polished sections desired.

1. Gravel. Sandy gravel. Sand. Impregnation in the field desired.	1. Method of Brown and Patnode (*in situ*) (2.3A), method of McMullen and Allen (2.3A), gel method [not suitable for thin sectioning (2.3C)]..
2. Gravel. Sandy gravel. Sand. Samples, cores.	2. Dry the material and follow part A of this table. For wet and moist samples: method of McMullen and Allen (2.3A), method of Reineck (2.3D) or gel method (2.3C) (not suitable for thin sectioning).

Table II.1. Key to the Different Impregnation Methods Applicable for Certain Types of Sediment with Regard to Moisture Content, Size and Shape of Sample, Results Desired, and Time Available

3. Gravel, sandy gravel and sand with less than 10% clay. Samples and cores.

   3. Point B.1 can be tried. One also can try to dry the samples and then follow the directions given in part A. Small samples: method of Tourtelot (2.3B) (not suitable for thin sectioning), method of Reineck (2.3D).

4. Gravel, sandy gravel and sand with more than 10% clay. Samples and cores.

   4. Since drying will result in forming of cracks, it is better to take small samples and apply the method of Reineck (2.3B) or the method of Tourtelot (2.3B). The latter method is not suitable for thin sectioning.

5. Silts. Samples and cores.

   5. See point B.1 or B.2.

6. Silty clays and clays. Samples and cores. Only a few days available between collection and investigation.

   6. Method of Tourtelot (2.3D). Size of samples should not exceed 5 x 5 x 3 cm. Small-sized thin and thick sections can be made.

7. Silty clays and clays. Samples and cores. Some weeks available between collection and investigation.

   7. Method of Reineck (2.3B). Hard results can be obtained of samples which do not exceed 2-3 cm in thickness.

C. Partially Consolidated Sediments

Thin sections desired for petrographical analysis as well as for sedimentary structure studies. Thick sections and/or polished surfaces may be required.

These methods are rather imbedding techniques than impregnation techniques; however, enough material may be impregnated to allow wetting.

1. Badly consolidated sediments, excluding fine silts and clays. Loose grains and/or very small pieces. Only petrographical analysis desired.

   1. Sample has to be air-dry. Method of Moore and Garroway (2.4A), open method (2.4A).

2. Badly consolidated sediments, excluding fine silts and clays. Good permeability. Sections desired.

   2. Sample has to be air-dry. Open method (2.4A), vacuum application (2.4C) in general, or the Shell method (2.4C) in particular.

3. Badly consolidated sediments, excluding fine silts and clays. Low permeability. Sections desired.

   3. Sample has to be air-dry. Method of Shell (2.4A), vacuum pressure application (2.4D; see 2.2D).

4. Badly consolidated sediments, fissile. Cracks need to be filled rather than sediment itself.

   4. Sample has to be air-dry. Gluing method (2.4B), open method (2.4A), method of Shell (2.4C), temperature application (2.4E): epoxy, lakeside or Canada balsam.

5. Shale, dry.

   5. See point 4.

6. Shale, moist.

   6. Temperature application (2.4E): Carbowax (see 2.3B).

7. Coal, air-dry.

   7. Gluing method (2.4B).

8. Cuttings, air-dry.

   8. Method of Shell (2.4C), vacuum-pressure application (2.4D, see also 2.2D), temperature application (2.4E): epoxy, lakeside, Canada balsam.

9. Badly consolidated silts and clays. Air-dry.

   9. Method of Jongerius and Heintzberger (2.2A), method of Altemüller (2.2A), method applying Pleximon (2.2B). See part A of this table.

Table II.1.   (Continued)

10. Badly consolidated silts and clays. Moist.	10. Temperature application (2.4E): Carbowax (see 2.3B).

D. Imbedded Sediments

Thin or thick sections desired. Polished surfaces sometimes can be made without additional impregnation (see 2.5 and Chapter Four for imbedding).

1. Sediments coarser than silt. Air-dry.	1. See part A of this table: points 1, 2, 5, and 6.
2. Sediments coarser than silt. Moist.	2. Method of McMullen and Allen (2.3A), method of Brown and Patnode (2.3A), method of Reineck (2.3B). If samples can be made air-dry, see point D 1.
3. Sediments finer than silt. Air-dry.	3. See part A of this table: points 3, 4, 7-15.
4. Sediments finer than silt. Moist.	4. Method of Reineck (2.3B), method of Tourtelot (2.3B). If the samples can be made air-dry see point D 3.

E. Araldite and Lacquer Peels

Thin or thick sections desired.

1. Sediments coarser than silt with less than 5% clay. Clay laminae not over 2 mm thick.	1. Pour an unsaturated polyester resin mixture onto it. If necessary apply vacuum (of open method) (2.4A). Epoxy resins can be applied also (2.2D, 2.2E, 2.2F).
2. Sediments coarser than silt with more than 5% clay. Clay present as layers over 1 mm thick.	2. See E 1. Vacuum should be applied. If the clay does not impregnate properly use method of Jongerius and Heintzberger (2.2A) applying only slightly diluted mixture or mixtures given by Altemüller (2.2A), method applying Pleximon (2.2B).
3. Fine silty and clayey sediments, with or without sand.	3. Method of Jongerius and Heintzberger (2.2A), method of Altemüller (2.2A), method applying Pleximon (2.2B).

the resin. The liquid resins are unstable and after several months or years of storage (*shelf life* or *storage life*), they will set, through a gel phase, into the solid state (copolymerization); this occurs even at normal room temperature. Shelf life can be increased considerably by storing the resins in a cool, dark place in their original containers or by adding an inhibitor such as hydrochinone.Exposing to ultraviolet light, as well as to temperatures greater than 25°C (77°F), reduces storage life.

The forming of styrene bridges between the polyester chains for obtaining a spatial net structure (cross-linking) can also be influenced by adding a *catalyst* with or without an *accelerator*. The catalysts used for the copolymerization of unsaturated polyester resins are products which decompose rather easily, each having an unpaired electron, which give them the name "free radicals." Substances (catalysts) that are capable of supplying radicals, thus creating active centers from which the attachment of monomer molecules will start, are organic peroxides, persulfates, ozonides, azo compounds, hydrazines, diazonium salts, and amine oxides (Noury & Van der Lande, not dated a). Since the free radicals which are formed become interwoven into the growing polymer molecules, the term "catalyst" for these compounds is wrong, according to the classical definition of a catalyst: the reaction must leave a catalyst unchanged. A better name is *initiator* (Conix, 1950). Both terms are used in industry.

Since the commonly used organic peroxides are fairly stable at room temperature, energy has to be added in the form of heat or radiation in order to facilitate the decomposition of the initiator into free radicals. For curing an unsaturated polyester resin rapidly at room temperature or at temperatures up to 100°C (212°F), it is necessary to add an accelerator to break the oxygen bridge of the peroxide (Noury & Van der Lande, not dated b, 1961a, 1962a).

Various materials can act as an accelerator.

For this type of work the most important ones are soluble metal soaps such as cobalt-octoate and cobalt-naphthenate, tertiary amines and lauryl mercaptan derivative.

Initiators (catalysts)

The type(s) and quantities of the initiator(s) and accelerator(s) used depend on the working temperature, the desired curing time, type and amount of resin, type and amount of filler, and the color of the final result. Since type and number of initiators and accelerators and their applications are very extensive, it is necessary to confine this discussion to those products used in impregnation and imbedding (Chapter Four) techniques at room temperatures, with or without a postcuring at higher temperatures (see also Fig. 2.3).

Benzoyl peroxide is normally used in combination with tertiary amines such as dimethyl aniline. This peroxide in a powdered form with 25% water is unsuitable for the polymerization of unsaturated polyester resins. Benzoyl peroxide, in the form of a white paste dissolved in dibutyl phthalate or other vehicles, is known under the names Luperco ABB (Wallace & Tiernan, 1962a), Lucidol B-50 (Noury & Van der Lande, not dated c, 1964a), Garox QZA (Ram, 1961). The benzoyl peroxide content is about 50% and has an active oxygen content of 3.3 to 3.6%. The same peroxide compounded with tricresyl phosphate is manufactured under the names Luperco ATC (Wallace & Tiernan, 1962b), Lucidol C-50 (Noury & Van der Lande, 1964a), Garox BZP (Ram, 1963a) and Nuodex B.P. paste (Nuodex, 1961a), and it is also a paste.

The disadvantage with pastes is that they are more difficult than liquids to measure in small quantities. Also, the combination of benzoyl peroxide and a tertiary amine turns the resin yellow under the influence of time.

From the ketone peroxides the methyl ethyl ketone peroxides (MEK) and the cyclohexanone peroxides are the most common initiators used in combinations with cobalt accelerators. Lupersol DDM, Lupersol Delta and Lupersol Delta-X are methyl ethyl ketone peroxides in liquid form dissolved in dimethyl phthalate manufactured by Wallace & Tiernan (1962c, d, e); Butanox M-50, Iso-butanox M-50, and Iso-butanox M-60 are made by Noury & Van der Lande (not dated d, 1956, 1964a); Garox MEK comes from Ram (not dated a), and Nuodex M. E. K. peroxide and Nuodex polycure M. E. K. peroxide from Nuodex (1960, 1962).

Lupersol Delta-X decreases the gel time (see Fig. 2.2) considerably more than Lupersol Delta does, while the amounts of initiator and accelerator required for a desired gel time increase from Lupersol Delta-X to Lupersol DDM (Wallace & Tiernan, 1962e). Butanox is much more active than Iso-butanox. When a longer gel time is desired one can decrease the amount of Butanox. However, increasing the amount of Butanox causes too long a curing time. It is therefore better to use Iso-butanox. The Nuodex polycure M. E. K. peroxide is also more active than the Nuodex M. E. K. peroxide.

The cyclohexanone peroxides are available in powder, paste, or liquid form. The powder is not suitable for this work since it contains water. Wallace & Tiernan (1962f) manufacture Luperco JDB-50-T, which is a cyclohexanone peroxide dissolved in dibutyl phthalate. It is a thick white paste with an active oxygen content of 5.8%; it easily dissolves in resin. From the Cyclonox series (Noury & Van der Lande, not dated e, 1964a, b) the author uses the products Cyclonox B-50 (paste with dibutyl phthalate) and Cyclonox LTM-50 (solution in triethyl phosphate). The LTM-50 is more active than B-50.

Both groups of ketone peroxides can be used in combination with metal soap accelerators for curing unsaturated polyester resins at room temperature. These peroxides are not detonable since they contain a phlegmatizing agent such as dibutyl phthalate or dimethyl phthalate.

Cumene hydroperoxide can also be used. It is a very stable peroxide at room temperature and can be stored in a cool dry place for a long time without noticeable decomposition. This peroxide can be used for slow impregnations, which can be placed in an oven afterward for final curing. It can be obtained from Hercules Incorporated under the name Hercules Cumene Hydroperoxide (not dated a,b), and as Trigonox K-70 from Noury & Van der Lande (not dated f, 1964a). Cumene hydroperoxide retards gelling of the resin mixture when used in combination with one of the ketone peroxides. It does not cure the resin at room temperature.

All peroxides should be stored away from direct heat, open flames, sunlight, all acids, fine-powdered metals, accelerators, and easily oxidized materials. Some of these products are

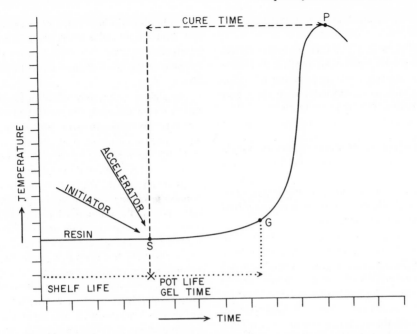

Fig. 2.2. General polymerization curve of an unsaturated polyester resin: (S) starting point of the polymerization reaction; (G) gelling point; (P) peak exotherm is the point of maximum heat evolution. (Modified after Jongerius and Heintzberger, 1962, 1963; Scott Bader, 1965.)

toxic and may cause damage to the eyes and skin after only short contact.

Accelerators

The most widely used accelerators in this field are the soluble cobalt soaps Co-octoate and Co-naphthenate with cobalt percentages of 1 or 6%. Cobalt-octoate has an advantage over cobalt-napthenate: it can be formulated more accurately because the cobalt-napthenate may also contain other cobalt compounds (Jongerius and Heintzberger, 1963). Nuodex manufactures cobalt-naphthenate which contains 6% metal cobalt in mineral spirits (Nuodex, 1961b). From Noury & Van der Lande (1961a, 1962, 1964, not dated b) other types of accelerators are available: accelerator NL-49 is a cobalt-octoate of 1% Co in dioctyl phthalate, NL-51 is similar but contains 6% Co, and NL-50 is a cobalt-octoate with 1% Co dissolved in styrene.

A disadvantage of these metal soaps is that they give a strong color to the final product. Depending on the amount of accelerator, the composition of the resin used, and temperature during the curing, the color may become (light) yellow, light red to violet, yellow-brown, or green (Table II.2). For impregnation purposes this is not important, but for thick imbeddings a colorless end product is necessary.

These cobalt accelerators are used in combination with ketone peroxides and cumene hydroperoxides. Mixtures of polyester with cobalt accelerators are not as stable as those without, depending on the type of resin used and on the storage conditions. The use of cobalt accelerators has a favorable influence on the stickiness due to air inhibition of the cured resin. For this reason these accelerators are sometimes used in combination forms with other accelerators.

When dealing with benzoyl peroxide, a 5% solution of dimethyl aniline in dimethyl phthalate (NL-63 of Noury & Van der Lande) is used. At least one part, and preferably two parts, of this accelerator is required for every two parts of initiator. A strong discoloration of the cured polyester resin, especially under the influence of light, will result.

It is better to use a lauryl mercaptan accelerator (NL-70 of Noury & Van der Lande) for imbeddings since it is a clear colorless liquid. Under the influence of UV-light no more pronounced discoloration with this accelerator is shown than the resin shows without the accelerator.

Besides the mentioned initiators and accelerators, a few others have been applied. Since the type is not always mentioned by the manufacturer they will be described under their own names without remarks (Table II.2).

Mixing

Warning: Never add a cobalt accelerator directly to an organic peroxide: the mixture may decompose violently due to the tremendous exothermic reaction; serious fire may result. Tertiary amine accelerators should never be mixed directly with benzoyl peroxide.

"The order in which peroxides and accelerators are added to the polyester resin depends on the type of the peroxide applied. Applying liquid peroxide, one prefers to add the accelerator first and after that the peroxide, as the mixture polyester + accelerator has a longer pot-life compared with the mixture polyester + peroxide.

On the other hand, when a solid type of peroxide (powder or paste) is used it is preferable to dissolve the peroxide in the polyester first, and then to add the accelerator. The dissolving of the solid type of peroxide in the polyester requires some time and the presence of an accelerator might then cause a premature reaction" (Noury & Van der Lande, 1961a).

Polymerization Curve

Polymerization of the unsaturated polyester resin, dissolved in monostyrene, at room temperature takes place only when both an initiator and an accelerator are added. The exothermic process can be divided into a number of steps, and a polymerization curve (Fig. 2.2) can be made by plotting temperature against time. A rather slow temperature rise occurs until the *gel-point* is reached. During the *gel-time* or *pot-life,* the mixture remains in a liquid state, but from this point the gelling will begin, as is noticeable by a sudden rise in temperature. Actually these two terms are not completely identical. The pot-life is the period during which the mixture can be processed and is therefore shorter than the gel-time. "The *hardening time* is the time from the setting of the resin to the point when the resin is hard enough to allow the molding or laminate to be withdrawn from the mold" (Scott Bader, 1965). The *peak exotherm* indicates the maximum temperature found during the polymerization process. It may take days or weeks before the product reaches *maturing time* and acquires

its full hardness, chemical, and physical properties. Maturing usually takes place at room temperature; however, this time can be shortened by *postcuring* at a higher temperature (50 to 80°C).

The shape, width, and height of the polymerization curve depends on several factors: type and quantity of resin, size and shape of cast, type and amount of initiator(s) and accelerator(s), operating temperature, and the type, amount, and condition of the fillers.

Cywinski (1960, Table 13) gives differences in polymerization rate of several plastics. The more cubic the container in which the resin cures, the smaller the total surface, and consequently the less the exchange of heat. Curing in laminae of a thickness up to 1 cm scarcely raises the temperature above 60 to 80°C.

The operating temperature strongly influences the rate of polymerization. A rise of 5 to 8°C, during the hardening of Vestopal resins with cyclohexanone peroxide and cobalt naphthenate at room temperature, reduces the gelling time approximately 50% (Hüls, 1961, p. 21).

As discussed earlier, the type and amount of the initiator strongly influence the curve. Mixtures of initiators sometime have advantages: they may react as synergistic, intermediate, inhibitive, or as other types of activity in a resin (Harrison et al., 1962). More data can be found in Farkas and Passaglia (1950), Mageli et al. (1959), Noller et al. (1961), and others.

Fillers

Type and amount (volume %) of fillers are very important for our work since the samples act as fillers. Quartz decreases the polymerization rate slightly and carbon black slows it down considerably. The delaying effect of fillers has many possible causes. A delaying effect to a liquid resin implies a dilution of reactive double bonds (Berndtsson and Turunen, 1954). Electrolytes and traces of different metals in the fillers can slow down the reaction. Initiators and accelerators can partly be absorbed on the surface of fillers (Maltha, 1957). Even very small amounts of water will decrease the rate of polymerization (Scott Bader, 1965), and therefore it is very important that the samples to be impregnated are absolutely air-dry (Jongerius and Heintzberger, 1963).

Resins

More extensive than the number of initiators and accelerators is the number and types of resins manufactured, although only a small percentage is suitable for application in our fields. The author will therefore discuss only the few types he has experimented with (Fig. 2.3, Table II.2).

A resin should have the following properties to be suitable for impregnation and/or imbedding applications:

1. The resin should be cured at room temperature. Postcuring in an oven is possible.

2. The resin should produce a hard product which is not tacky, and should have the proper properties for sawing and grinding purposes.

3. Small amounts of accelerator must be sufficient to start the polymerization process.

4. The color of the resin should not interfere with later analyses. For imbedding purposes the polyester must be colorless.

5. The resin/initiator/accelerator mixture must cure in such a way that cracks do not form too easily.

6. The resin should allow a strong dilution with styrene monomer or other liquid in order to obtain the desired viscosity.

7. A low viscosity of the resin is required, especially for imbedding purposes.

8. If possible, the resin should not be too water-sensitive.

9. The polymerized resin mixture should have a refractive index which approaches that of Canada balsam and should produce no difficulties in microscopical investigations. Especially for microscopical investigation, the proper type of resin should be continually used.

10. The final result should resist ultraviolet rays.

The Chemische Werke Hüls manufactures a series of unsaturated polyester resins under the trade name Vestopal (Hüls, 1961). Vestopal is diluted with styrene monomer, is flame resistant, and is slightly yellow in color. The resin should be stored in a cool (below 20°C), dark place. Containers should be cleaned with solvents such as ketones or chlorinated hydrocarbons. Once Vestopal has hardened it cannot be dissolved. When it is left for some time in contact with acetone or methylene chloride, it will decompose gradually.

Altemüller (1956, 1962) introduced Vestopal H, which has a viscosity of about 11 cp. Since a cobalt accelerator alone does not affect the resin, it can be added before the initiator. The volume shrinkage is about 8%. Postcuring may take 2 to 3 weeks at room temperature, a few days at 30 to 40°C, and a few hours at 70 to 80°C.

In the United States, Plaskon 951 resin from the Allied Chemical Corporation is used very successfully. Its viscosity is about 400 cp, its color is a faint yellow, its density is about 1.13, and can be dissolved in acetone. The same company manufactures a two-component system under the name 6605 Glasskin resin (similar to Plaskon 9407). This light pink colored resin, with a viscosity of about 475 cp, contains 34% free styrene and a cobalt accelerator (Fuller, 1957); catalyst No. 6606 should be used. This is a 60% solution of methyl ethyl ketone peroxide in dimethyl phthalate (Lupersol DDM, Delta and Delta-X; Butanox and Iso-butanox; Garox MEK; Nuodex M. E. K. peroxide and polycure M. E. K. peroxide). To 1 gal (U.S.A.) of 6605 Glasskin resin, ¾ to 1 fluid ounce of 6606 catalyst (100 cc resin: 0.6 to 0.8 cc catalyst) is added for curing at room temperature. Without fillers or additional styrene monomer, the resin mixture gels in 30 min at a constant temperature of 21°C (70°F).

The Pittsburgh Plate Glass Company manufactures an unsaturated polyester resin, dissolved in styrene monomer, under the name Selectron 5003. Its specific gravity is 1.11 to 1.13, and its viscosity is 600 to 750 cp at 25°C (77°F). For temperatures from 71 to 125°C (160 to 260°F) benzoyl peroxide is the preferred initiator, whereas at room temperature cumene hydroperoxide, methyl ethyl ketone peroxide, etc., are recommended (Pittsburgh, 1958; Curray, 1955).

The resin Paraplex P-43 from Röhm & Haas Company, Germany, is very suitable (Curray, 1955). It is a 70% polyester concentration in styrene monomer, and has a viscosity of 2500 cp at 25°C (77°F); its specific gravity is 1.148 (Rohm & Haas, 1962). Its storage life, when stored in its original container, at 25°F, is at least 6 months.

The products Crystic 189 LV and 191 E from the Scott Bader & Co. Ltd. (1965) have already been discussed in Chapter One (see Table 1.3). Catalyst paste H is a cyclohexanone peroxide and accelerator E is a cobalt soap.

Chapter One also discusses the products SR 19038 and DSR-19098 from BXL Plastics Material Group Ltd. (formerly Bakelite Ltd.; 1958a,b, 1962, 1963). Catalyst Q. 17447 is a methyl ethyl ketone peroxide in dimethyl phthalate, and accelerator Q.17448 is a solution of cobalt naphthenate with 1% Co.

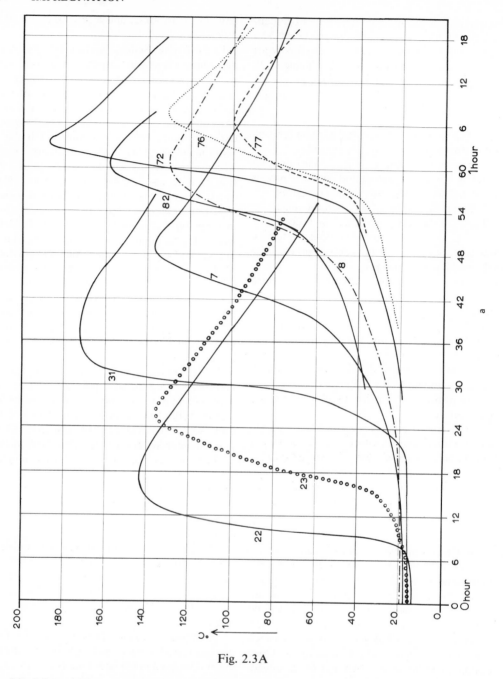

Fig. 2.3A

Fig. 2.3. Polymerization curves of a number of resin mixtures presented in Table II.2. The numbers in italics at the horizontal axis (c) refer to the curves given as dashed lines.

The Synres Chemical Industry manufactures an unsaturated polyester resin under the name Synolite. Two types, Synolite 336 and Synolite 711, have been applied. The first has more color, a long stability (pot life), and a little longer gel time than Synolite 711. The viscosities at 25°C are 6 to 8 and 8 to 11 poises, respectively (Synres, 1963, 1967).

The Synthese N.V. produces a resin known as Setarol. Setarol 3001 has a viscosity of 8 to 10 poises and Setarol 3120 a viscosity of 14 to 16 poises at 25°C. The former has a lighter color

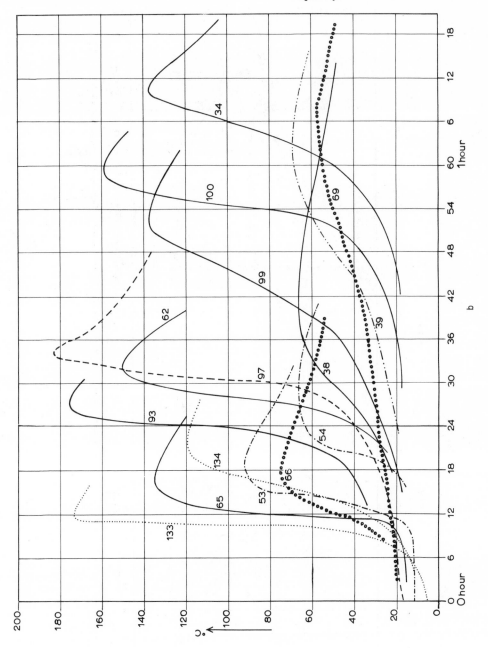

Fig. 2.3B

than the latter. Number 3001 can be used for imbeddings. The shrinkage is 6.1 and 4.9%, respectively (Synthese, 1962, written communication, 1963).

Five unsaturated polyester resins, which are colorless, have been successfully used in imbedding techniques (see discussion of Imbedding).

Lamellon 230, one of the resin series manufactured by Scado-Archer-Daniels, has a gel time of 5 to 6 min and a minimum cure time of 6 to 8 min. The viscosity of this resin is 17 to 20

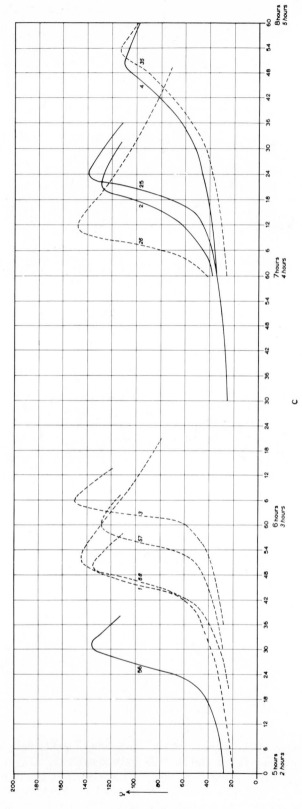

Fig. 2.3C

oises at 20°C (68°F). Intiator Hobilon C liquid) and accelerator Hobilon Z should be used with this resin (Scado, 1965, not dated a).

Frencken's Polyester 5132 with initiator FF 1964 and accelerator FF 1697 produces water-clear results. The linear shrinkage of the cured resin is about 2%, and about 7% in three dimensions (Frencken, 1962).

Under the name Bio-plastic, a pre-accelerated polyester resin can be obtained at Ward's Natural Science Establishment, Inc. Their assemblage kits are very useful, especially for a single imbedding of geological or biological material.

The plastic group of the north French coal district has a series of unsaturated polyester resin under the name Norsodyne. Norsodyne 50 with methyl ethyl ketone peroxide and cobalt naphthenate is a very successful product for imbeddings. To avoid any coloration due to the cobalt accelerator, they manufacture an initiator (C6) and colorless accelerator (LM/5) (H.B.N.P.C., 1959, not dated).

Reichhold's 32-032 Polylite unsaturated polyester resin is pre-accelerated. MEK peroxide is needed only to get the curing reaction started. The surface of the cured plastic is not tacky and it is very smooth.

Although these resins are primarily used for imbedding purposes, they can also be used for impregnation techniques.

Inhibitors

Inhibitors are added in order to increase the pot-life or gel-time of resin-peroxide mixtures, resin-accelerator mixtures, and resin-peroxide-accelerator mixtures at room and at elevated temperatures. An addition of 1% of the inhibitor NLC-1 (Noury & Van der Lande, not dated g) lengthens the gel-time of a mixture at least by a factor of 2, and it will not influence the final cure. Therefore, when a short curing time is required, consequently resulting in a short gel-time, the gel-time can be lengthened by adding an inhibitor.

The most common inhibitors are phenols, phenol formaldehyde resin dust, sulphur, rubber, copper and copper salts, most forms of carbon black, and methanol (Scott Bader, 1965, p. 14).

Yellowing by Ultraviolet Light

The influence of ultraviolet rays on resins has already been mentioned briefly. Under the influence of these rays, the cured resin may discolor, become sticky, or deteriorate. The ideal U.V. absorber should absorb the destructive, high-energy, ultraviolet rays and re-emit this energy as nondestructive wave lengths. It should have additional properties such as light stability, durability, low or no color, no odor, heat stability, low volatility and chemical stability, nonextractability and nontoxicity (Noury & Van der Lande, 1961c). Of the NL series the type NL/1 is most suitable for application in polyester resins. A dosage of 0.1 to 1% will be sufficient, since most of the resins contain an absorber for U.V. rays. For impregnations, which are normally stored in boxes, the yellowing or deterioration problem is negligible, but for imbeddings, which are normally stored open, the yellowing and deterioration can be very serious.

Frencken (1962) recommends the addition of 2 drops methyl-violet, dissolved in liquid, to 100^g resin. This primarily prevents yellowing due to the influence of the accelerator.

Thixotropy

The addition of 0.5 to 5% thixotropy powder, based on weight of the resin, can be used to prevent the flow of resin mixtures on vertical or inclined surfaces. The Noury thixotropy powder is a silica compound (almost a chemically pure, moisture-free silicon dioxide with a particle size of 0.0115 to 0.020 μ). High-viscosity resins do not impregnate well with thixotropy. Adding thixotropy powder to a low-viscosity resin will not decrease the impregnating properties. In this field of work, this can be useful for repairing surfaces of impregnated blocks by applying a thin coat to the polished surfaces of samples.

Thixotropy powder can be mixed easily with all known resins. If fillers are to be used with the thixotropy powder, it is advisable to add the fillers first (Noury & Van der Lande, 1961d).

Dyes

For certain applications it may be desirable to color the resin. Manufacturers produce a series of organic pigments of dyes of which small quantities (transparent dyes 0.01 to 0.2%, pigments 0.3 to 5%) normally give sufficient color to the polyester. The amount necessary depends on the application, type, and color of the organic material. The desired amount can only be found by trial and error and first should be checked on a cured sample. The basal layer of an imbedding absorbs the dye easily and may become too intense in color (for further information see Chapter Four). In Fig. 1.22 pigments are used to make the very thin clay layers more distinct.

Table II.2 Data about Mixtures of Different Resins, Initiators and Accelerators.
Nearly all experiments have been carried out in small tins (diameter 5 cm)
with quantities of 100 cc without using a thermostat. Room temperature 20°.

Number	Resin Type	Resin cc	Initiator Type	Initiator cc	Accelerator Type	Accelerator cc	Sty. (cc)	NLC 1 (cc)	Sand (cc)	Gelling Days	Gelling Hrs	Gelling Min	Peak °C	Post Cure	Tacky	Crack	Color	Milky	S.c.	Graph Fig.2.15
3	Ve	100	MEK1	2	Co-oct	1					3		150°			x	v.f.rose			x
4	–	100	–	1	–	1					7		109°				f.rose			x
5	–	100	–	1	–	0.5					12						colorl.			
24	–	100	–	0.5	–	0.5					6	30	138°	50°		x	colorl.			
25	–	100	–	2	–	1		1								x	f.yel.			x
26	–	100	–	2	–	1	0.5				4		148°			x	f.yel.			x
27	–	100	–	1	–	1		1			12						yellow			
2	–	100	MEK2	2	–	1					7		129°			xx	v.f.rose			x
16	–	100	–	2	Lauryl	1					4		35°	50°		xx	v.f.yel.			
21	–	100	–	2	Di.an	1					4			50°		xx	red-br.			
1	–	100	Cy.p	2	Co-oct	1					2		145°			x	f.rose			x
104	–	100	Cy.1	2	–	1					1	30					colorl.			
105	–	100	–	1	–	1					2						f.rose			
106	–	100	–	1	–	0.5					4						colorl.			
9	–	100	Ben	1	–	1		1						50°		x	f.rose			
10	–	100	–	0.5	–	0.5					12		25°	50°		x	colorl.			
18	–	100	–	2	Lauryl	1						8	144°				colorl.			
22	–	100	–	2	Di.an	1						14	135°				red-br.			x
23	–	100	–	1	–	1						6					ochre			x
101	–	100	–	2	–	1		1									red-br.			
103	–	100	–	1	–	0.5						15					f.red-br.			
107	–	100	–	0.5	–	0.5						40					yellow			
108	–	100	–	0.5	LM-5	0.5		1									yellow			
7	–	100	C 6	2	–	1					1		137°				f.yel.			x
8	–	100	–	1	–	1					1		130°				f.yel.			x
11	–	100	–	1	–	0.5					1		107°				colorl.			
28	–	100	1694	1	1697	0.5					3		26°	50°			f.yel.			x

No.		%													Color			
29	–	100	–	0.5	–	0.3	–					1			colorl.			
30	–	60	MEK1	2	Co-oct	1	40						25			v.f.rose	xx		x
31	–	85	–	4	–	2	15							174°		rose	xx		x
73	–	80	–	2	–	1	20					1		176°		f.rose	x		x
72	–	60	–	2	–	1	40					1		189°		f.rose	x		x
71	–	60	–	1	–	0.5	40					5		35°		colorl.			
70	–	60	–	0.16	–	0.08	40								50°	colorl.	x		
75	–	50	–	2	–	1		50		>3								x	
77	–	40	–	2	–	1	10	50		>3		1		100°		v.f.yel.	x		x
74	–	30	–	1	–	0.5	20	50		>3				25°		v.f.yel.	xx		x
76	–	45	Cum	2	–	1	30	25		>3		1		131°					
6	–	100	–	2	Lauryl	1				>3					50°	v.f.yel.			
19	–	100	–	2	–	1				>3					80°	v.f.yel.	xx		
102	–	100	–	2	Di.an	2				>3					50°	yellow	x		
13	–	100	MEK1	2	Lauryl	1				>3					50°	colorl.			
14	–	100	–	1	–	1						30		46°	50°	colorl.			
15	–	100	–	1	–	0.5						5		31°	50°	colorl.			
20	–	100	–	2	Di.an	1				>3		4			50°	yellow			
17	–	100	Cy.p	2	Lauryl	1				>3		5			50°	v.f.yel.	xx		x
34	PI	100	MEK1	2	Co-oct	1						1		138°		f.red-br.			x
35	–	100	–	1	–	0.5						4		112°		red-br.			x
36	–	100	–	1	–	0.5				>3		12				colorl.			
55	–	100	–	0.5	–	0.5			1			5		137°	50°	v.f.yel.			x
56	–	100	–	2	–	1										v.f.yel.			
57	–	100	–	2	–	1			0.5			2		129°		f.yel.			x
58	–	100	–	1	–	1			1			12				v.f.yel.			
44	–	100	–	2	Lauryl	1				>3						colorl.			
45	–	100	–	1	–	1				>3						colorl.			
46	–	100	–	1	–	0.5				>3					50°	f.yel.			
51	–	100	Di.an	2	Di.an	1				>3		1		135°	50°	red-br.	x		
33	–	100	MEK2	2	Co-oct	1						1				f.rose			x
47	–	100	–	2	Lauryl	1				>3					80°	colorl.			
52	–	100	Di.an	2	Di.an	1						1			50°	red-br.	x		
32	–	100	Cy.p	2	Co-oct	1								132°		f.rose			x

99

Table II.2 (continued)

Number	Resin Type	Resin cc	Initiator Type	Initiator cc	Accelerator Type	Accelerator cc	Sty. (cc)	NLC 1 (cc)	Sand (cc)	Days	Hrs	Min	Peak °C	Post Cure	Tacky	Crack	Color	Milky	S.c.	Graph Fig. 2.15
48	–	100	–	2	Lauryl	1				>3				80°			colorl.			
40	–	100	Ben	1	Co-oct	1								50°			f.rose			
41	–	100	–	0.5	–	0.5								50°			v.f.yel.			
49	–	100	–	2	Lauryl	1				>3				80°			v.f.yel.			
53	–	100	–	2	Di.an	1						12	92°				f.ochre			x
54	–	100	–	1	–	1						20	66°				yellow			x
38	–	100	C 6	2	LM.5	1						20	67°				colorl.			x
39	–	100	–	1	–	1						20	69°				colorl.			x
42	–	100	–	1	–	0.5						30	67°				colorl.			
37	–	100	Cum	2	Co-oct	1								50°			f.rose			
50	–	100	–	2	Lauryl	1				>3				80°			colorl.			
59	–	100	1694	1	1697	0.5				>3				80°			v.f.yel.	x		
60	–	100	–	0.5	–	0.3				>3				80°			v.f.yel.	x		
62	–	85	MEK1	4	Co-oct	2	15					20	150°			xx	rose			x
61	–	60	–	2	–	1	40				1		143°			x	rose			x
63	–	40	–	4	–	2	60					30	145°				v.f.rose			x
124	–	100	–	2	–	0.16					1						f.brown			
122	–	100	–	1	–	0.15						35					rose			
123	–	100	–	0.5	–	0.1					2						v.f.br.			
125	–	100	–	0.25	–	0.08					7						f.yel.			
128	–	100	–	4	–	0.17	50					30				xxx	yellow			
132	–	150	–	2	–	0.15	100					35				x	v.f.yel.			
127	–	150	–	1.4	–	1.1	125					50					f.purple			
131	–	150	–	0.8	–	0.07	100			3							f.yel.			
126	–	150	–	0.4	–	0.033	100			5							v.f.yel.			
88	La	200	C 6	4	LM.5	2					12						colorl.			
87	–	200	Hob.C	6	Hob.Z	2						15	176°			x	f.ochre			x
86	–	100	–	4	–	1						10	170°			xx	f.rose			x
96	–	100	1694	4	1697	0.25						30		80°			v.f.yel.			
92	–	50		1		0.5	50													

No.															
91	–	80	C 6	4	LM.5	2	20	125°	30					colorl.	x
90	–	80	–	2	–	1	20	36°	30				x	colorl.	x
89	–	60	MEK1	2	Co-oct	1	40	182°	30	3		80°		f.rose	x
64	K	100	–	2	–	1		158°	30	1			x	gr.-yel	x
129	–	100	–	0.5	Co-nap	0.1			30	2			xxx	yel.	
130	–	100	–	2	–	0.1				1			xx	yel.	
93	–	80	C6	2	LM.5	1	50	176°	15					yel.	x
65	–	100	Ben	2	Di.an	1	20	136°	11					red-br.	x
68	S.01	100	MEK1	2	Co-oct	1		137°	30	2			x	f.rose	x
69	–	100	C 6	2	LM.5	1		58°	30					colorl.	x
66	S 20	100	Ben	2	Di.an	1		74°	10	2	x			red-br.	x
67	–	100	MEK1	2	Co-oct	1			30			80°		yel.	
97	Ful	100	6606	2				185°		1			xxx	red-br.	x
98	–	100	–	1				173°	30				xxx	red-br.	x
99	No	100	C 6	2	LM.5	1		137°	30				x	colorl.	x
100	–	100	MEK1	2	Co-oct	1		158°	40					f.br.	x
78	Fr.	100	1694	1	1697	0.5	20	130°	15	1			x	red-br.	x
85	–	80	–	2	–	1		34°	30					ochre	x
79	–	100	MEK1	2	Lauryl	1		149°	30	1			x	red-br.	x
81	–	40	–	2	Co-oct	1	60	40°	30					colorl.	x
82	–	60	–	2	–	1	40	158°	30	1				rose	x
83	–	85	–	4	–	2	15	159°	30				x	red-br.	x
80	–	100	C 6	2	LM.5	1		152°	10					purple	x
84	–	80	–	2	–	1	20	163°	15				x	purple	x
94	–	80	MEK1	2	Di.an	1	20	174°	3				x	red-br.	x
95	–	100	1694	4	1697	0.25			15		(x)			brown	
133[a]	336	100	MEK1	4	Co-oct	1		195°	6					colorl.	x
134[a]	711	100	–	4	–	1		138°	9				x	l.ochre	x
135[a]	S 03	100	–	1	Co-nap	0.17			5						
136[a]	–	100	–	1	–	0.05			12						
137[a]	–	100	–	1	–	0.03			21						

101

Table II.2 (continued)

Number	Resin Type	cc	Initiator Type	cc	Accelerator Type	cc	Sty. (cc)	NLC 1 (cc)	Sand (cc)	Gelling time Days	Hrs	Min	Peak °C	Post Cure	Tacky	Crack	Color	Milky	S.c.	Graph
138[a]	P 43	100	–	1	–	1				1		25								Fig. 2.15
139[a]	–	100	–	2	–	1						30								
140[a]	–	100	–	1	–	0.6					4	40								
141[a]	–	100	–	2	–	0.6					3									
142[a]	189	100	H	4 g	E	4 g						10	160°				f.yel.			Fig. 1.48
143[a]	191	100	–	–	–	–						15	160°				colorl.			Fig. 1.48

Legend

Resin: Ve = Vestopal H; Pl = Plaskon 951; La = Lamellon 230; K = K-46; S 01 = Setarol 3001; S 20 = Setarol 3120; Ful = Fuller Glas-skin 6605; No = Norsodyne 50; Fr = Frencken 5132; 336 = Synolite 336; 711 = Synolite 711; S 03 = Selectron 5003; P 43 = Paraplex P 43; 189 = Crystic 189; 191 = Crystic 191.

Initiator: MEK1 = methyl ethyl ketone peroxide (Butanox M-50); MEK2 = methyl ethyl ketone peroxide (Iso-Butanox M-50); Cy.p = Cyclohexanone peroxide, paste (Cyclonox B-50); Cy.1 = Cyclohexanone peroxide, liquid (Cyclonox LTM-50); Ben = Benzoyl peroxide, paste (Lucidol B-50); C 6 = catalysateur C 6; 1694 = Frencken FF 1694; Cum = cumene hydroperoxide: Hob.C = Hobilon C; 6606 = Fuller catalysator 6606.

Accelerator: Co-oct = Co-octoate, 1% Co (NL-49); Lauryl = lauryl mercaptan derivative (NL-70); Di:an = dimethyl aniline (NL-63); LM-5 = accelerator LM-5; 1697 = Frencken FF 1967; Co-nap = Co-naphtenate, 6% Co; Hob.Z = Hobilon Z.

Styrene: Styrene monomer; *NLC 1* = inhibitor NLC 1; *Sand* = fine sand.

Peak °C: Peak temperature in °C.

Postcure: Postcuring in the oven at 50°C for 2 days and 1 day at temperature (C°) as indicated.

Tacky: Sample still tacky, even after postcuring in the oven.

Sample Cracked: x = 1–2 cracks, xx = 4–7 cracks, xxx = completely cracked into small pieces (about 1 cm³).

Color: v = very, f = faint, l = light, colorl. = colorless, yel. = yellow, br. = brown, gr. = green.

Milky: Milky appearance; that is, not transparent.

S.c.: Sand concentrated on bottom.

Graph Fig. 2.15.: Gel-time temperature curve given in Fig. 2.15.

[a]Data obtained from manufacturer.

The operator should take care not to use too much pigment since this may result in a tacky surface of the cured resin or in a grainy result of the cured polyester. The pigments should be applied in paste form to ensure even dyeing. Powder should be dissolved in styrene monomer before it is mixed with the resin. Some colorants delay the hardening process.

Experiments have been made with dyes (as paste) from Synthese N.V., Ram Chemicals (Ram, 1962a) and DuPont de Nemours. The last-mentioned manufacturer produces Luxol dyes which are also soluble in nitrocellulose lacquer (see Fig. 1.22).

2.1B Epoxy Resins

Liquid epoxy resins also can be used for impregnating nonpure clayey sediments. These resins are based on the reaction of bisphenol-A and epichlorohydrin. Epoxy resins are superior to other resins in the following properties: no volatile loss during curing, little shrinkage, resistance to solvents and chemicals, chemical inertness, acceptance to a wide range of fillers and pigments, great toughness and shock resistance, high adhesion properties to almost any surface, and a high degree of moisture resistance (Ciba, 1964a; Dow Chem., not dated a,b).

A disadvantage of the epoxy resins, compared to the described unsaturated polyester resins, is that they usually have to be heated to decrease their viscosity (see Araldite Peels), although reactive diluents can also be used to lower the viscosity, e.g., Ciba's reactive diluents DY 021, DY 022, DY 023. This decrease in viscosity restricts their application to nonclayey sediments. However, when epoxy resins can be used, results can be obtained in a few days.

D. E. R. 332, liquid epoxy resin from the Dow Chemical Company, proved to be a good epoxy resin. Its viscosity at room temperature is 6400 cp, which can be lowered considerably by adding phenyl glycidyl ether, butyl glycidyl ether, or styrene oxide (Dow Chem., not dated b, p. 17). The final product is virtually colorless when proper hardeners are used, such as meta xylylene diamine. Hardening will take place at room temperature or at high temperature, depending on type of hardener.

Scotchcast No. 3 from the 3M Company has a slight color and does not completely harden. The curing process takes a long time at room temperature, and the use of an oven causes an overdarkening in the final color (Coleman, written communication, May 1963). Its pot-life at 23°C is 3 to 4 days, the color clear amber, and the viscosity at this temperature is 1000 cp. The most favorable curing temperature varies between 77° and 120°C. At 120°C the viscosity goes down as low as 20 cp, remaining there for about 20 min before gelation and curing begin. Total curing time at 120°C temperature is 1 to 2 hours. The mixture should be composed of two parts A and three parts B by weight. Both parts are liquids (MMM, not dated a, b, c).

2.1C Molds

Several materials can be used as molds in which sample and resin stay during the hardening process of the resin. The operator must take care not to apply materials that cannot withstand high temperatures (see Table II.2 and Fig. 2.3). Porous materials and products with rough surfaces require extensive precautions and are generally not suitable for molds (see Mold Release).

The most simple molds are made of pieces of plate glass mounted together with cellotape. Inner as well as outer sides should be taped to prevent leaking. Mixtures with a long life should not be cured in such molds because the cellotape may slowly dissolve.

Sturdy cardboard and wooden boxes lined with sheet plastic make inexpensive molds and can be used successfully if the plastic does not dissolve in the resin mixture.

Metals produce suitable molds. However, since metal is rather rigid, it is necessary to incline the sides at an angle of about 70° so that the upper surface is larger than the lower one; in this way the cast can always be removed.

If special shapes or sizes are desired, the operator can use special molding materials to make his own mold. Products have been invented for making molds of irregular shapes, and these are used by paleontologists for duplicating fossils. Some of these products produce rather thin-walled molds and consequently require sturdy frames. Successful experiments have been carried out with the following products.

Clear polyurethane molding compound, CS 3501 from the Chem Seal Corporation, is normally supplied as two components, A and B. Both parts have an amber color. The mixing ratio is part A:part B = 100:25; the mold release time is 24 hours at 25°C (77°F) or 2 to 3 hours at 82°C (180°F); the curing time is 8 hours at 82°C (180°F) or 5 days at 25°C (77°F); and the volume shrinkage is 3%. In order to obtain good adhesion, CS 3501 should be applied only to clean dry surfaces. For binding to metal, neoprene and polyvinyl chloride, special primers, must be applied. Since part A absorbs moisture from the air it should not be exposed to air. For dense, void-free compounds it is recommended to degas this component. When dealing with small quantities, part A must be heated to about 82°C (180°F) on a hot plate and then degassed in a standard laboratory desiccator connected to a vacuum pump. A vacuum of 3 mm or less of mercury for a period of 10 to 15 min will be sufficient. Because component B tends to crystallize, it can be warmed up prior to mixing, to 99°C (210°F), and then degassed. After both parts are degassed and cooled to room temperature, they can be mixed thoroughly using a metal stirrer. After mixing, the mixture should be degassed again (Chem Seal, 1961, 1963, not dated).

The Chemical Products Corporation manufactures a product for dip coating under the name Chem-o-sol. This plastisol is a one-component system, a dispersion of polyvinyl chloride resin in suitable liquid plasticizer. At room temperature the viscosity is 5000 to 10,000 cp, and the fusing temperature is 163° to 191°C (325 to 375°F). Only a tank large enough for dipping the object and an oven are necessary. The author made several molds of screen wire coated with Chem-o-sol. The chemical resistance is good except when exposed to solvents and products like acetone, amyl, ethyl, and butyl acetate, benzene, chloroform, ketones, lacquers, and toluene (Chem Prod., not dated a, b; Kaufman and Jackson, 1962; Allison, 1962).

Silastic RTV (Room Temperature Vulcanization) Silicone Rubber products from Dow Corning Corporation consist of fluid rubber and a catalyst or curing agent. The two parts must be blended before curing at room temperature can take place. Viscosity, working time, and cure rate depend on the particular Silastic RTV rubber selected and the amount of catalyst. The three types, Silastic RTV 502, 588, and 589, have viscosities of 50,000, 30,000, and 30,000 cp, respectively. RTV 502 is slow-flowing, whereas 588 and 589 have the consistency of corn syrup. The working times are 10 min, 3 hours, and 4 hours, the cure times 30 min, 24 hours, and 24 hours, respectively, at 25°C (77°F). A special Silastic RTV thinner can be added. About 10% by weight mixed with the Silastic RTV lowers the rubber viscosity by as much as 75%.

Silastic 502 RTV rubber vulcanizes to a rubbery solid in about 30 min. The working time can be influenced considerably by varying the amount of catalyst 502 (Table II.3).

Catalyst 502 contains stannous octoate in a set of fillers and can be obtained from Nuodex Products Company under the name Nuocure No. 28. The catalyst should be handled carefully since it can cause skin or eye irritation. It is difficult to measure rather small quantities, so it should be diluted with Silastic RTV thinner or with xylene. This catalyst-thinner combination should be thoroughly stirred before adding to the fluid rubber because the catalyst tends to separate from the thinner when left standing. Addition of thinner has no significant influence on the working time or on the ultimate physical properties of the cured rubber if no more than about 10% by weight of thinner is added. It should be added to the Silastic before adding the catalyst. Sufficient vulcanization takes place in sections of about 3 mm thickness over a period of 24 hours at 25°C (77°F) and 50% relative humidity. Complete vulcanization is finished in about 4 to 7 days (Dow Corning, 1963e).

Table II.3. Influence of the Amount of Catalyst 502 on the Working and Curing Times of Silastic RTV 502 Rubber

Grams Catalyst per Pound (450 g) Rubber	Grams Diluted (20%) Catalyst per Pound Rubber	Working Time at 77° F (25° C)	Curing Time at 77° F (25° C)
4.5	22.5	2 min	10 min
2.3	11.5	10 min	30 min
1.2	6.0	30 min	90 min
0.6	3.0	60 min	3 hours

After Dow Corning (1963e).

Remarks: 2.3 g catalyst per pound rubber is equivalent to 10 drops catalyst per 100 g rubber.

Silastic 588 RTV rubber, together with catalyst 588, which contains dibutyl tin dilaurate, can also be lowered in viscosity be adding Silastic RTV thinner (10% by weight reduces the viscosity to 25%). Working time depends on the amount of catalyst. Temperature also influences the working time (Dow Corning, 1963f).

Silastic 589 RTV rubber cures in about 24 hours in unlimited thicknesses. Like 588, it is excellent for making molds for casting epoxies, polyesters, and other plastics (Dow Corning, 1963g).

Silastic RTV mold release is supplied in pressurized spray cans. It need not be used to release epoxy and polyester resins (Dow Corning, 1964b). To adhere the Silastic to various substrates, Silastic 1202 RTV primer must be applied (Dow Corning, 1964a). Molds can be repaired easily by using Silastic 732 RTV silicone rubber (Dow Corning, 1963d).

Dow Corning room temperature vulcanizing silicone rubbers can also be obtained in precatalyzed form. The moisture of the air acts as a catalyst to start the curing process. It can be useful for repairing small holes in molds and also for closing metal bindings of some apparatuses as discussed in Chapter Six. A special primer is necessary (Dow Corning, 1963a, b, c; Ram, not dated b, c).

Recent developments resulted in improvements of the RTV mold-making rubbers. Silastic A RTV mold-making rubber is a highly pourable rubber which is available with three catalysts (1, 2, and 3). Catalyst 1 is light blue in color and consists of pretested dibutyl tin dilaurate in paste consistency. The normal mixing ratio is 10:1 (Table II.4). Catalyst 2 is a light tan, pretested stannous octoate in paste form, while paste 3 is a yellow-colored paste containing lead octoate. It is used for heat-accelerated curing (Dow Corning, 1966a). The recommended 10:1 mixing ratio allows 5 hours working time and 5 to 45 min curing time at 300°F depending on material thickness. The catalysts will burn skin or irritate eyes. Silastic B RTV makes a firm, resilient, white rubber (Dow Corning, 1966b) which can be mixed in a 10:1 ratio with the same catalysts (Table II.4). Silastic C RTV is a red, nonflowing, paste-consistency rubber, for which only catalysts 1 and 2 are suitable (Dow Corning, 1966c), whereas Silastic D RTV can be used with all three catalysts and will produce a firm, resilient, red rubber with exceptional heat stability (Dow Corning, 1966d; Table II.4). Difference in color between rubber and catalyst makes it easy to determine when the two are uniformly mixed. The mentioned mold release agent and primers can be also applied when working with the more recent developed types of rubber.

Table II.4. Data on Silastic A, B, C, and D RTV Mold-Making Rubbers at 77°F. (25°C), Unless Otherwise Indicated

Type RTV Rubber:	A			B			C			D		
Mixing Ratios:	5:1	10:1	20:1	5:1	10:1	20:1	5:1	10:1	20:1	5:1	10:1	20:1
Viscosity in poises		140			300			8000			300	
Color		white			white			red			red	
Specific gravity		1.14			1.38			1.47			1.50	
Catalyst 1:												
Working time in hours	1	2½	4	1½	2½	4	2½	4	8	2½	3½	3½
Cure time in hours	18	24	48	12	24	36	18	24	48	18	24	36
Catalyst 2:												
Working time in min	5	20	50	5	45	60	20	40	60	4	10	20
Cure time in min	60	120	240	40	120	180	120	180	360	45	60	120
Catalyst 3:												
Working time in hours		5			4						4	
Cure time in min at 300°F												
film ⅛ in. thick		5			5						5	
film ⅛–¼ in.		10			10						10	
film ¼–½ in.		15			15						15	
film ½–1 in.		20			20						20	
film 1–2 in.		45			45						45	

After Dow Corning (1966 a, b, c, d).

2.1D Mold Release Agents

In spite of poor binding between most molds and cured products, it will still be difficult to separate the two if special precautions are not taken..

Stiff molds from glass or metal with smooth surfaces should be covered with a "release agent," a product which forms a thin film. Rough surfaces and porous molds must first be covered with a product to smooth the walls. Waxy products, plastic sheet, flexible products such as CS 3501, Chem-o-sol, and Silastic RTV Silicone Rubber can be applied for smoothing purposes. These flexible materials, used internally or as a complete mold, do not require mold release agents.

Several types of mold release agents are produced: (1) film-forming barrier coats; (2) internal releases; and (3) wipe-on or spray types. The film-forming barrier coats are used as sealers against plaster, plastic, and plywood or other rough-surfaced or porous molds. Ram Chemicals produces a cellulosic type under the name Garalease 915 (Ram, 1958, 1962b, 1963b) and a polyvinyl alcohol solution named Plastilease 521B (Ram, 1959, not dated d). Other materials have been mentioned previously.

Internal release agents are applied for continuous production runs and are used directly in the resin. They are soluble in uncured liquid polyesters but incompatible with the cured resin, and consequently they migrate to the mold interface during curing. These can be obtained from Ram Chemicals under the name Molgard (Ram, 1962b, 1963b, not dated d).

The third mentioned group of release agents is used when the mold surfaces are smooth. The number of trade names of these products is very extensive. Those most frequently used are the wax types, usually carnauba wax, transformed into a paste. Noury & Van der Lande (1961b) suggests using their NL/1 as an underlayer on the mold. A solution of polyvinyl alcohol in water (NL/2 of Noury & Van der Lande, 1961b) can be spread or sprayed over the wax layer. When dealing with molds of plate glass, a polyvinyl alcohol solution is most often used. Excess solution is poured into the mold, which is turned and tilted in all directions to wet bottom and sides thoroughly, after which the mold is turned upside down to drain before it is let dry in normal position.

Ortholeum 1962 lubricant assistant is a mild film and antiwear additive recommended for use in lubricating oils and greases (DuPont, 1959). It is used as a mold release agent when dealing with metal molds for imbedding slices of unconsolidated sediments in plastic, and also for making a cover layer of the imbedded slice on plate glass. With a piece of clean cloth the mold is covered in a thin uniform film of Ortholeum 162. A brush should be used to cover the corners of the mold (Seppic, not dated).

Talcum powder can be used very successfully as a release agent when the hardening takes place at room temperature (Synres, written communication, 1966). The molds used should be made of glass or well-polished metal.

2.2 AIR-DRIED UNCONSOLIDATED SEDIMENTS

Most sediments can be dried artificially without cracking. Sandy samples may be placed in an oven to have the water evaporate; when clay bands are present, or when clay forms the majority of the deposit, the samples must be wrapped in a plastic foil or sealed with material that allows the water from the sample to escape slowly, or freeze-drying can be applied (see Chapter Six).

2.2A Impregnation under Vacuum

Impregnating samples with unsaturated polyester resins can be carried out successfully only when the samples are air-dried and when the resin penetrates the pores under vacuum. Altemüller (1956, 1962) and Jongerius and Heintzberger (1962, 1963) use the resin Vestopal-H, which has a viscosity of about 11 poises. The other unsaturated polyester resins mentioned give equivalent results. Due to differences in viscosity of all resins, different dilutions are necessary to obtain comparable fluids. Styrene monomer is added to the resin to decrease its original viscosity.

Altemüller impregnates small soil samples of which normal-sized thin sections are made with thicknesses down to 15 μ, while Jongerius and Heintzberger deal with soil samples of brick size (15 x 8 x 5 cm) from which large-sized thin sections are obtained. The quantities used to make the resin-styrene-initiator-accelerator mixtures vary between these authors. During curing, most of the styrene monomer evaporates. Since the volumes of the impregnating samples differ considerably, and consequently the distances the styrene molecules have to travel to the surface, the total curing time takes Altemüller about 6 to 8 days, whereas Jongerius and Heintzberger have to wait over 6 weeks.

Method of Jongerius and Heintzberger

Jongerius and Heintzberger (1962, 1963) use the cyclohexanone peroxide Cyclonox LTM-50 of Noury & Van der Lande as initiator and NL-49 (Co-octoate, 1% Co) as accelerator.

Four air-dry samples, each 15 x 8 x 5 cm, are placed next to each other on one of the 8 x 15 cm sides in a cardboard box (24 x 17 x 11 cm) lined with plastic foil. The distance between the samples is about 1 cm, and also a 1 cm space is left between the samples and the sides of the box. Labels with sample dates are marked at the exposed sample sides. The authors state:

"The bottom of the box is virtually evenly filled with soil, which is very important for a successful hardening. In fact, if the samples did not fill the box uniformly it would contain large empty spaces which would obviously become full of plastic during the impregnation. Such large plastic nuclei polymerize more rapidly than the plastic in the soil samples. As a result so much heat may be liberated that the polymerization in the soil samples is accelerated as well, thereby causing disadvantages. In practice four large samples can conveniently be treated together. With a view to continuous production of thin sections, simultaneous impregnation of a number of samples is required. This can be done by treating the samples separately in series, but this involves the regular cleaning of much glassware (taps, tubes, etc.). This applies much less to the procedure above. At the same time the impregnation of more than four or five large samples in a single block is not recommended. This would demand not only very large impregnation equipment, but it would also make the hardened blocks too large and heavy to handle."

Slager (written communication, January 6, 1967) places the samples on strips of plastic to allow impregnation through the bottom. He prefers a selection of the samples according to clay content, percentage of organic material, etc., and varies the styrene monomer concentration with regard to the permeability.

*Impregnation Conditions.*Several conditions must be fulfilled in order to obtain complete impregnation and proper handling of the samples (Jongerius and Heintzberger, 1962, 1963; Altemüller, 1956, 1962):

1. A high vacuum is required during impregnation.
2. The viscosity of the impregnating resin must be low.
3. Excess monomer styrene must evaporate in a reasonable amount of time, since a high styrene content reduces the polymerization rate of the resin and adversely affects properties of the result, such as shrinkage, modulus of rupture, and its resistance to various chemicals. This may result in difficulties during grinding of the sections, while the shrinkage may cause birefringence in the cured polyester resin. Under crossed nicols this birefringence can be observed as light margins around structural elements and in cavities.
4. The application of low amounts of initiator and accelerator results in a long gelling time of the resin, which is necessary to allow the excess styrene monomer to evaporate. A long gelling time also allows a low exotherm peak (Table II.2 and Fig. 2.3) and consequently decreases the risk of internal stresses which form cracks and/or birefringence.
5. Low concentrations of accelerator produce better hardening than high concentrations and also prevent strong coloration of the cured resin.

Procedure. A cardboard box with four samples is placed in the cylinder of the vacuum impregnation apparatus (Figs. 5.3 and 5.4). The O-ring and the plexiglas top are placed in position, and the resin funnel is put in one of the cylinder-cover holes in such a way that the resin will not drip on any of the samples. The other holes of the plexiglas top are plugged. Vacuum, obtained by an electrical driven vacuum pump, is now applied to the vacuum cylinder to evacuate all air from the samples.

Just before or after the application of vacuum, the impregnating mixture is mixed. A total of 2500 cc is needed (for composition see Table II.5 A, B). Next the mixture is poured into the resin funnel, keeping the valve closed. Vacuum is then applied also to the resin mixture.

After approximately 20 min evacuation of air from both sample and impregnating liquid should be completed and the lower tap of the funnel should be opened to allow the resin mix-

Table II.5. Composition of the Resin Mixture for the Impregnation of Clays Below 22°C (A), above 22°C (B) and for Replenishing (C)

	A (<22°C)	B (>22°C)	C
Vestopal-H	1500 cc	1500 cc	1000 cc
Styrene monomer	1000 cc	1000 cc	0
Cyclohexanone per.	4 cc	2 cc	2.5 cc
Co-octoate	2 cc	1 cc	2.5 cc

After Jongerius and Heintzberger (1962, 1963).

ture to drip into the cardboard box. This should take about 7½ hours to reach the top of the samples. This slow process prevents air, still present in the sample, to become trapped by the resin. After the samples are filled with the resin, the tap underneath the funnel is closed, and the whole is left overnight. During this period the vacuum will disappear slowly because of leakage.

The following day the cylinder can be opened and the remaining mixture, still present in the funnel, is added to the samples. According to Jongerius and Heintzberger, the samples are then covered by 1½ cm liquid.

The funnel and its tap should be cleaned thoroughly to prevent curing of the remaining liquid. The O-rings should be replaced since it will take 1 to 2 days for them to return to their original shape.

Curing. The box with samples and liquid is placed in a continuously running fume-cupboard to promote evaporation of excess styrene monomer, thereby causing the liquid level to fall fairly rapidly and exposing the samples for several days. As long as styrene monomer is evaporating it is necessary to "top up" or submerge the samples completely. This replenishing, using the mixture as given in Table II.5C, should be repeated every 2 or 3 days. When all the excess styrene is evaporated, the resin starts curing and the box with samples is left for a few more weeks in the fume-cupboard. During this period there is a very gradual gelling, followed by the hardening. As soon as this stage is reached the box can be stored in a cool place for some additional weeks to harden thoroughly. The surface may stay sticky due to air inhibition. This hardly affects the polymerization process. Postcuring in an oven at a temperature of about 40 to 50°C for 2 to 3 days completes the process, and the stickiness may disappear. If the cardboard box is transferred directly from the fume-cupboard to the oven it is possible that cracks and birefringence form by internal stresses.

The block with samples is now removed from the box, the plastic foil pulled off, and the samples sawed separately with a large diamond saw.

Conclusions. Jongerius and Heintzberger (1962, 1963) found that a low-polyester viscosity alone may not be sufficient to completely impregnate a sample. Some of the styrene monomer evidently penetrates fairly rapidly into the sample, followed slowly by the polyester. These authors became aware of this when they discovered that the sample centers were moist within a comparatively short time after the impregnation began, and that only the edges of a sample were found to have been penetrated with sufficient polyester. As a result a very long gelling time is necessary.

"It might be concluded from the foregoing that some of the excess monostyrene is absorbed in the soil without being released again afterwards and even that the gradual fall in the liquid level during the time in the fume-cupboard is not caused by an evaporation of the monostyrene, but by its continued absorption in the soil. If this were, however, the case, the concentration of the monostyrene in the samples might become so high that it would prevent hardening of the blocks" (Jongerius and Heintzberger, 1963).

Daily weighings of a box with samples and mixture during the fume-cupboard period showed a fairly considerable decrease in weight per day. This was in accordance with the weight of the volume of styrene monomer equal to the daily decrease in volume. The total loss of weight during this period was about equal to the weight of the styrene originally added. From this it can be concluded that all styrene monomer initially present in the samples gradually disappears again.

Method of Altemüller

Because of the samples impregnated by Altemüller are about 15 x 20 x 10 mm, he does not need a large vacuum apparatus. Altemüller (1956, 1957, 1962) continued the earlier work of Kubiena (1937, 1938, 1942) and introduced Vestopal-H of the Chemische Werke Hüls in the field of soil science.

Procedure. Each sample is placed in a small tin foil container. An apparatus is constructed from six thick-walled glass tubes. A vacuum connection runs through their rubber stops, as do high cylindrical separation funnels, which serve as impregnation-liquid containers. A vacuum of 15 mm is applied to the samples for 15 min before the Vestopal mixture (Table II.6A) is slowly added to the samples. The samples remain in the vacuum environment after all the liquid has been added until no more air bubbles rise. The tin foil containers with samples are then taken out and stored. After about 3 to 4 days, replenisher (Table II.6B) should be added. Polymerization can be quickened by placing the containers in an oven at a temperature of 30 to 40°C for about 1 or 2 days. If cracks are formed the operator should use lower amounts of initiator and accelerator as indicated in Table II.6A. If complete hardness is desired, it is recommended to leave the sample untouched for an additional week after removing it from the oven.

Conclusions. Table II.6C and D show that Altemüller uses a more diluted mixture than Jongerius and Heintzberger, but the amounts of initiator and accelerator are twice as high. Altemüller favors a faster gelling time and risks a higher peak exotherm temperature. However, since his containers are much smaller, heat exchange is much easier. No difficulties are encountered when dealing with sandy samples; however, with clayey samples there may be insufficient resin penetration. To compensate for this lack of penetration, Altemüller uses lower amounts of initiator and accelerator. The replenisher used by Altemüller (Table II.6B) has a slightly slower gelling time, similar to that used by Jongerius and Heintzberger (Table II.5C).

Remarks. Altemüller (1962) discusses the impregnation results of different types of soils. Sandy soils allow thorough impregnation; however, increase of clay matrix decreases impregnation possibilities. Further decrease of the liquid viscosity in clay samples may be successful. Altemüller found that lime-rich soils are relatively easy to impregnate, but soils with a soapy appearance, and poor in lime, often demand a double impregnation. Soils containing organic matter may discolor due to the dissolving of organic material. This may result in a longer gelling time, and therefore organic soils should be impregnated wih mixtures containing a high initiator content to quicken gelation and slow down solution. Salts, present in the sample, absorb part of the initiator and consequently prevent curing of the mixture.

Method of Curray

Curray (1955) reported briefly on the impregnation of sand collected with thin-walled stove pipes, or with tin cans, tops and bottoms removed.

Impregnation Liquids. Curray uses Selectron 5003 resin from the Pittsburgh Plate Glass Company, or Paraplex P-43 from Röhm & Haas Company. Styrene monomer is used to decrease the viscosity of the impregnating liquid. The proper consistency can be determined only by trial and error on each particular sand. The initiator (½ to 1%) can be a methyl ethyl ketone peroxide, or a cumene hydroperoxide, and the cobalt naphthenate from Nuodex is used as accelerator.

Procedure. The sample container is pushed into the sand without twisting. It should not be

Table II.6. Composition of the Resin Mixture for Impregnating Samples According to Altemüller (1962): (A) Amount and Composition Used by Altemüller; (B) Replenisher; (C) Composition According to Jongerius and Heintzberger; (D) Composition Used by Altemüller Recalculated to the Amount of Resin Given under (C).

	A		B		C	D	
Vestopal-H	300	cc	100	cc	1500 cc	1500	cc
Styrene monomer	250	cc	0	cc	1000 cc	1250	cc
Cyclohexanone per.	1.65	cc	0.3	cc	4 cc	8.25 cc	
Co-octoate, 1% Co	0.8	cc	0.15	cc	2 cc	4	cc

filled over ⅔ or ¾. The removed samples are left in the container, placed in a shallow box, and allowed to dry.

When the sample is ready for impregnation, the resin is mixed and carefully poured into the unfilled top of the core tube twice. The second filling is necessary to ensure complete saturation of the upper part of the sample. To prevent air inhibition of the resin at its upper surface, the gelled surface can be covered with wax paper, cellophane, or a liquid such as glycerine.

If a long sample has to be impregnated, the sample can be turned over after the resin has cured, and if the bottom is not impregnated, sand can be dug out, creating a small reservoir, and the polyester resin mixture can be poured again. Allow the sample to dry as long as possible before pouring the resin mixture.

Postcuring can be done in an oven to speed the curing process.

Discussion. The Curray method has one disadvantage in that the impregnation is carried out from above with the risk of trapping air. If this method is used, better results can be obtained with the use of the vacuum equipment as described in Chapter Four.

Another way to impregnate the stovepipe with sand is to place the lower end of the pipe on a piece of mosquito wire (screening) on top of which part of a nylon stocking is placed. The container with this bottom is placed in a large pipe and supported by small metal or wooden blocks. In this position air ventilation aids in drying of the sample, while it decreases the risk of trapping air in the plastic since the dripping resin pushes the air out. The blocks are removed and the sample placed directly in the box when the resin runs freely out of the pipe, and the operator is then certain that the sample is impregnated.

Impregnation from below (under vacuum or not) can be carried out when a few pipes are placed in a box that is as high as the samples. The box should be lined with plastic or tin foil; the outer sides of the sample containers are painted or sprayed with a mold release agent to facilitate later removal. The samples should be air-dry. The lower ends of the pipes can be closed by pieces of nylon stocking and mosquito wire. The pipes should be placed on thin strips of wood or cardboard. Resin mixture is then poured slowly into the box as described in the method of Jongerius and Heintzberger. Holes drilled into the metal container facilitate impregnation, but the operator must be sure that the drilling operation does not disturb the sample.

2.2B Impregnation under Vacuum Using Pleximon 808 (Plexigum M 7466)

Method of Borchert

Borchert (1961a, 1962, 1963, 1964) tried several methods for impregnating soil samples in order to make thin sections. He concluded that Plexigum M 7466 (now called Pleximon 808) has advantages over other products such as Vestopal-H. His thin sections are not made thinner than 35μ.

Pleximon is built from acid methacrylate esters, which are inflammable. The influence of light and heat may cause pleximon to polymerize; it therefore should be stored in a dark, cool place. Catalyst 20 (formerly called catalyst 7466) should be added at the rate of 10% per volume to induce fast hardening at room temperature (Röhm & Haas, not dated).

Borchert collects the sample with an iron ring 6 cm high and 10 cm in diameter (see Chapter Six). The ring is opened in the laboratory, the sample thinned down to a thickness of 2 cm, and the side rubbed off carefully with a needle cutter. Next, the sample is left to dry, which takes from a few days to a few weeks, depending on type of soil. The operator should prevent the sample from forming cracks. When the humidity is too high, the polymerization of the water-soluble impregnating liquid is influenced, and consequently the hardening is incomplete.

Impregnation. The impregnation is carried out in a home-made cellophane bag. The bottom of the bag is sealed, the sample placed in it, and the mixture added. This mixture of Pleximon 808 and catalyst 20 (10:1) should be stirred well for 3 min before it is poured into the bag. The mixing should be done under the hood of a fume-cupboard to prevent gasses from escaping into the working area.

A series of numbered bags with samples are placed in a vacuum desiccator with a vacuum of 20 mm Hg. The samples will now be saturated under the formation of foam within 25 min. The bags are then removed from the desiccator and further filled with the Pleximon 808-catalyst 20 mixture. The bags are each closed with a broad clamp (Fig. 2.4) to prevent air from entering the bag. This air-tight closing is important for a good gelling, since air will weaken the mass. The samples are put aside to allow them to gel, and in a period of 3 to 4 hours the contents of the bag become hard. The process is exothermic. After the whole has cooled it is placed in water to allow the cellophane to loosen easily from the contents.

Kuron and Homrighausen (1959) suggest evacuating the sample prior to adding the Pleximon mixture, after which a Pleximon 808-catalyst 20 at a ratio of 13:1 is added during a period of 15 min. The desiccator is now filled with nitrogen gas which has passed a solution of pyrogallol to absorb all oxygen present (Fig. 2.4). The sample is left in the nitrogen atmosphere for 35 min; then it is removed and placed in the fume-cupboard. It will take about 3 hours to get the sample well hardened.

Sawing, Polishing, and Mounting. The sample is cut in two disks with a diamond saw and one-half of the sample is ground and mounted with Plexigum M 354 on a 4 mm thick plexiglas plate. The mounted half can now be ground to a thin section. These techniques are similar to those described in the discussion of Thin Sections; the only difference is the mounting medium and the base plate. The powder Plexigum M 357 should be mixed with the liquid catalyst Pleximon 804 in a 2:1 ratio. This mounting medium should be hardened under pressure. The mounting should be finished within 15 to 20 min.

Method of Werner

Werner (1966) uses Pleximon 808 to impregnate his freeze-dried, silty-clayey Recent sediments. His reasons are the low viscosity of the mixture, short curing time, good cutting, grinding, and polishing properties, no color, good tenability, and a favorable refractive index of 1.51. The disadvantage is the volume decrease, which, however, has a negligible effect on his samples (60 x 40 mm in section and 4 to 5 mm thick since the shrinkage mainly occurs at right angles to the fluid surface). The mixture also has a tendency to curl, which can be prevented by immediate mounting on the object holder. Heating of the samples for about 24 hours at 60 to 70°C is favorable for a proper polymerization (Werner, written communication, May 1967).

The following procedure is followed by Werner:

Impregnation. Freeze-dried samples should be impregnated as soon as they are dry to prevent damage of the powdery-voluminous slice of which the cohesion is just enough to prevent collapsing. They also are very sensitive to humidity, which will create cracks.

Werner impregnates several samples at the same time in a vacuum desiccator 28 cm in diameter. He describes a distribution arrangement that enables the mixture to reach all samples. Recently Werner (written communication, May 1966) accelerated his freeze-drying technique

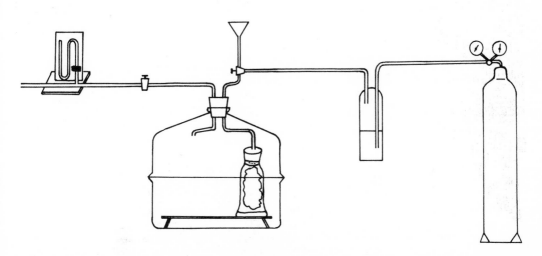

Fig. 2.4. Schematic presentation of an apparatus for the impregnation of small soil samples. For explanation see text. (Redrawn from Kuron and Homrighausen, 1959.)

by using aluminum foil as sample containers. These containers can be placed in a large tray that fits the impregnation desiccator (Fig. 2.5).

Placing a label on top of each sample is recommended. Data can be engraved in small pieces of tin, which will be visible later through the impregnation medium.

Vacuum is applied to the samples for 20 to 30 min by means of a filter pump, after which the mixture (10 parts Pleximon 808 and 1 part catalyst 20 by volume) is applied slowly to the samples. The vacuum is left for about 20 min after impregnation is finished. Nitrogen is then let into the desiccator through a rubber balloon to release the vacuum partly. The operator should wait ½ hour before letting air in to release all vacuum since oxygen prevents polymerization.

After approximately 30 min the polymerization has progressed enough to allow oxygen to contact the mixture. The samples are removed from the desiccator to facilitate heat escape. At this time the resin is not yet hard and therefore allows easy cleaning of the equipment.

About 2 hours later, the polymerization is completed. The single sample trays can be separated easily by breaking the small resin bridges between them. The samples can be removed from their aluminum trays by tearing the aluminum off.

Cutting and Mounting. Lower as well as upper sides of the sample are covered with resin. Instead of grinding so much away that the slices can be mounted, Werner glues the backs of two slices together with Pleximon 804, a fast-hardening glue, and then cuts the glued sample again in two along this seam, using a rocksaw blade of 4 mm thickness.

Plexiglas pieces of 2 mm thickness act as object glasses. The sediment slices are mounted on it with Pleximon 804, and the majority of the sample is cut off.

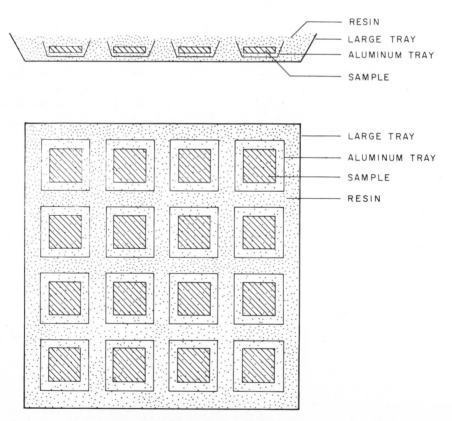

Fig. 2.5. Metal tray containing aluminum foil trays with freeze-dried samples and the impregnating liquid. The large tray has to fit the vacuum unit. The aluminum trays are not placed against each other to facilitate later removal. The single samples can now easily be separated by breaking the small resin bridges. (Redrawn from a letter from Dr. F. Werner, Geol. Inst., Kiel, Germany, April 1967.)

Thin Sectioning. Grinding can be carried out on a planing machine, using oil instead of water as a cooling agent to prevent the clay minerals from swelling. A "thick" section can be made to study primary and secondary structures. However, a "thin" section is required for microscopical determinations. The operator can also mount the slice which he cuts off prior to grinding on the plexiglas to make a thick and a thin section.

2.2C The Shell Method for Impregnating Cores of Unconsolidated Sediments

At the Shell Development Company, Exploration and Production Research Division, in Houston, Texas, a method has been developed to impregnate 3 in. diameter cores with an unsaturated polyester resin (Plaskon 951) under the influence of vacuum and pressure. This method, described by Ginsburg et al. (1966), is an addition to the methods just described. For soil scientists this method is not suitable when the samples must be dried at high temperatures.

The authors mentioned state:

"The impregnation of soft sediments has three main advantages: 1) The slabbed cores show many structures and textures that are invisible or indistinct in broken or sliced soft-sediment cores; 2) multiple slabs of the same core permit a serial, three-dimensional examination of sedimentary structures, particularly those produced by organisms; 3) impregnation and slabbing simplify storage, provide specimens for sectioning, and make the cores readily available for study."

Procedure

In 1958 one of the four authors, Moody, developed a technique for which standard laboratory equipment and a pressure chamber are required to impregnate core sections up to 6 in. (15 cm) long. Cores are extruded in the laboratory, cut into sections of 15 cm, and allowed to air-dry for several days. Next, the core pieces are placed in a low-temperature oven (93°C, 200°F) for the first few hours and then raised to 177°C (350°F) for final drying. In order to secure satisfactory hardening of the impregnating liquid, the sample should be absolutely moisture-free. Therefore the core pieces are transferred from the oven to a desiccator, and the moisture is checked by the color of silica gel. When discoloration occurs, it is necessary to place the samples back into the oven.

In the next step the sample is placed in an aluminum pan, which is put into the vacuum desiccator. An open reservoir funnel is placed on the vacuum desiccator to allow the resin to flow out freely. The chamber is evacuated, and then the mixed resin is introduced slowly. At least ½ in. liquid should cover the sample. Air is let in very slowly, and the pan and sample are then transferred to the pressure chamber. Nitrogen gas is used for pressurizing the chamber. One atmosphere pressure is sufficient to drive the resin in the pore space. Constant pressure is kept for 15 to 30 min and then is relieved slowly. The sample is then removed and allowed to gell and cure overnight. It is not necessary to apply the pressure treatment to sand and gravel samples since nearly complete impregnation will occur in the vacuum chamber.

A special vacuum chamber was designed by Bernard and Daigle to facilitate longer cores, with a total length of 3 ft. It consists of an open-end pipe which can be closed by a plexiglass door. To secure an air-tight closure an "O" ring and four bolts are used. A vacuum tube and a funnel are connected to the upper side of the horizontally placed, cylindrical chamber. A vapor trap is placed between the vacuum pump and the impregnation chamber(s). Loose sand cores are commonly wrapped in fiberglass cloth to prevent the core from falling apart during the impregnation process.

Curing and Cutting

If the impregnated core sticks to the mold, curing has not been completed. Curing can be completed by allowing the cores to set in the sunlight for 24 hours. Postcuring can be carried out in an oven, but care should be taken to avoid cracks which form because of overheating and exothermic reaction (see Table II.2).

The short, 15-cm-long cores can be cut on any standard diamond saw. However, for the previously mentioned 3-ft-long cores a standard stone saw mounted on rails is required (see Chapter Five).

If the cut section is still soft in areas where the resin is not yet dry, it should be placed in the sun or in a low-temperature oven to complete curing.

Discussion

To study the sedimentary properties on the cut slab we can coat its surface with mineral oil.

The authors state that this method has been used successfully for calcareous and terrigenous sediments. Sandy cores are impregnated easily and excellent results have been obtained for all but the finest calcareous clay-sized sediments. Large samples of terrigenous clays have not yet been impregnated satisfactorily.

In fine-grained materials the shrinkage, during drying, distorts the textures considerably, but macro sedimentary properties are less affected. Some cores may not completely impregnate due to incomplete drying. Peat- and organic-rich deposits are difficult to impregnate due to reactions between the resin and the organic material. Therefore organic-rich cores should be saturated first with cobalt naphthenate diluted with mineral spirits.

If certain laboratory analyses are to be carried out cutting the core in half directly after it has been extruded and impregnating only one-half of the core is recommended.

A vapor trap is placed between vacuum pump and vacuum chamber to protect the pump from being affected by resin vapor and water. Dry ice and isopropyl alcohol or acetone are used as cooling agents in the trap. If the oil in the pump is changed frequently, it is not always necessary to use a trap.

A distinct advantage of an impregnated core over a nonimpregnated one is that the cut faces show more details of the sedimentary structures and other properties (Fig. 2.6). By cutting the slabs in a series, a three-dimensional sedimentary structure can be obtained.

Materials and Impregnating Liquids

Ginsburg et al. (1966) use two sizes of aluminum pans. The small ones are 18.1 x 8.6 cm at the top and 16.5 x 7.3 cm at the bottom (7⅛ x 3⅜ in. and 6½ x 2⅞ in., respectively). The depth is 10.1 cm (4 in.). The pans for the 3 ft core sections are 8.9 x 94 cm (3½ x 37 in.) at the bottom and 11.5 cm (4½ in.) high. A mold-release agent (Allied Chemical Company's Polylease 77) is used.

The authors use Plaskon 951 resin manufactured by the Allied Chemical Company. Approximately 0.03% of cobalt naphthenate (6% CO) is added as accelerator, and 0.3% methyl ethyl ketone peroxide as initiator. The result obtained from this mixture produces a transparent, colorless product which cures without cracking in large castings (see Table II.2).

About two parts of resin per foot of core, and 1 ft^2 of fiberglass cloth per foot of core are required.

2.2D Impregnation under Vacuum and Pressure with Epoxy Resins

At the Coastal Studies Institute of the Louisiana State University epoxy resins are applied to impregnate unconsolidated samples. Vacuum as well as pressure is used to get the resin into the sediment pores (see Shell Method).

Material and Equipment

Coleman (written communication, 1963a) obtained good results with a mixture of 65% epoxy resin (D.E.R. 332 from Dow Chemical), 25% butyl glycidil ether, and 10% diethylenetriamine (Dow Chem., not dated b; Cavanaugh and Knutson, 1960).

The impregnation takes place in the vacuum-high pressure apparatus as described in Chapter Five (Fig. 5.8). The samples have to be small since the diameter of the apparatus is limited. Small waxed cups are used as sample containers.

Procedure

The samples to be impregnated are placed in the waxed cups and allowed to dry. When dry, enough resin mixture is added to completely cover the samples. The cups are placed on the sample rack, which is then lowered into the apparatus.

The head is placed on the pressure cylinder and clamped firmly. Vacuum is now created in the apparatus. Either the valve or the cylinder head or the valve in the vacuum pipe is used to increase the vacuum gradually in a period of about 3 to 4 min. This is done to avoid excessive boiling of the plastic in the cups. Allow the vacuum to operate for about 40 min in hot weather. Cool surroundings make it necessary to keep the vacuum much longer since the viscosity decreases with increase of temperature. During the vacuum period, air migrates from the pore spaces.

When the vacuum period is finished, the valve in the vacuum line is closed and the vacuum pump is shut off. Pressure is now applied to the cylinder. As source, pressure bottles with nitrogen should be used. Allow a pressure of 91.4 to 98.4 kg/cm^2 (1300 to 1400 lb/in.^{2}) for 20 min in hot weather or 30 min in cool weather.

Next, the pressure is released very slowly to avoid excess boiling of the resin and to prevent the samples from cracking. Allow 10 to 12 min for this process. The valve on top of the cylinder head should be used to release pressure after the

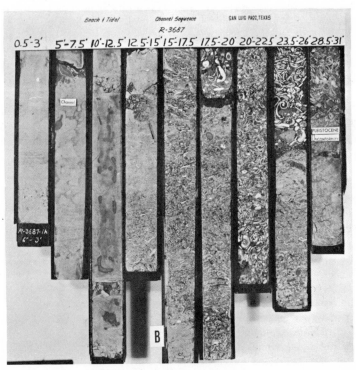

Fig. 2.6. Comparison of the same (A) nonimpregnated with impregnated and cut (B) cored sections of beach and tidal channel deposits from the northwestern Gulf of Mexico. (Courtesy of Shell Development Company, Exploration and Production Research Division, Houston, Texas; Ginsburg et al., 1966.)

cylinder gauges and the valve in the pressure pipe are closed.

The cups with samples are finally removed and stored for curing. At room temperature curing may take a couple of days, but it will take only 2 hours in an oven at 110°C (230°F).

Remarks

This viscous resin mixture is unable to penetrate sediments consisting of almost pure clay, and therefore only the surface of such sample may be impregnated.

Some difficulties may be encountered when dealing with coarse and loose sediments. The chip to be impregnated tends to break up with the boiling that occurs during the application of vacuum and the release of pressure. However, this can be prevented by "shoring up" the chip with a few drops of resin around the edges before impregnation. This breaking can also be prevented by wrapping the chip in pieces of mosquito wire, or cheesecloth if the sample is a little more coherent. Sediments consisting of coarse silt or fine sand can simply be soaked in resin without application of vacuum and pressure.

This procedure can also be applied to other epoxy and unsaturated polyester resins.

2.2E Impregnation under Vacuum; Curing in Oven

A combination of low viscosity, long pot-life, and quick cure has been developed in Scotchcast resin no. 3 of the 3 M Company (MMM, not dated c). The viscosity at 25°C (77°F) is initially around 1000 cp and will remain below 5000 cp for about 3 to 4 days at room temperature. The cure time is 1 to 2 hours at 121°C (250°F), 5 to 7 hours at 93.3°C (200°F), 12 to 16 hours at 76.6°C (170°F), and 28 to 35 hours at 65.5°C (150°F). The viscosity can be lowered by raising the temperature: 90 cp at 78.8°C (120°F), and 20 cp at 121°C (250°F)—gelation starts after about 20 min. However, some pot-life will be lost.

Scotchcast resin No. 3 is furnished in two liquids, part A and part B. The mixture is made of 40 parts by weight of part A and 60 parts by weight of part B. Thorough mixing is necessary. Heating the mixture facilitates the removal of excess air.

Procedure

The sample must be air-dry (Coleman, written communication, 1963a). After the specimen has

been placed in a container, the mixture is added until the chip is well covered. In order to decrease the viscosity, the parts can be mixed under heat. The container is next placed immediately in a vacuum apparatus, and vacuum applied immediately. Repeated cycles of vacuum and release facilitate impregnation and penetration.

After the impregnation, the container with soaked sample is placed in an oven. If the sample is not covered completely with mixture, more liquid should be added.

Remarks

Curing at room temperature takes a long time and does not completely harden. Oven curing results in a hard product; however, the color darkens.

2.2F Impregnation at High Temperatures

As stated in the section on epoxy resins, the viscosity of such resins can be lowered by raising the temperature. The samples should be dry and heated up to temperatures of 120°C (250°F). This normally eliminates all samples containing clay since they will crack.

Pure, coarse-silty, and sandy sediments can be impregnated very quickly with epoxy that can be cured in the oven. The procedure has already been discussed in the preceding section (Scotchcast resin no. 3 of the 3M Company) and in the discussion of Araldite Peels in Chapter One.

The sample, while still in its container, is dried in an oven at a temperature of about 105°C (221°F) for at least 24 hours. Next the impregnation mixture is prepared and poured onto the sample. The total quantities should be doubled or tripled with regard to the amounts given in Table I.6. Curing can be carried out at a temperature of about 120°C (250°F) for a time indicated by the manufacturer. After the sample has cooled, it can be sawed, polished, etc.

2.3 MOIST AND WET UNCONSOLIDATED SEDIMENTS

When time is limited, or when conditions make it impossible to air-dry samples before they are impregnated, it is necessary to apply impregnating materials that harden in the presence of water or that are soluble in water or in a diluted liquid that has replaced the water beforehand (see 1.3B, 1.5A, and 1.5B). The glue peel methods can be followed here if one keeps in mind that there is very little penetration and that the impregnating material is too soft for section purposes. A disadvantage of water-soluble products is that the operator cannot use water during the cutting and grinding operations (J. Bouma, 1965).

The method of McMullen and Allen utilizes materials for impregnation which can be sawed and ground easily. The only restriction is that only sands can be impregnated. Reineck uses a rather time-consuming method for impregnating wet clays, but he obtains good results. Wax, as applied by Tourtelot to impregnate clay, will result in soft products of which thin sections can be made. Transferring the free water within the sediment into a gel can be done without difficulties as long as permeability is good, as is the case with sands. This method should be applied when impregnating cores of which only surface sections have to be studied.

2.3A Moist Sands Impregnated in the Field

Method of McMullen and Allen

McMullen and Allen (1964) collected samples with a Senckenberg sample box (Chapter Six). Products as well as application are discussed in Chapter One. Compositions, as given in Table I.3, should be tested to see if enough time is allotted for impregnation before gelation begins. Care should be taken to pour enough liquid onto the sample to avoid insufficient filling of pores in the upper part. If the exposed surface is not of importance, excess mixtures can be used. If it is desired to preserve the surface as well as the body of the sample (see Fig. 1.49), one can spray the surface after the poured liquid starts gelling, or the operator may start by spraying several times. Once the liquid is hard, the sample is turned over and excess liquid is poured onto the base.

Binding the grains with the mentioned BXL Plastics Materials Group Ltd. and Scott Bader products is usually sufficient for preparing thin sections. If there are still loose grains, one can use an unsaturated polyester resin to fill all remaining holes before sawing and grinding activities are carried out.

Method of Brown and Patnode

Brown and Patnode (1953) developed a method for impregnating materials above or under water. The technique does not apply to low-permeability sediments such as silts and muds.

An injection apparatus is clamped onto a core barrel (Fig. 2.7). The device, as developed by these authors, is for use in relatively shallow water. In Fig. 2.7 the most essential elements of construction are given.

"Prior to sampling, the piston rod assembly is clamped in position by means of the split collar (2). The volume of the cylinder is regulated by the position in which the piston rod (1) is clamped. The instrument is inverted and plastic, to which the catalyst and accelerator have just been added, is poured into the body of the cylin-

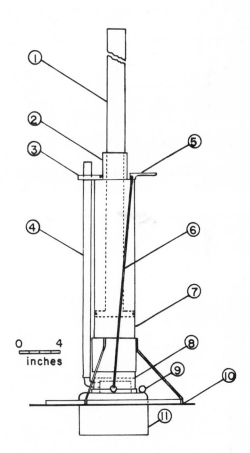

Fig. 2.7. Schematic presentation of a plastic injection sampler: (1) piston and rod; (2) piston rod guide and clamp; (3) vent clamp and piston rod release; (4) vent; (5) bail clamp; (6) bail; (7) cylinder; (8) gasket; (9) ring for attaching rope and float to facilitate recovery of core barrel; (10) adjustable support flange; (11) core barrel. (Redrawn from Brown and Patnode, 1953.)

der (7). The end of the cylinder is closed by inserting the rubber gasket and plug (8) in the end of the cylinder. The core barrel (11) is positioned on the gasket and clamped against it by exerting tension on the bail (6) by means of the bail clamp (5). The instrument is then righted and the core barrel (11) pushed into the sand body to a depth controlled by the support flange (10). Water that is displaced from the core barrel by the sand, escapes through the vent tube (4). This vent is essential to prevent disruption of the core when inserting the core barrel. When the support flange (10) is firmly seated on bottom, the lever arm (3) is raised, releasing the piston rod clamp (2) and closing the vent tube (4). Pressure is then applied to the piston by means of the rod (1). The piston forces the plastic downward from the cylinder. The plastic passes through a small orifice in the top of the plug (8) and through a large hole in the top of the core barrel. The plastic then displaces the water in the top of the core barrel above the support flange and distributes evenly over the top of the sand. Continued pressure forces the plastic through the porous sand driving out most of the pore water. When all the plastic is driven from the cylinder and the piston rests on the plug (8), the bail release (5) is lifted. This frees the core barrel from the rest of the assembly which is raised, leaving the core barrel in the sand. A rope and float previously attached to the ring (9) serves to mark the location of an underwater core and facilitate its recovery" (Brown and Patnode, 1953).

All types of cans can be used; their volume depends on the size of the cylinder (7). An opening is cut in the bottom of the can to let the resin pass, while two eye-bolts are inserted near the rim to attach the bail. An O-ring between the piston and cylinder acts as a seal and enables the piston to move easily.

The obtained samples are dried prior to impregnating them with an unsaturated polyester resin. Selectron 5001 from the Pittsburgh Plate Glass Company was found to be very satisfactory. Methyl ethyl ketone peroxide (1.5 to 1.0%) is used as initiator and cobalt naphthenate (0.05% with 6% Co) as accelerator. Oxygen and water inhibit the curing; under water, a preflush with acetone or alcohol is very effective.

When the Selectron 5001 is to be stored for some time, it is necessary to use more initiator. The authors successfully removed underwater samples within 1 to 2 hours after injection when

the formation temperature was between 10 and 21.1°C (50 and 70°F). They did not observe deleterious effects on leaving the samples in position in the formation for an indefinite period. One should therefore let the resin cure in place for 24 hours or more.

The largest core mentioned by Brown and Patnode (1953) was 17.2 cm (6¾ in.) in diameter and about 183 cm (6 ft) long.

2.3B Impregnation of Non-dry Clays

If the clay samples are still moist they can be hardened with a wax soluble in water or prehardened with a preservative and then hardened with an epoxy resin.

Method of Tourtelot

Mitchell (1956) developed a method for impregnating moist shale samples which transverses thin sectioning into routine work. Mitchell was chiefly concerned with the relation of the fabric of clays to their engineering properties. He used samples that contained their natural moisture since the drying effect on fabric was unknown. Tourtelot (1961) called attention to this geologically unknown technique.

The impregnating material used is a high molecular weight polyethylene glycol compound, known under the name Carbowax, a product of Union Carbide Corporation. This company manufactures a series of carbowax products of different molecular weights. Carbowax 6000 proved to be best suited for this type of work. The average molecular weight is 6000 to 7500 and it melts at 60 to 63°C (140 to 145°F). Its solubility percentage by weight in water of 20°C is 50 (Union Carbide, 1962). The cooled, impregnated sample has a hardness comparable to that of talcum.

Procedure. The moist sample is impregnated with Carbowax 6000 by placing the specimen in a cup of molten wax at a temperature of about 60°C (140°F). Carbowax is easily soluble in water, and, consequently, the wax can move into the sample by diffusion. For this process a period of 3 days is required at a temperature of 60°C.

Since the samples may tend to fall apart it is necessary to wrap them either in mosquito wire or in gauze (Fig. 2.8a). After 3 days the cup with molten wax and sample(s) is taken from the oven and the samples are removed and hung over the cup to allow free dripping and cooling.

The cooled sample can be sawed and ground to a thin section. Water cannot be used for cooling, and Mitchell, as well as Tourtelot, uses kerosene for sawing and grinding. A petroleum solvent should be applied to remove the kerosene from the slice before it can be mounted with an unsaturated polyester resin onto a glass slide. The cover glass is also cemented with polyester resin. Reasonably thin sections can be obtained in this way. It will be difficult to make them really thin because of the softness of the wax, but a thickness of 40 μ can be easily obtained (Fig. 2.8*b*).

Remarks. The effects of the water replacement by wax on the clay are believed to be negligible since Mitchell (1956) did not observe any volume changes. He feels that his thin sections give an accurate picture of the clay fabrics.

The wax crystallizes in feathery aggregates, which makes uncertain the study of very small grains such as the interior of foraminifera or in other rock cavities, but the aggregates do not interfere with studies of grains larger than about 50 μ or with analyses of the orientation of clay minerals (Tourtelot, 1961).

The Carbowax reacts with pure montmorillonite in bentonite, but Tourtelot did not find any reaction in rocks that contain up to 50% montmorillonite associated with other clay minerals.

The main disadvantage of the Carbowax is that it is not isotropic. The birefringence of the wax varies from slide to slide, but it has second-order yellow and red interference colors in most sections of standard thickness (Tourtelot, 1961). This variation might be due to the slight differences in water content of the samples.

It is necessary to wrap the collected samples in a plastic bag and two sheets of aluminum foil immediately after collection to protect the samples from a rapid loss of moisture and from falling apart. If the samples are too dry it is impossible to impregnate them with Carbowax.

During grinding it is necessary to use a more fluid slurry of kerosene and carborundum powder than is used for thin-sectioning hard rocks. The pressure used during grinding should also be less to prevent the abrasive material from being imbedded in the rock.

If the slice tends to break up after sawing, it can be repaired by dipping it into molten wax. If the sample contains hard fragments, which make it impossible to get undisturbed thin sections, the specimen should be impregnated under vac-

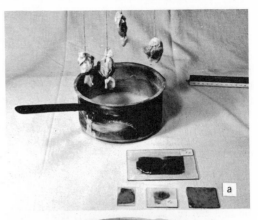

Fig. 2.8. (a) Four wet clay samples (subrecent river clays, excavation in the Rhine near Wijk bij Duurstede, The Netherlands) are impregnated with Carbowax 6000 (method of Tourtelot, 1961). They hang above the impregnation pan. In front, two thick sections and two not yet finished slices (on both sides of the smallest thin section). Diameter of the pan is 16 cm. (b) Photograph of thick section (S6) of clay impregnated with Carbowax 6000.

uum with a polyester resin. Normally this does not penetrate far and Tourtelot (1961) recommends the use of an aerosol spray to grind the flat surface first and to spray it again until no more resin is absorbed, and then to regrind the surface. Before applying the spray, the specimen should be washed in a petroleum solvent to remove all kerosene.

Method of Reineck

Reineck (1963c) developed a technique for making thick sections of clayey and sandy clayey materials. To prevent the forming of desiccation cracks, he prehardens the sample prior to real impregnation. Samples could be transformed by this method into projectable sections of 5 x 5 cm.

Materials. Prehardening is carried out with a preservative Arigal C from the Ciba Company.[3] It is a white, hygroscopic powder used in industry to prevent cellulose-destroying organisms from attacking materials such as canvas, tarpaulin, and sail (Ciba, not dated p). According to Müller-Beck and Haas (1961), 25 g of Arigal C is dissolved in 100 g of water at a temperature of about 80°C (176°F). Later a special catalyst is added.

Mosquito wire is used as a backbone for the sediment sample to prevent it from falling apart, and a fast-drying nitrocellulose lacquer of plaster of Paris is applied to cover the sample sides.

The second impregnation step results in hard samples and is carried out with Araldite Casting Resin F (CY 205)[4] and Hardener HT 902 or 905, both from the Ciba Company (Section 1.6; Reineck, 1958b, 1962; Bouma, 1964b).

Procedure. A mosquito wire frame is made (Fig. 2.9a) and pressed into the sediment sample which has been precut with a knife. Upper and lower face of the sample are smoothed as much as possible, applying the electro-osmosis effect (see Chapter Six).

The frame is made of a strip of wire, 220 mm long and 10 mm wide. It is bent around a mold with sides of 50 mm. Both ends are bent outward and sealed together with thin wire. This tail should always point in the same direction, thus fixing the position of the sample.

The fenced sample is next covered on all sides with a fast-drying nitrocellulose lacquer (Fig. 2.9b). As soon as the film is dry it is perforated with a needle. It is also possible to place rust-proof needles in the four corners and to imbed the sample in plaster of Paris. The needles serve as marks for later sawing (Fig. 2.9c).

The sample is next placed onto a plate covered with a piece of filter paper. It is then dipped and left in the Arigal C solution for 5 days. Rei-

neck (1963c) uses about 150 cc Arigal solution for one sample of 5 x 5 cm, 1 cm thick.

After 5 days a catalyst solution is added; it is stirred constantly; an amount of 10% by weight is needed. A mixture of 37.5 g Arigal C in 150 cc water requires 3.7 cc catalyst. After about 48 hours, depending on room temperature, a gray, slimy deposit becomes visible, indicating the end of the pre-impregnation. The sample, still on its plate, is removed from the solution and placed in a polyethylene bag. The bag is closed tightly and placed in an oven for 48 hours at a temperature of 65°C (149°F). Warming the catalyst results in a fixing of the Arigal solution.

The glass container in which the Arigal impregnation took place can be cleaned with a diluted solution of HCl.

After 48 hours the oven is turned off and not opened until the sample has cooled down. The specimen is now taken from the polyethylene bag and dried for 48 hours at room temperature, after which the sample is placed in a desiccator.

When dealing with sandy clay samples which are easy to impregnate, the prehardened sample can be placed in a tray of metal or other material and a mixture of Araldite casting resin F (CY 205) and Hardener HT 902 can be poured onto it. A period of 24 to 36 hours at a temperature of 120°C (248°F) is sufficient to finish the curing. The resin mixture has a composition of 100 parts by weight of CY 205 and 65 to 70 parts by weight of HT 902 (87 cc resin and 75 g hardener). The mixture is made while stirring at a temperature of 100 to 120°C (212 to 248°F) until a clear liquid is formed.

Clayey samples have to be impregnated with a mixture of Araldite casting resin F (CY 205) and Hardener HY 905 (Ciba, 1963, 1964b, not dated c) at a temperature of 40°C (104°F). The composition is made of 100 parts by weight of resin and 100 parts by weight of hardener. The impregnation process will be accelerated if a vacuum desiccator is used which can be heated to 40°C. The vacuum should be 65 to 45 cm mercury for a period of 30 min to 1 hour. Next the sample is transferred to an oven and left for 16 days at a temperature of 40°C. After this period the resin mixture is tough, designating completion of the impregnation. For final curing the temperature is raised to 120°C (248°F) for 48 hours, allowing the sample to be fully hardened.

The hardened sample is sawed to a size of 40 x 40 or 45 x 45 mm. Since the mosquito wire will be cut off, it is important to make a saw-mark at

---

[3]For American equivalents see "Remarks" at the end of 1.6.

[4]American equivalents are given in Chapter One.

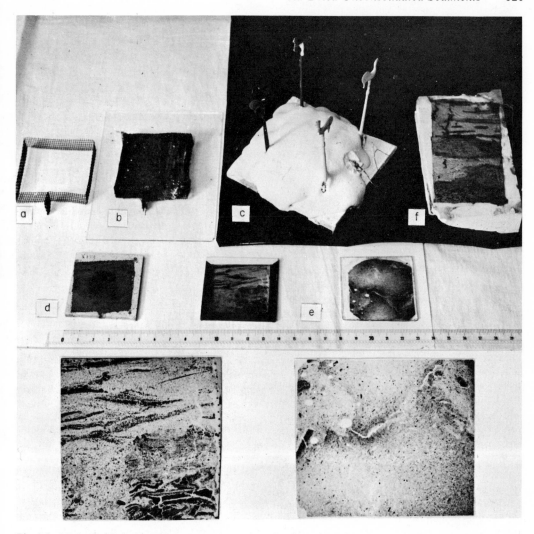

Fig. 2.9. Method of Reineck to impregnate wet sands and pelites with Arigal C and Araldite F: (a) frame of mosquito wire; (b) frame with sediment is covered with lacquer; (c) frame with sediment is imbedded in gypsum. The tail of the frame is visible at the right-hand side. The four pins indicate the corners of the frame; (d) slice of impregnated material; (e) two thick sections of 5 x 5 cm made from impregnated pelite. The left one is ready for projection; (f) poorly impregnated sample. Only the darker areas are well impregnated. Lower photographs show the thick sections of sandy pelites collected outside the excavation in the Haringvliet, Delta Project, The Netherlands.

the place where the tail of the wire was. Sawing and grinding can be carried out as for normal thin sections. The slice can be thinned down so that the areas with the finest materials become transparent enough for projection (Fig. 2.9d, e).

Remarks. If the procedure is followed accurately, excellent results can be obtained. If too much gypsum is used, the impregnation is often not homogeneous, and nonimpregnated parts

may result (Fig. 2.9f). The whole technique is somewhat time consuming, but it is the best method known for hardening fine-grained sediments without the formation of cracks.

2.3C Gel Method

When wet samples have a porosity that allows easy transport of water, this water can be bound to a gel, thus stabilizing the position of all grains.

Wet cores can be impregnated in this way, as can dry dune sands.

Impregnating Materials

The American Cyanamid Company produces the AM-9 Chemical Grout, which represents a new concept in the fields of soil stabilization and grouting. It is applied to seal off the flow of underground water into oil wells, drill holes, basements, tunnels, mine shafts, coffer dams, sewer pipe joints, caissons, dams, and dikes. Material with grain sizes ranging from 10 to 0.009 mm can be stabilized (Cyanamid, 1961).

AM-9 Chemical Grout is a white powder, often consisting of friable lumps. It is very soluble in water, ethanol, and methanol, and is insoluble in gasoline, kerosene, and oils. A 10% solution at 20°C (68°F) has a viscosity of 1.2 cp, a density of 1.04 and a pH of 4.5 to 5.0. Solid AM-9, as well as the solution, can be absorbed into the body by passage through the unbroken skin. Since it contains acrylamide, care should be taken and contacted areas should be washed immediately in running water.

AM-9 is a mixture of two organic monomers—acrylamide and N, N'-methylenebis-acrylamide—in proportions which produce very stiff gels from dilute, aqueous solutions when properly catalyzed. The process of gelation is a polymerization—cross-linking reaction.

Many catalysts and mixtures of catalysts can be used to gel AM-9. For normal use DMAPN (B-dimethylaminopropionitrile) is applied as an activator for the reaction. It is a somewhat caustic liquid of which 1 g is equivalent to 1.16 cc. Care should also be taken to avoid contact with the skin. If DMAPN is accidentally introduced into the eyes, immediate washing for at least 15 min with plenty of water is necessary and medical treatment must be sought.

Ammonium Persulfate (AP) serves as initiator, triggering the gelling reaction, and consequently is the last material to be added (Karol, 1964, 1966b). It is a strong oxidizing agent and should be handled only while wearing protective glasses and using rubber gloves.

Potassium Ferricyanide (KFe) acts as an inhibitor to control the reaction. It is a nontoxic, reddish, granular material.

Factors Affecting the Gel Time

(Cyanamid, 1961; Karol, 1962, 1963a, b, 1966b).

AM-9 *Concentration*. Decreasing the percentage from 15 to 5% causes only a slight increase in gel time.

DMAPN, KFe *and* AP *concentration.* Changes in the concentration of one or all components of the catalyst system strongly affects the gel times (Figs. 2.10 to 2.12). Too much KFe or too little AP or DMAPN will produce weak gels or none at all. The lower limit for DMAPN is about 0.4%, or AP 0.25%, while in normal cases not more than 0.035% KFe should be used.

Temperature. The gel time lengthens markedly with decreasing temperature and vice versa (Figs. 2.10 and 2.11). A rough rule of thumb is that the gel time is cut in half if the temperature goes up 5.5°C (10°F).

pH. pH also influences the mixed solution. Up to 1-hour gel time, the pH should be in the range of 7 to 11. Below a pH of 6.5 the gel times may become long and indefinite.

Air. Air decreases the gel time. After stirring the solution, a saturation of air is common. In Fig. 2.12 and in Table II.7 data are given on well-stirred mixtures.

Metals. Iron, copper, and copper-containing alloys have an accelerating effect on the gel time. The application of DMAPN, however, makes possible the presence of standard equipment. Aluminum paint prevents iron tanks from influencing the processes. Aluminum, stainless steel, plastic or rubber containers, pipes, and valves must be used to handle solutions of AP.

Sunlight. Sunlight may cause local gelation of AM-9 solutions, due to ultraviolet rays.

Inhibitors. Most of the ordinary polymerization inhibitors such as hydroquinone, oxygen, ferric ions and sodium nitrite can stop or slow down the gelation. However, these materials cause the formation of weak gels, which cannot be improved. KFe is applied as an inhibitor when supplied in very low percentages (Figs. 2.10 to 2.12, Table II.7).

Salts. Salts often have an accelerating effect on the gelation time (Fig. 2.12, Table II.7) (Drake, 1962). They also increase the gel strength. Some of them, such as calcium chloride, decrease the rate at which water is lost from gels under dehydrating conditions.

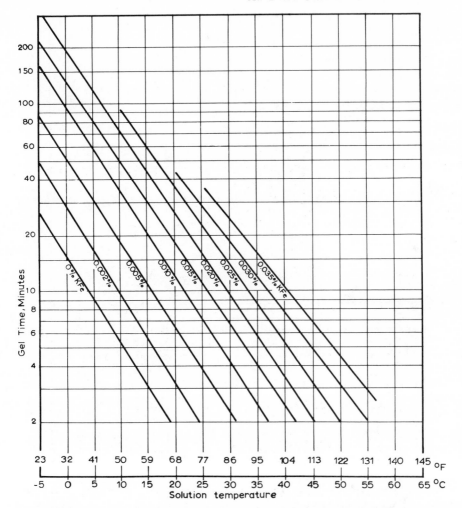

Fig. 2.10. Get time-temperature curves of AM-9 mixtures. The composition of grout solution by weight is: AM-9: 10%; catalyst DMAPN: 0.4%; AP: 0.5%; and KFe as indicated. (Redrawn from Cyanamid, 1961.)

Particulate Materials. Such materials as clay types slow down the gelation to some extent. Therefore, if such fillers are present, the concentration of DMAPN should be increased.

Control of Gel Time

Figures 2.10 to 2.12 and Table II.7 present a number of examples of different mixture systems where temperature and/or components vary.

Before applying this gel method the operator should make a test in order to find out the influences of the water and salts in his water and sediment.

Instead of using DMAPN one can use sodium carbonate (Na_2CO_3) to a pH of 10 (Fig. 2.11).

Effects of AM-9 Gels on Samples

Karol (1961, 1965a, b, 1966a) reports that stabilized soils retain their properties for a minimum of at least 10 years, which was the length of the experiments at that time.

AM-9 gels contain about 90% water, which is integrated in the space lattice of the gel structure. When the gel is stored under low-humidity conditions, the water will evaporate slowly, and consequently the gel will shrink. However, when placed under conditions of 100% humidity, water will be re-absorbed until the original volume is reached. These effects are not important when dealing with sediments whose grains touch one another. Voids filled with gel may rupture when the gel is completely dried out.

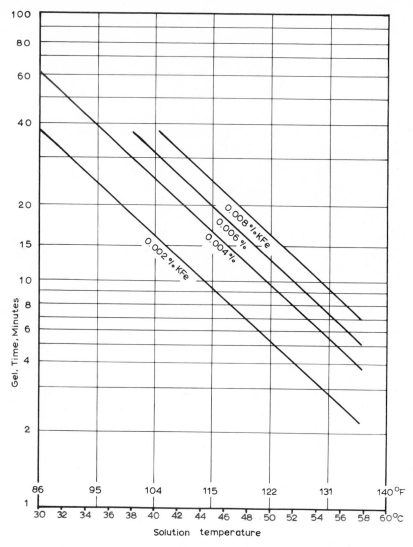

Fig. 2.11. Gel time-temperature curves of AM-9 mixtures. The composition of grout solution by weight is: AM-9: 10%; catalyst Sodium Carbonate to pH 10; AP: 1.0%; and KFe as indicated. (Redrawn from Cyanamid, 1961).

Loose, fine-grained sediments and porous deposits like diatomites will be subject to volume shrinkage when the gel dries (Fig. 2.13). Normal sandy cores do not show this effect.

Coloring of the AM-9 Gel

Tracing the path of the gel may be facilitated by coloring the AM-9 solution. Calcocid Uranine B-4315 of the American Cyanamid Company will not affect the gel strength or the induction period when used at a concentration of not more than 20 ppm. For lower concentrations the aid of an ultraviolet lamp is necessary.

The bright green color of the uranine in solution will fade to an orange shade shortly after gelation (Cyanamid, 1961).

Types of Applications

Some applications were attempted utilizing AM-9 gel. On board ship, during a cruise, the top of a box sample was covered while still in the box, using a fast-gelling AM-9 mixture to stabilize the surface for later vertical sectioning and to prevent loss of water due to evaporation. Since the gel loosens water under low humidity conditions, the top of the box had to be well

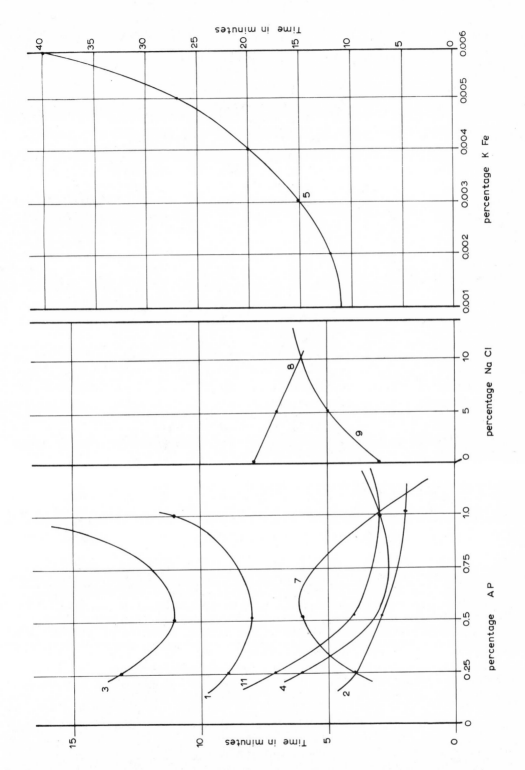

Fig. 2.12. Gel time-composition curves of AM-9 mixtures. All curves are drawn through three points. The numbers refer to Table II.7. Three groups of AM-9 mixtures are presented, from left to right: percentage AP, percentage NaCl, and percentage KFe. The experiments were carried out at room temperature without any temperature control.

Table II.7. Variation in Gel Time of AM-9 Solutions by Different Amounts of Components

Number	Water (cc)	NaCl (g)	Sea Water[a] (cc)	AM-9 (g)	DMAPN (g)	Wter[b] for AP (cc)	AP (g)	KFe (g)	Gel Time (min)
1a	79	–	–	10	0.4	10	0.25	–	9
b	79	–	–	10	0.4	10	0.5	–	8
c	79	–	–	10	0.4	10	1.0	–	11
2a	79	–	–	10	0.8	10	0.25	–	4
b	79	–	–	10	0.8	10	0.5	–	3
c	79	–	–	10	0.8	10	1.0	–	2
3a	79	–	–	10	0.4	10	0.25	0.001	13
b	79	–	–	10	0.4	10	0.5	0.001	11
c	79	–	–	10	0.4	10	1.0	0.001	23
4a	79	–	–	10	0.8	10	0.25	0.001	6
b	79	–	–	10	0.8	10	0.5	0.001	3
c	79	–	–	10	0.8	10	1.0	0.001	3
5a	79	–	–	10	0.4	10	0.5	0.002	12
b	79	–	–	10	0.4	10	0.5	0.003	15
c	79	–	–	10	0.4	10	0.5	0.004	20
d	79	–	–	10	0.4	10	0.5	0.005	27
e	79	–	–	10	0.4	10	0.5	0.006	40
6a	79	–	–	10	0.1	10	0.5	0.001	X[c]
b	79	–	–	10	0.2	10	0.5	0.001	X
c	79	–	–	10	0.3	10	0.5	0.001	21
7a	79	–	–	10	1.2	10	0.25	0.001	4
b	79	–	–	10	1.2	10	0.5	0.001	6
c	79	–	–	10	1.2	10	1.0	0.001	3
8a	79	5	–	10	0.4	10	0.5	–	7
b	79	10	–	10	0.4	10	0.5	–	6
9a	79	5	–	10	0.8	10	0.5	–	5
b	79	10	–	10	0.8	10	0.5	–	6
10a	79	5	–	10	0.4	10	0.5	0.001	12
b	79	5	–	10	0.4	10	1.0	0.001	X
c	79	5	–	10	0.4	10	0.25	0.001	13
11a	79	5	–	10	0.8	10	0.25	0.001	7
b	79	5	–	10	0.8	10	0.5	0.001	4
c	79	5	–	10	0.8	10	1.0	0.001	3
12	169	–	169	40	1.4	40	2	0.032	4.5
13	328	–	–	40	1.4	40	2	0.032	16
14	41	–	41	10	0.345	10	0.6	0.008	1.5
15	246	–	82	40	1.4	40	2	0.032	8.5

[a]Water collected at the beach off Scripps Institution of Oceanography, La Jolla, California.
[b]The water used to dissolve is the same water as used with the AM-9 in that experiment.
[c]X indicates that no complete gelation took place.

sealed with two or three layers of thin plastic sheet and aluminum foil taped to the box.

Beach sand samples, collected with the Senckenberg box, were impregnated and cut with a wire for direct structural examination.

Sands collected with plastic or metal tubes were stabilized with AM-9 gel. Impregnation from above, with the container base open and resting on mosquito wire, is not always successful; thin, "impermeable" films may prevent the passage of the solution.

Some types of laminae can influence the vertical movement of the solution when impregnation is done from the top or the bottom of the core. However, extruded cores are easier to impregnate since impregnation can be done from the side. To prevent the core from flowing away, two half-cylindrical plastic tubes are placed around the core in such a way that on the upper side a groove is left open.

By injecting the solution through a perforated pipe into the beach sand, it is possible to obtain a

Fig. 2.13. Samples impregnated with AM-9 solutions. Front row left: diatomite completely shrunk. This also applies to the sandy pelitic sample on the right side. Second row, from left to right: coarse sand core impregnated from above; artificial sample in tray (# 15) with layers of coarse and fine sand and iron hydroxyde containing fine sand at the bottom. The sample is broken at the joint of fine sand on coarse sand. Impregnation from the side. Third row, from left to right: beach sand collected with Senckenberg box. The impregnation is not uniform; artificial core of medium sand with a layer of iron hydroxyde containing fine sand in the middle. The core is broken along the joints. Impregnation from above; beach sand (medium sized) with a coarse sandy laminae along which the core (# 10) is broken. Impregnation from above. Beach near San Diego, California, U.S.A. Upper row, artificial cores standing upright, from left to right: coarse sand (# 12) impregnated from above; medium sand (# 13) underlain by fine sand with iron hydroxyde resting on coarse sand. Impregnation from above; fine beach sand with a nonimpregnated area. Impregnation from above.

more or less cylindrical impregnated sand mass. The outer shape is normally irregular due to differences in permeability. It is difficult to remove the impregnated body since the strength is rather low.

Dry dune sands can be successfully impregnated. An area is slowly soaked with water and gel solution is poured onto it. By leaving the area undisturbed overnight, enough water will escape to dig out the impregnated body. The upper part of the sediment must be discarded, for it is destroyed by the pouring and the transport of water.

2.3D Method of Replacement

Wet samples can be impregnated with materials that do not withstand water. It is a very time-consuming technique since water must be replaced by another liquid which does not resist the impregnating material.

The replacement should be carried out slowly and often must be repeated. For example, the water can be replaced by styrene monomer after which a long-gelling mixture of an unsaturated polyester resin is added, which, in turn, must penetrate.

The total time involved depends on the size and porosity of the sample. The author never experimented with it since faster methods (previously described) exist. Slager (written communication, January 1967) came to the conclusion that the results are poor, even if liquids other than styrene are applied.

2.4 PARTIALLY CONSOLIDATED SEDIMENTS

Poorly consolidated deposits such as shale, coal, and other fissile sediments are often difficult to saw and grind because they break easily or fall apart. Prior to these working activities such samples must be impregnated and/or imbedded (see Section 2.5). The choice depends on whether thin sections or polished surfaces are desired. We discuss here some techniques for hardening samples before thin sectioning.

In order to decrease the amount of impregnating liquid needed and to speed up the gelling and curing processes, it is advisable to take the smallest parts of the samples as possible or to cut them prior to impregnation.

2.4A Open Methods

The simplest method is to place a sample in a container and pour impregnating liquid beside the sample. To facilitate the escape of air bubbles, care should be taken that the uppermost part of the specimen is not covered by the liquid. The operator should first cut a smooth plane at right angles to the specimen and place this flat face on some bars to allow free entrance to the liquid. Depending on the type of sediment, the operator may dilute the impregnating liquid, but primarily the open method should be used for sandy or other porous materials applying nondiluted unsaturated polyester resins of low viscosity. Normally it is not necessary to impregnate the complete sample but only a few millimeters or centimeters to make a thin section at right angles to the stratification.

When it is desired to impregnate a part parallel to the bedding, the face should be cut or smoothed prior to the impregnation and, if possible, placed on bars. Laminae that are difficult to penetrate may give improper saturation (see also Shell method with Araldite D).

Experiments were carried out in which the face to be impregnated was oriented upward and the liquid was poured onto the face. This requires more resin and there is a risk that the liquid will move too far downward, leaving the upper portion more or less unfilled. Repeated pouring then makes the method time consuming.

Method of Moore and Garraway

Moore and Garraway (1963) use polymethyl methacrylate, an acrylic compound manufactured under the name Stellon by Claudius Ash, Sons and Co., Ltd. The polymerization can be carried out quickly at room temperature. The authors use this method to make thin sections of loose materials. It cannot be used too successfully to impregnate undisturbed samples.

Materials. Frosted microscope slides; flat wire about 2 mm wide or strips of thin metal; small crucible for mixing the acrylic; small spatula; plasticine for positioning the mold and to prevent leakage of the slurry; petroleum jelly (Vaseline); Stellon; nonpigmented powder; and clear fluid.

Procedure. A rectangular mold, about 28.5 x 19 mm (1⅛ x ¾ in.), is made using the flat wire. The mold is placed on the frosted side of the object glass and banked with plasticine to prevent leakage and air bubbles from forming prior to hardening. Petroleum jelly is smeared over the glass and inner side of the mold and serves as mold release agent.

Two parts of dry sediment are mixed with 1½ parts Stellon powder. Once this step has been taken, the process must be carried out to completion. Add a slight excess of Stellon liquid to the powder and stir with a spatula for 25 sec, no longer.

The sediment slurry is now poured into the mold and the surface is smoothed with the spatula. Polymerization starts immediately at a room temperature of about 20°C (68°F). Complete hardening takes about 1 hour. Care should be taken that the slurry is not disturbed or moved during hardening.

The hardened wafer can now be removed from its mold by tapping the wire gently with the spatula. The smooth side of the wafer can now be ground with 500-grit carborundum powder until an adequate surface has been cut into the grains.

The wafer is then cleaned in water, left to dry, and mounted on the frosted side of an object glass, applying 2 drops of Stellon mixture: 1 part of Stellon powder and 1 part Stellon liquid plus a very slight excess liquid. The mixture should be stirred for 30 sec. Excess mixture is shaken from the wafer prior to mounting. Care should be taken not to trap air bubbles. Pressure can be applied to the wafer during the to and fro movement to remove surplus slurry and any air bubbles. A small weight can be placed on the wafer during the 1-hour cementing period.

The wafer can now be ground normally. When the section is thin enough it can be placed in water for 10 min, after which the very thin film can be removed and washed in water. The size can be trimmed with a pair of scissors.

The thin film, when dry, is mounted with Canada balsam on a 35°C (95°F) hot plate. The cover glass should be warmed to 55°C (130°F). Slides prepared by this technique are chemically and physically stable. The polymerized polymethyl methacrylate does not discolor with time or with changes in temperature. Moore and Garraway (1963) have succeeded in making petrographic slides of unconsolidated sediments which ranged in size from argillaceous silts to mixtures of coarse sand and pebbles. The refractive index of Stellon is slightly less than that of quartz; the authors suggest checking the refractive index of each new order of Stellon.

2.4B Gluing Method

Sediments such as sandy pelitic deposits containing fine lamination tend to break along laminae planes. These samples should normally be imbedded in plastic or gypsum before cutting. However, the cut face often shows too many cracks to enable the operator to grind the surface or even to polish it. Especially for making acetate peels, it is better not to impregnate the specimen.

In such instances a glue is poured or painted onto the face, with the excess later removed by grinding. At the Geological Institute in Utrecht, Glyptal Cement number 1276 from General Electric, Insulating Materials Department, is used. Since the viscosity of this product is about 9500 cp at 25°C (77°F) (General Electric, 1963) the cement can be diluted with acetone (1 part Glyptal and 3 parts acetone). The application is repeated about three times.

2.4C Vacuum Application

Instead of using the open method, the impregnation can be carried out under the application of vacuum, thus facilitating the resin's movement into the sample. The method then does not differ from the method of Jongerius and Heintzberger, the method of Altemüller, or the Shell method.

Shell Method for "Impregnating" Shales and Cuttings

At the Koninklijke Shell Exploratie en Produktie Laboratorium at Rijswijk (Z.H.), The Netherlands, the epoxy resin Araldite casting resin D (CY 230) and hardener HY 951 are used to imbed and "impregnate" shale pieces and cuttings.

Araldite D casting resin (CY 230) is a slightly yellow liquid and hardener HY 951 a yellow-green liquid (Ciba, 1963, not dated d). A mixture of 100 parts by weight of resin and 9 to 10 parts by weight of hardener starts gelling after about 1 to 1½ hours at a temperature of 20°C (68°F). Curing at this temperature takes about 14 hours, whereas at 100°C (212°F) it takes about 20 min (Table II.8). Using 6 parts by weight of hardener, the mixture can be used during 2 to 2½ hours. The specific gravity of the resin at 20°C is 1.15, that of the hardener is 1.0. It is very important to mix both components thoroughly.

Procedure. The sample pieces are placed in a small galvanized tray (25 x 21 x 10 mm), after which the mixture is poured onto it. If the sample is porous, the operator should apply vacuum to it for about 20 min to impregnate the pieces as much as possible. The tray with sample is left overnight to cure. It is advisable to lengthen the curing and not to saw and grind the sample the next day.

Table II.8. Curing Times of a Mixture of Araldite D Casting Resin and Hardener HY 951 at Different Temperatures

Temperature (°C)	Minimum Curing Time
20	14–24 hours
40	5– 7 hours
70	1– 3 hours
100	10–30 min
130	5–10 min

After Ciba (not dated d).

In general, this method is more an imbedding than an impregnation method, depending on the permeability of the sample.

2.4D Vacuum and Pressure Application

Vacuum as well as pressure can be applied to impregnate badly consolidated and fissile sediments. This method is very convenient when dealing with small samples. For details the reader is referred to Section 2.2D. Epoxy resins as well as unsaturated polyester resins can be applied.

2.4E Temperature Application

Impregnation with the help of temperature is an easy and rather fast method. Raising the temperature lowers the viscosity of epoxy resins and melts all types of resin that are solid at room temperature.

This method is very suitable for sandy materials or samples with good permeability values. For dense specimens improper impregnation may result, but with small chips or thin slices of sediment good results can be obtained. Several techniques are known, but only few will be discussed.

Epoxy Resins

Epoxy resins have been discussed in previous sections and in Chapter One. Since curing of some epoxies occurs only at higher temperatures after many hours or days, care should be taken to apply the method only to samples that are not disturbed at those temperatures.

Lakeside

Lakeside No. 70 C is a thermoplastic transparent cement in bar form that melts quickly above 80°C (176°F) to form a uniform thin film. It is a product from Courtright & Co. The cement is colorless in thin films and has the same refraction index as Canada balsam. Accurate heat control is essential. Below 135°C (275°F) the cement is somewhat viscid and may form films of uneven thickness, which do not give full transparency. Above 155°C (311°F) inadequate adhesion and some bubbling may be encountered. The cement may start burning if temperatures above 155 to 160°C are prolonged. It is very difficult to remove cement which is burned or fused to a surface. Fuming nitric acid will eventually dissolve the Lakeside, but only slowly, and even then not always completely (Courtright, not dated a).

Alcohol is the most common solvent. Good results are obtained, but the method is very slow. Acetone evaporates much more quickly, but it may leave an emulsion of microscopic bubbles. If these are objectionable they can be removed by gentle heating. Xylene and toluene can also be used. A 5% (by weight) solution of borax in boiling water leaves no residue after about a 10-min exposure.

This type of Lakeside is used primarily for mounting slices of samples on glass for thin sectioning, but the cement can also be used for impregnation. The side to be mounted on an object glass should be as smooth as possible. The specimen is placed on a hot plate, the smoothed side facing up and leveled horizontally as much as possible. The hot plate should be set a little above 140°C (284°F). As soon as the sample

reaches the right temperature, the cement bar is smeared over the surface. This is repeated a few times until the Lakeside no longer penetrates the sample. The sample is removed from the hot plate and allowed to cool. The face can now be ground further. The impregnation depths are often not more than 1 or 2 mm, which is the reason why the sample surfaces should be as flat as possible.

Vernicolor

Vernicolor S. A. Meilen near Zürich, Switzerland, supplies a resin called Vernicolor for the preparation of thin sections (Frei, 1947; Jongerius, 1957). Like Lakeside No. 70 C, it must be heated to become fluid. The application of Vernicolor is more or less like that of Canada balsam. The manufacturer has announced that the fabrication of the product will be discontinued.

Canada Balsam

Canada balsam was one of the favorite products for the preparation of thin sections. Raising the temperature results in a decrease of its viscosity. Care should be taken that it does not become too hot since it may become brittle and turn dark in color near its boiling temperature. Van Straaten (1951) used a mixture of Canada balsam and shellac, whereas Hickman (1956) and Kurotori and Matsumoto (1958) apply Canada balsam only.

At the Geological Institute in Utrecht, the Canada balsam is dissolved in xylol until a good fluidity is obtained. This mixture is heated in an iron spoon or small container and the specimen is placed in it for a few minutes. The specimen often is precut and preground. Since an open gas flame often is used, care must be taken that the mixture does not catch fire. The resulting impregnation is more or less similar to that obtained with Lakeside.

Carbowax

Tourtelot (1961) applies the Carbowax 6000 to impregnate not only moist clays (see 2.3B) but also shales that still have their moisture content. For shales it is absolutely necessary that they keep their moisture after the sample has been collected. Wrapping them in aluminum foil and placing them in plastic bags directly after collection is necessary.

2.5 IMBEDDED SEDIMENTS AND ARALDITE PEELS

Many sample specimens are either not well consolidated or more or less fissile, which makes it impossible to cut them directly because many pieces will break off, resulting in incomplete surfaces. A number of techniques are known to prevent a sample from falling apart. A gluing technique previously was discussed.

2.5A Imbedding with Gypsum (plaster of Paris)

Gypsum imbedding is one of the most inexpensive techniques known. Since a layer of gypsum can withstand only small forces it is necessary to make the cover rather thick (Fig. 2.14). The thickness depends on the size of the specimen and its tendency to break.

A base layer of gypsum is poured into a box. As soon as the gypsum becomes a little stiff, the sample is placed on it. It this way it is possible to press the sample slightly into the gypsum base to ensure a proper contact. Prior to placing the specimen, the sample should be slightly wetted. The next step is to pour a thin gypsum slurry around and on top of the sample. As soon as the gypsum begins to harden, the number and orientation are marked into the gypsum. When the gypsum body is hard, the sample can be taken out of its container and is ready for sawing and polishing.

Remarks. At the Geological Institute at Utrecht polyethelene trays are often used as containers. When they are too large, a piece of cardboard is taped inside to give the working area its proper size. A brick is placed against the cardboard to keep it in place.

Prior to grinding, the pieces are cemented better by applying an unsaturated polyester resin mixture or Glyptal 1276 from General Electric (Fig. 2.14).

2.5B Imbedding with an Unsaturated Polyester Resin or an Epoxy Resin

When dealing with small samples (10 cm or smaller) it is not too expensive to use synthetic materials for imbedding. The advantage of these products is that they are more fluid than gypsum and consequently enclose the sample better and fill more holes (Fig. 2.15). In 2.4 some impregnation methods were discussed which are essentially imbeddings (see 2.4A and 2.4C).

Dollé (1959) imbeds pieces of coal in unsaturated polyester (Norsodyne 50) before cutting (Fig. 2.15). This method is employed for many specimens prior to cutting an X-ray slice (see Chapter Three).

Materials

It is important only for imbeddings to know what the final color of the cured resin is to be. The best resin mixture is one that cures rather fast but does not crack. Therefore it is important to work with small quantities and to build up thin layers (not over 15 mm thick), which have enough exchange of heat. From Table II.2 and Fig. 2.3 it can be observed that suitable mixtures are, for example, the numbers 7, 8, 22, and 23 with Vestopal H, numbers 53 and 54 with Plaskon 951, number 65 with K-46, number 133 with Synolite 333, number 66 with Setarol 3120,

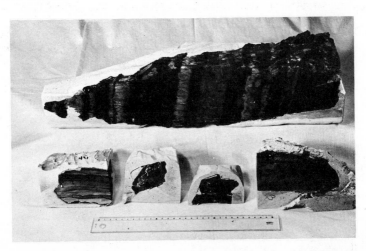

Fig. 2.14. Samples imbedded in gypsum and then sawed. Upper row: a shale with carbonate concretions in levels, Upper Devonian sill deposits, Eibach, Germany. Front row, from left to right: Middle Buntsandstein, Bad Salzdetfurth, Lower Saxony, Germany (J. A. Broekman); three Eocene clayey limestones with different carbonate contents from Navascues, Northern Spain.

Fig. 2.15. Samples imbedded in unsaturated polyester resin (Plaskon 951). From left to right: a broken claystone collected from the head of the La Jolla Canyon, California, and a fine-grained Eocene sandstone from the same area; Eocene sample from the coast north of La Jolla, California; three pieces of anthracite from Belgium. After the samples were imbedded completely they were cut with a diamond-impregnated saw.

numbers 86, 87, 89, 91, 92, and 96 with Lamellon 230, numbers 99 and 100 with Norsodyne 50, and number 139 with Paraplex P-43.

Different molds can be used. However, a plastic sheet, a mold-release agent, or aluminum foil must be used with all of them to prevent the cured resin from sticking to the mold.

Procedure

A base layer of about 1 cm thickness is first made. As soon as curing starts, the specimen is put in place. Depending on the shape of the lower plane, the sample rests in place or must be kept in position with cellotape. Each consecutive layer of resin mixture should not be poured before the last one is cured so completely that its temperature has dropped below 40°C (104°F). The mold should be covered with a glass plate, a sheet of plastic or aluminum foil, or with a piece of paper to prevent dust from falling in. It is also important not to wait a long period of time (weeks) between the pouring of two layers to prevent the forming of bad contacts (see 4.5, Fig. 4.23b).

The sample should be covered completely with resin mixture. If the surface remains tacky, the imbedded sample can be placed in an oven for a day at a temperature of 40 to 50°C (104 to 122°F). A tacky surface can also be removed with acetone.

The imbedded sample can now be sawed or a slice can be cut (Fig. 2.15). For thin-sectioning it is often necessary to impregnate only a few millimeters.

2.5C Araldite Peels

As mentioned earlier, Araldite peels often are still porous, which indicates that not all cavities and pores have been filled with the Araldite mixture. Cutting and grinding an Araldite peel occasionally results in the loss of grains.

An ideal combination for the analysis of sedimentary structures is to make an Araldite peel first in order to obtain a relief picture of the structures present. After the peel has been photographed, undiluted polyester resin mixture, with a curing time of at least two hours, is poured onto the back of the peel. This must be repeated until the resin no longer soaks in. The result obtained is very hard and is ideal for making large-sized, thin sections (Basumallick, 1966).

2.6 POLISHED SURFACES AND THIN OR THICK SECTIONS

For the study of sedimentary structures, a polished surface often shows these properties more clearly as a rough face or a face which is only cut. In certain instances it is even necessary to etch the polished surface and/or stain it (see Section Four).

Normally a thin section is too thin for sedimentary structures to be observed well, and when the object is to study those properties, a "thick section" can be made. Such sections normally have a thickness ranging from 75 to 30 m, depending on the average grain size.

2.6A Polished Surfaces

As just mentioned, the use of "polished" surfaces is very important. Under polished we include not only smooth surfaces obtained by the application of fine abrasives but also surfaces that are treated with synthetic materials to give them a polished appearance.

Samples

Very hard samples can be polished only with different types of abrasives and polishing paste to obtain the desired result. Examples are quartzites, hard and dense limestones, and marmors. Porous or loose-grained specimens seldom produce good results. In industry certain waxes are often used to obtain luster. It may even be necessary to impregnate the sample to at least a few millimeters to prevent grains from falling from the surface during grinding.

Sawing and Grinding

When large samples are dealt with it is important to cut the specimen in such a way that sawing marks are of minor order since it is very difficult and time consuming to grind them away. The specimen should therefore be mounted very tightly to a sturdy cutting machine table.

Grinding can be carried out in several ways. The size of the cut face should be smaller than the size of the grinding disk, the width of the polishing belt, or the size of the planing machine table, unless hand polishing wheels are used. Different grades of abrasives (carborundum powder) should be applied from coarse (grade 360 to 400) to fine (grade 600 or 800 to 1000). If fewer grinding disks are available than the number of carborundum grades, the operator *must* clean disk and specimen well before starting a finer grade. During grinding it is necessary to clean disk and specimen from time to time to prevent loose grains from making scratches.

Finishing. Only dense specimens can be polished with polishing paste to give them the luster desired. Normally polishing is very time consuming, especially for specimens of more than a few centimeters in diameter. The easiest way is to grind the sample as far as possible so that cutting marks are removed. Wax or lacquer applied may produce the polishing effect.

Stonemasons often use a special "polish paste" to obtain a higher luster. A product manufactured by Reek (Zaandam, The Netherlands) is in general application in Holland. It is a combination of natural wax types, such as prime-yellow, carnauba wax, white beeswax, ozocerite, and different binders to obtain a stiff wax. The wax is applied to the section with a tissue. Excess wax can be removed with fine steel wool. The wax must be thoroughly rubbed to obtain the desired luster.

A synthetic lacquer or a nonviscous polyester resin mixture can be painted onto the dry, clean surface. The lacquer should be diluted. An aerosol spray may also be applied. Many samples such as sandstones and shales absorb the lacquer. Therefore the painting should be repeated until the lacquer no longer soaks in. The operator may increase the viscosity of his fluid.

Since there are always spots that will absorb less mixture than others, the result will be mottled. The surface can be cleaned with acetone, after which one or two final coats or sprayings should be applied. For many sandstone samples it is difficult to obtain a real luster when the specimen is observed from the side. The varnish acts like water on surfaces of dense materials and, consequently, details stand out more clearly.

Instead of using products mentioned in Chapter One, an air-drying, two-component paint, manufactured under the name Metakote by Sikkens (1965), is applied at the Geological Institute at Utrecht. This product has a very high resistance to chemicals and solvents and has strikingly good mechanical properties, such as hardness and impact and abrasion resistance. Two hours after application, the Metakote is dust-free; at the same room temperature another 12 hours is required for complete hardening. At 120°C this hardening process takes ½ hour. The gloss is high and the flow excellent. The hardness after complete thorough-hardening is like a nail. It can withstand temperatures up to 150°C. Neither fresh nor sea water attacks the paint film. Of all types of chemicals, only alcohol, acetone, methylisobutylketone, and esters soften the coat slightly, but it recovers completely. The mixing ratios in parts by volume and in parts by weight are presented in Table II.9. The pot-life of the mixture at room temperature is at least 8 hours. Metakote is available in several colors, but for this work only colorless can be used. If the drying time of the last coat is longer than 24

Table II.9.   Mixing Ratios of Metakote Air-Drying
Two-Component Paints

	White and Colored Metakote		Colorless Metakote	
	By Volume	By Weight	By Volume	By Weight
Metakote paint	100	100	100	100
Metakote hardener	30	30	50	30
or				
Metakote brushing thinner	50	25	40	25
Metakote spraying thinner	150	100	130	100

After Sikkens (1965; written communication, August, 1966).

hours, it is absolutely necessary to polish this coat thoroughly before applying the next coat. Drying and thorough-hardening should take place at a temperature above 10°C. The degree of moisture is not significant (Sikkens, 1965). The brush should be cleaned several times a day in Metakote Thinner, to which about 10% ice-vinegar or vinegar-extract has been added, or with Sikkens Detergent for brushes.

Remarks on Photography

Since no relief is present and mirror effects can be obtained from the coated surface, the photographer should be careful in placing artificial lights on the side of a sample. In many instances good photographs can be made. If the results do not show the desired details, glycerine may be used to wet the specimen surface. If in-sufficient results are still obtained, it may be worthwhile to try out other techniques such as etching, staining, or radiography.

2.6B Thin and Thick Sections

Numerous contributions to this subject are known. Only a few of them will be mentioned here.[5] However, it cannot be stated that they cover all aspects of these techniques.

General Procedures

Canada balsam, Lakeside No. 70 C, and unsaturated polyester resins are the most common mounting media at present. The first two require heat, unless the balsam is diluted with xylol.

First the sample is cut, after which the sawed section is ground with various grades of abrasives. If necessary, the sample is partly impreg-

[5]Ross (1926), Hanna (1927), Krumbein and Pettijohn (1938), Reed and Mergner (1953), Taylor (1960), and Milner (1962). In the field of micropaleontology many authors have contributed to thin sectioning of foraminifera, e.g., Emiliani (1951). When dealing with minerals and ores it is necessary to polish the surface. Bakelite, Lakeside, Canada balsam, or diallyl phthalate prepolymer are used as mounting media (Short, 1931; Krieger and Bird, 1932; Dunn, 1937; Kennedy, 1945; Meyer, 1946; Brison, 1951; Amstutz, 1960). Lamar (1950) and Lees (1958) etch limestone samples for 5 min in a 20% solution of acetic acid. Lees then mounts them with Lakeside No. 70 C, grinds them down to 30 μ, and etches the sections again in a 20% acetic acid solution. Hickam (1956) applies Canada balsam for the preparation of thin section of well cuttings.
Conkin (1956) uses Krylon 1303 instead of a cover glass and Canada balsam. The aerosol bomb is held about 30 cm away from the ground section and two or more thin coats are applied. The drying time is about 5 min. Excess lacquer can be removed easily with xylol. A cover glass can be used for protection, but it is not necessary.
The preparation of thin sections of coal is extensively described by Thiessen et al. (1938). Thin-sectioning of clays is discussed by Weatherhead (1940),who uses pyroxylin for impregnating the sample and Canada balsam as cement. Mitchell (1956), Tourtelot (1961), and Catt and Robinson (1961) also describe the preparation of thin sections of clays (see 2.3B). Of the number of authors contributing to the making of thin sections of sediments and friable rocks we particularly note Schlossmacher (1919), Ross (1924: bakelite varnish diluted with methyl alcohol and ether — 1:1, requiring 2 days for evaporation and baked in an oven for 2 days at 70 to 100°C; Ross also uses kollolith), Hanna (1927), Legette (1928: bakelite varnish and methyl alcohol using a slight vacuum for evaporation and kerosine or mineral oil for grinding instead of benzene), Fowler and Shirley (1947), Emery and Stevenson (1950: a mixture of the resin Castolite with thinner and hardener to impregnate beach sand prior to sectioning), Exley (1956: Bakelite: see McMullen and Allen, 1964), Debyser (1957), Thissen (1959), and Richardson and Deane (1961).
Read and Mergner (1953) use Canada balsam, kollolith, glycol phthalate, or bakelite for mounting. When dealing with water-soluble minerals, which are decomposed by water, they use oil or alcohol instead of water for grinding.

nated and ground again prior to mounting on an object glass. Care should be taken not to trap air bubbles when mounting the specimen. The best way is not to spread the mounting material evenly, but to leave it in a roll with a width similar to the length or width of the specimen surface. The flat face is held inclined with the lower end on the object glass. It then is moved a few millimeters to the center to form a real roll of cement. The specimen is now folded downward in a hingelike motion, thus pushing away excess cement and air bubbles. The specimen can now be moved a little to make the film of cement as thin as possible.

When dealing with slow-curing media such as unsaturated polyester resins and Canada balsam diluted in xylol, it is important to place the mounted sample on a horizontal table; next place a weight on top of the sample to prevent air from penetrating from the side; and, finally, place four rubber stops against the corners of the sample to prevent sliding (Jongerius and Heintzberger, 1962, 1963).

When the specimen is mounted onto the object glass, most of the sample can be cut off on a small-sized cutting machine. Further thinning down can be carried out on a grinding disk, planing machine, or diamond impregnated disk (see Chapter Five).

Dollé (1959) uses pieces of plate glass (60 x 49 mm and 6 mm thick) to which the object glass is mounted temporarily with Lakeside No. 30 L (Courtright, not dated b). This plate glass gives the operator an easier grip on the object glass (see further, Chapter Five).

Before a cover glass is mounted, it may be necessary to polish the thin section when dealing with metallic minerals and ores or to etch and/or stain the surface. Mounting the cover glass is carried out in the same way as described for mounting the specimen.

Large-sized thin or thick sections cannot be mounted with Canada balsam or Lakeside No. 70 C because of their large surface. The best method is to use an unsaturated polyester mixture with a short curing time (Table II.10).

To secure good mounting without air bubbles it is very important to use a well-leveled table and weights that fit the size of the cover glass. Rubber stops placed against the four corners prevent moving of the specimen or the cover glass.

Jongerius and Heintzberger (1962, 1963) do not place the section on the object glass, but they place the polished side of the specimen upward and cover it by a thin layer of resin mixture. After a thicker layer has been applied to one of the long edges, the glass is applied to the section in a hingelike motion.

When the proper equipment is present, there

Table II.10.   Composition of Cementing Material to Mount a Sample to an Object Glass and to Mount the Cover Glass

Vestopal-H	Initiator LTM-50	Accelerator NL-49
10 cc	0.2 cc	0.1 cc

The gelling time is about 1½ hours at room temperature. At temperatures above 22°C half the amount of both initiator and accelerator is used.

After Jongerius and Heintzberger (1962, 1963).

is no limit to the size of the section (Fig. 2.16). Very practical are sections with a size of 5 x 5 cm, which can be used directly as slides for projection.

According to Harbaugh (1953), a broken microscope slide can be repaired with Scotch tape. However, this tape has the ability to polarize light.

Remarks on Photography

There is a difference in the manner in which thin sections can be photographed. Micropho-

tography is a technique whereby small details can be made more clear by means of enlargement, whereas for the study of sedimentary structures, enlargements are not normally required (macrophotography, Fig. 2.16). It is important to use photographic equipment such as plate cameras with extension bellows or a miniature camera with reflex accessories (e.g., Visoflex housing) and telelenses to fill the size of the negative as much as possible.

As a light source for macrophotography a lightbox with matt glass and opalin glass is necessary to obtain uniform illumination.

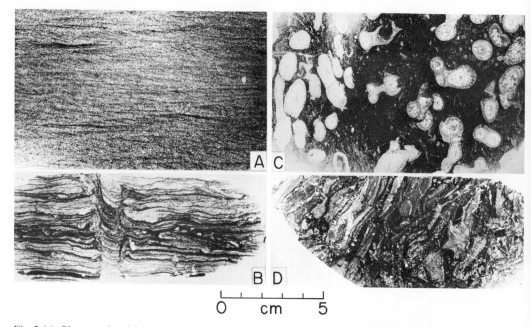

Fig. 2.16. Photographs of thick sections, using transmitted light. Not only the thickness and slight differences in the thickness of a section but also the type of material is important for the resulting photograph. (A) Maulbronner sandstone with small-scale foreset bedding. Maulbronnen, between Karlsruhe and Stuttgart, Germany (see X 26 of Fig. 3.29E). (B) Banded tuff (submarine) with burrow. Rhodes, Greece (Collected by E. Mutti, Esso Rep. Bordeaux, France) (see X 40 of Fig. 3.29R). (C) Coral limestone of Silurian age. Erratic boulder from Scandinavian origin. Hilversum, The Netherlands (see X 68 of Fig. 3.29M). (D) Reef debris from side of a bioherm (Devonian), Taillfer, Belgium.

2.7 CLEANING OF EQUIPMENT

It is generally rather difficult to clean equipment and tools when unsaturated polyester resins and epoxy resins are applied. The operator should use disposable mixing vessels such as cans.

Pipettes and measuring cylinders used for initiators and accelerators can be cleaned directly after use with a lot of water or with acetone. Special products such as P3-Nevol and RBS-25 Concentrate can also be used (see below).

Impregnation cylinders (see Chapter Four), tools, and other containers on or in which polyester resin mixtures are cured can be cleaned with acetone or with the special products mentioned. A closed bucket containing acetone, in which the equipment can be left for a day, is very practical. Occasional stirring of the acetone speeds up the dissolving process. Jongerius and Heintzberger (1962, 1963) clean all glass parts of their impregnation equipment in a solution of P3-Nevol in warm water. P3-Nevol is a slightly alkalic product. A concentration of 5 to 10% normally cleans glassware overnight (Henkel, 1961a, b).

RBS-25 Concentrate from Borghgraet (not dated; Hicol, not dated) is a concentrated cleaning agent of which a concentration of 2% is sufficient. The solution is alkalic. Equipment should be left in the solution for about 24 hours. Since it may be difficult to clean glassware, a warm solution should be used. Boiling may be necessary in some instances. The solution should be stirred from time to time. All tools should be rinsed in water thoroughly after cleaning.

Epoxy resins normally do not dissolve in any mixture. For cleaning the hardened resin from any equipment one may use glacial acetic acid with some methylene chloride as an additional softening and swelling agent. Also heavy chlorated hydrocarbons, chlorine, phenol, or dichlorethane can be used. These products do not dissolve the hardened material, but after a few days, they initiate a swelling which loosens the resin from the equipment.

All manufacturers provide data in their technical brochures on storage and handling of their products. If the rules given are followed, no danger to persons or buildings is involved.

Some of the products are harmless, other are inflammable, toxic, or decompose easily. The large variety of products and national, state, and local rules prevents a detailed description of all regulations. Each investigator should ask the manufacturer for further information on his products. There are special rules for bulk storage.

2.7A Storage

Unsaturated polyester resins and epoxy resins should be stored and kept sealed in their original containers. These products are more or less inflammable. Shelf-life of most polyesters is about 6 months if kept in a cool place (not above 20°C). Epoxy resins generally are more stable. Contamination and loss of volatiles can be prevented only when the containers are well-sealed (Hüls, 1961; Scott Bader, 1965; Dow Chem., not dated a; MMM, not dated b).

Accelerators and styrene monomer are inflammable and should be kept away from open fire. Their containers should be well-stoppered.

The initiators are mostly organic peroxides and (in general) are classified as hazardous chemicals. By nature, they are usually extremely inflammable and may decompose with explosive violence under certain conditions. Their stability improves in the series of liquid-pastes-solids. Small quantities are best stored in well-stoppered containers, and placed in a refrigerator, or in a dark, cool, fireproof area. For large quantities an extensive series of rules is necessary (Noller et al., 1961; Scott Bader, 1965; Hercules, not dated b; Nuodex, not dated; Wallace and Tiernan, not dated).

Initiator and accelerator should never be mixed directly with each other, since they can react with explosive violence.

Leftover peroxide should never be poured back into the original container; it should be destroyed. Small quantities can be destroyed by dissolving them in diluted alkaline, sodium metabisulphite, or other reducing solutions.

2.7B Handling Precautions

As stated in the rules given below, it is important to keep the working area as well as all its equipment clean, to have a good ventilation system, and to be certain that all spilled chemicals are cleaned up immediately (Arbeidsinspectie, 1964; Wallace and Tiernan, not dated).

1. The working space should be well-ventilated and kept clean.

2. Each person should have a free air space of at least 20 m³ of which at least 10 m³ is above a height of 180 cm.

3. For first aid protection, extinguishing equipment of the dry chemical or foam type should be available.

4. Open flames must be avoided. Electrical hotplates should be used instead of gas.

5. The surface of working tables should be made of a material that is not easily corroded by the chemicals used. It is recommended to cover them with paper which can be replaced regularly.

6. The garbage cans used should be kept closed as much as possible, preventing the escape of volatiles.

7. All glassware and other measuring and mixing equipment should be kept clean. Tin cans can be used for mixing and thrown away after use.

8. A good sink, with warm and cold water and with an elbow or a foot operation system, should be installed in the working area for personal safety.

9. A sprinkling system is recommended.

10. Different types of protective creams and solutions as well as safety goggles and gloves should be placed at hand.

11. All containers should be clearly labeled.

12. All containers should be kept closed. Spilled material must be cleaned up immediately, applying absorbing paper, towels, or mica powder (vermiculite, perlite).

13. Different, well-labeled, measuring cylinders and pipettes should be used for the different chemicals the operator deals with. The initiator and the accelerator especially must be kept separate to prevent explosive reactions and/or contamination.

14. Chemicals that evaporate easily should be handled in a running fume-cupboard since they all are more or less toxic.

2.7C Safety Precautions

Most of the products we are dealing with irritate the skin. Even people who are not necessarily sensitive to these chemicals can develop a hypersensitivity when contact with the skin occurs repeatedly. This can be avoided by wearing plastic or rubber gloves. Also, the use of barrier creams during and after the work is recommended. *Never should any of the chemicals be touched by bare hands*. Even small skin wounds should be treated without delay.

Some of the vapors, such as that from styrene monomer, are irritating to the eyes and respiratory passages. If a person is sensitive, he should wear a mask to protect the mouth and nose.

Peroxides are extremely irritating to skin and eyes. They can cause burns on the skin if not washed off immediately with plenty of warm water, followed by application of lanoline ointment. The operator should wear protective goggles when working with liquid initiators. If peroxide comes into contact with the eyes, it can cause serious injury if not treated immediately. The affected eye or eyes should be washed vigorously with plain water for at least 15 min. Preferably a 2% aqueous solution of sodium bicarbonate should be available for eye washing, after which washing with plain water is necessary. In all cases a doctor should be consulted after the first aid, since serious consequences may not become apparent for several days (Scott Bader, 1965). As a further and subsequent first aid treatment, a 5% solution of ascorbic acid can be used. Under *no circumstances* should the eye be treated with oil or with an oily solution, because such products aggravate the damaging effect of peroxides.

For measuring the small quantities of initiators the operator normally uses a pipette. If some of this peroxide is swallowed, vomiting must immediately be induced; this should be followed by the drinking of copious draughts of sodium carbonate in water. The stomach should next be emptied with a stomach pump.

All operators carrying out impregnation work should be instructed regularly. They should be required to pass a physical test (Van Luyt, 1964).

The present author has had good results with a number of special products from Stockhausen & Cie. In this type of work the use of the chemicals may be only for short periods, but if continuous use is necessary, it must be realized that professional illnesses can be very serious (Koehler, 1951; Hook, 1952; Barkow, 1957, 1959, Stockhausen, 1962a to f, 1963a, b, c, 1964a, b).

Since there are no universal barrier creams and cleaning agents, different products have to be applied in accordance with the chemicals used. In particular, there are special silicon-free

hand barrier creams; Arretil-R (when dealing with thinners, mineral oils and fats, and lacquers); Arretil-T, which is soluble in water (for asphaltic lacquers, dry pigments, fiber glass) (Stockhausen, 1962b); and Arretil-Qu (when dealing with thinners, styrene monomer, epoxy resins, oil paints, and lacquers). Hands should be washed and dried prior to the application of a barrier cream. Three silicon-free general barrier creams are Kosmosan, Stokolan, and Stoko-Emulsion. Kosmosan is a cream which primarily is applied after the work is finished. This cream should be used especially when dealing with lacquers or watery solutions (Stockhausen, 1962c). Stokolan (a greasy cream) can be used before and after the work (Stockhausen, 1962d). Stoko-Emulsion has special creams added to it and has a general application. It can be applied where the special barrier creams are not suitable or where one prefers an emulsion rather than a cream, for example, to protect the face (Stockhausen, 1962e).

Industrie-Preacutan is an alkali-free skin cleaner. It should not be used like soap (Koehler, 1951; Stockhausen, 1962f). Es-Te-Lackentferner is a special cleaner used when dealing with lacquers, unsaturated polyester resins, and epoxy resins (Stockhausen, 1963d). Besides cleaning, these products contain special skin-protection chemicals.

2.8 DISCUSSION

For several types of investigations in nonconsolidated or weakly consolidated sediments — such as polished sections, thin or thick sections of different sizes, grain orientation, and inclination, sedimentary structures, and the combination of petrographical investigations and location of minerals by staining — it is necessary to fill the pores between the grains with some artificial product in order to bind the particles together to one massive block without changing the position of grains.

A large variation in lithology and moisture content of samples and a discrepancy in the time allowed between sampling and investigation necessitate an extensive number of different techniques and, consequently, use of different artificial products.

As in making peels, the sands are the easiest material to work with since water can be removed without causing any disturbance. Pure clays, on the other hand, require a difficult and time-consuming type of impregnation.

The application of the different chemicals is sometimes not without danger to the health of the operator. However, by following a small number of safety regulations, no dangerous results need occur for any person.

Practically all methods discussed are rather easy to carry out, and the necessary equipment can be restricted to a minimum. It is important to realize that besides safe and accurate handling, a great deal of patience is needed. This still is not a 100% guarantee of good results. Experience is an important factor since the majority of samples never fit any division given. It is therefore important to restrict oneself to a few methods and to improve the personal procedure by experience.

CHAPTER THREE

RADIOGRAPHY

X-ray radiography, as applied to sediments, is a technique based on differential passage of X-radiation through a sample onto a special film. Differences in composition (density) from place to place through the sample cause differences in attenuation. The result and variation in the amount of radiation that reaches the film creates differences in photographic density.

It should be stated that this technique is not designed to replace other methods but rather to supply the investigator with an additional source of information. Often, because of the nondestructive nature, X-ray radiography can significantly enhance the application of other methods of study.

X-ray radiography was recently introduced into sedimentology by Hamblin (1962a). The method has already been applied for many decades in the fields of medicine, biology, paleontology, and industry (Phillips, 1963a). The explosive development in industry has made the applications of X-ray radiography in the study of sediments highly valuable.

At the present time only few contributions to this subject are known, but it seems likely that their number will increase rapidly in the near future (Hamblin, 1962a, 1963, 1965; Calvert and Veevers, 1962; Rioult and Riby, 1963; Bouma, 1963, 1964c, d, e, 1965; Bouma and Shepard, 1964; Bouma and Marshall, 1964; Bouma and Boerma, 1968; Oele, 1964; Farrow, 1966).

Radiography can also be used for other geological investigations, such as to observe microfaults and folds in samples, certain structures and textures in igneous and metamorphic rocks,

distribution of certain minerals such as ores, the presence and disturbing influence of fauna and flora, detection and distribution of fossils in a sediment, sedimentary structures preserved in soils, and distribution of pores through soils. It may be instructive even to make radiographs of ancient and recent sedimentary samples after paleomagnetic measurements have been carried out to find whether a low value of magnetism is due to the sample itself or due to reworking by burrowing organisms that disturb the primary orientation of the grains. For purely mineralogical purposes this technique is not applicable since homogeneous material or a composition of minerals with small variation in atomic number of its elements will not produce differences in absorption.

Once an X-ray unit is available this method is relatively inexpensive, easy to handle, not time consuming, and not directly dependent on size, type, shape, or condition of a sample. The geological material may be consolidated or unconsolidated, dry or wet, impregnated, inbedded or nonpreserved. In many cases one can first make a radiograph of a sample slice before taking small parts for grain-size analysis, thin sections, or impregnation. This procedure is the best method for unconsolidated fine-grained sediments, where it is difficult to make a thin section.

In general, one makes a radiograph of a thin sample slice, but it is also possible to take two films from cores and irregular samples, made under different angles, which can be observed under a stereoscope to study a three-dimensional picture.

X-rays, as well as gamma rays, can be used to

study sedimentary structures. Both types of radiation will be mentioned, but to date only the first is used. Differences in radiation that passes through a sample can be observed with the aid of a suitable fluorescent screen, but for research purposes a film has more advantages.

Every scientist who is interested in working with the technique of radiography should read all parts of this chapter carefully and should consult someone with experience in radiation before purchasing an X-ray apparatus or beginning radiographical work to avoid serious difficulties with regard to health and expected results. He is also referred to the literature published by various companies such as Gevaert, Ilford, and Kodak. Parts of this chapter are based on such contributions and also on papers from the *Encyclopedia of X-rays and Gamma Rays* (Clark, 1963). The text of this chapter may not be correct in every detail and it does not claim to be complete, but enough information is given to introduce the reader to the field of radiography. Figure 3.1 gives a general guide to the various subjects treated.

3.1 RADIOGRAPHIC SOURCES AND PROCESSES

X-rays are comparable to ordinary light; they are a dual form of particles of energy and electromagnetic oscillations. Their wave lengths, however, are much smaller, giving them some other characteristics: X-rays are not directly visible, they can penetrate materials which absorb or reflect visible light. Gamma rays are similar to X-rays in most properties and they can be distinguished by their source rather than by their nature.

Both types of radiation travel in straight paths at the velocity of visible light. In practice they cannot be directed or deviated by means of lenses or glass prisms. When radiation passes through a body, the beam is attenuated by various physical phenomena, for example, absorption and scattering. They exert an ionizing effect by liberating electrons from any material upon which they fall. The ionizing action may produce the following effects: fluorescence of certain materials; chemical effects such as the production of latent image; and biological effects in the sense of deterioration or destruction of living cells (Kodak, 1957).

The wavelengths of electromagnetic radiation can be expressed in meters, centimeters, milli-

meters, microns (1000 μ = 1 mm), in millimicrons (1000 mμ = 1 μ), in angstrom units (1 Å = 0.1 mμ = 10^{-8} cm = 3.937×10^{-9} in.), and in X units (1.000 X = 1.00202 Å) (Fig. 3.2).

The continuous spectrum of an X-ray tube extends over a wide range of wavelengths and is characterized by a short wavelength limit and an intensity distribution. The spectrum varies according to the voltage applied to the X-ray tube. Gamma radiation, however, consists of a number of specific wavelengths, which differ according to the nature of the radioactive source.

The wavelength λ (lambda) is normally given in angstrom units. The formula used to calculate the short wavelength limit of the continuous X-ray spectrum is

$$\lambda_o = \frac{12.35}{V} \quad \text{in Å}$$

in which V represents the value of the voltage.

3.1A X-Rays and Gamma Rays

X-rays are normally produced by an X-ray apparatus. The shorter the wavelength of this type of radiation, the higher their energy and the greater their penetrating power. "Hard" X-rays have short wavelengths and "soft" X-rays have relatively long wavelengths. *Quality of radiation* refers to the radiation's relative hardness or penetrating power.

To obtain information on the sedimentary properties of a sample the only ionizing action of importance is its chemical effect on the film emulsion. The photographic record obtained is called a *radiograph*. In fact, the radiograph is a kind of shadow picture; the darker regions present the more penetrable parts of the sample and the lighter regions the more opaque parts. If a print is to be made, it should be remembered that dark and light areas are reversed when compared to the radiograph.

Gamma rays are emitted from the disintegrating nuclei of radioactive material. Quality and intensity of this type of radiation can only be controlled by using various isotopes and by changing the distance source-film. In general, the gamma rays are a hard radiation. Some gamma rays are naturally emitted by radioactive isotopes such as radium and mesothorium. Others like cobalt-60, caesium-137, cerium-144, thulium-170, and iridium-192 are artificially produced (Kodak, 1957; Kodak Ltd, 1964a).

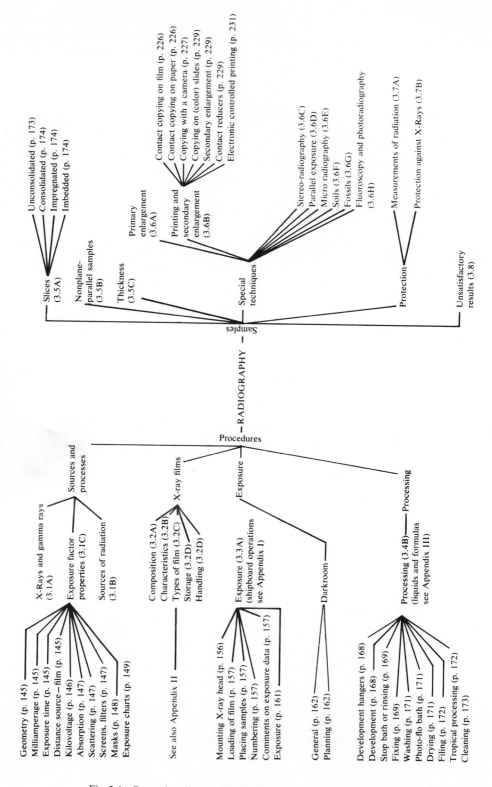

Fig. 3.1. General outline on the subjects discussed in Chapter Three.

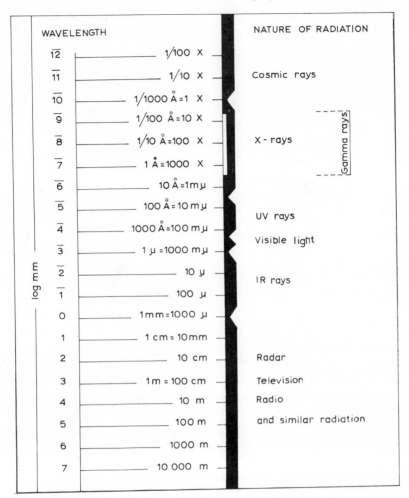

Fig. 3.2. Portion of the electromagnetic spectrum with different types of radiation. (Redrawn from Gevaert, not dated.)

3.1B Sources of Radiation

X-Rays

When electrons travel with high velocity and collide with matter, they produce X-rays. The normal X-ray tube consists of an evacuated glass tube. This vacuum is important because it allows the liberated electrons a patch of free travel. In the tube an arrangement of an incandescent filament is mounted, supplying the electrons and consequently forming the negative electrode, or *cathode,* and a positive electrode, called the *anode* (Fig. 3.3A).

The filament of the cathode is heated to incandescence by a current of several amperes. It frees electrons, which will move in any direction until a high electric voltage (*tube voltage*) is es-

tablished between the cathode and the anode. Under the effect of this voltage the electrons are propelled toward the anode (Bowman, 1963). The electrons are concentrated into a beam by means of a *focusing cup* (Gevaert, not dated; Kodak, 1957).

A target or anticathode (Fig. 3.3A) is built at the place where the electrons hit the anode. The sudden stopping of a small portion of the electrons at the surface of the target results in the generation of X-rays. Of the kinetic energy of the electrons about 99.7% is transformed into heat; the rest is converted into X-rays, when a small apparatus is employed. Only a small area of the target, called the *focal spot* (Fig. 3.3B) is hit by electrons. The dimensions of this spot are very important and should be kept as small as

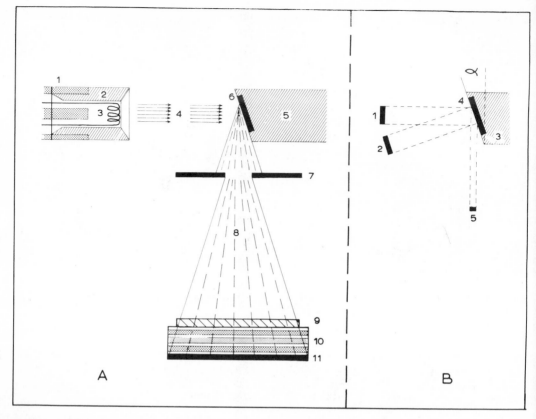

Fig. 3.3. (A) Schematic diagram of an X-ray tube and the setup for making a radiograph: (1) cathode; (2) focusing cup; (3) filament; (4) electron paths; (5) anode; (6) target; (7) diaphragm; (8) X-rays; (9) sample; (10) exposure holder with X-ray film; (11) sheet of lead. (B) Schematic diagram of the relation between the actual focus spot and the effective focus spot as projected from an anode with an angle α: (1) section of the electron beam; (2) surface of the actual focus spot; (3) anode; (4) target; (5) surface of the effective focus spot; α: angle of the anode. (Redrawn from Gevaert, not dated; Kodak, 1957.)

possible, since the smaller the focal spot, the sharper the radiograph can be made.

Manufacturers of X-ray tubes furnish data charts indicating the kilovoltages and milliamperages that can safely be applied at various exposure times. The life of a tube will be lengthened considerably if it is operated within the rated capacity.

The actual dimensions of the focal spot are not equal to the effective dimensions since the surface of the target is at an angle to the bundle of electrons (Fig. 3.3B). If the angle α between anode face and central ray is small, say 20°, the effective focal spot is only a fraction of its actual size.

Apparatus

For the purposes of this book, small X-ray units with a small focal spot are best to work with. The use of low voltages gives more details of a sample on the film, and since the samples are dead bodies (without movement) it does not matter if they are exposed to radiation for long periods (Rogers, 1963). By application of low kilovoltages and milliamperages one may use a relatively cheap X-ray unit, which is also easy to handle. For very thick samples (over 10 cm in thickness) it will be nevessary to work with a higher powered unit (over 100 kV), but for regular work in our field, dealing with slices of rock or cores and irregular samples up to 9 cm in diameter an apparatus with a maximum voltage of 100 kV is quite suitable.

It is beyond the scope of this work to discuss all X-ray units available, and the author will mention only those he has worked with.

At the Scripps Institution of Oceanography, La Jolla, California, radiographs were made with a General Electric Mobile "90-11" X-ray unit, set at 40 kV and 5 mA, with unfiltered tungsten radiation (Calvert and Veevers, 1962; Bouma, 1963, 1964c, d, e, 1965; Bouma and Marshall, 1964; Bouma and Shepard, 1964). The focal spot size is 1.5 x 1.5 mm. The head hangs in a horizontal tubehead support, which is part of the tubestand (for further information see General Electric, X-Ray Dept., 1961).

At the Geological Institute in Utrecht, the Netherlands, the sedimentological department uses a Philips-Müller Macrotank B. The size of the focal spot is 1 x 1 mm. The voltage range is continuously variable from 30 to 100 kV. The tube current ranges from 0 to 4 mA for all kilovolt values and is also continuously variable (Müller, 1959; Philips, not dated).

The Baltospot 140B from the Balteau Electric Corporation is also a portable unit consisting of two parts. The circuit is 140 kV and 5 mA, self-rectified. The focal spot is 1.5 x 1.5 mm (Balteau, not dated).

The Picker "Hotshot" 110 kV portable X-ray unit is equipped with a beryllium window, which allows emission of "soft" long-wavelength radiation for radiography of thin sections of low-density materials. The geological oceanography section of Texas A&M University successfully applies this unit on board ship as well as at the institute (Appendix I). The focal spot is extremely small, 0.5 mm; the voltage range goes from 10 to 110 kV (without settings); the mA control goes to 3.5 mA (without settings) for continuous work or to 5 mA at a 50% duty cycle 5 min "on," 5 min "off" (Picker, not dated).

3.1C Exposure Factor Properties

Geometric Principles

A radiograph is a shadow picture of an object that has been placed in the path of an X-ray beam, between the anode and the film. The geometric laws of shadow formation are identical for the penetrating radiation and for visible light.

In Fig. 3.4 five examples are given to demonstrate the geometric principles. Figure 3.4A shows that the shadow cast naturally shows some enlargement because the sample slice is often not in direct contact with the film. Since the focal spot is not a point but a small area, the shadow cast is never completely sharp. The shape of the shadow will depend on the angle between X-ray beam and film and/or the angle between sample and film.

Figure 3.4 illustrates that the following conditions should be fulfilled to obtain the sharpest, truest shadow of an object (Kodak, 1957; Kodak Ltd, 1959):

1. The surface of the source of the X-ray beam (focal spot) should be small, which means that it should be as close to a point as possible to avoid unsharpness (compare Figs. 3.4A and B).

2. The source of the X-rays should be as far from the object as is practical in order to obtain a maximum of sharpness (compare Figs. 3.4B and C). In both figures the sizes of the object and the focal spot and the distance between object and film are the same.

3. The object should be close to the film to make the penumbra as small as possible (Fig. 3.4C).

4. The X-ray beam should be directed at right angles to the film (compare Figs. 3.4A and D).

5. The object and the film should be parallel in order to obtain true-shaped results (compare Figs. 3.4A, D, and E).

The geometrical unsharpness is determined by the size of the penumbra (p) whose formation is illustrated in Fig. 3.4C. According to this figure,

$$p : b = f : (a - b) \quad \text{or} \quad p = \frac{f \times b}{a - b}$$

(Kodak Ltd., 1959).

Milliamperage, Time of Exposure, Distance from X-Ray Source to Film

The blackening of a radiographic image depends upon the amount of radiation absorbed by the emulsion of the film. Several factors influence this; for example, the amount of radiation emitted, the amount of radiation falling upon the sample, the proportion of this radiation that passes through the sample, and the intensifying action of screens.

The total amount of radiation emitted by an X-ray tube depends roughly upon the tube current, tube voltage, and time of energizing the tube. A change in milliamperage (tube current) causes a change in the intensity of the radiation emitted. The intensity is approximately proportional to the milliamperage.

Figure 3.5 shows spectral emission curves for an X-ray tube operated at two different tube currents—the higher is twice the lower. This implies that each wavelength is twice as intense in the high milliamperage beam as it is in the low milliamperage beam. However, both beams con-

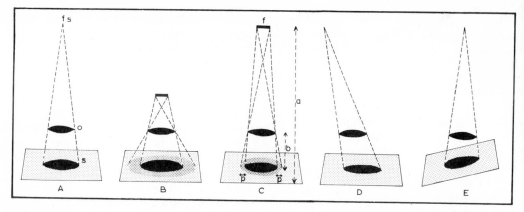

Fig. 3.4. General geometric principles of shadow formation. (A) The focal spot (fs) is a point; the distance between object (o) and film on which the shadow (s) is formed is very small compared to the distance fs-o. (B) The focal spot has a large surface and is near the object. (C) The focal spot has a large surface (f) and is far from the object (a − b > b); p = penumbra. (D) The axis of the X-ray beam is not at right angles to the film surface. Object and film are parallel. (E) Object and film are not parallel. (Redrawn from Kodak, 1957, Fig. 10; Kodak Ltd., 1959.)

tain the same wavelengths. Thus there is no difference in X-ray quality of penetrating power, implying the rule: *The total amount of radiation emitted by an X-ray tube operating at a certain kilovoltage is directly proportional to the milliamperage × the time the tube is energized.* This product is called exposure. It can be given algebraically: $E = Mt$, where E is the exposure, M the tube current, and t the exposure time. Radiographic exposure can be specified in terms of milliampere–minutes or milliampere–seconds, without stating the specific values of tube current and time.

X-rays diverge when they are emitted from the anode and they cover an increasingly larger area with decreasing intensity as they travel from their source (Fig. 3.6). In this example it is assumed that the total energy of the X-ray beam remains constant and that the X-rays passing through the aperture or diaphragm cover an area of 4 cm² on reaching the recording surface C_1 at a distance d from A. When the recording surface is moved another distance d farther from the anode, a total of 2d from A, the X-ray beam will cover a surface of 16 cm². This area is four times as great as that at C_1. The radiation energy per cm² at C_2 is only one-quarter of that per cm² at C_1. The "inverse square law" states this algebraically:

$$I_1 : I_2 = d_2^2 : d_1^2$$

where I_1 and I_2 are the intensities at distance d_1 and d_2, respectively.

Kilovoltage, Absorption, Scattering

The kilovoltage which is applied to an X-ray tube effects not only the intensity but also the composition of the X-ray beam. By raising the tube voltage, X-rays of shorter wavelengths and thus of higher penetrating power are also produced. Figure 3.7 shows spectral emission curves for an X-ray tube operated at two different kilovoltages, but with the same milliamperage. It can be seen that there are some shorter wavelengths in the higher-kilovoltage beam which are absent in the lower-kilovoltage beam. All wavelengths in the lower-kilovoltage beam

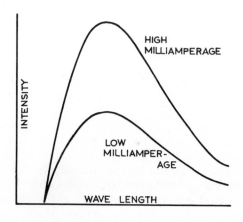

Fig. 3.5. Curves demonstrating the effect of a change in milliamperage on the intensity of an X-ray beam. (Redrawn from Kodak, 1957, Fig. 12.)

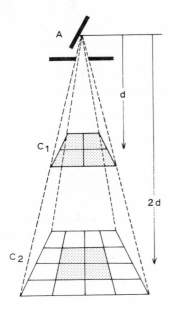

Fig. 3.6. Scheme illustrating the inverse square law. (Redrawn from Kodak, 1957, Fig. 14.)

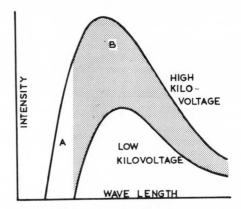

Fig. 3.7. Curves demonstrating the effect of a change in kilovoltage on the composition and intensity of an X-ray beam. (A) Area where shorter wavelengths become added to the X-ray beam by increasing kilovoltage. (B) Area where the intensity of the wavelength increases by increasing kilovoltage. (Redrawn from Kodak, 1957, Fig. 13.)

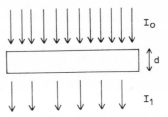

Fig. 3.8. Relation between the ingoing (I_0) and the outgoing (I_1) intensity of the radiation due to absorption. (Redrawn from Eering, 1965.)

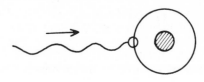

Fig. 3.9. Schematic presentation of the photoelectric effect. (Redrawn from Eering, 1965.)

are present in the more penetrating beam and in a greater amount.

Raising the kilovoltage increases the penetrating power of the X-rays emitted and increases, sometimes to a great extent, their intensity.

The disadvantage of an increase in the tube voltage is a decrease of contrast of the picture obtained.

When X-rays hit an absorber, some of the radiation is absorbed and scattered while the rest passes through. Attenuation of radiation in the object is the difference between the radiation intensity before reaching the object and after passing through this object (Fig. 3.8). This is expressed in the equation:

$$I_1 = I_0 e^{-ud}$$

where e = base of natural logarithms, d = thickness of the object, μ = linear attenuation coefficient, which depends on the atomic number and the density of the material and the wavelength of the radiation.

In working with low kilovoltages ($< 150 \, kV$) only two effects are of importance as far as the absorption is concerned (Eering, 1965):

1. *Photoelectric effect* (Fig. 3.9). A photon hits an electron. In this case it implies total absorption. The effect decreases with increase of penetration power of the radiation. The decrease of this effect is negligible within the reach of our X-ray apparatus.

2. *Compton effect* (Fig. 3.10). A photon hits an electron and becomes partly absorbed, after which it continues in another direction. This

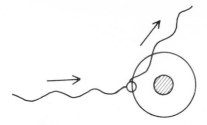

Fig. 3.10. Schematic presentation of the Compton effect. (Redrawn from Eering, 1965.)

effect (*scattering*) may be 20% smaller at 100 kV than it is at 10 kV.

These effects work against each other. The scattered radiation is nonimage forming, but it tends to obscure the contrast of the radiographic image. Another part of the energy in the original X-ray beam is spent in liberating electrons from the absorber. These electrons are unimportant radiographically, but those coming from lead foil screens can be very important.

The absorption of X-rays in a sample depends upon the thickness of the object, its density, and the atomic nature of its elements. The relative absorptions of different materials are not constant but vary with the kilovoltage (Kodak, 1957, Table IV). As the kilovoltage increases the differences between the types of material within the sample tend to decrease.

The investigator should use the lowest possible kilovoltage that is still enough to penetrate the sample well, in order to obtain most details in his radiograph.

Screens, Filters, and Masks

Since less than 1% of the energy of the X-ray beam is used to obtain the photographic image, it may be desirable to apply certain techniques to utilize part of the wasted energy, as long as it does not complicate the technical procedure. Fluorescent and lead screens are the two types that can be used (Fuchs and Corney, 1965).

Fluorescent screens contain fine powdered chemicals (such as calcium tungstate and barium lead sulfate) in a suitable binder coated in a thin, smooth layer on a special supporting cardboard. These chemicals have the ability to absorb X-rays and to emit light immediately. The luminous image is less sharp than the X-ray image and has a certain graininess (Kodak, 1957; Kodak Ltd, 1963a). There is little sense in using them below a voltage of 100 to 120 kV.

Lead foil screens minimize the effect of scattering on the image, because they absorb the soft scattered radiation more than the hard primary radiation. No intensifying effect is observed below a kilovoltage of 120 to 140 kV, although theoretically the amplification works from tube voltages above 88 kV.

The application of a sheet of metal makes it possible to filter heterogeneous radiation to make it harder. The filter is placed either as close as possible to the radiation source or immediately behind the object to be examined. In the first case it is also necessary to mask down the X-ray beam so that the smallest possible part of the filter is hit by the radiation, and it consequently forms a source of scattered radiation.

The effects of filtration are a decrease of the contrast of the image and a decrease of the veil from scattered radiation. The filter not only makes the radiation harder but it also adds a certain thickness to the object, with the result that the percentage of increase in thickness of thinner parts of the object is higher than that of thicker parts. This can be desirable when radiographing cores and irregular-shaped samples.

If the filter is placed directly behind the sample one obtains not only a decrease of the contrast but also a stronger absorption of the scattered radiation coming from the object than of the primary radiation. The filter must be absolutely free from scratches and irregularities since these become recorded in the radiograph and might cause interpretation errors.

If it is desirable to radiograph the edges of irregular objects also, and if it is difficult to use a lead mask, it may be advantageous to place a filter close to the source. Table III.1 shows how much of the original intensity is left after the addition of a filter.

Another method for reducing scattered radiation from external sources is to place a lead antiscatter cone around the window of the X-ray tube. The base of the cone should be such that the radiation beam just covers the object. Pieces of lead with different openings can be made which fit into the antiscatter cone when smaller samples are irradiated.

Some of the cassettes and exposure holders, in which the X-ray film must be placed, have a protective lead foil in the back side. In spite of this it is advisable to place an additional plate of lead underneath the cassette to make sure that scattering initiated by the table, the floor, and other materials does not reach the film.

Table III.1. Percentage of the Original X-Ray Intensity that
Remains after the Addition of a ¼-in. Steel Filter Placed
Near the Tube. The Steel Sample Has Been Radiographed
First without the Filter. In Both Cases the Exposures
Were Made at 180 kV

Region of the Sample	Thickness	Percentage of Original Intensity Remaining after Addition of Filter
Outside specimen	0 in.	under 5%
Thin section	¼ in.	about 30%
Medium section	½ in.	about 40%
Thick section	1 in.	about 55%

After Kodak (1957, p. 32).

X-Ray Exposure Charts

An exposure chart indicates the exposure (mA × time) required for different thicknesses of a given material to obtain a given density on a given type of X-ray film (Gevaert, not dated; Kodak, 1957). The application of exposure charts (see Section 3.5C) saves the operator time, sheets of film, and failures.

Exposure chart sets can be made by utilizing one type of X-ray apparatus, the same type of film, the same kilovoltage, and the same source-film distance. For each chart the variables are the type of material, its thickness, and the exposure.

The establishment of the chart requires a step-wedge, which is made of the material to which the chart relates. The author made step-wedges of sandstone and limestone, each wedge consisting of 10 steps with thickness differences of 3 mm between the steps. Using a given kilovoltage (in our case 50 kV), a Philips-Müller Macrotank B, Kodak Industrial X-ray Film Type AA, a distance between focal point and film of 1 m, one type of Kodak developer and no screens and filters, many exposures with different exposure factors were made until good radiographs were obtained from each step. The results were plotted on an exposure chart graph sheet with a linear horizontal axis referring to the thickness of the sample and a vertical axis with a logarithmic scale indicating the exposure (log exposure or mA·min).

It was found that the curve obtained is a straight line. Its position and inclination vary between different types of materials. All curves given in 3.5C are obtained by making exposures of two small slices of material (3x5 cm), 3 and 9 mm thick, instead of making the time-consuming step-wedges. The two values were plotted and the straight line through both points gives the desired curve.

Thickness, Shape, Type, and Condition of Samples

The advantage of the X-ray radiography technique is that it is not limited to shape, type, condition, or thickness of a sample within certain limits (see also Section 3.5).

Since the radiographic image is the result of the passing of radiation through a sample onto the film, which gives a type of projection in straight lines of all different parts encountered, it is clear that the sample should not be too thick if the investigator is to be able to make correct interpretations. The best way is to work with thin plane-parallel slices, for example, 3 to 12 mm thick; but with special methods radiographs can be made of thick slices, cores and irregularly shaped samples.

It is not of technical importance if a sample consists of quartz, calcite, or other minerals. However, the results obtained from many limestones are poor since variation in absorption between the different grains is often so small that no sedimentary picture is obtained (see Section 3.5C).

It makes no difference to the technique whether the samples are consolidated or unconsolidated, wet or dry, impregnated or not impregnated with plastic, imbedded or not in plastic or plexiglas or in another artificially made product as long as it is not of glass. Glass proved to absorb too much radiation.

3.2 X-RAY FILMS

The radiation that passes an object can be made visible on certain types of screens. However, it is better to use special films on which the image is permanently recorded and no danger exists that the investigator receives radiation while observing the image on a screen.

3.2A Composition

An X-ray film normally consists of seven layers (Gevaert, not dated). They are symmetrical in upbuilding. The fact that they have two sensitive layers makes the film twice as fast and twice as contrasty. A single emulsion layer of double thickness would require longer processing and drying and would permit the use of one lead screen or intensifying screen only.

The seven layers of the X-ray film are:

1. The support, which carries the other layers. Originally it was a cellulose acetate or tri-acetate (1x). Polyester supports are now being used.

2. A substratum to ensure that the emulsion coating adheres properly to the support (2x).

3. The emulsion or sensitive layer with a thickness of about 0.001 in. The emulsion consists of a special kind of gelatin containing countless infinitesimal crystals of silver halide (2x).

4. A protective layer of special hardened gelatin (2x).

3.2B X-Ray Film Characteristics
Transparency

$$\text{Transparency} = \frac{\text{intensity of transmitted light}}{\text{intensity of incident light}}$$

Opacity.

The degree of local opacity of a developed film depends primarily on the amount of metallic silver found locally in the emulsion coating. The opacity is the reciprocal of the transparency.

$$\text{Opacity} = \frac{\text{intensity of incident light}}{\text{intensity of transmitted light}}$$

Sensitometry

Sensitometry indicates the "relation between exposure and processing conditions on the one hand and film response on the other" (Corney, 1963). These characteristic or sensitometric or H and D curves (after Hurter and Driffield who first used the term in 1890) represent the relation between the logarithm of the relative exposure (absissa), which is independent of the kilovoltage, and the density (ordinate) (Fig. 3.12).

Relative exposure is used since there are no convenient units, suitable to all kilovoltages and scattering conditions, in which radiographic exposures can be expressed.

The curves can be used as the basis of calculations to increase the density from a measured density to a desired density, or to find the multiplying factor between two different films of which a similar density is desired, or to find the density one gets when dealing with a thinner or a thicker part of the sample when thickness and density of that part are known (for further information see Gevaert, not dated; Kodak, 1957).

Contrast

The contrast observed in a film increases if the difference in blackness of two points increases. From Fig. 3.11 it can be seen that if the difference between I_1 and I_2 is as great as possible, the inclusion is distinguished best.

$$I_1 - I_2 = \Delta I; \qquad \Delta I = I_0 (e^{-\mu d} - e^{-\mu_1 d_1})$$

(The material above and below the inclusion is neglected.)

I_o: Intensity of the radiation just before it enters the sample.

I_1: Intensity of the radiation after passing the object.

I_2: Intensity of the radiation after passing the inclusion.

d: Thickness of the object.

d_1: Thickness of the inclusion.

e: Base of natural logarithm.

μ: Linear absorption coefficient of the object.

μ_1: Linear absorption coefficient of the inclusion.

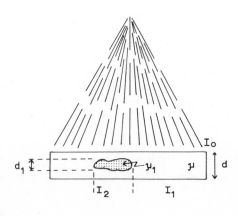

Fig. 3.11. Schematic drawing of the contrast idea. For key see text. (Redrawn from Eering, 1965.)

If the wavelength λ is high the factor ΔI will be high. It is therefore desirable to keep the kilovoltage low in order to get maximum contrast.

Negative Duplicates

For a number of X-ray films the characteristic curves do not continue with a steep slope but start to flatten and even turn down (Fig. 3.12). This characteristic of the curve can be used for making duplicates in negative shape. Some films are available which have already been exposed. If this is not the case one can continue the exposure (for contact printing) to go through the positive stage into the negative stage. However, many films are not suitable for this type of overexposure.

Sharpness of the Image

It is the intention of the investigator to get as much information as possible from a radiograph. The limits of observation depend upon:

1. Size of the inclusion within the object.
2. Contrast of the inclusion image.
3. Sharpness of the inclusion. Two types of blurring are known:

    (a) External unsharpness, which is due to the fact that the focal point has a certain surface. Figure 3.13 shows that x decreases in size by increasing F and decreasing F¹ for a certain focus surface f. Decrease of d is also favorable, and another important point is whether the inclusion is situated high or low in the object.

    (b) Grain size of the film. The silver halide grains in the emulsion (Kodak) are less than 1μ in diameter. The graininess, that is, the visual impression of grains of the silver deposit, observed in the processed radiograph (normal X-ray exposure) is due to random grouping of these developed grains. This, in turn, is due to statistical variation of the number of absorbed radiation quota from one place to another.

Influence of Developing Time on Contrast and Speed

For the sharpness of the image (internal unsharpness) it is of considerable importance to follow the rules given by the manufacturers with regard to developing the exposed film in order not to increase the graininess of the film (Fig. 3.14). In this figure it can be observed that when the developing time increases the sensitometric

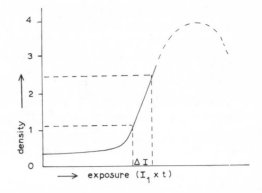

Fig. 3.12. Schematic drawing of a density curve showing the down-turning effect. (Redrawn from Eering, 1965.)

curve becomes progressively steeper, thus the contrast increases. The curves move to the left, which implies that smaller quantities of exposure are required for obtaining the same density and consequently the effective speed and the grain size of a film increase with increasing developing time.

However, the developer acts not only on the exposed silver halide crystals, but the nonexposed crystals also undergo a reduction, although far more slowly. The faint degree of gen-

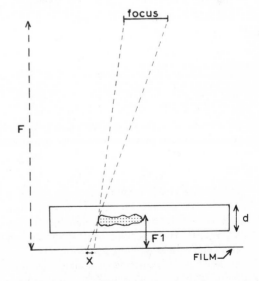

Fig. 3.13. Schematic presentation of the geometric unsharpness (x). (Redrawn from Eering, 1965.)

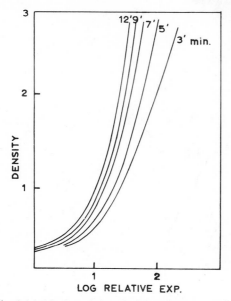

Fig. 3.14. Moving of the density curve of an X-ray film to the left due to increase of developing time. (Redrawn from Gevaert, not dated; Kodak, 1957.)

eral blackening is called "chemical fog." This fog increases as the developing time is lengthened and consequently the contrast decreases.

3.2C Types of Film

X-ray films are available in different sizes, mostly 25, 75, or 100 sheets in a pack. The packing methods of manufacturers differ slightly, but generally the investigator can choose among three types of packing:

1. The X-ray film sheets are placed in a box without protective sheets of paper in between — noninterleaved sheets.
2. The X-ray film sheets are each put in a folded sheet of paper before they are placed in a box — interleaved, or folder-wrapped sheets.
3. The X-ray film is enclosed in a folder of black interleaving paper sealed in a light-tight envelope — ready pack or envelope-packed sheets.

For these films it is not necessary to use a darkroom for placing the film in an exposure holder or a cassette, since the envelope serves this purpose. The envelope-packed sheets can be provided with or without lead protection.

Several film types are also available in rolls of different widths and lengths. These rolls sometimes are envelope-packed and are used in industry for outside work.

From normal photography it is known that there are differences in graininess of the films. The lower the graininess, the slower the film speed and the greater the enlargement can be. Fine-grained films give finer image details than coarse-grained ones. The same applies to X-ray films. Table III.2 shows a number of types from different manufacturers and their approximate correction factor compared to Structurix D10, D7, and S films from Gevaert. The films were processed according to the specifications of the manufacturers. These correction factors are relative since progress and improvements in the production of films change their speeds. The correction factor is a factor which must be applied to correct the times shown on the exposure charts (see also Appendix Two).

Some Eastman Kodak (U.S.A.) X-Ray Films. The author primarily has used Kodak Industrial X-Ray Film, Type AA. It has a fine grain and a high contrast. It is designed for the radiography of light metals at lower voltages and for 1000 kV and higher radiography of heavier steel parts. It can be used directly or with lead screens (Kodak, 1957, 1961, 1966).

Kodak No-Screen Medical X-Ray Film has nearly the same speed as Type KK and it has a "high" contrast. The author started with this type of film (see also Hamblin, 1962; Calvert and Veevers, 1962) but soon changed to Type AA, which is much more suitable for this type of work. In Fig. 3.15 the difference between the types No-Screen and Type AA can be seen (Bouma, 1963, 1964e). However, the details of the *Spartina* stem come out slightly more distinctly on No-Screen than they do on Type AA. Especially when the investigator needs the maximum detail from organic rests, such as plants, it might be worthwhile to try also the No-Screen type.

The relative speeds of the types M, AA, and No-Screen are 40, 150, and 950, respectively.

Kodak Ltd (Great Britain) X-Ray Films. Kodak Ltd. Kodirex X-ray film is a very high-speed medical and industrial X-ray film for use without intensifying screens. It has a high contrast (Kodak Ltd., 1964b, 1967).

Kodak Ltd. Industrex X-ray film type D is recommended for radiography without salt screens and for radiography with lead screens. Its speed is lower than the one of Kodirex by a factor 2. Its contrast is high (Kodak Ltd., 1964c, 1967).

Kodak Ltd. Industrex X-ray film type S is a fast film designed for use with salt intensifying screens of the calcium tungstate or zinc sulfide

Table III.2. The Exposure Factors of Several Films Compared to Structurix D 10, D 7, and S from Gevaert. These Data Were Calculated in 1964 and Should be Considered Approximate Values

		D10	D7	S			D10	D7	S
Gevaert	D 10	1	–	–	Kodak	AA	4	1	–
(Belgium)	D 7	4	1	–	(U.S.A.)	M	18	4.5	–
	D 4	15	4	–		F	–	–	–
	D 2	60	16	–		KK	0.6	0.15	–
	S	–	–	1					
Ilford	G	0.6	–	–	Kodak	Kodirex	0.8	–	–
(Great	B	1.8	–	–	(Great	Industrex D	1.8	–	–
Britain)	CX	4	1	–	Britain)	Cristallex	4	1	–
	F	12.5	3.1	–		Microtex	16	4	–
	A	–	–	0.8		Industrex S	–	–	2
Ferrania	ID	0.8	–	–	Kodak	Kodirex	1.4	–	–
(Italy)	IC	3	0.8	–	(France)	Definix	4.5	1.1	–
	Gamma	6	1.5	–		M	17	4.2	–
						Regulix	–	–	0.8
Dupont	504	4	1	–	Siderix	R	1.2	–	–
(U.S.A.)	506	5.5	1.4	–	(France)	N	3.4	0.9	–
	510	10	2.5	–		F	14	3.5	–
					Ansco	C	1	–	–
					(U.S.A.)	A	3	0.8	–
						B	8	2	–

(After Balteau, 1964).

type. When medium contrast is required this type of film should be used with lead screens (Kodak Ltd, 1963d).

Kodak Ltd. Crystallex X-ray film is a fine-grained film designed for work in which grain size must be at a minimum with a reasonable speed. It should not be used with salt intensifying screens, but it can be applied with lead screens. Its contrast is about equal to that of Industrex film type D (Kodak Ltd, 1962a, 1967). The Crystallex can be used to replace the U.S.A. Kodak film type AA.

Kodak Ltd. Microtex X-ray film is very fine-grained and has a high contrast (Kodak Ltd, 1964d, 1967). It does not differ too much from Eastman Kodak film type M.

The relative speeds of the types Kodirex, Industrex type D, Industrex type S, Crystallex and Microtex are 400, 200, 130, 100, and 25, respectively.

Gevaert-Agfa (Belgium) X-Ray Films. Gevaert Structurix D2 X-ray film is very fine-grained and is designed for the highest radiographic sensitivity and the finest detail. Its speed is low (Table III.2). It should be applied without salt screens; however, lead screens can be used (Gevaert, not dated, 1964; Agfa-Gevaert, 1966).

Gevaert Structurix D4 X-ray film has an average speed and a high contrast. It can be applied with lead screens, but salt screens should not be used.

Gevaert Structurix D7 X-ray film has a rather high speed and a high contrast. It should be used without salt screens, but it can be combined with lead screens. It is designed for thicker sections than can be radiographed with D4. According to Table III.2 it should be more or less equivalent to the Kodak types AA and Crystallex.

Gevaert Structurix D10 X-ray film has the highest speed of the series if it is used with calcium tungstate intensifying screens. Its contrast is high when such screens are applied (Gevaert, not dated, 1964).

Ilford (Great Britain) X-Ray Films. Ilford Industrial B X-ray film is a fast, high-contrast film for use with or without lead screens. Do not use it with salt screens. This is the general-purpose film for radiography (Ilford, not dated, 1958a).

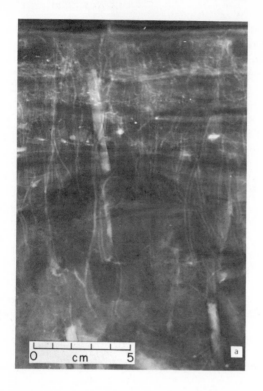

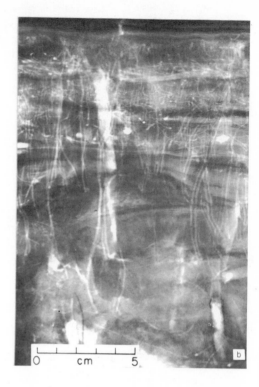

Fig. 3.15. Prints of radiographs made from clayey marsh deposits. For all three photographs the same sample slice has been used, only for (c) the upper ¾ has been removed. The more or less vertical running double lines are roots of *Salicornia*, the straight, thick, nearly vertical bands are stems of *Spartina*, and the more or less horizontal lenses are concretions of iron hydroxide. Mission Bay, San Diego, California, U.S.A. (a) Medical film, 90 kV, 15 mA, 2 sec (Kodak film). (b) Industrial film AA, 40 kV, 5 mA, 160 sec (Kodak film). (c) Industrial film AA, 40 kV, 5 mA, 45 sec (Kodak film). Distance source-film: 36 in. Slice thickness 20, 20, and 5 mm, respectively. (From Bouma, 1963, 1964e.)

Ilford Industrial G X-ray film is the fastest film of the Ilford industrial X-ray films. It is about twice as fast as type B. Its contrast is high. It is designed for use without salt screens, with or without lead screens. This film is recommended for thick objects (Ilford, not dated, 1961a).

Ilford Industrial CX-ray film has a medium speed, fine grain, and high contrast for use without salt screens, with or without lead screens. It has a much finer grain than type B and about half the speed. According to Table III.2 this CX film must be more or less equivalent to Kodak type AA, Kodak Limited Crystallex and to Gevaert type D7 (Ilford, 1959).

Ilford Industrial F X-ray film is a very fine-grained film with a high contrast and a speed of about half that of type C. It can be used with or without lead screens, but not with salt screens. Its high contrast renders it a film which gives the very best possible definition in radiographs (Ilford, not dated, 1958b).

Ilford Industrial A X-ray film is an extremely fast, medium-contrast film, primarily intended for use with salt intensifying screens. When applied without salt screens, or with lead screens, its speed is the same as type F, but its contrast is lower (Ilford, not dated, 1958c).

3.2D Storage and Handling

Storage of the X-Ray Film

X-ray films like other sensitized materials, slowly deteriorate with age. The first sign of such lower quality is visible as a higher chemical fog. Manufacturers who produce films give a number of rules that should be followed to avoid loss of film due to this deterioration (Gevaert, 1964; Ilford, not dated; Kodak Ltd, 1965a).

1. The stock of film should be kept at a minimum.

2. All boxes in stock should be opened in order of date of obtaining so the oldest films are used first.

3. The films should be stored in a cool place, if possible with a temperature below 10°C (50°F) if it is for a longer period (over 6 months) and at a maximum temperature of 20°C (68°F) if it is for a short storage. If the films are stored in a refrigerator, as may be the case in tropical areas, one should take the film box out of the refrigerator at least a few hours prior to use since the sudden change in temperature causes absorption of water from the air onto the films.

4. The storage place should have a relative humidity, within the range of 40 to 60%, since humidity has a bad influence on films. This is especially important if the hermetically sealed package of the box has been opened.

5. To avoid possible pressure marks on the films due to folds in the packing material, the boxes should be stored standing on edge and not stacked horizontally one on top of another.

6. The storage place should be free from any possible exposure to harmful gases such as formaldehyde, hydrogen sulfide, sulfur dioxide, ammonia, coal gas, industrial gases, motor exhausts, mercury vapor, and vapors of solvents and cleaners.

7. The unexposed films should be stored far away from any radiation source.

When the low kilovoltages are dealt with, as is done in this field, these precautions will not be difficult to adhere to.

Storage of the Radiographs

When the processing of the exposed films is carried out properly, including rinsing and drying, the radiographs may be considered permanent. All films are manufactured on a safety base, which means that their fire sensibility is not greater than that of normal paper (Kodak, Ltd, 1965a). A number of rules can be given for storage:

1. Radiographs should be stored in a rather cool, dry, and ventilated room. Ideal conditions are a temperature between 15 and 27°C (60 to 80°F) and a relative humidity of 40 to 50%.

2. Radiographs should be stored in a room which is free from harmful gases such as formaldehyde, hydrogen sulfide, sulfur dioxyde, ammonia, coal gas, industrial gases, exhaust gases, mercury vapor, and vapors of solvents and cleaners. A large filing cabinet should have ventilation with a filter for the incoming air.

3. Radiographs should not be stacked on each other, but the covers from the interleaved film sheets or manilla envelopes should be used as cover as soon as the radiographs are really dry.

4. On the covers the number of the radiograph and, if desired, other data should be written in order to avoid removing the radiograph from its cover too often.

5. The author uses a continuous numbering system, in spite of the type or series of investigation and the operator who carried out the radiographing (see Section 3.3).

6. Before putting the dried radiograph in its cover, the corners with the pinholes made by the developing hangers should be cut off to avoid scratching.

Handling of X-Ray Films

Since the operator will find his task of interpreting the radiograph difficult enough, he will not want to find spurious marks (artifacts) due to improper handling or processing (see also Section 3.8). Therefore never remove a film rapidly from cartons, exposure holders, etc.

X-ray films are sensitive to pressure, creasing, and kinking. Friction may produce an electric discharge giving rise to treelike marks known as "static" (Ilford, not dated).

All operations should be carried out properly, methodically, and without haste. A little extra time and care pays well because of the improved quality of the films.

Never touch the body of the film by hand; always take hold of it by the edges. This is important to remember during loading exposure holders and fastening them on a developing hanger.

3.3 EXPOSURE

Exposure means the passing of radiation through a sample onto a special film in order to create a latent image in the sensitive layers of that film. This discussion is confined to normal contact radiography (sample in close contact with the exposure holder) of plane-parallel slices of sediment. Other shapes and types of samples and special techniques are treated elsewhere.

The time the sample is exposed to radiation is the only period of the entire radiography technique during which the "dangerous" X-rays exist.

3.3A Exposure Room

The X-ray head should be placed in the exposure room, while the control panel should be positioned outside this room. Since one is dealing not only with the conical shaped X-ray beam but also with scattering of X-rays, caused by the object, table, floor, and walls hit by the beam, this room should have a lining of lead (see discussion of Protection). It is also possible to make use of the basement of building whose walls have such thickness that indirect radiation does not pass.

Besides the X-ray apparatus, with or without a special support, a table for the samples to be radiographed and a box with lead letters and numbers should also be in the exposure room.

The door opening should have a protecting lead screen at the side of the X-ray tube. This door will lead to another room where the control panel is placed and where the processing can be carried out. Anyone wishing to construct a small, useful, setup for this type of work may consult Fig. 3.19, which gives a plan for exposure-processing rooms. The door between both rooms has a mechanically operated safety switch which allows working of the X-ray tube only when the door is closed. A piece of lead-glass is placed in the wall between both rooms. This can be very important in spite of dead objects. It sometimes occurs that the operator is not completely sure whether he has carried out all the necessary manipulations. Instead of switching off the unit, so that he can go back to the exposure room, often he can observe enough through the lead-glass window.

In order to prevent disturbance at the wrong time, signal lights are placed above the outside door. One of the red lights goes on as soon as the X-ray tube is activated. Another can be switched on by the operator as an indication that he is processing radiographs. A third light indicates that someone is working in the normal darkroom which is adjacent to the processing room and which can only be entered through the light-lock built in the entrance of the processing room.

National, state, and local laws on protection and radiation work should be observed carefully, even when examples of setups given in this book are followed.

Mounting of the X-Ray Head

The author has constructed a special support for greater diversification in application (Fig. 3.16). The X-ray head can be placed at the base of the support directing its beam upward. For stereoradiography and parallel exposure techniques the head can move over a distance of 12 cm from the front to the back and the head can turn over an angle of maximum 10°.

On the base of the support two round stands are mounted, along which a horizontal table-top frame can be moved vertically. On top of this frame a plexiglas plate is placed on which the samples to be radiated must be laid. The vertical movement of the table is necessary in order to change the distance between focal spot and film.

Loading of the Film

The loading of the X-ray film is a simple procedure, which requires only a little practice to be carried out easily in the darkroom. The loading should be done near the cupboard or lead box where the film is stored. The "dry" table (see discussion of Loading Area) should only be used for this work to prevent any contact of developing liquids with the unexposed film.

There are a number of ways to place a cover around the X-ray film to protect the film against daylight and lamplight.

1. Ready-packed or envelope-packed films can be used directly since the envelope serves as light protector.

2. Cassettes in which intensifying or lead foil screens can be placed may be used. When screens are used, the film sheet must be placed in the cassette without the interleaving paper. The best way to load the cassette is to open it when it lies on the "dry" table. While using proper illumination only, a film sheet still in the interleaving paper, is carefully drawn out of the box. Holding the sheet between thumb and finger, the operator pulls apart one side of the paper and puts the film in the cassette while holding the remaining paper in his hand. In this way the film is not touched (Kodak, 1957, Fig. 47; Ilford, not dated, Figs. 4, 5).

3. X-ray exposure holders contain a thin sheet of lead at the back to minimize the effects of back scatter. They are very convenient and easy to load in the dark. The exposure holder has a front and back of sturdy paperboard or plastic. A duplex paper (or plastic) envelope attached inside folds around the film to exclude light. The film is placed in the open holder. Next, the large flap of the envolope is turned down so that it is flat on the film, and then the two long side flaps, followed by the bottom flap, are folded over the large flap. The cover is now turned over it (Kodak, 1957, Fig. 48) (Fig. 3.17).

4. Rubber or plastic envelopes consist of two covers, each closed on three sides, which fit over each other, thus protecting the film against light. They are made of rubber or plastic and are flexible. The operator must therefore be very careful when using them, being sure that the film does not bend, thus preventing "artifacts." It is advantageous to work with different sizes in order to use the smallest amount of film.

Placing of the Sample

When the exposure holder is placed on the table of the X-ray head support with the X-ray beam directed downward, one should be careful that the right side of the holder is facing the X-ray tube. The sample slice(s) are now placed directly on the holder. A sheet of lead (1 mm) is placed underneath the holder to be sure that no back scattering hits the film. If only part of the exposure holder is covered by the sample, for instance, when two or more exposures are to be made of the same sample on the same film but with different exposure times, the rest of the exposure holder is covered with a sheet (1 mm) of lead. Also, the sample is framed by lead (see Fig. 3.18).

The part to be radiated is now centered on the support table and the sample is ready for the exposure. It is advisable to use a numbering system before the exposure is made (as discussed later).

If the X-ray beam is directed upward, first the sample is placed on the table and numbers and letters are laid upside down. The whole sample is now surrounded by lead, and next the exposure holder (front side toward the tube) is placed over it and covered by lead.

Numbering

Since the operator normally makes more than one exposure in a series, it becomes difficult to remember later all data concerning type of film, number of sample, kV used, exposure time, etc. It is therefore advisable to number the films immediately before the exposure.

To accomplish this one uses a set of lead letters and numbers, available from dealers of films and X-ray apparatus. The author mounted the lead letters and numbers between two pieces of tape (MMM masking tape No. 202) to grip them more easily. The corresponding number or letter is written on the tape (upper side). A special case holds the complete set (six of each) and a number of arrows made of lead. These arrows are very useful when one is dealing with oriented samples.

If the operator wants to have more indications on the film, he can compose complete data in lead next to the samples (Fig. 3.18).

Comments on Exposure Data

The author has found it very convenient to use only one system of numbering when a small group works with the equipment. In most cases it will take a long time before the numbers become too long and thus impractical to compose.

For every exposure a sequential number is used even if more than one exposure is made on one film. This is necessary particularly for the

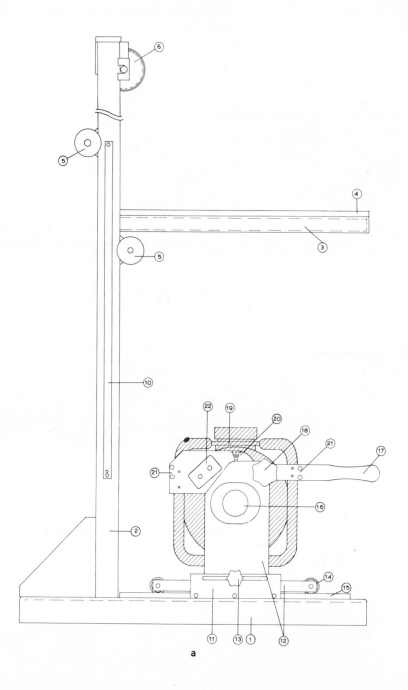

a

Fig. 3.16. Simplified drawing of the X-ray head support in which the X-ray head can move horizontally and turn over small angles about a horizontal axis. The table with plexiglas top can be moved up and down: (1) base plate; (2) round stands; (3) table supports; (4) plexiglas top; (5) guiding wheels; (6) end disk of table moving part; (7) wire between table moving part and table support; (8) brake block; (9) handle; (10) scale, zero point at the height of the focal spot; (11) sled fastening plate; (12) sled to which the tank unit is mounted; (13) knob for

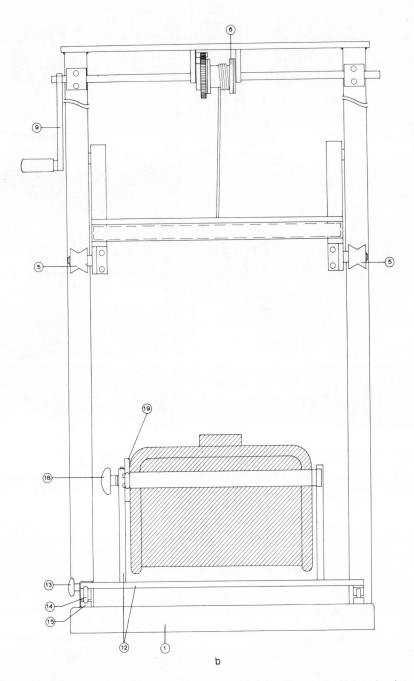

b

securing position of the sled; (14) sled wheel; (15) sled wheel rail; (16) axis around which tank unit can turn; (17) handgrip for turning tank unit about horizontal axis; (18) fastening knob; (19) frame attached to tank unit with screws (21). Handgrip (17) forms part of this frame. The frame hangs in the sled (12) at the axis (16); (20) scale for amount of horizontal turning; (21) screw; (22) window in the frame for the two connections of the cooling circuit. (Constructed from author's design at the workshop of the University of Utrecht, The Netherlands.)

Fig. 3.17. Method of loading a Kodak exposure holder. (1) The exposure holder is opened, a sheet of film is removed from the box, the cover is removed from the film, and the film is inserted carefully in the exposure holder. (2) The large flap is closed. (3) The side flaps and then the small end flap are closed. (4) The large outer covering flap of the holder is closed and the three open sides are taped to prevent the holder from falling open.

comments on exposure data. When more than one sample is irradiated at the same time we often give, besides the number of the film, short indications on the upper side of each sample. In cases of only one sample the film number is always placed at the top side of the sample.

The comments on the exposure data are written in a simple exercise book in which columns are drawn. The following headings can be used:

1. Number. This means the film number.
2. Date. Date of exposure.
3. Operator. Initials of the operator. This facilitates answering any questions that may arise later.
4. Investigator. The name of the investigator.

This is practical since the one who made the sample does not always carry out all laboratory work.

5. Sample data. In this column necessary data can be written to enable rapid location of the sample. Also the abbreviations in lead can be explained here.

6. Orientation. The direction in degrees in which the lead arrow points.

7. Film. Type of film and size can be indicated here.

8. kV. The kilovoltage used.

9. mA. The milliamperage used.

10. Time. Time in minutes and seconds of the exposure. This can be of importance, together

with columns 8 and 9, if something happens with the X-ray tube.

11. Remarks. This can be used for the type of sample, its thickness, type of radiography technique (slice, core, irregular samples, enlarged exposure, stereoradiography, etc.). It often serves as a reserve for column 5. Also a note can be given in this column if the exposure was wrong and the reason for it.

This comment on exposure data often proves useful when questions arise or when the operator remembers he had similar samples some time ago. By consulting those data he knows at once the best exposure he should use to get good results.

Exposure

When the exposure holder with lead underneath and with the sample on it surrounded by numbers, letters, arrows, and sheets of lead are all in the right position, the operator must be sure that no other exposure holders with sensitive materials are left in the exposure room. As soon as all has been found to be in order the real exposure can be made. The exposure must be carried out according to the data written in the exercise book.

Before the next exposure is made the first one must be removed. The same precautions should be followed. It was found that only a few days are necessary to pick up the real routine. On the other hand, there is no use at all if the handlings can be carried out more rapidly, since it is necessary for most X-ray apparatus that the period of rest, and consequently of cooling, must be at least as long as the exposure time. This is to prevent overheating of the tube when no water cooling is applied.

3.4 PROCESSING

The size, arrangement, and equipment of the darkroom will depend primarily on the frequency of use, the amount of sensitive material to be processed within one series, the number of operators working in it, the total floor space available, and its location in relation to other rooms. We consider here only a small processing room in which one or two operators can work and of which the equipment is little more than a minimum requirement.

For a large setup the investigator should read the data sheets mentioned in this discussion or he should request the booklet "Professional

Darkroom Design" from Kodak Ltd (Kodak Ltd., 1962d). The best way is to consult special darkroom building manufacturers such as Donka, who know all safety regulations and local by-laws and who can get the most economic use out of a special situation (for shipboard operations see Appendix I).

The author has experience with his Kodak equipment only, although his darkroom design is based on excursions, discussions with specialists, and the consulting of literature. A number of requirements should be fulfilled if a certain space is converted into a darkroom (Kodak Ltd., 1963b):

1. The darkroom must be completely light-proof with no trace of stray light coming in under doors, through window crevices, or through light-locks.

Fig. 3.18. Example of a complete exposure setup. Underneath the exposure holder a protective sheet of lead is placed. On top of the exposure holder the sediment slices are placed together with some indications in lead letters and numbers. The arrow points to the top of the samples. The number in the upper right corner (128) indicates the film number. The samples are surrounded by thin lead foil screens placed in black paper to prevent scattered radiation from reaching the film.

2. The darkroom must be located well away from X-ray apparatus or other radioactive sources, or it must be properly screened therefrom.

3. The ventilation must be sufficient.

4. Drainage, running water (preferably hot as well as cold), and means of heating must be provided.

5. The darkroom must be dry, easy to clean, not located at the side of the midday sun, which may increase the temperature too much, and it should be away from rooms and other buildings where harmful gases are produced. For normal processing the temperature should be such that the baths are not warmer than 21°C (70°F). In warm climates an air conditioner is necessary.

6. Electricity must be present in order to have enough white lights and connection points for special lights, heaters, etc. All electrical plugs should be grounded.

7. The arrangement of the equipment should be in two groups — "dry" and "wet" parts — and it should be set up so that one can work efficiently. Easy access to plumbing and pipe work is recommended.

8. Access to the darkroom should be possible from a corridor and not from another room. The entrance should be built as a light-lock.

9. No darkroom should be smaller than 2.50 x 1.80 m (8 x 6 ft).

The entry should be constructed as a light-lock. There are several ways to make one, but it is preferable to make it with two doors and not with curtains. Curtains may not always close enough and they are dust collectors.

Adequate ventilation is very important. This can be achieved with a mechanical device. A nonmechanical ventilation unit may be insufficient to obtain enough ventilation (Kodak Ltd., 1962d, 1963b). The inner sides of all ventilations pipes, openings, and ventilators should be matt-black to ensure that they act as efficient light-traps.

3.4A Darkroom Planning

As stated earlier, the darkroom should have a "dry" and a "wet" part, well separated. The areas with their equipment will be described first; the handlings are discussed separately.

The dry area, in our design, consists of two parts: an exposure control part and a loading part with storage facilities. The wet part also consists of two parts: a processing part and a drying part (Fig. 3.19).

The darkroom should have one or more central white lights, all operated by one switch system. The best lighting systems are the indirect ones since they give an even illumination and less shadow effects than direct lights.

Exposure Control Part

The control panel of the X-ray unit is placed at one side of the dry part of the darkroom. This is very easy when a tank model is used. When the cover is removed all switches and meters are directly accessible.

Above the control panel instruction sheets are hanging, one with the operation instructions of the X-ray apparatus, and one with the general safety instructions (Table III.3) based on general protection rules and the local by-laws. The exercise book for the comments on the exposure data is in front of the control panel box or in a little drawer underneath. Above the panel a white-light lamp with switch can be mounted. The lead-glass window is not placed here but more off to the side since the control part is placed as far as possible from the X-ray table.

Underneath the table part for the control panel a lead shelter can be planned (two vertical walls and one top part). Each sheet of lead is glued between two pieces of fibrous wood. The top part rests inside the vertical parts on wooden strips to ensure a radiation-tight unit. This shelter is an additional protection against radiation for the X-ray films since the source is nearby.

Loading Area

This part of the dry bench is used for loading and unloading of films, for fastening them on a developing hanger, and for cutting off the corners of the dried films. As cover for the bench, melamine plastic materials such as Formica are recommended.

Against the wall, processing hanger brackets, one pair for every size developing hanger, are mounted as storage. Also, a safelight is placed high. The distance from safelight to bench must be at least 1.20 m (4 ft). At shorter distances it should not be directed straight down.

On one side an open cupboard with shelves is placed as storage for exposure holders and cassettes. Each type and size should have its own shelf. Underneath the bench should be drawers for the storage of different items (scissors, cleaning brushes). Underneath this drawer a waste basket can be placed. Another drawer can be used for the temporary storage of the interleaving paper covers during the processing period,

Table III.3. General Safety Instructions for Persons Working with Ionizing Rays

1. Follow exactly all indications given for your own protection.
2. Carry your laboratory coat and your own film badge when you are doing radiographic work.
3. Only persons appointed by the head of the institute are allowed to work with X-ray apparatus. They have to take care that nobody is near the X-ray source when his presence is not absolutely necessary.
4. The control panel is locked. Only the signers have a key. After finishing the work the operator should lock the control panel and return the key.
5. Be sure that the set-up for taking a radiograph is right and that no persons are present in the X-ray room before sliding door is closed, and consequently the electrical circuit is closed.
6. Realize that each object hit by X-rays—including the human body—scatters radiation to all sides. Never expose yourself nor your hands to the direct radiation beam.
7. All necessary data must be noted in the comment on exposure data after each exposure.
8. Report on your radiographic work every day to one of the signers.
9. Take care that you store your film badge in a cool, radiation-free room.
10. Remind the head of the institute (department) to replace the film badges regularly and to ask for a periodical medical check-up.
11. Report all irregularities to the signers, and the medical center must be consulted at once if there is a slight chance that one might have been exposed to radiation.

and a third drawer, placed near exposure holder cupboard, for a temporary storage of a box of unprocessed films. This drawer should be light-proof.

It is practical to place two differently shaped light switches above this bench, one for the safelight and one for the central white light(s).

Processing Area (Fig. 3.19)

The wet part of the darkroom should be kept as dry and clean as possible to avoid the formation of artifacts on the films. The equipment used for processing can be obtained from the manufacturer or from special companies, but they can be made also in one's own workshop (for materials see Kodak Ltd, 1964e).

Hard rubber tanks, manufactured by Kodak (France), with a capacity of 10 liters, can be used as developing tanks. However, the washing tanks should be 20 liters. A set of four small tanks and one large tank (developing, rinsing, fixing, washing, photo-flo), with pieces of wood between to prevent the upper sides from touching, forms a good unit. With two iron bands all tanks can be mounted into one set (see also Appendix One).

A small wet table with Formica cover and a small raised front side is placed at normal height (standing height) next to the sink and can drain into the sink. Mixing and measuring of developers and fixers can be done at this table. Underneath this table a low rack is placed for storing the bottles of liquids.

Next to this wet table a cupboard or rack is placed for the storage of small-sized processing equipment and for bottles and boxes with undiluted chemicals. Funnels, reserve thermometers, measuring cylinders, and stirring rods should be placed on a shelf that is covered with a glass plate or with Formica. The chemicals also have their own shelves.

Since the sink is placed near a corner of the room the walls can be used for several purposes. Just left of the developing tank a small shelf is mounted on which the developing timer is set. Next to the timer an electrical heating rod is placed.

A little more to the right two differently shaped safety switches are mounted, one for the central white light(s) and one for the safelight. A small illuminator, with an opalin glass window of 10-cm width and 48-cm height, can be placed next to the developing tanks. An electrical outlet is placed next to the illuminator, with a switch at the side of this light box. The illuminator is very practical since the films, having just passed the fixing bath and washed for a minute, can be checked against this light to see if they are right or must be redone.

Underneath the illuminator three watertaps are mounted. One tap is for normal use, to the second one a hose is connected permanently for the water supply into the washing tank and for cleaning the tanks, while the third one is a mixing tap for cold and hot water. Two thermometers and two tank stirrers are hung at the wall, one for the developing tank and one for the mixing tank. Also, a few long brushes and a few plastic tubes with different diameters for emptying and cleaning the tanks and two instruction

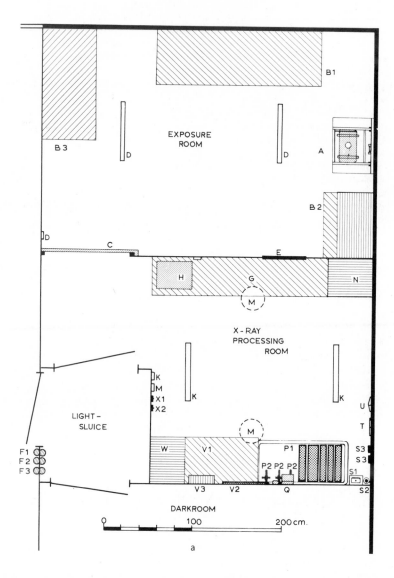

Fig. 3.19. Detailed layout of the exposure and the processing rooms. *Exposure room*: (A) X-ray table with X-ray head; (B 1) table for samples; (B 2) table with box with lead letters and numbers; (B 3) table for samples; (C) sliding door with safety switch; (D) white lights and switch; (E) lead glass. *Processing room*: (F) light sluice with signal lamps [(1) red light starts burning when X-ray tube is operating; (2) red lamp, indicating the operator works in the processing room; (3) red lamp indicating the operator works in the dark room]; (G) table; (H) control panel X-ray unit with plug connection; (I) instruction sheets; (J) drawers; (K) white light and switch; (L) lead shield for placing unexposed X-ray films; (M) safe light and switch; (N) cupboard; (O) developing hangers; (P 1) sink; (P 2) water taps; (P 3) soap holder; (P 4) brushes for cleaning the developing tanks which are placed in the sink; (Q) illuminator; (R) plug connection; (S 1) timer; (S 2) heating rod; (S 3) tank stirrers; (S 4) thermometers; (T) instruction sheet; (U) towel; (V 1) wet table; (V 2) draining rack for processed films still hanging in the developing hangers; (V 3) shelf; (W) cupboard; (X 1) room thermometer; (X 2) hygrometer; (Z) wastebasket.

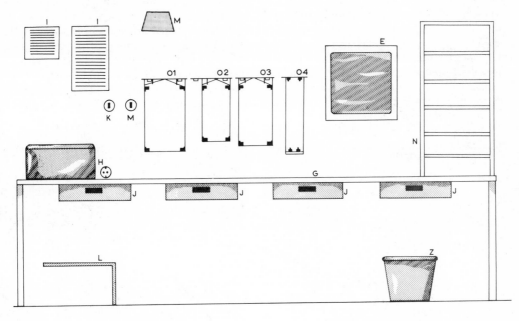

b

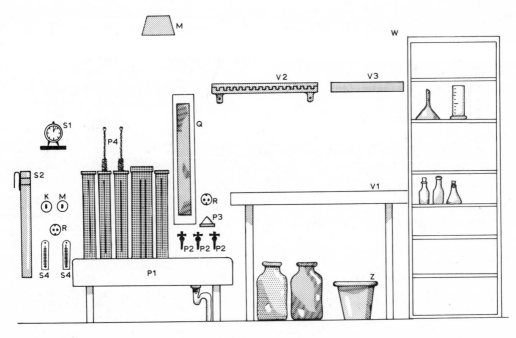

c

165

sheets, glued on cardboard (Fig. 3.20, Tables III.4, III.5) are hung here. One sheet gives directions and graphs for the preparation of the developer and fixing baths and gives some developing time-temperature graphs; the other gives a checklist for the fixing bath.

On this wall, between the dry and wet parts, a set of towels and paper towels with a wastebasket underneath can be placed.

A room thermometer and a hygrometer are hung against the side of the light-lock, where differently shaped switches for the central lights and safelights are also placed.

Drying Area

Films that have been washed must be dried. When using developing hangers, the film can be left attached to them and the hanger can be placed in a special drying rack. In this way the films will hang at right angles to the wall, they will not touch each other, and they will hang slightly inclined, which ensures that all water drops run to one point. The drying rack is mounted above the wet table at a height the smallest operator can reach.

A drying cabinet with a heating element and forced ventilation is very convenient. Drying

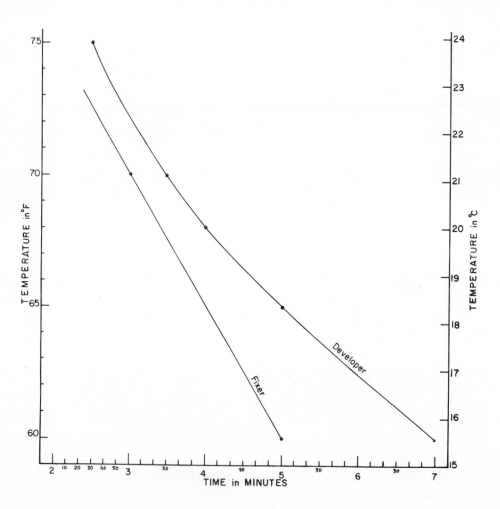

Fig. 3.20. Developing and fixing curves for Kodak DX-80 developer and Kodak "Unifix" fixer (for instructions see Table III.4).

Table III.4. Instruction Sheet for the Preparation of Developer (Kodak DX-80), Replenisher (Kodak DX-80R), and Fixing Bath (Kodak "Unifix" powder) (For developing and fixing times see Fig. 3.20)

Kodak DX-80 Developer

Dilute one part DX-80 with four parts of water.

Temperature	Developing time
60°F (15.5°C)	7 min
65°F (18.3°C)	5 min
68°F (20°C)	4 min
70°F (21°C)	3½ min
75°F (24°C)	2½ min

Kodak DX-80R Replenisher

Dilute one part DX-80R with four parts of water. Take care that the developing tank is always filled to the same level by adding replenisher to it.

Kodak "Unifix" Fixing Bath

Package no. 1 must be dissolved in 2.25 liters of water.
Package no. 2 must be dissolved in 9 liters of water.
Package no. 3 must be dissolved in 22.5 liters of water.
Dissolving temperature should not be higher than 27°C.

Fixing times: 3 to 5 min for temperatures ranging from 60°F (15.5°C) to 70°F (21°C). See also Fig. 3.20.

The activity decreases during use and consequently the fixing times must be lengthened.

Number of Sheets of 17 x 14 in. Fixed in a Tank of 10 liters	Fixing Time in Minutes at 65°F (18°C)
0–13	4
13–25	5
25–36	7
36–50	9
50–62	11
62–75	15

Table III.5. Checklist for the Fixing Bath (Kodak "Unifix" powder). From this List It Can Be Read Whether It Is Necessary to Lengthen the Fixing Time (compare with Table III.4)

Date	Initials of Operator	Film sizes being developed and fixed						Remarks
		10-24 cm	24-30 cm	10-48 cm	25.4-20.5 cm	27.9-35.6 cm	35.6-43.2 cm	

Conversion Table for Different Film Sizes

Film Size	Number of Sheets Equivalent to One Sheet of 14 × 17 in.
10–24 cm (3⅞–9½ in.)	6.9
24–30 cm (9½–11¾ in.)	2.3
10–48 cm (3⅞–18⅞ in.)	3.5
25.4–20.5 cm (10–12 in.)	2.1
27.9–35.6 cm (11–14 in.)	1.7
35.6–43.2 cm (14–17 in.)	1

time is cut down to 20 to 30 min, while the closed cabinet prevents dust from falling upon a film.

3.4B Processing

X-ray film is sensitive to blue light, and until the film is processed it should be handled in light from which the blue rays are missing. The author uses a Beeheve Safelight lamp with a "Wratten" number 6 B safelight filter for direct lighting (Kodak).

The darkroom illumination should be tested by placing a piece of film at a dry table in normal position of working. In this position the film is exposed to the illumination for 1 min, except for a small strip. A strip of the film is now covered with a piece of opaque cardboard and the rest of it is exposed for another minute. Cover a part of the film again and continue for about 6 to 8 min for the last strip. The film sheet is now developed and fixed in the normal way. A comparison of the strips will show that the film can be exposed for some time to the darkroom illumination without harm. If the strips become fogged, one has to make a check on the light-proofness of the darkroom; fogging may also be caused by special illumination placed too close to the table (Kodak Ltd, 1964e).

Loading the Film into Developing Hangers

The exposed film must be taken out of the envelope, cassette, or exposure holder and immediately placed into a suitable hanger for processing operations. There are two types of hangers: the channel type and the tension type. The tension hanger has advantages over the channel and will be the only one described (Kodak, 1957; Ilford, not dated; Kodak Ltd, 1962b).

The tension-type hanger has two fixed clips near the bottom corners and two tension ones at the top side. First the film is attached to the bottom ones and then to the top clips (Fig. 3.21). A little practice will be sufficient to render the job routine. One should take care that the pins of the clips are not too near the edge of the film to avoid tearing, and the pins should be pushed well through the film. Care should be taken not to touch the emulsion with the fingers, except in the film corners.

Development (see Fig. 3.21 for general procedure)

The unexposed silver halide crystals of the film are not affected in a developer solution, but from the exposed crystals the silver is liberated and deposited as tiny metallic grains which form the black image (Kodak Ltd, 1962b).

With a stirrer the developer is well stirred in order to obtain a homogeneous liquid in the tank. Also the developer is brought to level by adding a replenisher to it (see below). Next the temperature is measured, after which the length of the developing time can be read from the graph (Fig. 3.20). (It is better to use a thermostatically controlled bath and a constant developing time.)

The exposed film should now be mounted in a developing hanger and lowered carefully into the developing tank. The preset clock should then be started (Fig. 3.22).

If the film is not moved, different concentrations will form in the thin film of developer touching the film, which results in so-called "flow marks" at the radiograph. To help the slow process of diffusion, the films must therefore be moved up and down for a few seconds directly after immersion. This brief agitation should be repeated after 30 sec and then every minute.

The degree of agitation influences the rate of development. Always applying the same system of development will ensure comparable results as far as it concerns this step of the whole radiography technique.

Every developer has its own characteristics; therefore one should avoid frequent changes of developers. The best temperature is 20°C (68° F), but a few degrees lower or higher is permissible. A change in temperature affects the degree of developing intensity and consequently the time of development. It is therefore advisable to use a time-temperature chart (Fig. 3.20), which can be made from the data provided by the manufacturer. Normally the temperature should not be below 16°C (60°F) and not above 24°C (75° F). For higher temperatures the reader is referred to the Tropical Processing discussion.

Each time a film is developed some of the active chemicals are used up and at the same time some soluble bromides are left behind. With time this will noticeably retard development. Each film leaving the developing tank takes a little liquid with it and consequently the level of the tank falls. The tank must be "topped up" regularly with a replenisher solution to keep the level at the right height and maintain the activity of the developer.

Developer and replenisher should be stored in

Fig. 3.21. Attachment of a sheet of exposed film in a tension developing hanger. (1) The film is removed from the exposure holder. In this case, the cover was left to protect the film during loading the exposure holder. (2) The film is attached to the fixed clamps at the bottom part of the developing hanger. (3) The film is attached to the upper clamps. Note that the film is held only at the edges to avoid finger prints. (4) The developing hanger is loaded and ready for developing.

well-stoppered brown glass or polyethyl bottles and should be kept in darkness to prevent oxidizing by air and other changes influenced by light. Both liquids should not be kept in the developing tank longer than about 3 months before renewing. It is advisable to use a floating lid for the developing tank, and a normal lid to prevent dirt from falling in during the periods the tank is not in use. The period of 3 months can be somewhat longer if the temperature does not rise above 20°C (68°F) but should be shorter for higher temperatures.

If one develops series of films with a number of hours between the series it is necessary to check the temperature of the developer each time.

Rinsing or Stop Bath

After the developing time is finished, the film is removed from the developing tank and held above it for about 5 sec to let the liquid drain as much as possible. When a film is removed from a bath slowly, draining is almost unnecessary.

After draining, the film is placed in the second bath of plain water or a stop bath. The film should be agitated for about 30 to 60 sec, then taken out, drained, and transferred to the fixing bath.

Fixing

The undeveloped silver halide crystals, still in the emulsion, must be removed to prevent them from darkening under the influence of light. This is done in the fixing bath, which is a solution of sodium or ammonium thiosulfate, known as "hypo." The liquid also contains an acid, for example, acetic acid, a preservative such as sodium sulfite, and a hardening agent such as potassium alum (Kodak Ltd, 1962b). The acid neutralizes the alkali in the developer carried over in the emulsion. The sodium sulfite is necessary to prevent the hypo from being decomposed by

1. STIR SOLUTION

2. CHECK TEMPERATURE

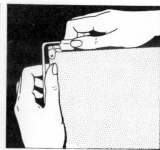

3. LOAD FILM ON HANGER

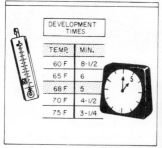

DEVELOPMENT TIMES	
TEMP.	MIN.
60 F	8-1/2
65 F	6
68 F	5
70 F	4-1/2
75 F	3-1/4

4. SET TIMER

5. IMMERSE FILM IN DEVELOPER

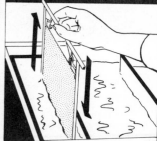

6. AGITATE FILM

7. DRAIN OUTSIDE DEVELOPER TANK

8. RINSE THOROUGHLY

9. FIX ADEQUATELY

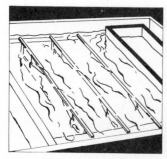

10. WASH COMPLETELY

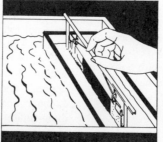

11. BATHE IN PHOTO-FLO

12. PLACE IN DRIER

the acid and the alum is added to harden the gelatin of the emulsion.

The fixing bath dissolves the undeveloped silver halide grains and the milky appearance of the film gradually disappears. This implies that the chemical activity decreases during use. Any fixing bath should be discarded when it fails to clear a film at a time twice that of a fresh bath. The film should remain in the fixer for *twice* the time required for it to clear.

A better method is to use a check list as is given in Table III.5. This list shows that after a number of films of the largest size the fixing time must be lengthened.

The time of fixing also depends on the temperature (see Fig. 3.20). To refresh the contact liquid against the film the film should be agitated regularly for 10 sec each minute.

When the fixation time is nearly finished the central white light can be turned on without any risk.

Washing

When the fixing time is over, the film, still in its tension hanger, is taken out of the fixation tank and drained. Next it is placed in the washing tank to remove all residual fixing liquid. If the hypo were allowed to remain it would slowly decompose and attack the image, causing it to become discolored and faded.

The washing tank should be equipped in such a way that continuously running water is secured with a minimum rate of 4 renewals/hour. The temperature of the water should not be less than 10°C (50°F) for a washing time of 30 min. For lower temperatures the washing is less efficient,

and the time should be increased considerably (for shipboard operations see Appendix One).

To prevent damage of the soft emulsion layer care should be taken that the water is clean and does not contain sediment particles. The author therefore uses a filter, which improves the quality of the water and decreases the risk of damage.

If the washing time is too short the risk of yellow-coloring of the film during time increases, and, consequently, the storage requirements will be greater (Kodak Ltd., 1965a).

Photo-Flo Bath (trade name)

The risk of drying marks may be prevented and drying may be hastened by rinsing the film, after it has been washed adequately, for approximately 30 sec in a wetting agent solution in the dilution recommended. This should be done expecially when the water is not absolutely soft. If no wetting agent is available one can use distilled water. The tank should be refreshed and cleaned at regular intervals.

Drying

Drying of the film can be done in two ways: natural drying or forced drying. Drying is very important for the quality of the finished radiograph. It should be carried out very carefully since touching of the wet film becomes visible afterwards. One also should take care that no dust becomes attached to the wet film since it is practically impossible to remove it. The film surface remains delicate until it is completely dry.

After drying has started the film should not be moved or shaken since water drops shaken from

Fig. 3.22. The ABC's of industrial X-ray film processing, as recommended by the X-ray Division of the Eastman Kodak Company. (1) Stir developer and fixer solutions to equalize their temperature. Use separate paddle for each bath to avoid possible contamination. (2) Check temperature of solutions with accurate thermometer. Rinse off each solution before checking next one. Adjust to 68°F if possible. (3) Attach film carefully to hanger of proper size. Attach at lower corners first. Avoid finger marks, scratches, or bending. (4) Set timer for desired period of development based on temperature of developer. See chart above for temperature and time. (5) Completely immerse film. Do it smoothly and without pause to avoid streaking. Start timer. (6) Immediately raise and lower hanger several times to bathe film surfaces throughly. Repeat at least once each minute. (7) When alarm rings, lift hanger out quickly. Then drain film for a moment. For fast drainage, tilt hanger. (8) Place film in acid rinse bath or running water. In an acid bath, agitate hanger for 30 to 60 sec; in running water, for 2 min. Lift from rinse bath and drain well. (9) Immerse film. Agitate hanger vigorously. Film should remain in fixer for twice required to "clean" it (when its milky look has disappeared), never less than 3 min. (10) Remove film to tank of running water. Keep ample space between hangers (water must flow over their tops). Allow adequate time for thorough washing—20 min or more. (11) If facilities permit, use a final rinse of Kodak Photo-Flo Solution to speed drying and prevent water marks. Immerse film for about 30 sec, drain for several seconds. (12) Place in drier, or rack in current of air. Keep films well separated. When dry, remove films from hangers and trim corners to remove clip marks. Insert in identified envelopes. (Redrawn from Eastman Kodak chart M4-26-8-62. Text also from that chart.)

the film hanger on to the partly dry emulsion are almost certain to cause drying marks (Kodak Ltd, 1962b).

Filing

When the film is absolutely dry it can be removed from its hanger. It is now ready for printing, filing, or direct study on an illuminator.

The film should be placed in its manila cover again. The data of the film can be written on this cover to avoid extra touching of the film. The corners must be cut off first. For further data the reader is referred to the discussion of storage of the radiographs in Section 3.2D.

3.4C Tropical Processing of X-Ray Films

Processing operations carried out at temperatures between 24 and 32°C (75 to 90°F) can only result in good radiographs when some special rules are followed. To obtain the correct contrast and to avoid swelling and softening of the film emulsion a fairly high concentration of a neutral chemical such as sodium sulfate should be added to the developer (Kodak Ltd, 1965b). At the same time a reduction of the developing and washing periods should be given.

It it is impractical to cool the solutions and to maintain them at a more normal level of temperature one should add 105 g crystals or 45 g anhydrous sodium sulfate (not sulfite) to 1000 cc of developer (Kodak D-19b or DA-19b. DA-19b is specially packed for the tropics) (Table III.6).

Table III.6 gives the corrected developing times for high temperatures based on standard developing times at 20°C (68°F). From this table a graph can easily be made to facilitate finding developing times of nonpresented temperatures.

It is absolutely necessary to use a stop bath between developing and fixing since time for draining after developing has too much influence on the developing. The film should be immersed for 3 min in the stop bath, during which one should agitate the film vigorously for the first 30 to 45 sec to avoid the formation of streaks. After an equivalent of seven 14 x 17 in. films per gallon (six 35.6 x 43.2 cm films per 4 liters) have been rinsed, the bath sould be replaced; otherwise scum markings will result (Kodak Ltd, 1965b).

The stop bath (Kodak SB-4) contains potassium chrome alum and sodium sulfate (see Appendix III). If excessive swelling is noticed, the stop bath may be made in a concentrated form, using only ¾ of the quantity of water given in the formula.

Fixing should be carried out in a solution of Kodak X-ray fixer powder or in a solution according to the formula Kodak F-5. Fixing time should not be less than 10 min.

Washing of the film must be done for at least 30 min in running water with a renewing of at least 4 changes /hour.

The last tank should contain a solution for preventing the formation of mold growth. The film should be rinsed for 3 min in an aqueous solution of ½ to 1% zinc fluosilicate to which Photo-Flo is added according to the recommended strength. Zinc fluosilicate is a poison which is fatal if swallowed even in dilute solution. The hands should be washed thoroughly after each contact with it and dust or spray from this chemical should not be inhaled (Kodak Ltd, 1965b).

Table III.6. Developing Times at High Temperatures for the Kodak D-19b or DA-19b Developer to Which Sodium Sulfate Is Added

Development Time (min) in Nonsulfated D19b or DA-19b at 20°C (68°F)	Calculated Development Time (min) in Sulfated D-19b or DA-19b			
	24°C (75°F)	27°C (80°F)	29°C (85°F)	32°C (90°F)
5	6	4½	3	2¼
7	8½	6½	4½	3¼
8	9½	7	4¾	3½
12	14	10½	7	5¼

(After Kodak Ltd, (1965b).

Cleaning of Developing Equipment

Cleanliness of the work improves the expected results and therefore a periodical cleaning is desirable.

The film hangers can be cleaned in a solution of 60 cc glacial acetic acid in 1000 cc water. Let the articles soak for an hour followed by scrubbing in clean water.

Deep tanks must be scrubbed thoroughly with clean water with a double-sided brush which is reserved for this type of work. It is advisable, especially during warm weather, to sterilize the developer tank in order to prevent bacterial growth. Bacteria may cause the sulfite to change to sulfide, which will then fog the undeveloped film.

A solution of sodium hypochlorite can easily be purchased, or it can be prepared in a fume-cupboard by adding a 10% solution of sodium carbonate to a 4% solution of bleaching powder until no more precipitate is being formed. The mixture is allowed to settle and the clear solution of sodium hypochlorite is decanted.

Tanks which have been sterilized must be thoroughly washed before use (Kodak Ltd, 1962b).

3.4D. Processing Liquids and Formulas

All manufacturers provide data to enable the investigator to make his own tank liquids. It would take too much space to give all the formulas and anyone interested in more details should request the data sheets mentioned in Appendix III or contact the manufacturer directly.

3.5 TYPES AND SHAPES OF SAMPLES

The radiograph presents an image of the absorption differences within the sample. The specimen is a three-dimensional body, whereas the radiograph is two-dimensional. In thick samples more characteristics will usually occur on top of each other, resulting in a confused radiograph (Fig. 3.23).

Preferably, detailed investigations should be carried out on thin slices. However, this does not exclude all applications of radiography to thicker samples, but it should be kept in mind that small properties in particular within a sample may not be visible on a radiograph when there is little difference in absorption between particle and mass. Increasing sample thickness decreases the possibility of such properties becoming visible. The application of X-ray radiography to thicker samples therefore should be considered as a general and preliminary investigation (scanning method), from which considerable information can often be obtained for further detailed analyses. These include sampling for grain-size determinations, petrology, mineralogy, petrochemistry or paleontology on fossils that are large enough to be visible. Scanning is also important, especially when stereoradiography is applied, for the selection of small samples which have to be impregnated (thin sections) or radiographed. Different types of rocks – sediments, metamorphic, igneous – are presented in Fig. 3.29 A-X and graphs given on the exposure times in relation to thickness.

3.5A Slices of Sediments

Unconsolidated Samples

Radiographs of plane parallel slices of unconsolidated sediments can be made without any difficulty as long as the operator does not destroy the material during slicing. Any ruptures created in the sample appear as black lines on the radiograph and often interrupt interesting parts. Clayey sediments should be cut using the electro-osmosis application, while sands can be sliced with a wire. Sediments with inclusions should be prepared with a knife, spatula, or brush.

To prevent the sample from falling apart, the operator can use plexiglas trays in which the specimen fits well (see Fig. 3.18). The height of the tray determines the thickness of the slice; the sides keep the material together better. The author used the following procedure for cores. Cut the core in half (Fig. 1.13) and smooth the cut surface. Place a tray on top of one half, turn the sample over, causing it to fall in the tray. The internal dimensions of the trays for 2½- and 3-in. cores, respectively, are 42 x 6 (7½) x 1 cm or 16½ x 2⅜ (2¹⁵⁄₁₆) x ¹³⁄₃₂ in.) The lengths of these trays fit a 14 x 17-in. sheet of film leaving space for lead numbers and letters. For smaller cores, strips of plexiglas are placed inside the tray. The protruding part of the core can be sliced off easily. If cutting marks are made they have to be removed since differences in slice thickness of both sides of the mark may show up on the radiograph as different grey zones. The mark also can be removed by smoothing a

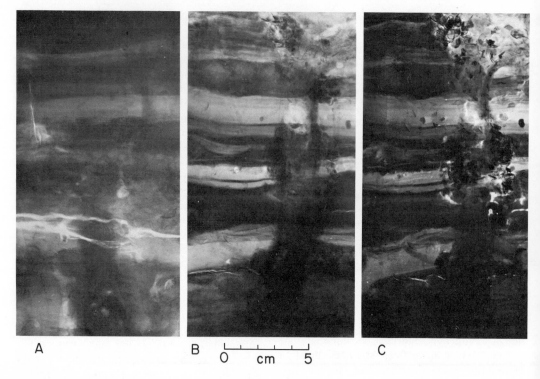

A   B   C

Fig. 3.23. Prints of radiographs showing the influence of the thickness of a sample on the sharpness of the image. The slice has been collected from a box sample. The original slice (A) was 7 cm thick, it was then thinned down to 4 cm (B), and next to 1 cm (C). Focal distance 39 in., Kodak Industrial X-ray film type AA, 5 mA, 100 KV, and 1½ min (A), 100 KV and 30 sec (B), and 60 KV and 30 sec (C). Sample collected near end of the jetties off Galveston, Texas, outside the channel at a depth of 24 ft of water. Note the increase in sharpness by decrease of slice thickness. Also notice the crab (?) boring, which is hardly visible in the radiograph made from the thickest slice.

larger area so a gentle decrease in thickness is formed, resulting in a gentle increase in blackening. However, a note should be made to avoid misinterpretations on the lithology. Smearing effects do not show up in the radiograph.

Consolidated Samples

Consolidated sediments are denser than unconsolidated ones, and a longer exposure is needed to obtain a density similar to that of an equally thick slice of unconsolidated material (see also cores).

The slices also should have smooth surfaces, and they should be plane parallel to facilitate interpretation.

Impregnated Samples

Sediments which are impregnated with an artificial product can be radiographed without any restriction. A few examples are given in Figs. 3.29E and F (X 129 and X 146). The artificial

medium absorbs radiation, but it is less than that of the sample. Very slight differences in absorption within the sample may blur, but the author did not observe this effect on the few samples of which parts were impregnated.

Imbedded Samples

Sediments that are imbedded react toward X-rays as do impregnated ones. The advantage of imbedded slices is that any risk of drying out and cracking is eliminated. The disadvantage is that an easy entry to the samples is eliminated.

To avoid unsharpness it is important that both surfaces of the slice are as smooth as possible and that the underlying or the overlying plastic layer is not too thick (see Fig. 3.4). The thinnest plastic layer should be placed in contact with the exposure holder. Figure 3.29G (X 151) presents 1-mm thick slice of tonstein (imbedded in Norsodyne to prevent breaking during transport).

3.5B Nonplane-Parallel Specimens

The main characteristics of nonplane-parallel samples with regard to radiography are the absolute thickness of the thickest part of the sample and the variation in thickness from place to place. The absolute thickness determines whether the X-ray unit can penetrate the sample. Since consolidated samples are denser than unconsolidated ones, the maximum thickness a certain X-ray unit is capable of penetrating will differ for both types. The X-ray units discussed in this chapter all are suitable for radiographing consolidated cores with a diameter of 9 cm (3½ in.), which was the thickest core handled by the author.

Complete Cores and Half-Cores

Cores form the majority of the group of nonplane-parallel samples to be radiographed. The cores can be of consolidated material, but most of them are unconsolidated and with or without a certain type of liner (Klingebiel et al., 1967). The author made radiographs in 1965 of shallow marine cores that were still in their thin-walled zinc tubes. Stanley and Blanchard (1967) report successful results with 6-cm diameter cores in plastic liner or wrapped in heavy paper or cloth.

The author found that by placing more zinc-lined cores next to each other in order to cover a 14 x 17-in. film, the outer sides of the tubes cause some type of reflection, which in turn causes a halo effect that interferes with the sides of the neighboring cores. The effect could be prevented by placing strips of lead between the core tubes.

Plastic, paper, or PVC liner does not cause any undesirable effects on the radiograph, and for scanning effects there is no objection to leaving the core in its liner. Saran Wrap (trade name for a very thin plastic foil made by Dow Chem.), however, used to prevent the imbedding sand (see below) from sticking to the core, always wrinkles. These wrinkles become visible on the radiograph.

Since each core has a variation in thickness, the absorption through the middle is maximal, and it is minimal along the edges, resulting in overexposure underneath the sides. Furthermore, the visible part of the radiograph becomes much smaller than the diameter of the core (Fig. 3.24). This effect can be eliminated somewhat by imbedding the core in loose, fine-grained sand in a plexiglass tray (Fig. 3.25). The author tried different types of sand and obtained the best results with well-sorted, fine sand (3.26).

Coarser sand shows up as grains. The less heavy the minerals in the sand, the better the result. This sand absorbs part of the radiation without creating an image and therefore decreases the effect of overexposure of the thinner parts of the core. It is important to have no, or very little loose sand underneath the core since this increases the total sample thickness and requires a higher voltage radiation penetration. On board ship the loose sand is replaced by a sand mold (impregnated) to speed up the process (see Appendix one).

The loose sand is packed very poorly and its absorption per unit thickness therefore may be considerably less than that of the core, especially for very fine-grained sediment or consolidated samples. To increase the effect of the imbedding sand, its thickness should increase with the decrease of core thickness. Good results have been obtained by imbedding the core first, as indicated in Fig. 3.25A, and then placing a cylinder of the same diameter as the core on top of the core and pouring sand around this cylinder to about half its height. The cylinder can be removed, leaving a countercylinder behind (Fig. 3.25B). This removal can be carried out better once the sample is located underneath the radiation source. When many samples have the same diameter it is worthwhile to make a mold.

A mold should have a minimum thickness underneath the maximum thickness of the core, but some material is needed to keep the mold together. The author obtained this effect by using a sheet of plexiglas as bottom. The binding between plexiglas and sand-plastic mixture is ascertained by drilling holes through the plexiglas in rows between the cores. Through the holes wood screws are placed prior to pouring the sand-plastic mixture.

Several attempts were made to obtain a good, smooth mold for half-cylindrical cores for routine scanning work on board ship (Appendix one). Poor results were obtained by wrapping Saran Wrap around PVC cylinders, which were then placed in a box. A sand-plastic mixture was poured around them to the desired height. Since the PVC pipes were placed rather closely together, it was difficult to stir the mixture between them. Air bubbles that became trapped were visible on the radiograph. Wrinkles in the Saran Wrap were also visible.

Better results were obtained by using a sturdy box, covered on the inside with a mold release agent. The box is larger than the size of required mold. PVC pipes are cut to a length to fit the

A   B

0   cm   5

Fig. 3.24. Prints of radiographs from a core not imbedded [thus having variation in thickness from the center to the side (A)] and of the same core imbedded in fine-grained loose sand (B). Note the difference in width obtained. Core 66-A-13-25E, 705 cm below surface, diameter 2½ in., Mississippi fan at a depth of 960 fathoms, focal distance 36 in., Picker "Hotshot" X-ray apparatus, Kodak Industrial X-ray film type AA, 7 mA, 80 KV, 5 min.

box. They are filled with sand and sealed well on both ends. The next step is to cover them with mold release agent. The plexiglas should be shorter on one side than the box. Holes are drilled through, and screws are placed. A little grease keeps them in place and prevents leakage. A small strip of plexiglas is placed to fill the gap next to the plexiglas bare plate (Fig. 3.27). This is later used as a place for lead letters, numbers, and arrows. A rather thick sand-plastic mixture is now poured into the box and air bubbles are removed. The PVC pipes can then be placed with a heavy weight on top to prevent them from floating or moving. Excess mixture must be removed. Once the mixture is cured thoroughly, the pipes, box, and plexiglas strip can be removed. Next a radiograph is made to examine the purity of the mold. It can then be trimmed to size.

For complete cores, it is better to have the molds, as previously described, in two pieces.

Irregularly Shaped Samples, Noncores

This section primarily refers to hand specimens of consolidated materials. Radiography can be used as scanning technique to determine direction and location for detailed work (thin sections), for analyzing spread of inclusions, etc.

Good results can be obtained by placing the sample with its "flat" side down in a plexiglas tray in which some sand has been poured. The sample is twisted several times to insure a proper fill of the sand in the holes. The specimen should be in contact with the bottom. The tray can then be filled with sand while the sample is covered. Tapping against the sides of the tray increases the setting of the sand. Sometimes the upper surface of the sand can be made higher at spots where the sample becomes thin.

3.5C Thickness of Sample

Increase in sample thickness requires increase in exposure time and often an increase in kilovoltage for radiation penetration. Unsharpness of the image normally increases too, depending on type and scale of sedimentary structures present (see Fig. 3.23).

The direction of sedimentary structures such as lamination planes compared to the direction of the X-ray beam can be very important (see also Ruault, 1966). An example may demonstrate this. Core 66-A-13 25 E contains well-developed parallel lamination below 4 m depth. The lamination planes are at right angles to the core and slightly curved along the sides. Several

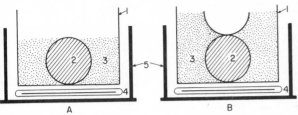

Fig. 3.25. Schematic presentation of two types of imbeddings of a core in fine-grained (impregnated) sand. (A) The top of the sand is flush with the top of the core. (B) The top of the sand has a countercylindrical shape. The situation in (B) is important when dealing with consolidated sediments which are denser than the sand used for imbedding: (1) box made from plexiglas; (2) core; (3) sand; (4) exposure holder with film; (5) protecting lead.

slices of different thickness have been cut at right angles to the stratification and an equivalent group has been cut at an angle of about 25° to the stratification (Fig. 3.28). Radiographs made of the first group show a slight decrease in sharpness with increased slice thickness. Radiographs of the obliquely cut slices do not reveal any information, except when the slice thickness does not exceed the thickness of the laminae. The absorption caused by a dark laminae is different from the absorption caused by a light-colored laminae, but in the obliquely cut slices any absorption difference is eliminated by underlying laminae (Fig. 3.28B).

It may not be automatically concluded that a sample is nearly or completely homogeneous when its radiograph reveals little or no data at all, since absorption differences between various properties may be beyond detection possibilities or the superposition of characteristics may eliminate the absorption differences that could otherwise be detected.

Stepped Wedge for Exposure Charts

In industry, exposure charts are often made by a series of radiographs of a built-up pile of plates consisting of a number of steps (Carson, 1964). The wedge is radiographed at several

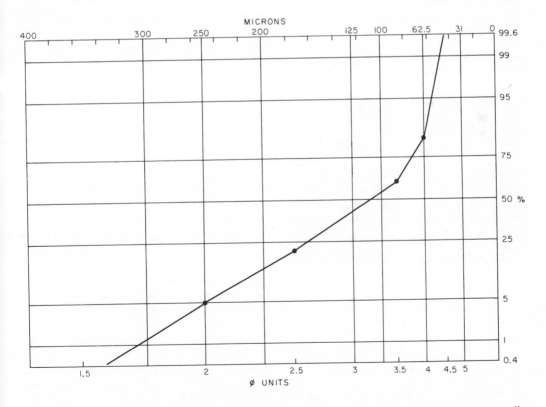

Fig. 3.26. Grain-size distribution on linear-probability paper of the sand used for imbedding, loose as well as impregnated. No material coarser than 1.5φ and finer than 4.5φ was present.

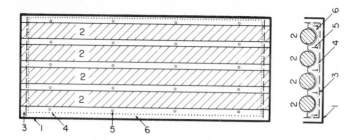

Fig. 3.27. Schematic drawing of how to make a sand-plastic mold for radiographing half-cylindrical cores. For complete cores two of these molds are necessary. The sand used is that presented in Fig. 3.26. A sturdy box (1) is covered with mold release agent. Along the short sides strips (3) are mounted to keep the PVC tubes (2) in place. Inside the strips fits a plexiglas plate (4: dotted line) in which holes (5) are drilled for screws. At the front side the plexiglas leaves an opening (6), which later gives space for lead letters and numbers.

different exposure times at each of a number of kilovoltages. Processing has to be carried out under identical conditions. The author made 10 steps of each of three samples (two different sandstones: Fig. 3.29E and F, X 121 and X 152; one limestone: Fig. 3.29L, X 34) with thickness intervals of 3 mm. It proves that the curves at the exposure charts were straight lines when dealing with one and the same kilovoltage. Dif-

ferent kilovoltages result in a very slightly curved line (Kodak, 1957, Fig. 32).

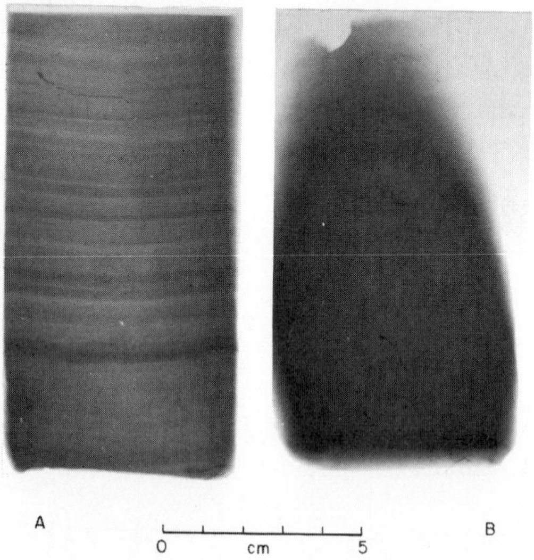

Exposure Charts for Various Samples: Sediments, Metamorphic, and Igneous

Since the making of stepped wedges is very time consuming and requires a lot of film, nearly 200 samples were collected, of which two slices were made. From all hard rocks a 3 mm slice and a 9 mm slice were cut and roughly polished. The unconsolidated sediments normally were slices in thickness of 1 and 2 or 1½ cm. An average of three different exposures was necessary to obtain radiographs with equal density. The densities were not measured but compared by eye, which is a very rough and inaccurate method. However, near all colleagues operate the same way. Since the technique of radiography also is successful for selection purposes (overall picture, two- and three-dimensional spread of properties, thin sectioning) in metamorphic and igneous rocks, some of these non-sediments have been radiographed too. A few imbedded and impregnated sediments have been incorporated (Table III.7). Figure 3.29 presents a selection of the samples radiographed. Black and white photographs as well as prints of radiographs are given along with the exposure charts. Some curves run up to an exposure close to 1000 mA· min, which is due to extension of the line through both measuring points. A higher kilovoltage is necessary to cut down the exposure. Normally 50 kV has been used for these tests, but lower kilovoltages are recommended when dealing with very thin slices. Figure 3.29 is made so extensive to allow the reader to obtain a good radiograph in one or very few trials.

Fig. 3.28. Prints of radiographs of the same sample (see Fig. 3.24), but cut at different directions with regard to the bedding. (A) The cut is made vertically, which is more or less at right angles to the lamination. The lamination is visible. (B) The cut is made about 30° off the vertical. This sediment slice also is 25 mm thick. The laminae do not come out separately because they overlap.

Table III.7. Directive to Fig. 3.29 Presenting the Different Groups of Sediments, Metamorphic, and Igneous Rocks Radiographed

Type of Sample	Sediments	Metamorphic	Igneous
Recent sands	A		
Recent muds	A, B		
Soils	C		
Conglomerates	D		
Sandstones	D, E, F, J		
Gypsum	G		
Diatomite	G		
Tonstein	G		
Marls	H		
Oolites	H, I, R		
Limestones	I, J, K, L, M, N		
Dolomite	J		
Marmorized limestones	O, P		
Marbles	P, Q		
Volcanics	R, S		
Travertine	R		
Gneiss		S, T	
Schist		S, U	
Quartzite		T	
Slate		T	
Serpentine		T	
Iron-ore		U	
Granite			U, W
Porphyry			U, V
Andesite			U
Rhyolite			U, V
Eclogite			V
Gabbro			V
Amphibolite			V
Diorite			V
Diabase			W
Labradorite			W
Hooibergite			W

Note. Letters refer to parts of Fig. 3.29.

Fig. 3.29A-W. Exposure charts for various geological samples, mainly sediments. Almost every curve is determined by two points corresponding to a thickness of about 3 mm and 9 mm. Both slices have been made from the same sample and radiographs were made until one was obtained with a density favorable for making photographic prints. Density has not been measured and consequently the curves are not exactly comparable. Radiographic data: distance focal spot-film: 1 m; Kodak Industrial Film Type AA; Philips-Müller Macrotank B X-ray apparatus; milliamperage normally used, 4 mA; 50 kv unless otherwise indicated.

For each sample presented on the exposure charts, normal photographs (indicated by a number) and contact prints of radiographs (generally at the right-hand side of the corresponding photograph) are given (Fig. 3.29A.1 corresponding to Fig. 3.29A.2, etc.). The thickness of both slices of each sample have been measured accurately before being plotted on the exposure chart.

From the series it can be seen that it is impossible to predict whether the radiograph will reveal better data than the black and white photograph. The samples are arranged in groups whenever possible, and within each group a number of examples are given to enable the investigator to make a decent exposure choice for his material. Many of the samples have been obtained from marble dealers (Hessel's Natuursteenhandel, N. V., Amsterdam, The Netherlands; Fèvre & Cie., Paris, France; Joh. Jansen & Zn., Utrecht, The Netherlands), and therefore many of the names are trade names. It was not always possible to obtain the location and stratigraphic data. Several samples were received from colleagues from the geological institute at Utrecht, The Netherlands.

The reader should be aware that dark parts on the radiograph prints represent dense parts in the sample, which have higher absorption capacities for X-rays than do light-colored parts.

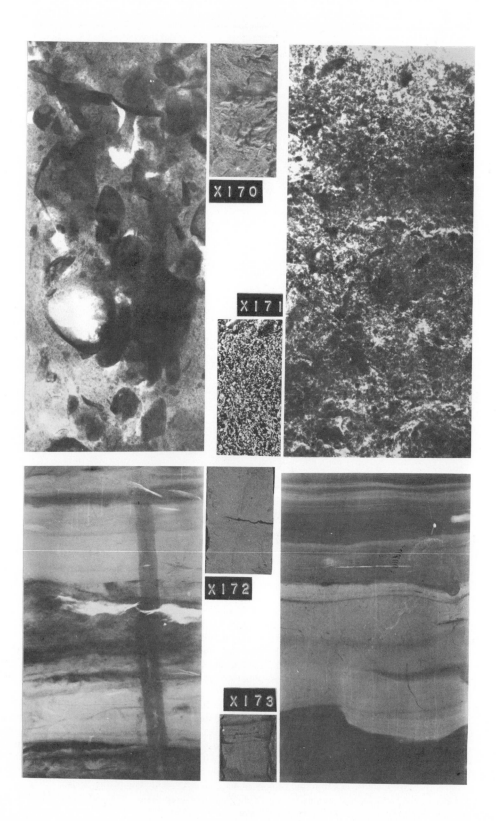

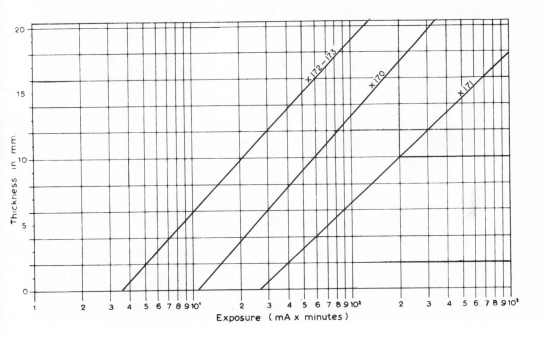

Fig. 3.29A. Exposure chart for various marine sands with shells and marine sandy muds, collected from the shelf of Surinam, South America, during the O. C. P. S. program, 1966, on board Hr. Ms. SNELLIUS. For explanation see text. Voltage used: 45 kV.

X 170: Very sandy mud with bryozoa and shells. Station D 34, Long. 7°08′N, Lat. 55°54′W. Water depth 84 m. Sample from 99 to 110 cm below surface. Piston core, central part.

X 171: Calcarenitic sand. Station F 44, Long. 6°33′N, Lat. 56°31.5′W. Water depth 33 m. Sample from 0 to 12 cm below surface. Box sample.

X 172: Sandy mud and mud in alternating bands, which are not visible on a normal photograph. A vertical burrow runs through the sample. The white parts in the middle are from the crack. Station A 14, Long. 6°14′N, Lat. 55°20′W. Water depth 19 m. Sample from 61½ to 70 cm below surface. Piston core, central part.

X 173: Sandy mud with sand laminae and scour and fill structure. Station C 21, Long. 7°03.5′N, Lat. 55°40′W. Water depth 70 m. Sample from 15½ to 24 cm below surface. Piston core, central part.

Note. It was not possible, due to the graininess of the film, to bring the black and white pictures back to their original size.

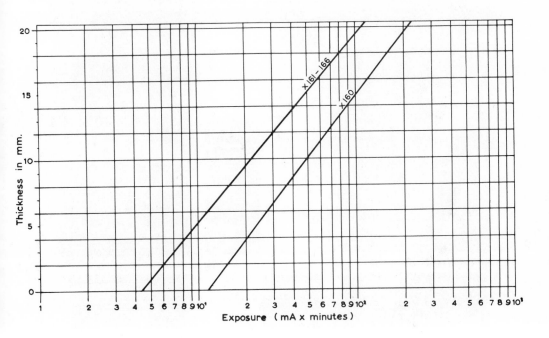

Fig. 3.29B. Exposure chart for various marine muds, collected from the shelf off Surinam, South America, during the O. C. P. S. program, 1966, on board Hr. Ms. SNELLIUS. For explanation see text. Voltage used: 45 kV.

X 160: Sandy to slightly sandy mud with an abundance of burrows, mainly going in vertical and horizontal directions. Station C 18, Long. 6°14′N, Lat. 55°50′W. Water depth 27 m. Sample from 45 to 57 cm below surface. Piston core, central part.

X 161: Very slightly sandy mud with some shells. The vertical scratches at the radiograph are artificial. Station A 11, Long. 7°08,5′N, Lat. 55°08.5′W. Water depth 77 m. Sample from 54 to 66 cm below surface. Piston core, central part.

X 166: Very weak mud from a mud bank along shore. The lower part of the sample reveals lamination with nests of shells on top. The upper part has some very indistinct banding. Station F 48, Long. 6°06′N, Lat. 56°36′W. Water depth 13 m. Sample from 51 to 56 cm below surface. Box sample.

Note. The curves are based on one point (thickness), but enough information was available to draw lines. It was not possible, due to the graininess of the film, to bring the black and white pictures back to their original size.

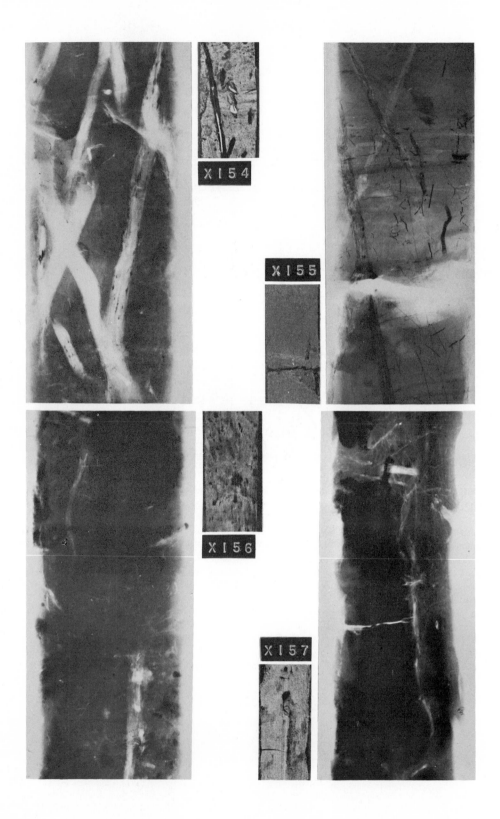

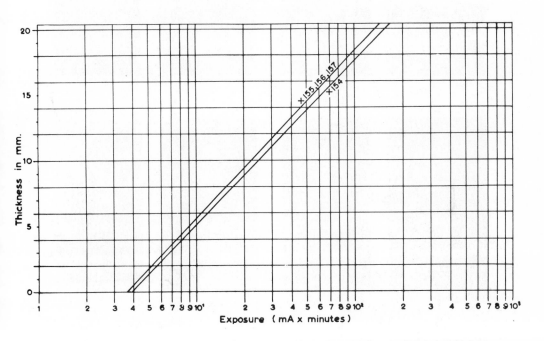

Fig. 3.29C. Exposure chart for various soils, collected by Dr. S. Slager. The curves are based on one point (thickness), but many radiographs made gave enough information to draw lines. For explanation see text. Voltage used: 45 kV.

X 154: Brackish clay, about 60% fine material, with roots of *Parwa*. Demarara marine clays. Northern part of Surinam, near Paramaribo, S. America: 18 to 27½ cm below surface.

X 155: Half "ripened" clay, about 60% fine material, which is only slightly desalted. Swamp vegetation, Demarara marine clays. Northern part of Surinam, near Paramaribo, S. America: 57 to 67 cm below surface.

X 156: Of the same core as X 155. Depth 179 to 188½ cm below surface.

X 157: "Ripened" clay bottom, about 60% fine material. This bottom is nearly completely desalted. Swamp environment. Demarara marine clays. Northern part of Surinam, near Paramaribo, S. America 55 to 65 cm below surface.

Note. It was not possible, due to the graininess of the film, to bring the black and white pictures back to their original size. Left and right of the prints of X 154 and X 155 are reversed when compared to the radiographs.

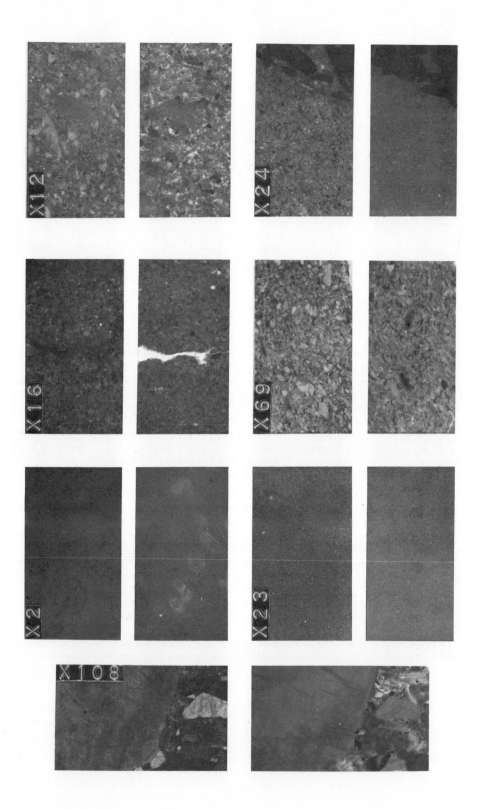

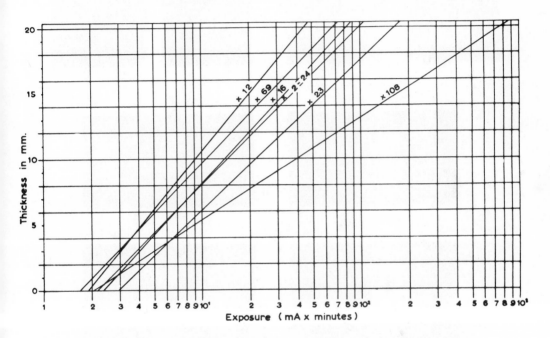

Fig. 3.29D. Exposure chart for some coarse sandstones and conglomerates. For explanation see text.

X 12: Conglomeratic sandstone with pores. Saxonian (Permian). Lodève, France (Kruseman, 1962).

X 24: Quartzitic sandstone with shale inclusions. Maastrichtian (Cretaceous). Stiegelbach near Adelboden, Switzerland (Bouma, 1962).

X 16: Conglomeratic sandstone with coal and clay pebbles. Coal mines near Hénin Liétard, Pas de Calais, France. (Courtesy P. Dollé.)

X 69: Coarse sandstone (Grès à Helmintoides). Series à Helmintoides. Province of Imperia, Italy. (Courtesy M. Vanossi.)

X 2: Scolites sandstone. Erratic boulder of Scandinavian origin (age presumably Cambrian). Donderse Veld, Drente, The Netherlands.

X 23: Red-colored, fine-grained sandstone. Saxonian (Permian). Lodève, France. (Kruseman, 1962.)

X 108: Conglomerate. Saxonian (Permian). Lodève, France. (Kruseman, 1962.)

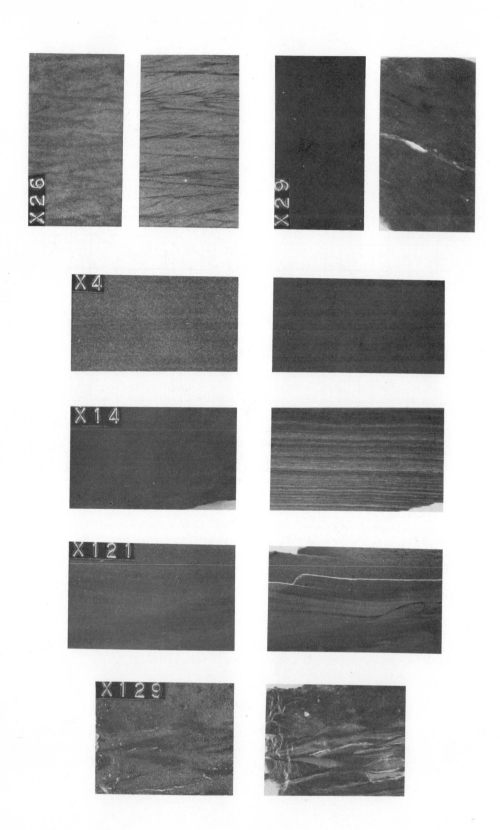

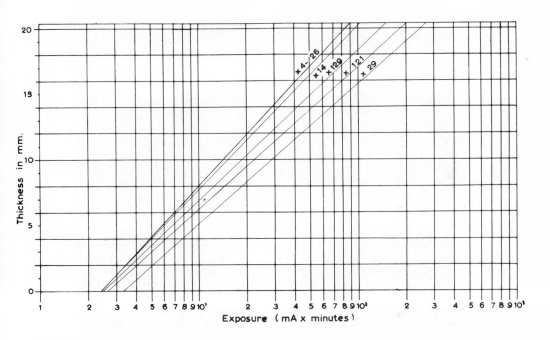

Fig. 3.29E. Exposure chart for various sandstones. For explanation see text.

X 26: Maulbronner sandstone. Maulbronnen, between Karlsruhe and Stuttgart, Germany.

X 29: Glauconitic sandstone. Middle Cretaceous. N. W. of Sospel, Alpes Maritimes, France.

X 4: Sandstone. Devonian. Coal mines near Hénin Liétard, Pas de Calais, France. (Courtesy P. Dollé.)

X 14: Sandstone with more coal flakes on laminae planes than X 4. Devonian. Coal mines near Hénin Liétard, Pas de Calais, France. (Courtesy P. Dollé.)

X 121: Sandstone. Turbidite (interval c-d). Eocene. Peira-Cava area, Alpes Maritimes, France. (Bouma, 1962.)

X 129: Sandy marsh, impregnated with Arigal C. Recent. Wilhelmshaven, Germany.

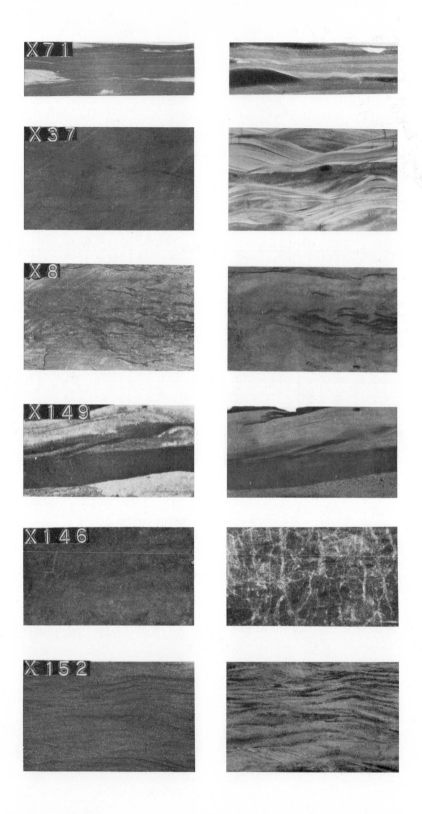

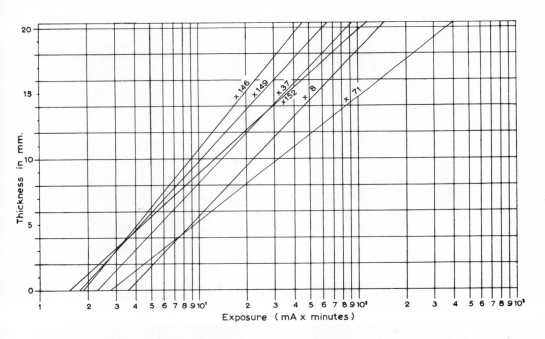

Fig. 3.29F. Exposure chart for some sandstones. For explanation see text.

X 71: Coblenz quartzite with flaser structure. Coblencian (Devonian). Coblenz, Germany.

X 37: Wave ripple lamination (upside down) in red sandstone and sandy pelite. Upper Saxonian (Permian). Lodève, France. (Kruseman, 1962.)

X 8: Fine-grained pelitic sandstone with clay streaks. Saxonian (Permian). Lodève, France. (Kruseman, 1962.)

X 149: Sandstone. Vollpriehausen Wechselfolge (Middle Buntsandstein). Bad Salzdetfurth, Germany. (Courtesy J. A. Broekman.)

X 146: Very sandy marsh, impregnated with Araldite F (CY 205) and hardener HT 902. Recent. South of Buren, Ameland, The Netherlands.

X 152: Fine-grained sandstone. Coblencian (Devonian). Alken near Coblenz, Germany.

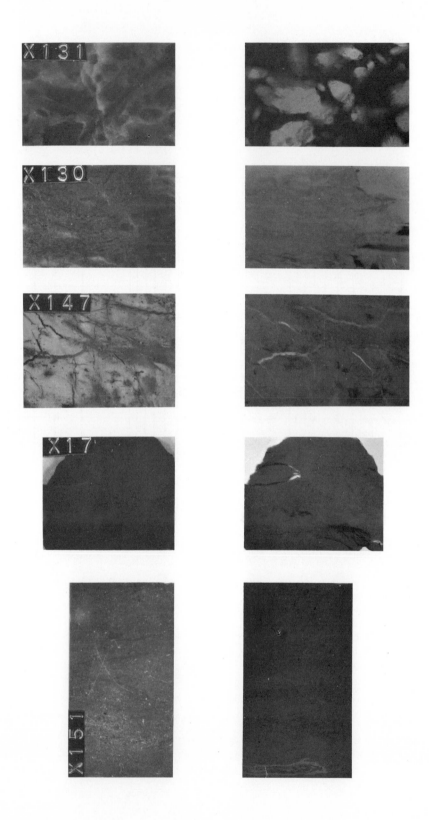

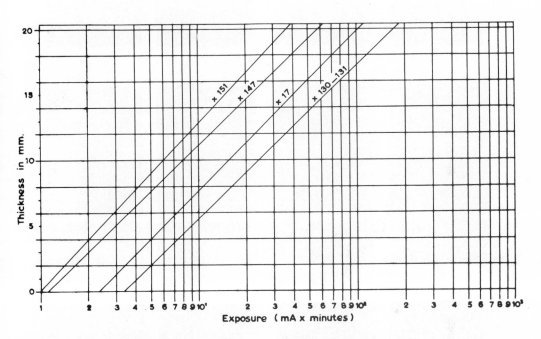

Fig. 3.29G. Exposure chart for gypsum, diatomite, and tonstein. For explanation see text.

X 131: Gypsum: flower gypsum. Lower gypsum-anhydrite zone. Middle Muschelkalk (Triassic). North of Gundelsheim, near Heidelberg, Germany.

X 130: Gypsum: banded gypsum. Lower gypsum-anhydrite zone. Middle Muschelkalk (Triassic). North of Gundelsheim, near Heidelberg, Germany.

X 147: Diatomite. Riss-Würm interglacial. Hollerup, Jutland, Denmark (see Fig. 1.37).

X 17: Section of a tree found in shale above a coal seam. Namurian (Carboniferous). Argenteau, Belgium.

X 151: Tonstein (1 mm thick) imbedded in Norsodyne 50 (plastic). The tonstein layer belongs to the coal seam Maxence. Coal mines, department Pas de Calais, France. (Courtesy P. Dollé.)

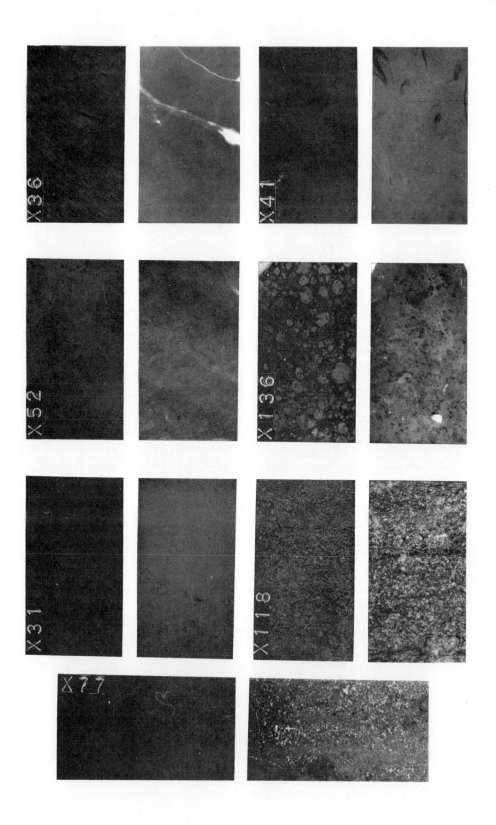

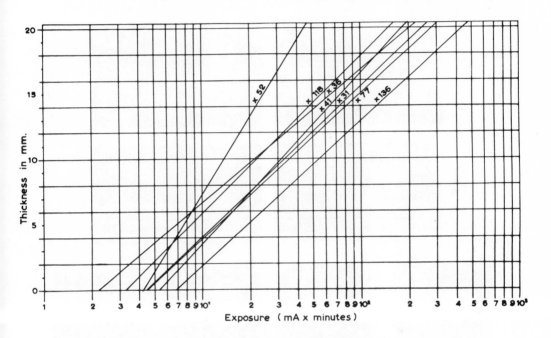

Fig. 3.29H. Exposure chart for various marls and oolites. For explanation see text.

X 36: Marl. Marnes Priaboniennes, Priabonian (Upper Eocene). Peira-Cava area, Alpes Maritimes, France. (Bouma, 1962.)

X 41: Marl with shell fragments. Marnes Priaboniennes, Priabonian (Upper Eocene). Lucéram, Peira-Cava area, Alpes Maritimes, France. (Bouma, 1962.)

X 52: Marl. Miocene. Malta.

X 136: Coarse oolite with glauconite. Gemmingen, Germany.

X 31: Oolite. Miocene. Basin of Paris, France.

X 118: Coarse oolitic limestone. Lias rubané. Bernais (Meuse), France.

X 77: Oolite. Marbrier near Baumont, France.

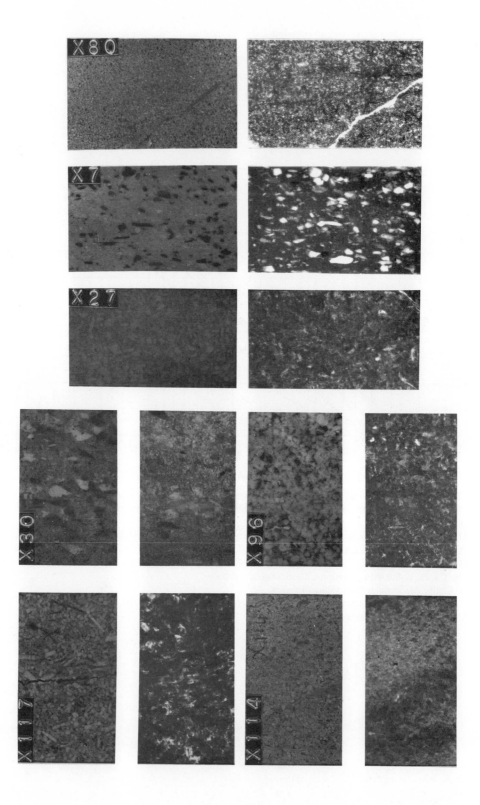

196

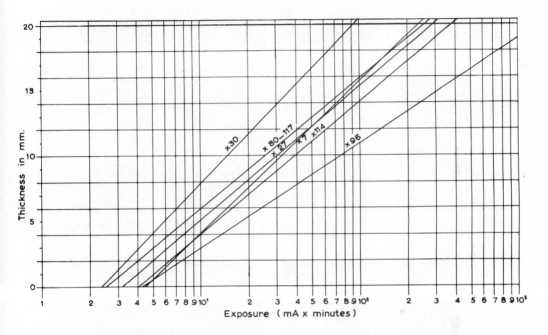

Fig. 3.291. Exposure chart for some limestones and oolitic limestones. For explanation see text.

X 80: Oolitic limestone. Lias choix. Barnais, France.

X 7: Limestone with calcareous pebbles. Tertiary. Near Salzburg, Austria. (Van Hinte, 1963.)

X 27: Pouillenay jaune (breccious limestone). France.

X 30: Breccious limestone with marl and shale pebbles. Maastrichtian (Cretaceous). Stiegelbach, near Adelboden, Switzerland. (Bouma, 1962.)

X 96: Pouillenay rouge (breccious limestone). France.

X 117: Coarse-grained fossiliferous limestone: Briantuille rosé. Meuse, France.

X 114: Coarse-grained fossiliferous limestone: Longchant Veine D. Yonne, France.

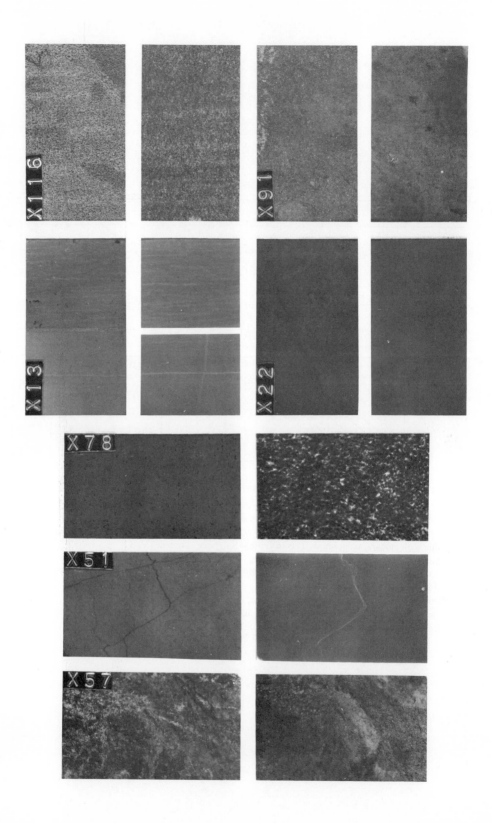

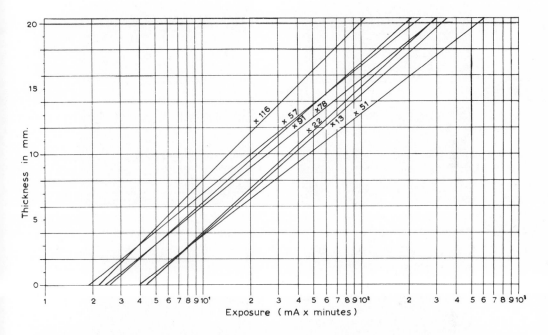

Fig. 3.29J. Exposure chart for calcareous sandstone, dolomite, and limestone. For explanation see text.

X 116: Calcareous sandstone: Saint Vaast fin. Oise, France.

X 91: Dolomite. Anrechte, Germany.

X 13: Dense lithographical limestone. Malm. Solenhofen, Germany.

X 22: Sandy fine-grained limestone with fossil remains. Jurassic. Alpes Maritimes, France.

X 78: Coarse porous calcarenite: Artiges jaune. France.

X 51: Very fine-grained white limestone. Upper Cretaceous. Priabona, Italy.

X 57: Gehlenite limestone. Badberg, east of Oberbergen, Germany.

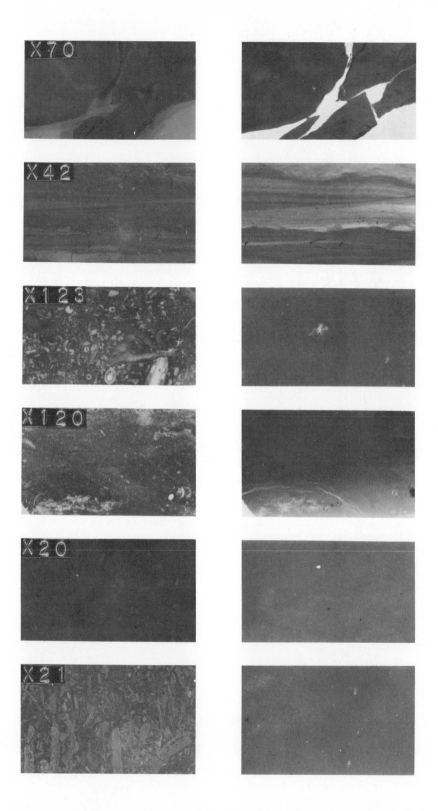

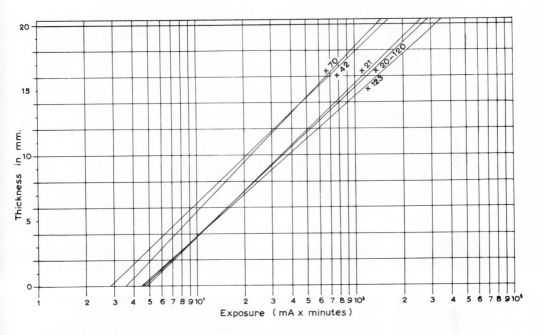

Fig. 3.29K. Exposure chart for various limestones. For explanation see text.

X 70: Griotte limestone. Upper Devonian. Tor-dal, S. Pyrenees, Spain. (Courtesy C. P. M. Frijlinck.)

X 42: Clayey fresh water limestone. Wealden. Smokejack, England.

X 123: Crinoidal limestone. Middle Devonian, Belgium.

X 120: Algal limestone. Couvenian (Devonian). Bioherm near Couvin, Belgium.

X 20: Breccious limestone with crinoid remains. Carboniferous. East Belgium.

X 21: Fossiliferous limestone. Devonian. Coal mines near Hénin Liétard, Pas de Calais, France. (Courtesy P. Dollé.)

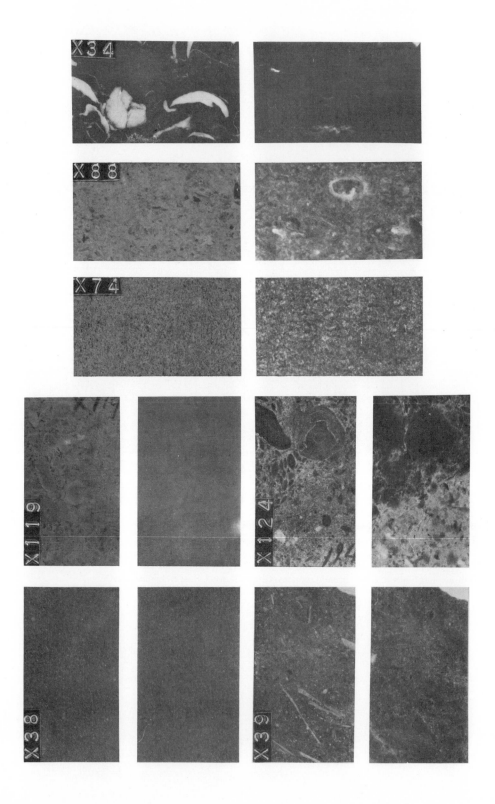

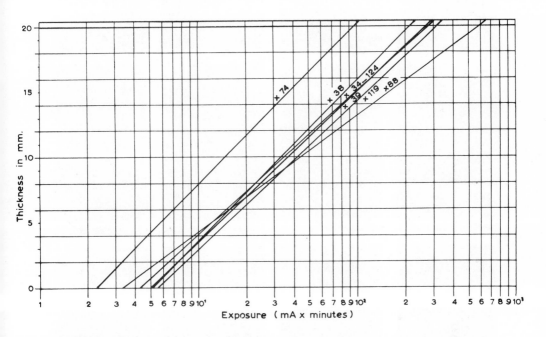

Fig. 3.29L. Exposure chart for various limestones. For explanation see text.

X 34: Limestone with brachiopod remains. Visean (Carboniferous). Ourthe Valley, Belgium.

X 88: Fossiliferous limestone: Splitska. Yugoslavia.

X 74: Bryozoan limestone: Jaune à grains. Hauteroche, France.

X 119: Coquina: Musauzy Capusine T. France.

X 124: Nummulitic limestone. Lower Eocene. Southern central Pyrenees, Spain.

X 38: Limestone with small nummulites. Eocene. Near Salzburg, Austria. (Van Hinte, 1963.)

X 39: Nummulitic limestone. Calcaire Nummulitique (Lower Eocene). Lucéram, Peira-Cava area, Alpes Maritimes, France. (Bouma, 1962.)

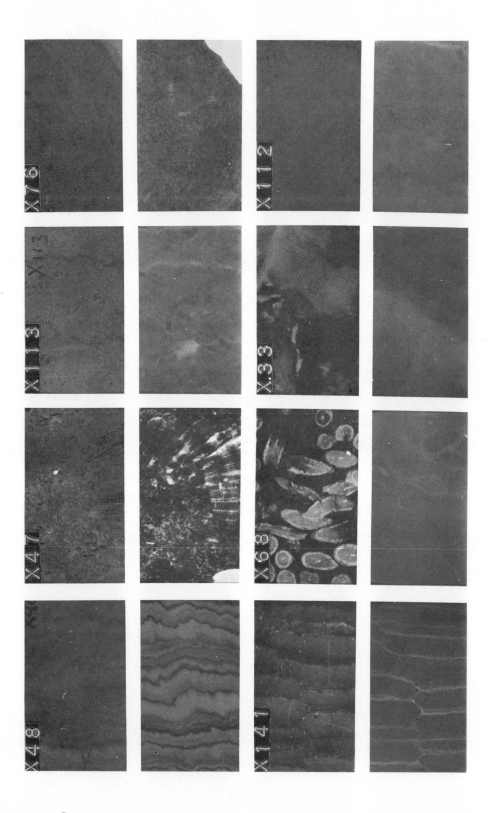

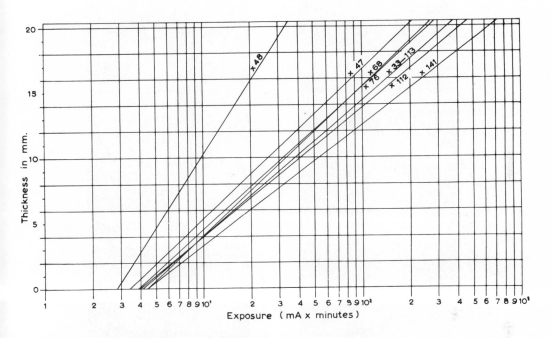

Fig. 3.29M. Exposure chart for some types of limestone. For explanation see text.

X 76: Limestone with small calcareous particles. Varzion Pointillé, France.

X 112: Limestone: Villefort ramagé B. Yonne, France.

X 113: Limestone with marmorized fossil remains: Villefort fleuri. Yonne, France.

X 33: Coral limestone. Devonian bioherm. Taillfer, Belgium.

X 47: Coral limestone. Miocene. Malta.

X 68: Coral limestone of Silurian age. Erratic boulder of Scandinavian origin. Hilversum, The Netherlands.

X 48: Dark-colored algal limestone. Miocene. Jaen, Andalusia, Spain.

X 141: Marly algal limestone. Visean V III b (Carboniferous). South Belgium.

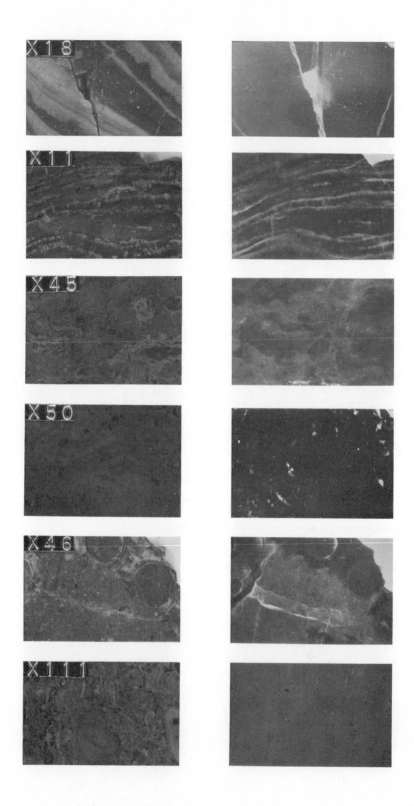

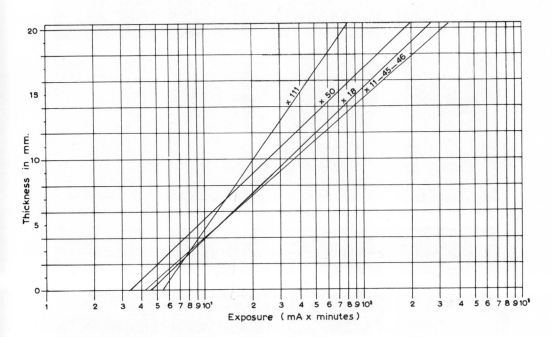

Fig. 3.29N. Exposure charts for algal limestones. For explanation see text.

X 18: Banded algal limestone. Visean (Carboniferous). Belgium.

X 11: Stromatopora limestone. Visean (Carboniferous). Belgium.

X 45: Algal limestone. Eocene-Oligocene. Priabona, Italy.

X 50: Algal limestone. Oligocene-Miocene. Malta.

X 46: Algal limestone. Campanian (Cretaceous). Pyrenees, Spain.

X 111: Algal limestone: Marmor Chanteuil, Bleu foncé (+). Yonne, France.

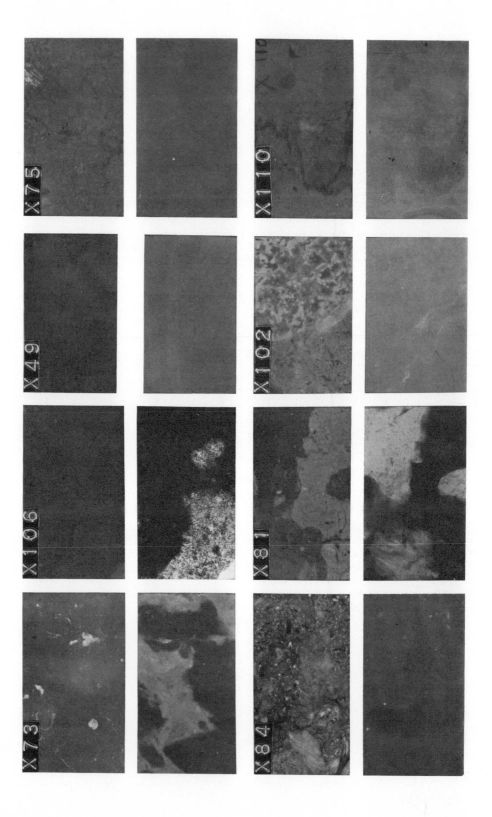

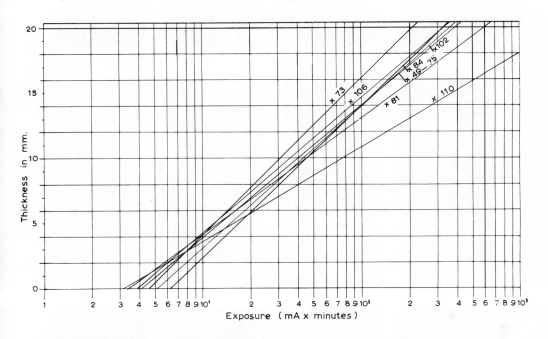

Fig. 3.29O. Exposure chart for various marmorized limestones. For explanation see text.

X 75: Marmorized limestone: Rose anemone. France.

X 110: Marmorized limestone: Marmor Chanteuil, Clair rubané. Yonne, France.

X 49: Marmorized limestone (dense, fine-grained). Upper Cretaceous. Priabona, Italy.

X 102: Marmorized limestone: Napoleon, France.

X 106: Marmorized limestone. Vauzion, France.

X 81: Marmorized limestone: Valorite F. France.

X 73: Slightly marmorized black limestone: Noir français. Nord, France.

X 84: Marmorized limestone: Montana. Yugoslavia.

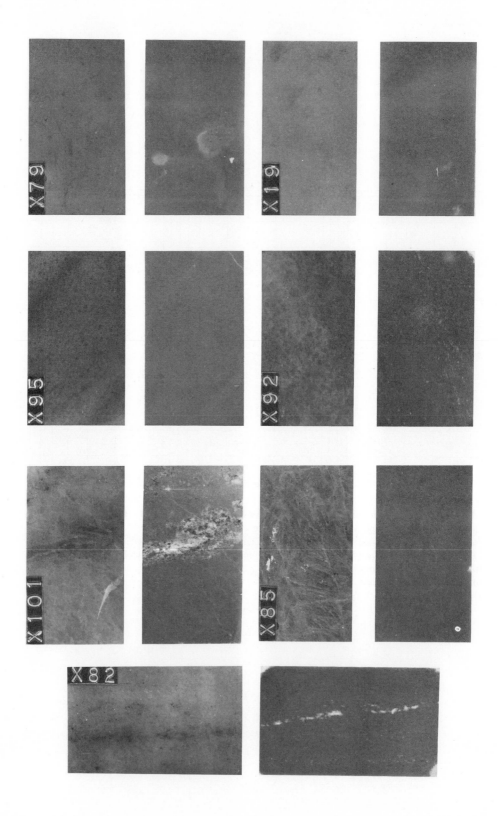

210

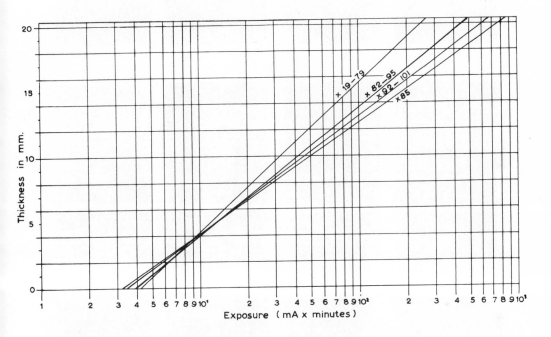

Fig. 3.29P. Exposure chart for various marbles. For explanation see text.

X 79: Brown-colored, highly marmorized limestone: Mabre doré. Valore, France.

X 19: Fine-grained white marble: Albast. Remich, Luxembourgh.

X 95: Medium-grained white-grayish marble, slightly layered: Drama. Greece.

X 92: Coarse-grained light gray marble. Switzerland.

X 101: Coarse-grained marble of light-green color: Verde Viana. Portugal.

X 85: Very coarse white marble: Filippi. Greece.

X 82: Coarse-grained, slightly banded, white marble: Cristallino 12. Norway.

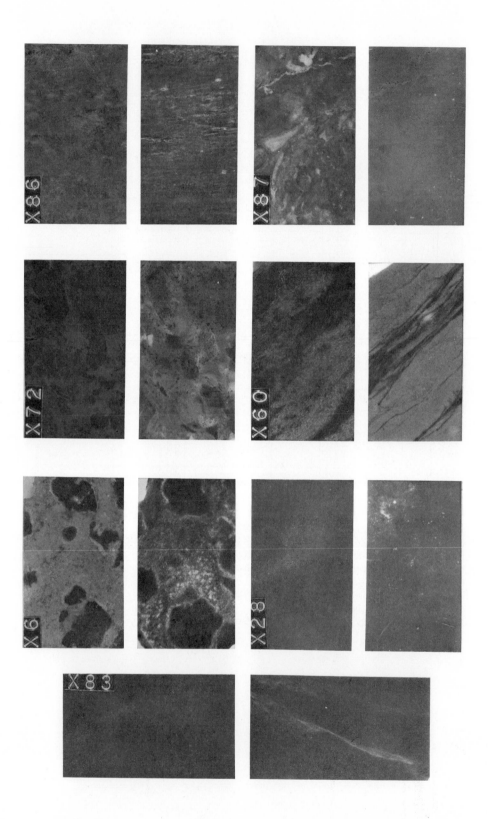

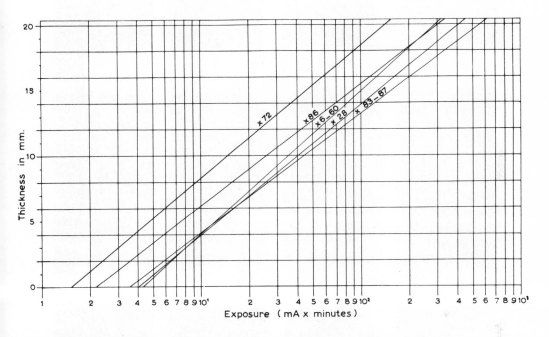

Fig. 3.29Q. Exposure chart for some marmorized limestones and marbles. For explanation see text.

X 86: Green marble: Verde Suède. Sweden.

X 87: Red marble with white calcitic veins which are partly recrystallized fossils: Rouge Royal. South Belgium.

X 72: Breccious marble: Vert Antique. Greece.

X 60: Crystalline limestone with contact minerals. Lles Aristot area, S. E. Pyrenees, Spain.

X 6: Garnet marble. Southwestern contact zone of the Panticosa granite, Spain.

X 28: Dense, light-brown marmorized limestone: Roman travertin. Bagni di Tivoli, Italy.

X 83: White, fine-grained marble with rose-colored bands: Rosé aurora. Portugal.

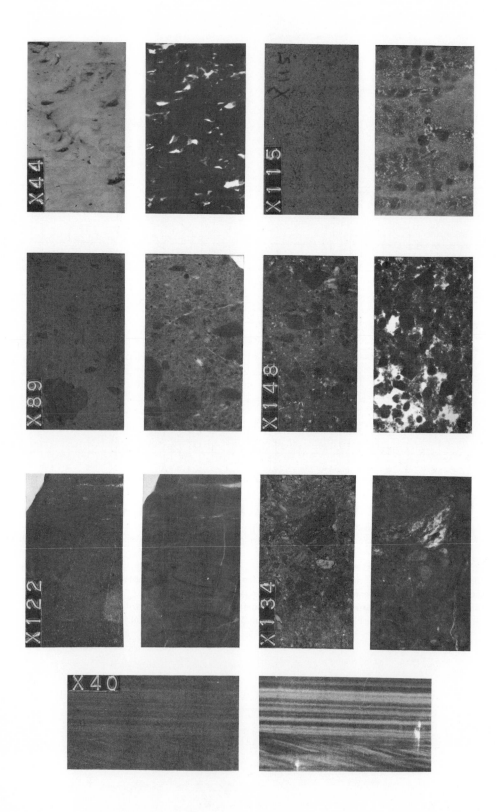

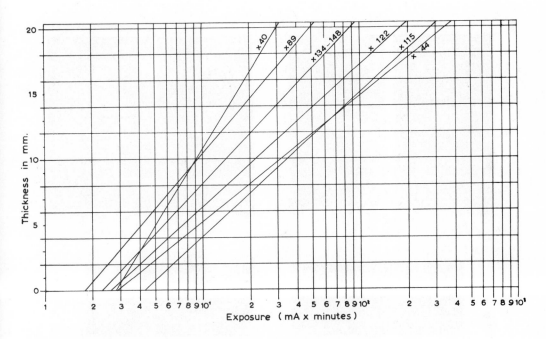

Fig. 3.29R. Exposure chart for volcanic materials, travertine, and oolite. For explanation see text.

  X 44: Travertine, Italy.

  X 115: Oolite: Baumont Dur. Charentes, France.

  X 89: Tuff. Lerchen, Germany.

  X 148: Lava. Vesuvius, Italy.

  X 122: Banded tuff. Northern Jutland, Denmark.

  X 134: Suevite. Das Ries, Henisfahrt, S. Germany.

  X 40: Banded tuff. Rhodos, Greece. (Courtesy E. Mutti.)

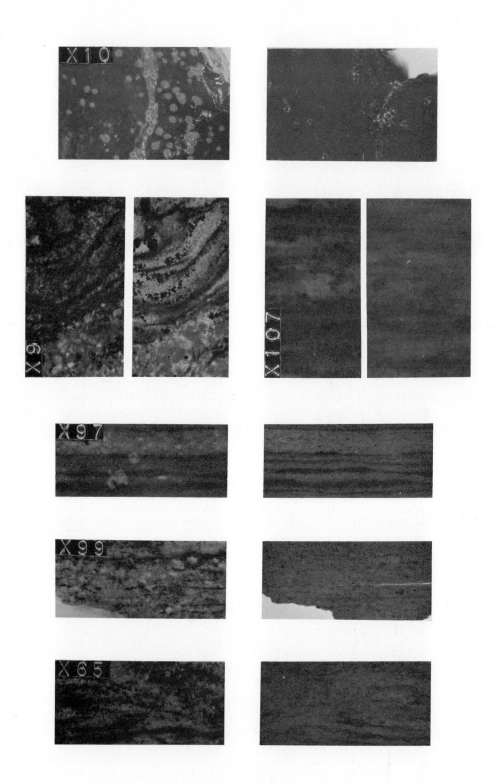

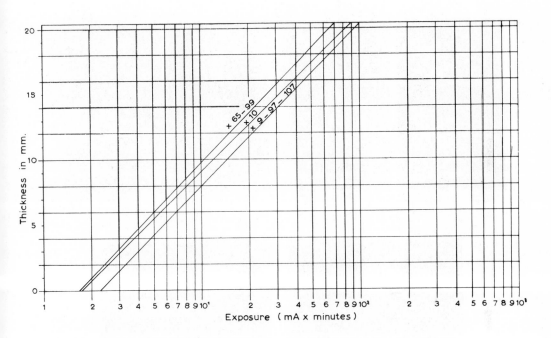

Fig. 3.29S. Exposure chart for some volcanic and metamorphic rocks. For explanation see text.

X 10: Obsidian. Iceland.

X 9: Granite gneiss. Erratic boulder of Scandinavian origin. Donderen, Drente, The Netherlands.

X 107: Gneiss. Zébraro/rayé. Northern Italy.

X 97: Gneiss: Polar quartzite. Snodal, Sweden.

X 99: Quartzitic gneiss. Brundal, Italy.

X 65: Chlorite schist. Hills along the river Mera near Chiavenna, Italy.

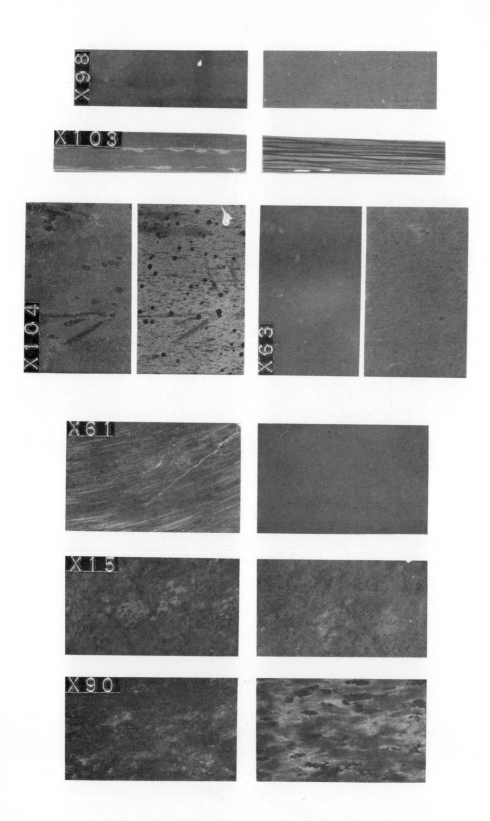

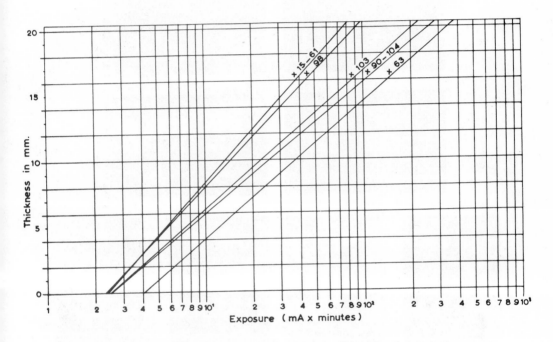

Fig. 3.29T. Exposure chart for different types of metamorphic rocks. For explanation see text.

X 98: Micaceous quartzite. Doredaj, Sweden.
X 103: Slate: black Finmarken. Norway.
X 104: Spotted slate: Sell royal. Norway.
X 63: Gneiss. Schapbach, Steinwasen (Schwarzwald), Germany.
X 61: Talc schist (with cutting marks). Pretoria, The Transvaal, S. Africa.
X 15: Gneiss with lindgrenite. Inspiration mine, Globe, Arizona, U.S.A.
X 90: Serpentine. Lugano area, Switzerland.

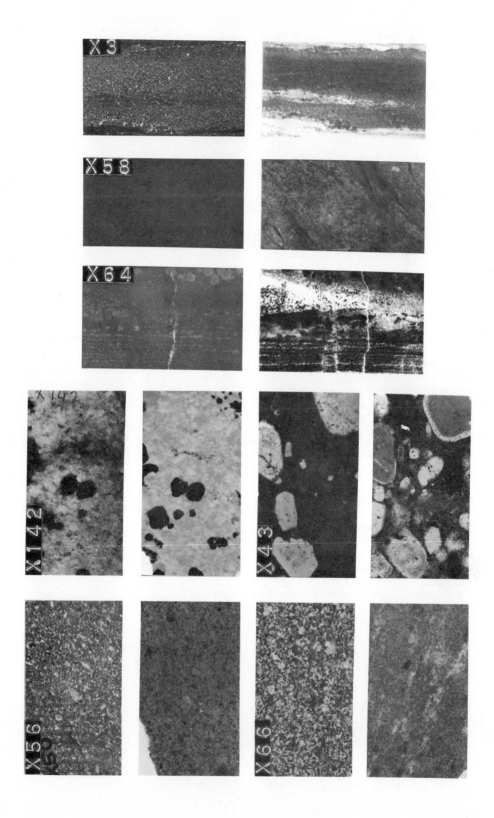

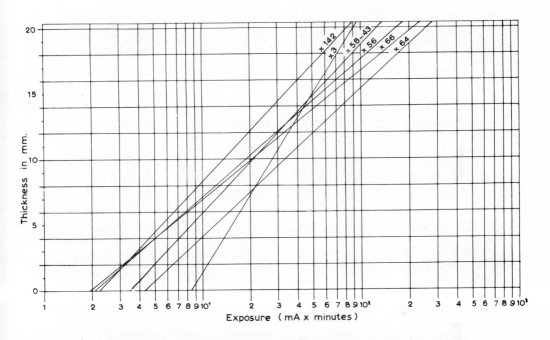

Fig. 3.29U. Exposure chart for various metamorphic and igneous rocks. For explanation see text.

X 3: Metamorphosed sedimentary iron ore with reorientation and growth of minerals (mainly pyrite). Lower Greenschist facies. Mine near Caldonazzo, east of Trento, Italy. (De Boer, 1963.)

X 58: Hornfels. Granite quarry near Nieder Ramstad, near Darmstad, Germany.

X 64: Chlorite-schist. Hills along the river Mera near Chiavenna, Italy.

X 142: Granite-pegmatite. Erratic boulder of Scandinavian origin. Texel, The Netherlands.

X 43: Rhomben porphyry. Erratic boulder of Scandinavian origin. Donderse veld, Drente, The Netherlands.

X 56: Hornblende andesite. Grosze Rosenau, Siebengebirge, Germany.

X 66: Rhyolite with quartz crystals. Gibbie's Pass, Lyttelton, New Zealand.

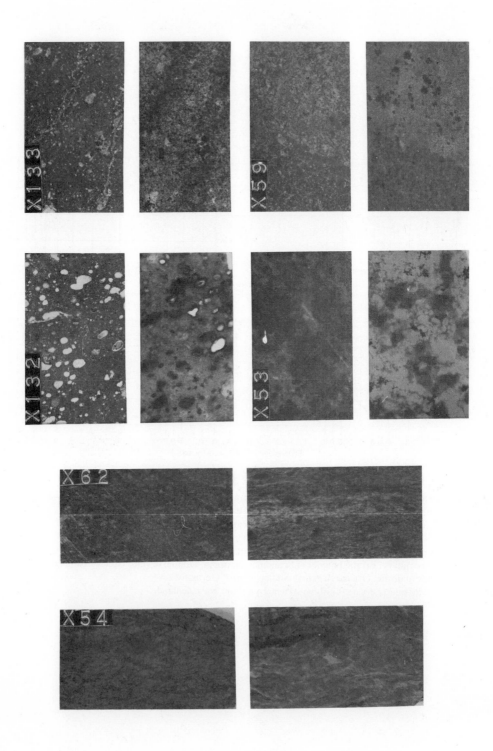

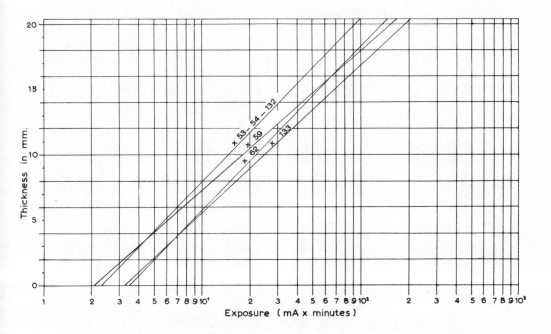

Fig. 3.29V. Exposure chart for some igneous rocks. For explanation see text.

X 133: Rhyolite. Rotliegendes (Permian). Schloszböckelheim, Mainzer Basin, Germany.

X 59: Eclogite. Edle Krone, Tharandt, Saxony, Germany.

X 132: Almond porphyry. Rotliegendes (Permian). Schloszböckelheim, Mainzer Basin, Germany.

X 53: Flaser gabbro. Hartenberg, near Rosswein, Saxony, Germany.

X 62: Amphibolite. Haslach, Schwarzwald, Germany.

X 54: Diorite. Unter Engadin, Graubünden, Switzerland.

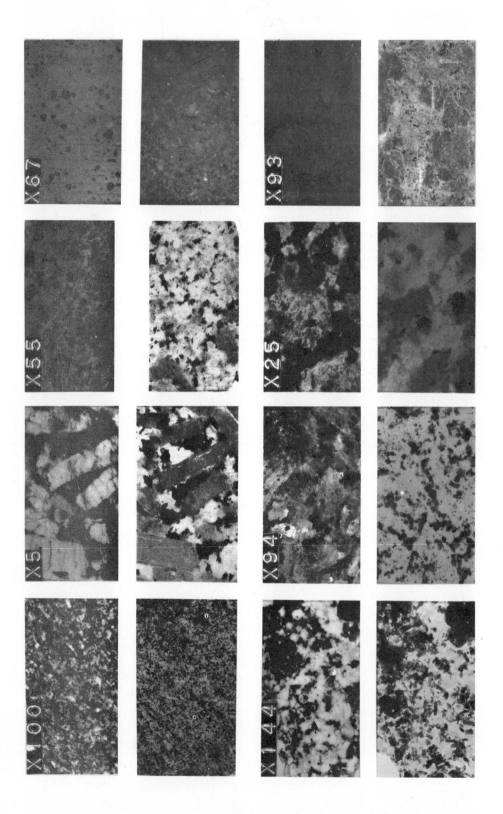

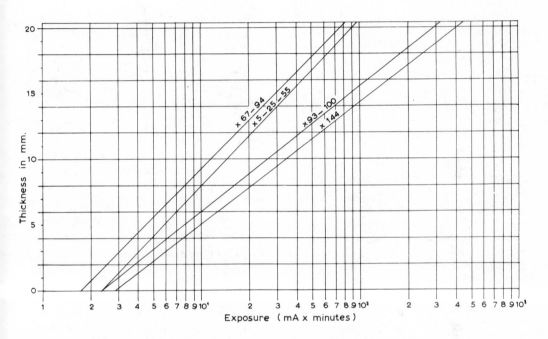

Fig. 3.29W. Exposure chart for various igneous rocks. For explanation see text.

X 67: Quartz-orthoclase granite (rhyolite). Erratic boulder of Scandinavian origin. Hilversum, The Netherlands.

X 93: Diabase. Hessen, Germany.

X 55: Amphibole-biotite granite. Vagney, Vosges, France.

X 25: Pegmatite: Red Balmoral. Finland.

X 5: Tourmaline pegmatite. Cape Cornwall, England.

X 94: Labradorite. Ojama, Finland.

X 100: Fine-grained granite. Finland.

X 144: Hooibergite. Hooiberg, Aruba, Caribbean Sea.

3.6 SPECIAL TECHNIQUES

A number of techniques beside the method of contact radiography as used in Section 3.5 are used in the field of industrial radiography. Some of these may be of interest to geologists.

The radiograph can be used as a negative, and therefore prints can be made. Even enlargements can be produced. It is also possible to obtain an enlarged radiograph, but enlarging diminishes the sharpness. If a radiograph is too dark, it has to be bleached before any print can be made of it. Since some radiographs contain overexposed and underexposed parts next to each other, one seldom can use them for copying work unless such instruments as electronically controlled printers are used.

Since all properties present in the sample are projected onto a two-dimensional sheet of film, one is restricted in the thickness of the sample (a confused picture, of course, is senseless) (Halmshaw, 1966). For thick samples it is recommended to apply stereoradiography. This method can also be used, as can the parallax exposure technique, to localize the depth of a certain property within the sample.

Small details can also be radiographed by means of microradiography, which enables the investigator to obtain results from very thin samples or from materials that have low X-ray absorption. Special single-coated films are made for this method.

3.6A Primary Enlargement

When the object is not placed in contact with the film, a penumbral unsharpness will result (see Fig. 3.4C). The thickness of the exposure holder and of the plexiglas tray, if used, can be neglected. Real penumbral unsharpness is obtained when the sample is placed high above the film in order to obtain a direct enlargement based on the same principle as the reverse square law (Fig. 3.6). In Fig. 3.30 it can be observed that the penumbral unsharpness increases to 1 mm if a sample is placed 30 cm above the film with a focus-film distance of 60 cm. This figure clearly demonstrates that the primary enlargement must be kept small and that the film-object distance should only be a small part of the focus-film distance. Figure 3.31 gives three examples. The first one (A) is the print of a normal contact radiograph. The distance between film and object is less than a few millimeters due to the cover of the exposure holder. The

film-focus distance is 100 cm. Figure 3.31B shows the same sample radiographed with an object-film distance of 20 cm; and Fig. 3.31C shows the radiograph of this 3-mm-thick slice with a film-object distance of 50 cm.

For this method we use a piece of plexiglas (15 x 10 cm and 3 mm thick) which fits in a holder. To the holder is mounted a ring with setscrews which fits around one of the round stands mounted on the base of the support (see Fig. 3.16).

Primary enlargement can be used to a certain extent. In many instances it is better to make enlargements of the radiographs (Fig. 3.31D).

3.6B Printing and Secondary Enlargements and Reducers

Copies of radiographs can be made in several ways using monochrome materials, or by the use of reversal color film (Kodak Ltd, 1965e). For making prints it is desirable to have radiographs that do not contrast too much. Otherwise, it will be necessary to reduce the darkness of some parts or the whole film.

Contact Copying at Film Material

As mentioned in Section 3.1B (Negative Duplicates), numerous films have characteristic curves that turn back at a certain moment (see Fig. 3.12). Such a "copy radiograph" is very useful when radiographs are used for examining purposes, as is done in industrial radiography.

Copy radiographs also can be made through the stage of a "reversed-tone radiograph" or "intermediate positive," both of which have dark and light reversed from the original radiograph. Instead of using paper material one can use certain types of transparent films.

Contact Copying at Paper

At the Scripps Institute of Oceanography contact prints on paper were made of the 17 x 14 in. radiographs made from sample slices collected with the large box samples. Only radiographs that did not have large contrast differences in many spots produced good results (Bouma, 1964e, 1965, 1969a,b).

A sheet of photographic paper was laid with its sensitive side up on a piece of black paper on a table or the floor of the darkroom just underneath the central white light. The radiograph was placed on top and covered with a plane-parallel glass plate. All this was carried out with the darkroom illumination. Next the central white

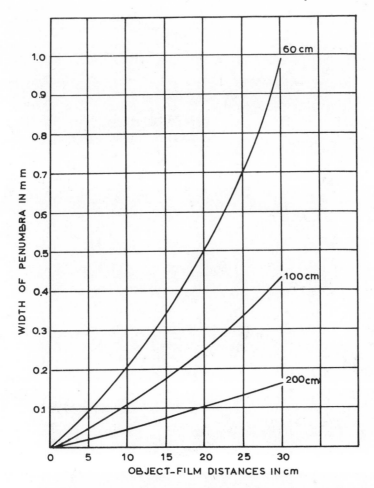

Fig. 3.30. Graph presenting the penumbral unsharpness. Three focus-film distances (200, 100, and 60 cm) are given, and there is a variation of object-film distances up to 30 cm. (Redrawn from Kodak Ltd., 1959.)

light was switched on for the time necessary, which had been determined before.

For further reproduction, negatives of 18 x 12 cm and also miniature size were made of the glossy prints. Prints made from these negatives were used for publication (Bouma, 1963, 1964e, 1965, 1969a, b; Bouma and Marshall 1964; Bouma and Shepard, 1964).

The disadvantages of this method are that too many steps are necessary to obtain a final reduced-sized, print and the obscure details of the radiograph disappear.

Small-sized radiographs can be printed at once, as were the 3 x 5 cm samples in Fig. 3.29. In spite of having many radiographs on one sheet of film, they were printed separately to obtain the highest quality. Instead of piling up paper, radiograph, and glass plate on the floor, paper and film were placed in a printing-frame, and a white lamp mounted on the side of the darkroom was used as light source.

This method has an advantage, especially for large prints, in that a long exposure time is required, therefore making it possible to black off lighter parts of the film by moving a mask above it. This can only be done if one or two parts or a side needs more or less light.

A smooth, glossy bromide paper is recommended; this is available in a range of contrasts.

Copying with the Help of a Camera

This certainly is a good method, but since sedimentologists are used to work with and to publish prints instead of the radiographs them-

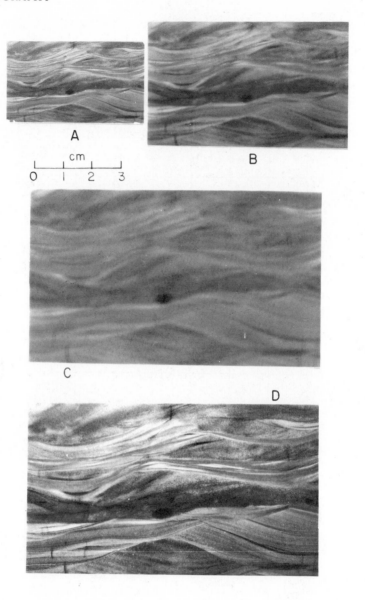

Fig. 3.31. Example (X 37 of Fig. 3.29F) of prints of radiographs with primary and secondary enlargement: (A) Print of original contact radiograph of a 3-mm-thick slice. (B) Primary enlargement of A with film-object distance of 20 cm. Film-focus distance 100 cm. (C) Primary enlargement of A with film-object distance of 50 cm. Film-focus distance 100 cm. (D) Secondary enlargement of radiograph made from A.

selves, it is not useful in this field, unless reversal films are used or contact negatives are made from the reversed-tone transparencies.

It is ideal to have a sturdy setup and to use a camera with ground-glass focusing. This may be a large plate camera with a long bellows extension or a single-lens reflex (miniature) camera with extension rings.

An illuminator is used as light source, against which the radiograph is mounted. All illuminated areas beyond the edges of the film should be masked off to prevent flare in the camera.

The camera must be perfectly rigid during the period of exposure; its optical axis should be at right angles to the center of the radiograph. A rail on the table on which the illuminator is placed or a sturdy tripod or stand will hold the camera (Kodak Ltd, 1965e, Fig. 2). With an exposure meter held near the position of the camera-film the light can be measured.

When working with roll film or miniature film it is recommended to make trial exposures such as the reading obtained from the light meter, one stop lower, and one stop higher. If very light parts are present on the radiograph, they can be blocked during part of the exposure, using a mask that is moved over it (for further details see Kodak Ltd, 1965e; Thorpe and Davison, 1944; Brownell, 1951; Gibson, 1951, 1962).

Copying at Projection Slides

With the camera method one can get positive pictures on transparent material which can be used directly as projection slides.

One can also use color films for copying radiographs. Reversal color films have to be used if the investigator wants a copy in original dark-light relation of the radiograph, whereas negative color film should be used for positive copies. The advantage of color film is its sensitivity to weak nuances in blackening which is somewhat higher than that of black and white film.

If the contrast of the radiograph is considered too high to be copied satisfactorily, one should give the film a short pre-exposure. The amount of pre-exposure depends on the contrast of the radiograph and the intensity of the light source used. As a guide, however, the exposure should be such that the pre-exposed portion of the film should be just perceptibly lighter than the unexposed border; additionally, the combination of the pre-exposure and the main exposure should reduce the contrast to an extent where detail in the densest regions of the radiograph is satisfactorily recorded in the copy. If this technique is employed, the camera must, of course, be capable of permitting double exposures (Kodak Ltd, 1965e). The use of certain filters may correct color differences between the types of opal glass of the illuminator and the film base.

Secondary Enlargements

When radiographs are not too large they can be placed in a photographic enlarger, as can normal negatives, and prints can be made. Figure 3.31D gives a secondary enlargement of the same size as C and of the same radiograph. Since no penumbral unsharpness is dealt with, magnificent enlargements can be obtained. This may be desirable for studying details. The only restrictions are the size of the radiograph, since it has to fit the enlarger, and the graininess of the radiograph. If large enlargements are desired, the use of very fine-grained X-ray films is recommended.

Chemical Action for Reducing Contrast of Radiographs

It has been mentioned previously that radiographs can be too contrastive to make good prints. Besides working with masks, which are moved over the lighter parts to throw more light on to the darkest parts or to pre-expose a film or a sheet of photographic paper, one can try to lower the contrast of a radiograph with chemical solutions (Tables III.8 to III.13).

The best results are obtained with radiographs that are only a few days old, since dried radiographs are difficult to handle. Unless information can be obtained from an experienced photographer, the proper reducer to use must be found by trial and error.

Table III.8.   Kodak R-1 Persulfate Reducer, a Superproportional Reducer for Overdeveloped Negatives of Contrasting Subjects

Ammonium persulfate	60 g
Sulfuric acid (concentrated)	3 cc
Water	1,000 cc

(After Kodak Ltd (1962c).

Note. The sulfuric acid should be added to the water very gradually under constant stirring. For use, take one part of this stock solution and two parts of water. When the reduction is complete immerse the negative in an acid fixing bath for a few minutes, then wash and dry it.

Table III.9. Permanganate Reducer

Stock solution A	
Potassium permanganate	52.5 g
Water	1,000 cc
Stock solution B	
Sulfuric acid (concentrated)	32 cc
Water (cold)	1,000 cc

(After Kodak Ltd (1962c).

Note. Add the acid to the water very gradually under constant stirring. This is a reducer for general use. It reduces density without loss of contrast. The stock solutions keep well but the mixture should be used immediately. For use, take 1 part of solution A, 2 parts of solution B, and 64 parts of water. When the negative has been sufficiently reduced, immerse it in a 2% solution of sodium bisulfite to remove the stain. Fix the negative in a fresh acid-fixing bath for a few minutes, after which it must be washed thoroughly. It is very important to wash the negative thoroughly before this permanganate reducer is applied, since irridescent irremovable scum may appear on the negative after drying.

Table III.10. Farmer's Reducer

Stock solution A	
Potassium ferricyanide	75 g
Water	1,000 cc
Stock solution B	
Sodium thiosulfate (hypo) (crystalline)	240 g
Water	1,000 cc

(After Kodak Ltd (1962c).

Note: This is a cutting reducer for correcting overexposed negatives and to clear shadow areas. Solutions A and B should not be mixed until use, since they will not keep long in combination. For use, take 1 part of solution A, 4 parts of solution B and 27 parts of water. Pour this solution at once over the negative and watch it closely. The action can be best controlled when a white dish is used. Once the reduction is sufficient, the negative should be washed thoroughly before drying.

Table III.11. Kodak R-4b, Two-Solution Farmer's Reducer

Solution A	
Potassium ferricyanide	7.5 g
Water	1,000 cc
Solution B	
Sodium thiosulfate (hypo) (crystalline)	200 g
Water	1,000 cc

(After Kodak Ltd (1962c).

Note. This two-solution reducer has the advantage of giving almost proportional reduction, thus correcting for overdevelopment by lowering contrast. Treat the negative in solution A with uniform agitation for 1 to 4 min at 18 to 24°C (65 to 75°F), depending on the degree of reduction desired. Immerse the negative next in solution B for 5 min and wash thoroughly. The process may be repeated if more reduction is desired. For the reduction of general fog, one part of solution A should be diluted with one part of water.

Table III.12. Kodak R-5 Proportional Reducer

Stock solution A	
Potassium permanganate	0.3 g
Sulfuric acid (10% solution)	16 cc
Water	1,000 cc
Stock solution B	
Ammonium persulfate	30 g
Water	1,000 cc

(After Kodak Ltd (1962c).

Note. The sulfuric acid should be added to the water very carefully and with constant stirring. This reducer gives a proportional reduction of the contrast of the negative. It treats high-contrast more intensively than low-contrast parts. For use, take one part of solution A and three parts of solution B. When the reduction is sufficient, clean the negative in a 1% solution of sodium bisulfite. The negative must be washed thoroughly. In order to obtain consistent results use distilled water (or water free of iron) for making up the stock solutions.

Table III.13. Superproportional Reducer (Compare with Table III.8.)

Ammonium persulfate	25 g
Sulfuric acid (10% solution)	10 cc
Water up to	1,000 cc

(F. Linschoten, personal communication, Dalco N.V. Photographic Industry, The Netherlands).

Note. The persulfate has lost its chemical activity if it does not give a cracking noise during dissolving. With this reducer the lightest parts become somewhat more transparent, the medium dark spots become reduced proportional and the dark part will be reduced strongly.

Electronically Controlled Printing

Since many radiographs are very contrastive and have a wide variation in grey shades, it is very difficult to obtain good prints in spite of masking parts with or without prior use of a contrast reducer. The help of a device that automatically corrects the density variations, called an Electronic Contrast Controlled Printing System (St. John and Graig, 1957; Hamblin, 1965), is the only way to obtain a 100% good print. Milligan Electronics Ltd. (England) now manufactures the Rank Cintel Ltd. electronically controlled printers, copiers, and enlargers (Cintel, not dated, 1961). The devices are used for photogrammetry, radiography in medicines and industry, press photography, and block making. The conventional type of exposure lamp is replaced by a high-intensity, cathode-ray tube. The tube produces a small spot of light due to the bombardment of a phosphor screen by an electron beam. This spot of light is projected through the radiograph and paper onto a photoelectric cell. This cell records the quantity of light which provides a voltage to control the intensity of the light from the cathode-ray tube. The light spot is much smaller than the negative, which allows the instrument to scan over the radiograph. Each scanning line is traced at a linear rate, and total exposure is achieved by causing the spot to transverse the entire area of the radiograph in a series of parallel lines (about 300 sequential lines, which overlap to insure complete exposure). In Bouma (1964e, 1965) several self-made prints are given, mainly of submarine canyon samples. Figure 3.32 presents one of these prints and also the electronically controlled print made by Milligan on a Milligan Electronic Contrast Controlled Printer.

3.6C Stereoradiography

A three-dimensional X-ray picture of a sample can be obtained by the method of stereo-

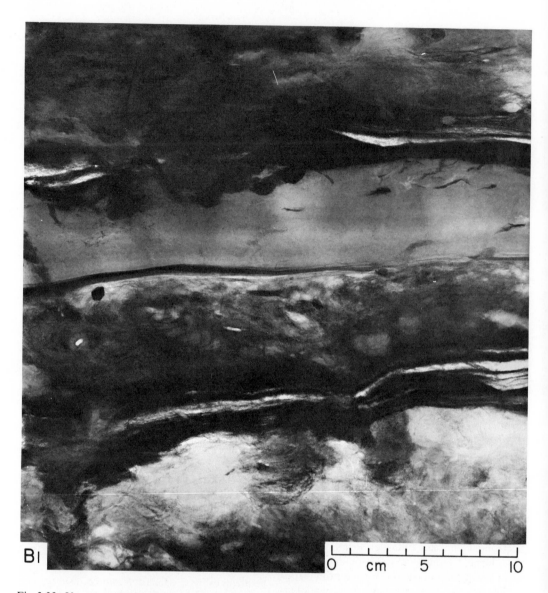

Fig. 3.32. Uncorrected print (B 1) and electronically controlled print (B 2) of the same radiograph. Radiographs with high contrast were chosen to demonstrate the difference between printing techniques. Box sample San Juan #46; 1035 fathoms, Pescadero Canyon. General Electric Mobil X-ray unit "90-II," focal distance 38 in., 50 mA, 40KV, 2⅓ min, Kodak Industrial film type AA, slice thickness ½ in. Alternation between muddy sand (with irregular lamination, burrows, type of loadcasts and pseudonodules) and clay with few burrows (?). Bottom part a sandy mud with mottled structure. Exposure times for printing: uncorrected, 16 sec; corrected, 44 sec. (Print made as gift by Milligan Electronics Ltd.)

232

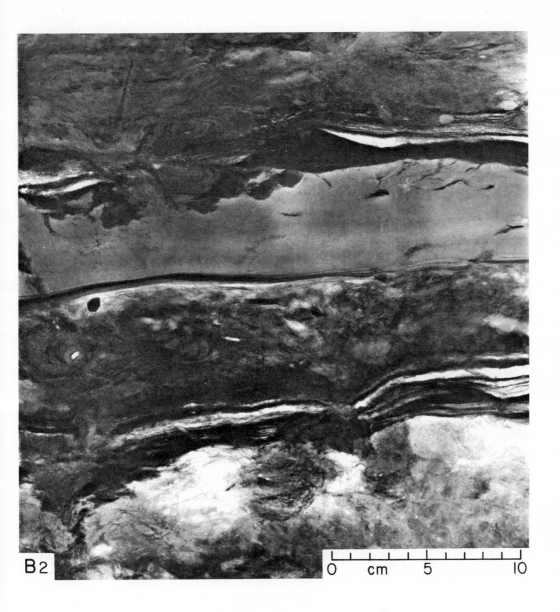

Fig. 3.32 (Continued)

radiography. This technique allows the determination of the relative position and depth of particles, the spread of inclusions, and the study of the three-dimensional shape of sedimentary structures and of pores.

Two similar exposures are made on two sheets of film; the only difference is that the position of the radiation source with regard to the sample is moved over a certain horizontal distance (Fig. 3.33) (Kodak, 1957; Eering, 1965). In practice it is easier to move the sample over a distance of 6 or 8 cm at right angles to its longest axis. The author applies the following technique to all his samples (primarily cores or slices of cores): place the middle of the sample exactly underneath the center of a vertical X-ray beam and mark this spot in such a way (slight dent) that it does not become visible on the radiograph. Move the sample 4 cm to the right parallel to itself for the first shot and then 8 cm to the left for the second exposure. Besides the usual code in lead letters and numbers, an R is placed for right and an L for left. The radiographs can be studied on a lightbox under a stereoscope, or prints can be examined (Fig. 3.34).

Since it is difficult to know what is up and what is down in a sample, it is advisable to place a lead mark (for example, B) underneath the specimen and another mark on top.

The sideway movement has to be carried out very precisely to obtain the same stereographic effect throughout the whole sample. Special attention must be given to the second move since another film has to be placed underneath. Non-plane-parallel samples should be imbedded (see preceding section) to get optimum results.

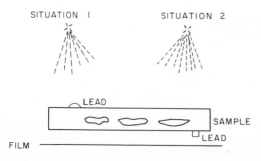

Fig. 3.33. Schematic presentation of the stereoradiography principle. First an exposure is made according to situation I and next a similar exposure according to situation 2. (Redrawn from Eering, 1965.)

Many stereoradiographs have been made. Experiments with core 66-A-13 #25E (Fig. 3.28) show that good results can be obtained when the laminations in the complete core, half-core, or slice are at right angles to the core axis, whereas no stereo effect could be collected from the oblique cut slices.

Similar stereo effects were obtained from samples of which just a normal radiograph was made, and where the specimen was positioned directly underneath the center of the radiation beam. Either the radiograph proved to be good and to present interesting aspects for a stereoscopic study, or second exposure was made easily by placing the sample 6 to 8 cm to the left or the right of the center. No differences could be detected between a normal stereoscopic set where the sample was placed at the right-hand side of the beam axis, as well as on the left-hand side, and a set with one vertical exposure and one oblique exposure.

The advantage of this procedure is that if a radiograph is processed directly and the specimen left in place, no measurements on horizontal movements have to be made for the first exposure. Once the radiograph indicates that a stereo effect is desirable, a second exposure can easily be made.

3.6D Parallax Exposure (double exposure method)

The parallax exposure method can be applied to locate the position of a certain property within a sample (Gevaert, not dated; Kodak, 1957; Kodak Ltd, 1964g; Eering, 1965). For this method the X-ray source is moved over a short horizontal distance without changing the vertical distance between source and film. Two exposures are made: if for the first exposure only $\frac{3}{5}$ of the total exposure is used, then for the second position the remaining $\frac{2}{5}$ is used. The radiograph thus consists of two overlapping images.

Different methods may be applied to obtain the double exposure effect (see also Kodak Ltd, 1964g and references cited). Only two techniques will be given here.

Measuring Distance from Object to Film

It is necessary to place the sample directly on the film or its cassette and to measure the horizontal movement of the X-ray source very accurately (Fig. 3.35). If M is the horizontal movement of the X-ray source, B_1 and B_2 the beams

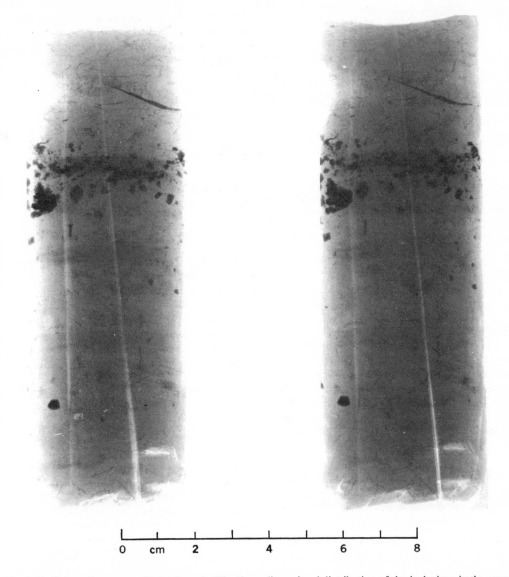

Fig. 3.34. Prints of a stereoradiography pair. The three-dimensional distribution of the inclusions in the upper part stands out clearly when the prints are observed under a stereoscope. Beaumont clay (Pleistocene), E. Texas coast, sample E 66–73 #13; depth 3002 cm, round core with diameter 1 ¾ in., core imbedded in sand for exposure; Picker "Hotshot" X-ray apparatus, focal distance 39 in., 4 mA, 80 KV, 3 min. (Sample received from Prof. S. J. Buchanan, Texas A&M University.)

that are effective with respect to the enclosure, E and D the distance from source to film, and S the distance of both images obtained from the enclosure E, the following equations can be given:

$$\frac{D-d}{M} = \frac{d}{S} \quad \text{or} \quad d = \frac{D \times S}{M+S}$$

in which d is the distance from the enclosure to the film. Suitable values for D and M are 90 cm (36 in.) and 30 cm (12 in.), respectively (Kodak Ltd, 1964g).

When the enclosure is linear, the source shift should always be at right angles to the long axis of the enclosure. It should be kept in mind that d is the distance from the enclosure to the film,

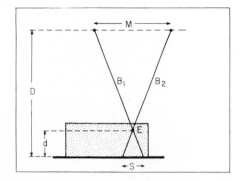

Fig. 3.35. Schematic presentation of the parallax exposure method (measuring distance from object to film). For explanation see text. (Redrawn from Kodak Ltd., 1964g, Fig. 1).

and consequently if a plexiglas tray or a film cassette is used, their thicknesses are incorporated.

Marker Method

Two small lead markers, preferably wires, are placed in contact with the sample, one on top and one at the bottom. The sample need not be in absolute contact with the film or cassette. The tube shift again is 10 to 12 in., but it need not be measured accurately. Figure 3.36 shows that the displacement X_3 on the film caused by both exposures of enclosure E lies in between the displacements X_1 and X_2 of the upper marker (M_1) and lower marker (M_2). The lead markers should be placed parallel to the major axis of the enclosure, and the tube shift should be at right angles to this major axis.

When d_v is the distance of the enclosure above the base of the sample and d is the thickness of the sample, the following equation can be used to calculate d_v:

$$d_v = d\,\frac{X_2 - X_3}{X_2 - X_1}$$

3.6E Microradiography

If details of a sample are required which are too small to be seen or to be studied with the naked eye, and if necessary enlargements create too great a degree of graininess in the film (Section 3.6B), the investigator can apply the principle of microradiography. Soft radiation, generated in the range of 5 to 50 kV, is normally used. The film usually is single-coated and finer grained than ordinary X-ray films (Engström and Lindström, 1951).

Goby (1913) reported contact microradiography as a new application of X-rays. This contact method is well known in biology and medicine (Lamarque, 1936; Bohatirchuck, 1942, 1961, 1963; Barclay, 1947; Engström, 1949; Mitchel, 1963) as well as in industry (Kodak Ltd, 1962e; Andrews and Johnson, 1963).

Projection microradiography is a primary enlargement method comparable to a normal microscope and an electron microscope. However, the apparatus is very complicated, and its application to the study of sedimentary structures must still be proven. Information can be obtained from Cosslett and Nixon (1955), Ong Sing Poen (1963), and Jongbloed (1965).

Fig. 3.36. Schematic presentation of the parallax exposure method (marker method). For explanation see text. (Redrawn from Eering, 1965.)

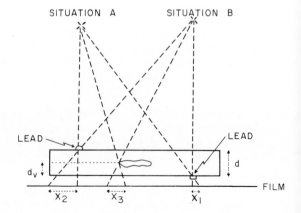

Contact Microradiography

Extremely good contact between specimen and film or exposure holder is required. A vacuum exposure holder may produce the best results, but a mechanical jig can be used as well (Kodak, 1957). The operator must be sure that any material placed between X-ray tube and sample has no marked structure and that its absorption is very low. Certain sheet plastic materials contain chlorine, which may absorb soft radiation.

The focus-film distance usually ranges from 7½ to 30 cm (3 to 12 in.). The best setup is to pack a very thin (about 1 mm) sample slice with the film together in black paper and to place this on top of a sheet of lead to prevent back-scattering. A frame, which fits around the sample, is placed on top to avoid side-scattering. The single-coated film is developed normally and later used as negative in an enlarger or directly placed under a microscope or binocular. Enlargements up to about 12 to 35 times can thus be obtained (Fig. 3.37), although special films and developing techniques are necessary for higher magnifications (Kodak, 1957).

3.6F Soils

X-ray radiography can be applied very successfully to the field of pedology. Remnant sedi-

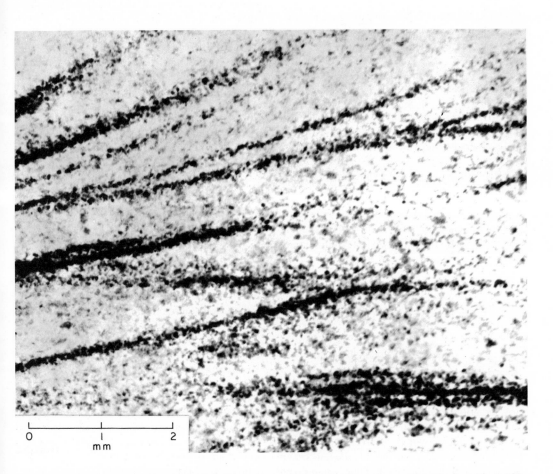

Fig. 3.37. Print of a contact microradiograph. The radiograph was made from a 3-mm-thick slice (see Fig. 3.29E, X 26) on Kodak Fine-Grain Positive Film. The sample was placed directly on top of the film inside the exposure holder to prevent penumbral unsharpness. Focal distance: 14 in., 5 mA, 40 KV, 4 min. The slice actually is too thick for microradiography. The radiograph was placed in the enlarger and a maximum enlargement was made. Exposure time 40 min, diaphragm of the enlarger two stops closed.

mentary structures can be detected and soil particles can be studied. Three-dimensional radiography is applicable as an additional tool when size and spacing of pores and cracks have to be investigated, especially when this technique is combined with a study on large-sized, thin sections (Slager, 1964, 1966). Charlesby (1963) reported on the effects of X-radiations on plastics and polymers. From his study it can be seen that "short" exposures are not harmful to these artificial products.

Radiography proves to be an enormous help when the investigator wants to study the disturbing influences of plant roots (see Fig. 3.29C).

3.6G Fossils

The application of X-ray has been known for many decades to paleontologists. Lehmann has been using stereoradiography since 1932, and Stürmer (1965) has already presented pictures obtained by microradiography. Many authors besides these have contributed to this technique. Hamblin and Van Sant (1963) and Farrow (1966) are writers who give excellent bibliographies on the use of X-ray radiography to the field of paleontology.

3.6H. Fluoroscopy and Photoradiography

Many other applications of radiography are known in industry, medicine, and biology. At the present time no publications on these methods have been published by the sedimentologists.

Fluoroscopy

The technique of fluoroscopy can be applied for scanning purposes. The X-ray image is not recorded on a film but can be observed visually on a fluorescent screen (Kodak, 1957). The advantages of this method are high-speed operation and low costs.

It is necessary to have a complete lead-lined cabinet around the X-ray apparatus to prevent any radiation from going toward the investigator. The image can be studied on a lead glass viewing window, which receives the image of the sample through a mirror from the fluorescent screen. The use of image amplifiers as an extension of fluoroscopy can be applied to increase the screen brightness.

The technique of fluoroscopy (Kodak, 1957) is generally said to have three limitations:

1. It is impractical to examine thick, dense, or high atomic number specimens.
2. The sensitivity of the fluoroscopic process is much smaller than that of radiography, which may give the operator false information since samples may look more homogeneous than they actually are.
3. There is no permanent record.

Photoradiography

Instead of using an X-ray film or a direct observation from a fluorescent screen, the image of the fluorescent screen can be photographed directly with a camera or even with a miniature camera.

The setup for photoradiography is partly similar to the one for fluoroscopy with the main difference that the mirror and the lead glass viewing window are replaced by a lead hood—the camera may be placed at this end (Kodak, 1957, Figs. 63 and 65).

This method certainly is cheaper than normal radiography, but the fine detail from the radiograph will be missing due to graininess and diffusion in the fluorescent screen and the graininess of the film.

3.7 PROTECTION

The most important consideration during radiographical work should be the operator's protection against radiation. Every country has its own precautionary rules, and therefore anyone who wishes to carry out radiography should consult these national codes and act accordingly. It is also desirable, and sometimes obligatory, to have a qualified radiation expert examine the radiation installation and the safety scheme (see further ICRP publication 3, 1960).

Any of the body tissues can be injured by overexposure to radiation. Blood, skin, eyes, and some internal organs are particularly sensitive. Tissues with cell division are most sensitive at the moment of dividing; for example, in bony tissue this would result in a change of the blood constitution. Mucus secretions covering stomach and intestinal canal are also very sensitive, and their overexposure will cause queasiness (roentgen sickness). Hair may fall out and skin may show burn marks.

Personal protection against radiation requires the use of special apparatus since the X-rays and gamma rays cannot be detected directly. For purposes of comparison, several units have been introduced—the roentgen, rad, and rem (International Commission on Radiological Units, 1950: Kodak, 1957; Adams, 1963a, b; Daniels, 1963; Gianturco, 1963; Roberts, 1963; Tsuda, 1963; Warrinkhoff, 1963).

3.7A Measurement of Radiation

Dosimeters are instruments to detect and measure the radiation at a certain place. Adams (1963c) divides the detectors of radiation into three groups: ionization chambers, counters (gas multiplication, proportional, flow proportional, Geiger-Müller, and scintillation counters), and photographic films (see also Philips, 1963b; Krebs, 1963; Potsaid, 1963; Ritz, 1963). The author has worked with only three detectors, used in Utrecht regularly.

The Geiger-Müller counter is a Philips P.W. 4014. It is actually based on the principle of ionization but is classified under counters. A meter indicates the equivalent of counts per second in milliroentgens per hour.

Generally this apparatus indicates values that are too high when working with X-ray tube voltages over 40 kV and too low for kilovoltages below 40 kV. Under 20 kV no radiation can be registered. The apparatus is operated by two buttons.

The ECKO ionization chamber has a chamber connected to an electrical (batteries) source and to the measuring part of the apparatus. In this chamber a metal plate is mounted; this is connected to the positive side of the electrical source. The chamber is filled with air (ECKO, not dated). The measuring principle of the dosimeter is similar to the definition of a roentgen.

The film badge is a third dosimeter every operator has to deal with when working with radiation sources, as is stated by local regulations. The film badge consists of one or two filmstrips with some filters packed in a lightproof cover. The films are personal and worn on the chest. At regular intervals they are developed and their photographic density is measured. The disadvantage of these dosimeters is that one never knows at a given moment how much radiation has been received (Kodak, 1957).

A small ionization chamber in the shape of a fountain pen also may be useful. When such an instrument is held against the light, one can read the scale. The setting at zero can be carried out by the operator. There are special types for work below 80 kV and others for radiography above 80 kV. Since the operator can measure the dose at any moment, its use is recommended if one has to deal regularly with X-rays.

3.7B Protection against X-Rays

In addition to using dosimeters and detectors of radiation, one has to take other precautionary measures (see Fig. 3.19).

The scientist working with a radiation outfit has to contact the local radiation authorities, be familiar with the local radiation rules, be certain that every person working with the outfit wears his film badge, take care of the regular change of film badges, and organize a regular medical examination. He must also be certain that the instructions (see Table III.3) are checked by the authorities and that they are placed in a convenient location. Every operator must know the basic principles of radiation and the safety precautions.

When working with low kilovoltages in the study of sediments, the best protection is lead. This metal combines high protective efficiency with low costs and is readily available. Other materials find application as protection against radiation. Structural walls of concrete or brick may afford considerable protection and therefore reduce thickness and costs of the lead required.

When lead is used care must be taken that the fastening leaves no leaks in the shielding. Adjacent sheets of lead must overlap, even if the sheets are welded together. The heads of nails or screws which pass through the lead should be carefully covered with lead (Kodak, 1957). Lead also can be glued between two pieces of thick plywood.

Besides precautions taken for floor, ceiling, and walls of the exposure room, extra precautions should be taken at all points where water pipes, electrical conduits, or ventilation ducts pass through the wall. Some examples are given in Fig. 3.38. Use of an antiscatter cone (Fig. I.1 in Appendix I) decreases the amount of scattered radiation enormously.

One has to be careful with lead protection when high kilovoltages are applied since lead-fluorescence (K-radiation) may result.

It is also possible to build a cupboard, covered with lead, around the X-ray unit. The total costs of a cupboard of about 1 x 1 and 2 m high will be less than the cost of the shielding of a larger

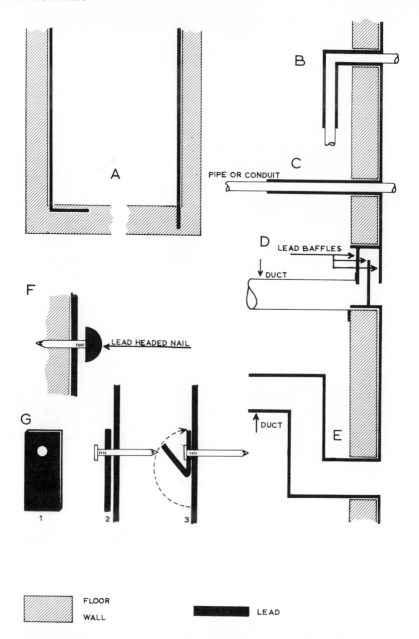

Fig. 3.38. Methods of handling protection against radiation of walls (A) with and without floor protection. Methods of shielding pipes (B), conduits (C), and ducts (D, E) when they have to go through a wall of the X-ray exposure room. Methods of sealing nail or screw holes in lead protection (F, G). (Redrawn from Kodak, 1957.)

room. Care must be taken that the door closes tightly, permitting no leaks. The disadvantage of such a cupboard, when placed in buildings, is that little space remains for work, but on board ships it may be of considerable advantage to use a protection cupboard.

An X-ray unit has to be operated with a voltage over 10,000 kV before any radioactivity can be induced in the material irradiated. Therefore *only* the *direct beam* and the *scattered radiation* during the period the X-ray unit is working is dealt with.

3.8 UNSATISFACTORY RESULTS AND THEIR POSSIBLE CAUSES

The interpretation of the image obtained from a sample often is very difficult. Very few publications give a description on the different characters visible on the radiograph (Hamblin, 1962, 1965; Bouma, 1964e, 1965). Besides the properties of a sample which may be found in the image, other patterns may be present which are due to artificial influences on the sample (Bouma, 1964e) or due to an error in one or more steps of the procedure of the X-ray technique (Kodak, 1957; Kodak Ltd., 1961; Gevaert, not dated). Very good examples can be found in the references cited. The most common ones will be discussed.

3.8A Density

A proper density for reproduction purposes normally gives the best radiograph on which most of the image characteristics can be detected. However, for certain parts with higher or lower absorption properties than the mass of the sample, it may be necessary to overexpose or underexpose the specimen.

A wrong density generally is due to overexposure or underexposure, overdevelopment or underdevelopment, no corrections made in developing time for exhausted developers or for developers which are too warm or too cold, or developing baths that have been incorrectly made.

A low density also may be due to material placed between the X-ray source and the film such as plate glass underneath a specimen or certain wrapping materials.

3.8B Contrast

Lack of contrast, as well as too much contrast, normally is due to the lithological differences of the sample. A change in the hardness of the radiation (lowering or raising the kilovoltage) sometimes may improve the resulting radiograph. Another possibility is to use a higher or lower contrast film.

Often very little can be done since the contrast represents the lithological properties of the specimen (Fig. 3.29). The use of contrast reducers or certain electronically controlled printers (Fig. 3.32) may give the best solution.

However, before reaching a solution, the operator should check his developer and developing procedure.

3.8C Sharpness of the Radiographic Image

There are many reasons why an image is not as sharp as it can be. When the distance between focal spot and film is too short, or the distance between specimen and film too large, or the size of the focal spot too large, an unsharpness may result (see Figs. 3.4, 3.30, and 3.31). Improper screens or a poor contrast between screen and film may cause an unsharp picture. Movement or vibration of the X-ray source, compared to the sample-film, causes a type of multiparallax exposure, which influences the sharpness of the image.

3.8D Fog

Different types of fog can be observed on some radiographs. The fog may be local or all over the image. The so-called gray fog can be due to defective safelights or general light leaks in the processing room; the film can be exposed too long or too close to a safelight. Also an excess of scattered radiation, improper storage of films with regard to radiographic protection, humidity, heat, or gasses can be a cause of this fog. Improper processing or deteriorated baths, or an exposure of the film to white light before the film is fixed properly, easily results in grey fog.

Marginal fog results from a defective closure of a cassette on a film holder.

Dichroic fog shows up as a yellow-green color by reflection and as a pink color by transmission. Causes may be a developer which is contaminated by fixer, improper rinsing between developing and fixing, which may cause an exhausted fixer, a prolonged development in an exhausted developer, or when a development continues locally in the fixing bath when films stick together.

An aging fog is a greyish reticulated fog which generally is an indication that the film is either too old or badly protected from humidity.

3.8E Yellow Stain

When development is carried out in an old and strongly oxidized developer, the film not properly rinsed before it goes into the fixing tank, or when the fixer is exhausted, a yellow stain can result which obscures details of the image.

This stain can also be produced weeks or months after processing if the film has not been fixed properly.

3.8F Density Differences Separated by a Distinct Contact

If a film is not immersed properly into the developing tank, especially when the operator stops lowering for a short time, differences in developing may occur on both sides on the contact line of the immersed film part and the dry film part. The same phenomena can occur when two films are brought into the developer at the same time while parts stick together.

The best procedure for developing is to have a bath filled far enough to keep the complete film in the solution, to immerse the film with an even and unbroken speed, and to insert no more than one film at a time.

3.8G Streaks

When developing hangers are not thoroughly cleaned and they are still wet when used again, drops of solution can run down over the film and cause chemical action prior to development. The result will be a number of contamination streaks on the radiograph.

Similar lines can also form when the films are not agitated regularly when in the baths. Especially during developing, the so-called bromide streaks are created when the bromide ions are released from the emulsion. Bromide is heavier than the developer, and when it sticks to the film it prevents proper development.

Drying streaks are produced when the film is dried slowly and irregularly. Careful draining (removing them slowly from a bath) can prevent these streaks.

3.8H. Clear Lines

Pressure lines are formed if something is drawn or written on the film while it is still in its paper cover. This can happen easily especially when dealing with ready-packed films. If the pressure is applied prior to exposure these lines are lighter than the surrounding regions. The author ruined some ready-packed films by writing "exposed" on the cover with a ballpoint. It proved that the paper became stuck to the film at places where the pressure, even light, had been applied (Fig. 3.39).

3.8I. Rugged Lines, Irregular Lines, and Dark Spots

Cracks in a sample, deep scratches, and thin parts in a slice allow more radiation to pass through and consequently produce dark parts on a radiograph (Bouma, 1964e) (Fig. 3.39). Very small thickness irregularities normally do not show up on an X-ray film when these thickness differences in the slice are neglectable. However, compared to the total slice thickness, cutting marks of ½ mm on a slice of 4 mm thick do show up.

3.8J Emulsion Damages

Scratches on the radiograph, especially when the film is still wet, are formed easily. Therefore the operator should be careful with handling the films during and after processing.

Dust particles easily stick to a wet or moist film and show up on a print as white lines (Fig. 3.39).

3.8K White Scum

When the water used for the developer, fixer, and rinsing baths is too hard, a milky scum appears on the film. If it is difficult to use softer water for all the baths, the operator can carry out a final rinsing in soft water, after the main rinsing. This soft water should be renewed regularly. After the soft-water rinsing a Photo-Flo bath (soft water) can be used.

3.8L Light Spots

Light-colored spots on a radiograph can be caused by stop bath or fixer solution splashed onto a film before developing.

"Crimp" marks are often half-circular in shape and due to sharp bending of the film while it is being inserted in a cassette or exposure holder.

Air bells are caused when air clings to the surface of the film after the film is immersed in the developer. They prevent an even developing. These spots can be prevented by tapping the top bar of the developing hanger a few times against the tank directly after immersing.

3.8M Dark Spots

Dark-colored "crimp" marks can be formed on a radiograph when the film is bent sharply during removal from an exposure holder or cassette.

Water or developer splashed onto a film before processing also results in dark spots on the radiograph.

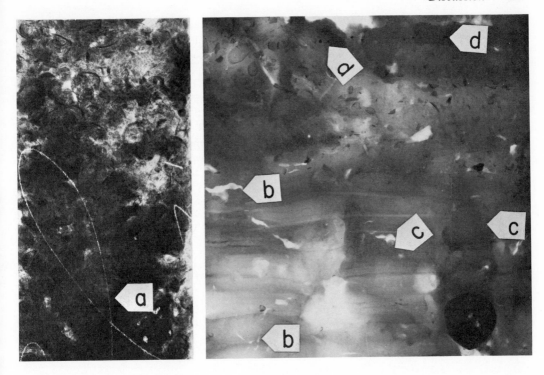

Fig. 3.39. Prints of parts of radiographs presenting some artificial mistakes. (A) Pressure line (a) created by writing with a ball-point on a ready-packed film. Sandy mud with abundance of shell remains. Surinam shelf, depth of water 29 m. Sample top 5 cm below surface. (B) Cracks in the sample (b), mainly due to slicing, form light spots on the print. Thinner parts of a slice (c) let more radiation pass and consequently come out darker on the print. Care should be taken not to confuse such spots with shading effects due to difference in lithology. Prints of dust particles and damage of the emulsion (d) also show up dark on the print. Clay with silt laminae and shell concentration on top. Surinam shelf.

3.8N Static Marks

When a film is removed too fast from the box or the exposure holder, electrostatic discharge figures may result on the radiograph. These marks can be of different shapes. Often they look like a star in a window with a central point and many branches going to all sides, they can also look like dendrites or like branch-shaped corals, or they can be treelike.

3.8O Fingerprints

Greasy fingers may touch the film prior to processing. Where grease patches stick to the film no developing can take place. These marks can be avoided easily by touching and handling the film only at the corners or, if necessary, along the edges.

3.8P Reticulation

A netlike structure can be formed at different scales if temperature differences exist between the processing baths and the washing tank. Often this structure appears as a microscopically fine structure and the image then appears as "grainy." The netlike structure, however, is quite typical when observed under a low-power magnifying glass.

3.9 DISCUSSION

This chapter along with Appendices One, Two, and Three reveal many aspects of X-ray radiography. It is an easy method, and when handled carefully by well-instructed operators, with safety operations checked by an expert, no harm to personnel or production should be encountered. However, it cannot be overstressed that radiation is harmful, and that protection is of primary importance.

The technique has no application restrictions as far as types and conditions of samples are concerned. However, it should be kept in mind that it is an absorption method, and consequently little or no information is obtained when the sample does not contain materials that have different absorption characteristics.

Radiography can be applied as a scanning technique as well as a method to study details of samples. In particular, the application of stereoradiography or microradiography often reveals additional information.

Section 3.8 presents a large number of possible mistakes and damages to the radiograph. However, when the whole radiographic procedure is followed properly, the chance of encountering one of these damages is very small.

CHAPTER FOUR

OTHER METHODS

This chapter is a composite of seven different groups of methods, some of them complex in character. Their short length is the main reason for this gathering of data in spite of the fact that little or no relationship exists between the various techniques. Four of the methods fall into the category of observation, two of them can be grouped under conservation, and one under preservation. They are outlined in Fig. 4.1.

Freeze Drying is treated as part of the Sample Preservation discussion in Section Six.

A short description of Infrared is inserted in this chapter in spite of some similarity to radiography.

Since some of the methods have a variety of submethods within their main title it is impossible to present all variations in Fig. 4.1. Details therefore can be found in the introduction to each section.

4.1 VAN STRAATEN TINS

Sections of wet Recent sediments seldom reveal sedimentary structures directly. The properties present are more visible when the material dries out. Van Straaten (1954, 1956, 1957a, b, 1959, 1965) uses this principle for his investigations on the structural and textural properties of sediment samples.

4.1A Equipment

Core holder, steel trays (Fig. 4.2), knife, spatula, electrical hot plate, and safety razor blade (see also "Comments"). For permanent recording: (miniature) camera, tripod or support, and two photolamps. For textural description: binocular and a light source for striking light.

The steel trays are made of ordinary plate material of a thickness of normal preserving tins.

The shape is rectangular. The dimensions depend on the size of the samples collected. Van Straaten (1954) uses trays which are 5 x 7½ cm in section and with edges of 1½ cm high. When dealing with round cores, the trays must not be wider than about 80% of the core diameter. For small piston and large Dachnowski cores, trays of 3 x 7½ cm, 1 cm high are recommended.

4.1B Samples

All shapes of samples, collected with coring devices which allow the obtaining of "undisturbed" samples, can be used for this method. Cylindrical cores must be extruded or the liner must be cut lengthwise. It is necessary to place the sample in a holder which has the proper shape (see Chapter 1.1B and Figs. 1.12-1.15). Cores can also be placed in a rack consisting of two wooden sides at right angles to each other (Fig. 4.3). It is necessary to have samples which are stiff enough to preserve their cylindrical shape.

4.1C Procedure

The technique here applies to cylindrical cores. The core is extruded or removed from its barrel into a half-cylindrical tube. An upper slice of the core is removed with a knife in order to obtain a flat surface with the same width as the metal trays. Several trays (with a series number at one of their short sides) are pushed into the core behind each other (Fig. 4.4). The number of each tray should point to the top of the sample to fix the position of each tray and the succession of the parts.

The trays are loosened from each other with a wet spatula or knife (Fig. 4.5) (see also Section 4.1F). The core pieces are turned over and the protruding part is cut off with a wet knife in such a way that approximately ½ cm still sticks out

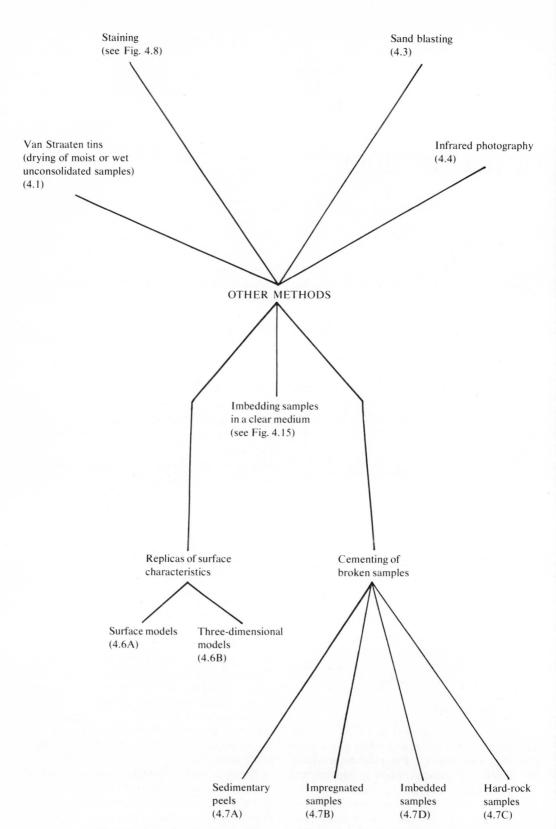

Fig. 4.1. General outline for the different techniques described in Chapter Four.

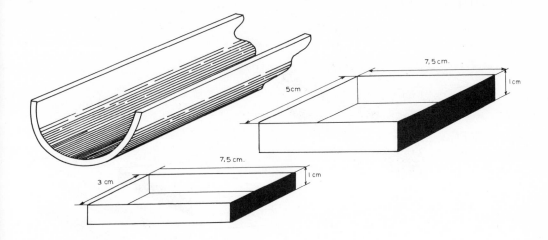

Fig. 4.2. Half-cylindrical core holder and two metal trays of different sizes used for temporary storage of a cylindrical core and for sampling small parts, respectively.

Fig. 4.3. Wooden rack for temporary storage of a core. The rack must be sturdy. Both inclined sides slope toward the center at an angle of 45°. This rack can be used only when the sediment core is cohesive enough to keep its shape.

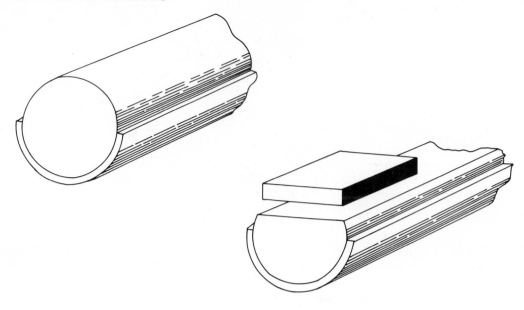

Fig. 4.4. After a slice has been removed from the core, the metal trays are pushed into the core behind each other.

above the edges of the tray (Fig. 4.6). The surface is smoothed with a knife as much as possible. The tray is now placed on the hot plate and heated until so much water is evaporated that air starts to penetrate; this is demonstrated by the "dry" color of the sediment at that state. The drying time varies for different materials and mainly depends on the moisture content, grain size, permeability, and thickness of the sample.

When this half-dry state is reached the surface is shaved off with a knife or a safety razor blade (see also Section 4.1F). The structures appear in every detail in this way. The scraping must be done very carefully to prevent the forming of cracks or breaking off of edges. The movement can best be carried out parallel to the bedding and from the sides to the center. The degree of dryness is very important since clayey layers will become smeared by the razor blade when the sediment is still wet, which results in a hiding of contacts. If the drying is too prolonged, cracks start to form.

4.1D Photography

Van Straaten found that some of the finest structures could be observed better on photographs made from the half-dried samples than by the human eye since photographic material has a greater sensitiveness to contrasts.

By using trays with a size of 5 x 7½ cm, one can take very good photographs with a miniature camera. Two lamps mounted in reflectors can serve as a light source. It is advisable to mount camera and lamps on supports, which in turn are mounted on a wooden board. The position of the tray is indicated on the board. With a measuring tape or a pointer the investigator can take care that the upper surface of the sediment always is at the same distance from the lens.

At the Geological Institute in Utrecht the "Reprovit II" from Leitz is often used for taking photographs. The advantage of this instrument is that the object can be placed in focus before a picture is made. Since the exposure time normally varies between ½ and ¼ sec, a delayed action shutter or a cable release must be used.

It is important to place the trays always with their number to one side. Using a fine-grained film, the investigator is able to make prints on true scale. By mounting the prints together one can obtain a continuous picture of the core (Fig. 4.7).

4.1E Microscopy

The tray with half-dried sediment can next be placed under a binocular for a fairly accurate grain-size investigation of the different depositional units, for the observation of fossils, etc.

Fig. 4.5. The metal trays are loosened from each other with a wet knife or spatula.

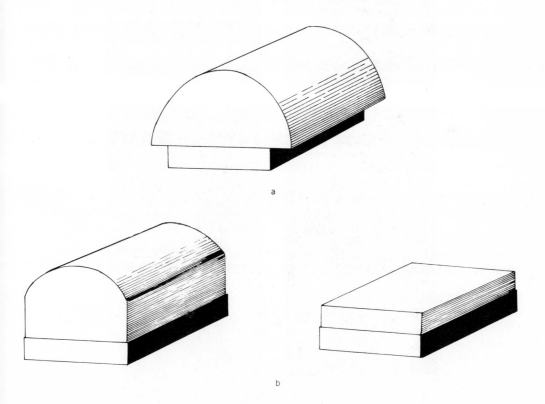

a

b

Fig. 4.6. The loosened core parts are removed from the core holder and placed upside down. The protruding sides and also all except about ½ cm of the material sticking out above the tray are removed. The surface of the sediment is shaved off carefully.

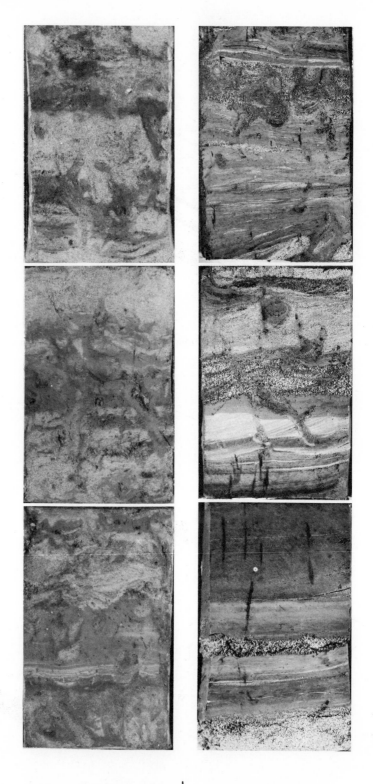

60-82 1/2 cm | 90-112 1/2 cm

4.1F Comments

Especially when dealing with clayey cores, the application of electro-osmosis to the cutting knife facilitates and speeds up the work. Also, smearing effects can be avoided, especially when slicing is carried out parallel to the bedding. It is the author's experience that shaving of the sample surface is seldom needed when it has reached the half-dry state.

4.2 STAINING

Staining is a well-known technique in petrology and soil mineralogy used to distinguish different light fraction components from each other.

To identify potash feldspar from plagioclase and quartz, the sample is etched with HF, then stained with sodium cobaltinitrite or malachite green (Gabriel and Cox, 1929; Keith, 1939a, b; Chayes, 1952; Jackson and Ross, 1956; Plafker, 1956; Rosenblum, 1956; Bailey and Irwin, 1959; Hayes and Klugman, 1959). Reeder and McAllister (1957) and Doeglas et al. (1965) use a buffered hematein solution after the cobaltinitrite solution treatment, whereas Bailey and Stevens (1960) dip the sample quickly in a barium chloride solution after the first coloring, followed by a covering of a rhodizonate reagent. Since these staining techniques are not yet applied in the study of sedimentary structures, no further details will be given here.

About 10 years after the famous publication of Gabriel and Cox (1929), etching and staining techniques were developed to identify calcite from dolomite and to distinguish different carbonate minerals (Rodgers, 1940; Friedman, 1959; Wolf and Warne, 1960; Wolf et al., 1967; Warne, 1962; Evamy, 1963; Chilingar et al., 1967). Katz and Friedman (1965) and Germann (1965) combined staining of carbonates with acetate peels, as described by Buehler (1948), Beales (1960), Lane (1962), Miller and Jeffords (1962), and Wolf et al. (1967).

Most of these staining methods can be applied to loose grains as well as polished or unpolished rock surfaces. Although these techniques have been developed for mineralogical purposes, they can sometimes also be used for the study of sedimentary structures (Weiss and Norman, 1960). By detecting variations in ferrous iron content, a differentiation of dolomite from calcite can be made (Evamy, 1963, Table I). Figure 4.8 presents a diagram of the various methods given in this section.

A very extensive description on the examination and analysis of sedimentary carbonates was published recently by Wolf et al. (1967).

4.2A Stained Acetate Peels of Carbonate Rocks

Katz and Friedman (1956) note "that many textural details of a stained polished surface cannot be satisfactorily detected under the binocular microscope. The color, moreover, is unstable and tends to fade, even when covered by a protective coating." Staining before making an acetate peel allows transfer of the actual layer of stain from the polished surface to a peel. Katz and Friedman (1965) describe the following methods.

Rock Preparation (see also Chapter One)

1. Cut the sample and grind the surface that is to be investigated.
2. The ground surface should be polished with a 800 mesh abrasive.
3. The polished surface must be thoroughly washed.
4. Etch with 1-9 diluted HCl. (The authors mentioned recommend 45 sec for pure dolomites, 30 sec for pure limestones, 25 sec for calcitic dolomites and dolomitic limestones, and 20 sec for carbonate-cemented sandstones.)
5. After etching, the surface is washed carefully, first in tap water and then in distilled water.

Fig. 4.7. Photograph of parts of core Van Straaten 277. The pictures were taken with a Leica camera as soon as the samples were dry enough to reveal the details of their texture and structures. The width of the samples is 5 cm. Core was collected in an "étang" of the Rhone delta, about 400 m south of Port St. Louis, France. The three core parts on the left-hand side come from the interval 60 to 82 ½ cm below surface, the three parts on the right-hand side represent the interval 90 to 112 ½ cm. Lithology, sedimentary structures, and burrowing activity are revealed clearly. (Courtesy L.M.J.U. Van Straaten, Geol. Inst., Groningen, The Netherlands.)

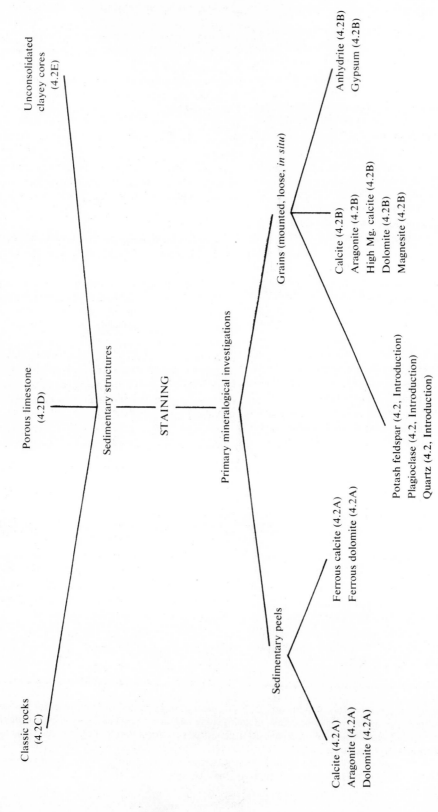

Fig. 4.8. Diagram of the different staining methods discussed in this section.

252

Staining with Alizarin Red S: Calcite

1. The solution is made of 1 g Alizarin red S and 2 cc concentrated HCl in 1000 cc distilled water (commercially available).

2. The solution is poured into a porcelain dish and warmed to 40°C.

3. The etched sample is immersed in the solution for exactly 4 min with the prepared surface up. Swirling the dish from time to time is recommended to keep the reaction going.

4. The solution is poured off, and the sample is washed with distilled water, the experimenter taking care that the water stream does not touch the delicate surface.

5. The sample is then immersed for 2 to 3 sec in acetone. A piece of acetate sheet is next placed on the stained surface and pressed firmly to it without sliding. If this is carried out correctly and quickly, no air bubbles will appear between the acetate sheet and rock.

6. After approximately 40 min the film can be removed and placed between two glass plates.

The imprints of calcite on the acetate film can be observed as a deep red color.

Staining with Potassium Ferricyanide: Ferrous Calcite and Dolomite

1. If a sample previously stained with Alizarin red S is available, it can be used after etching in diluted HCl for 5 to 10 sec (see Rock Preparation: Nos. 4 and 5).

2. The staining solution consists of 5 g potassium ferricyanide dissolved in distilled water containing 2 cc concentrated HCl. This solution is then diluted with distilled water to 1000 cc.

3. Staining and the preparation of an acetate peel are identical to the steps discussed in Staining with Alizarin Red S, nos. 2 to 6.

Any ferrous carbonate imprint will now turn blue. The deepness of the color reflects the Fe^{++} concentration.

This is a routine test for iron; it is described for dolomite containing ferrous iron (Friedman, 1959), but also for ferrous calcite (Evamy, 1963). To distinguish these minerals from each other, the reader is referred to Combined Staining Procedure.

Combined Staining Procedure: Calcite-Dolomite-Ferrous Calcite-Ferrous Dolomite

1. The sample is prepared as discussed in Rock Preparation and Staining with Potassium Ferricyanide, no. 1.

2. The solution is made by dissolving 1 g Alizarin red S and 5 g potassium ferricyanide in distilled water containing 2 cc concentrated HCl. The solution is diluted with distilled water to 1000 cc.

3. The staining procedure is carried out as discussed in Staining with Alizarin Red S. The colors received in the peel will be:

calcite	red
dolomite	colorless
ferrous calcite	bluish red
ferrous dolomite	blue

Staining with Feigl's Solution: Aragonite

1. The sample is prepared as discussed under Rock Preparation and Staining with Potassium Ferricyanide, no. 1.

2. The solution is prepared by adding 1 g solid Ag_2SO_4 to a solution of 11.8 g $MnSO_4 \cdot 7H_2O$ in 100 cc distilled water and then boiling. After cooling, the suspension is filtered and 1 to 2 drops of dilute NaOH solution is added. After 1 to 2 hours the precipitate is filtered off.

3. The prepared sample is immersed in this solution at room temperature for 2 to 3 min. Careful agitation from time to time is recommended.

4. The sample is removed and carefully washed with distilled water (tap water is not allowed since a white AgCl precipitate will result on the surface).

5. The acetate peel is prepared as discussed under Staining with Alizarin Red S. nos. 5 and 6. Aragonite stains black in this solution, whereas calcite and dolomite remain unstained (Friedman, 1959).

4.2B Staining of Some Carbonates and Some Sulfates

Friedman (1959), in his extensive review, describes two series of analysis to differentiate aragonite, anhydrite, calcite, high Mg calcite, dolomite, magnesite, and gypsum. He also lists in his tables a large number of organic dyes that can be used for staining dolomite, calcite, magnesite, and gypsum. Only a selection of this is discussed here (Figs. 4.9, 4.10).

Stain Solutions and Applications. (All dyes mentioned are commercially available.)

Alizarin Red S in 0.2% HCl. Dissolve 0.1 g of Alizarin Red S in 100 cc 0.2% HCl. The 0.2%

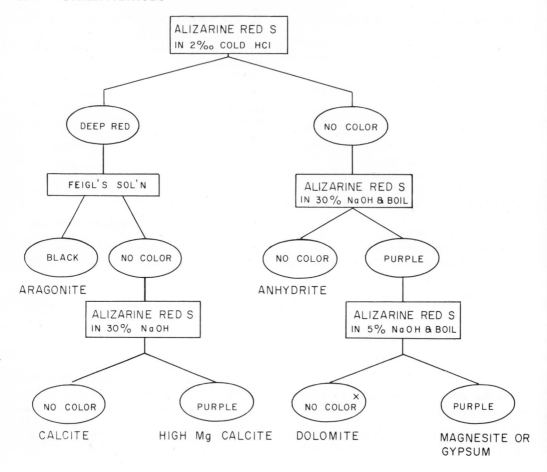

Fig. 4.9. Recommended staining procedure with Alizarin red S and Feigl's solution. (Redrawn after Friedman, 1959.)

HCl solution is made by mixing 2 cc of commercial grade concentrated HCl with 998 cc H_2O. Calcite is stained deep red within 2 to 3 min and dolomite is not stained except on excess exposure.

Alizarin Red S in 30% NaOH. Dissolve 0.2 g of the dye in 25 cc methanol by heating, if necessary. If evaporation occurs, replenish the methanol. Next add 15 cc of 30% NaOH and bring the solution to a boil. The 30% caustic soda solution is made by adding 70 cc H_2O to 30 NaOH. Immerse the specimen in the boiling stain solution. Anhydrite does not stain. Dolomite turns purple after 5 to 10 min. If the color of a specimen fades with time, it can be restored by immersing the sample in diluted NaOH. High Mg calcite also turns purple in this solution. However, it will obtain this color even if the solution is not boiled.

Alizarin Red S in 5% NaOH. The preparation of this solution is identical to that of the preceding one, except that a 5 % solution of NaOH is used. This caustic soda solution can be made by adding 95 cc H_2O to 5 g NaOH. This solution is not strong enough to give dolomite a strong stain. Magnesite and gypsum both turn purple. According to Megnien (1957), a 10% mercuric nitrate solution combined with a 1% nitric acid can be used to identify gypsum as it obtains a lemon-yellow color.

Feigl's Solution. The preparation is discussed in Staining with Feigl's Solution, no. 2. Aragonite stains black in this solution, whereas dolomite and calcite remain unstained.

Titan Yellow (Clayton Yellow) in 30% NaOH. The solution is made by dissolving 0.2 Titan yellow in 25 cc methanol by heating, if

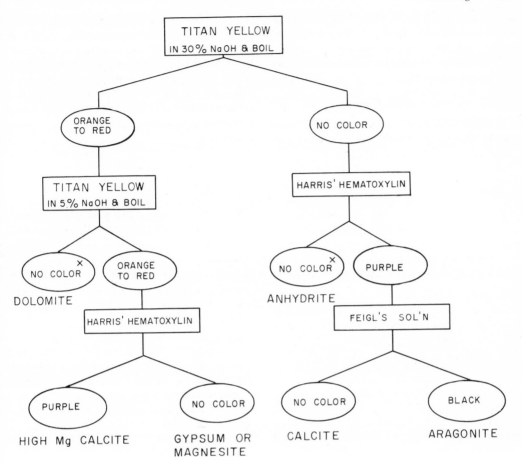

Fig. 4.10. Recommended staining procedure with Titan yellow, Harris' hematoxylin, and Feigl's solution. Friedman used very fine-grained high-magnesium limestones in his study. (Redrawn after Friedman, 1959.)

necessary. Methanol should be replenished if some is lost due to evaporation. Add 15 cc of a 30% NaOH solution and bring the solution to a boil. After immersing the sample in this solution for about 5 min the dolomite turns deep orange-red. Gypsum stains orange, and high Mg calcite stains orange-yellow.

Titan Yellow (Clayton Yellow) in 5% NaOH. Preparation of this solution is the same as the preceding. Data on this alkali solution are similar to Alizarin solution data. This stain solution stains the high Mg calcite, gypsum, and magnesite orange to red, whereas dolomite remains colorless or turns light orange.

Harris' Hematoxylin. The normal solution has some disadvantages, and Friedman suggests the following formula. Add 50 cc commercial grade Harris' hematoxylin to 3 cc of a 10% HCl

solution. This acid is made by adding 90 cc H_2O to 10 cc commercial grade concentrated HCl. The staining solution should be stirred well before use. Three to ten minutes are required to stain polished and unpolished surfaces. The more frequently the stain solution is used, the quicker the stain takes effect. Dolomite is not affected, whereas calcite obtains a purple color. High Mg calcite reacts with this solution as calcite does.

Method

1. It is not important if one is dealing with an unpolished or a polished surface. The face to be studied is etched in a 8 to 10% HCl solution for 2-3 min.

2. The sample is washed in running water.

3. The sample is stained by immersion in a staining solution. To follow the complete proce-

dures as given in Figs. 4.9 and 4.10, fresh chips can be used for each step (after being etched and washed). When only one sample is used, its stained surface should be rubbed off before applying the next test.

4.2C Staining Clastic Rocks

Hamblin (1962b) experimented with different organic dyes to accentuate obscure sedimentary structures in clastic rocks. Alizarin Red proved to be the best stain when used in strong solution.

Method

1. Cut a sample and smooth the surface.
2. Make a strong solution of Alizarin Red in water.
3. Apply the solution evenly over the surface with a small paintbrush or other suitable tool.

Clay present in a specimen will turn dark red, presumably due to absorption. The remaining material will turn light red in color. This method is only effective when a slight concentration of clay is available along laminae planes. Dispersed clay particles in sand or pure sandstones do not show any improvement after treatment.

Results

Different types of sediments have been treated by the author. All specimens were held under running water after the dye was dried in order to remove the excess dye. In some instances the dye came off so easily that nothing was left. Some specimens, slightly more porous, had to be rubbed by hand to remove excess dye. A Liassic limestone showed some improvement. The application of Alizarin proved to be superior to the use of India ink (See Section 4.21). A coarse coquinite and a carbonate containing tuff did not show any improvement, nor did a red-colored sandstone (Buntsandstein). A number of Eocene sandstones (Niesen flysch near Adelboden, Switzerland, and Peira-Cava sandstone, Alpes Maritimes, France) were tested. Only few of them showed up better after treatment, as did a Devonian sandstone from the north French coal mines.

4.2D Staining Porous Limestones with India Ink

West (1965) applied the principle of absorption of carbon particles from India ink to porous unfossiliferous limestones. A high contrast between more or less porous laminae in his pelsparites resulted. According to West this method also works well on coarser grained Portlandian bahamites and oolites, as it does on other porous granular limestones.

Method

1. The sample is cut, ground, and polished. The polishing is carried out with 600 grade carborundum powder.
2. The sample is washed thoroughly; take care that no carborundum is left in the pores.
3. The specimen is next etched for about 10 sec in a 20% solution of HCl.
4. The etched surface is then ground lightly several times on a glass or plexiglass plate, using a mixture of India ink and carborundum powder, grade 600.
5. Rinse the sample under a tap and rub the stained face to remove excess ink. If necessary, the procedure can be repeated.

Fig. 4.11. Staining porous limestone and a tuff with India ink. The parts on the left-hand side of the samples are not treated.

"In the case of allochemical limestones (Folk, 1959) the matrix is usually more porous than the allochems themselves, which may thus become more clearly distinguishable" (West, 1965).

Results

Different sediments were tried; the degree of improvement varied from good to none (Fig. 4.11). Dense limestones and graywackes were not affected at all. The Liassic limestones mentioned in the preceding section showed a minor result, the carbonaceous tuff and an oolite did

fine, whereas a noncarbonate containing Buntsandstein sample showed only slight improvement.

4.2E Staining of Unconsolidated Sedimentary Cores

Pantin (1960) experimented with several organic dyes to reveal lithological differences in core samples in order to identify sedimentary microstructures. His result was that the dye with a "molecular configuration which gives rise

Fig. 4.12. Air-dry unconsolidated samples (A, B) are stained with methylene blue (C, D). Lithological differences influence the amount of absorption. Core MT/1. Samples are about 2 in. wide. (Courtesy H. M. Pantin, 1960. New Zealand Oceanographic Inst., Wellington.)

to a blue or violet color also allows the dye to be rapidly absorbed on clay particles. This effect would tend to inhibit the penetration of the dye into fine-grained sediment, and would thus increase the contrast between the finer and coarser parts of a core sample." The best results were obtained with crystal violet, methyl violet 6 B, and methylene blue chloride.

Method

1. The core is dried or allowed to become air-dry.

2. A small flat surface is cut with a scalpel along its entire length.

3. By examining this fresh surface a number of representative parts (for example 2 x 3 in. when dealing with a 2-in. core) are selected for staining. Normally, enough transverse shrinkage cracks are available to predetermine the possible samples.

4. The sediment face has to be smoothed as much as possible. This can be carried out by running the specimen down a carpenter's plane, which is clamped upside down in a vice. If the sample contains large fragments, which may fall out and scar the face, one can moisten the surface and use a razor blade to smooth it.

5. The sample is then placed horizontally with its smoothed surface up.

6. The smooth face is moistened evenly with water, and a solution of dye is applied as evenly as possible. This can be carried out with a soft paintbrush, a pad of porous material soaked in the watery dye solution, or with an ink or small paint roller.

7. The dye is allowed to soak in for a few minutes, after which the colored surface is gently shaved with a razor blade or a scalpel. In some cases a better result is obtained if the specimen is first dried at 80°C.

The dye will soak into the sediment for a depth of only 1 to 3 mm, but in the more permeable parts it penetrates much deeper. Therefore the amount of material removed is very important. Variations in permeability can be distinguished by variations in the color of the shaved surface (Fig. 4.12).

Comments

1. The disadvantage of this method is that the sample has to be dry. Successful results were not obtained with fresh or moist cores.

2. Since the core must be dry, it is an excellent method for older, dried out cores, which have not been investigated for sedimentary structures.

3. Fine-grained homogenous cores did not allow the dye to soak in at all.

4. Only some fine-grained carbonate cores were available. Dry pieces showed some design, but no improvement could be obtained with this technique.

5. The cutting technique can be improved by the application of electro-osmosis. When the core is not completely dry a very smooth surface can be cut without any smearing effects. This shortens the time required with the shaving operation prior to staining.

6. Instead of a carpenter's plane, a set of sandpapers with different grades works well. A knife may be necessary to remove scratches.

7. This method proved to be more successful when dealing with fine-grained sediments in which some silt layers and/or sedimentation-erosion planes were available than by merely observing an ordinary dried surface.

4.3 SANDBLASTING

Hamblin (1962b) describes the application of sandblasting as a technique to accentuate obscure sedimentary structures in clastic rocks. The result obtained can be compared with the action from prolonged weathering.

The method is very simple. Hamblin uses a small, inexpensive, multipurpose air-cleaning, sandblast, and spray gun, which operates on an air pressure of 75 lb/in.^{2} (5.3 kg/cm^2) or more. This type of unit adapts to almost any source of air pressure. It works well from a standard laboratory air jet.

The actual sandblasting should be carried out under a laboratory hood; in this way there is no need for special respiration equipment, and it minimizes most of the inconveniences associated with ordinary sandblasting operations.

Hamblin found that the greatest structural detail can be obtained by using clean, unsorted sand of which the largest grain size is smaller than the average grain size of the sample. In less than 5 min the structural features can be etched out in moderately well lithified rocks.

The success of this technique depends on the degree of differential cementation along structural contacts. "Favorable results are not limited to sediments exhibiting differences in composition between structural planes but can often be obtained from even the most pure quartzose sandstones" (Hamblin, 1962b).

4.4 INFRARED PHOTOGRAPHY

Infrared photography has been applied to diverse scientific technological problems for many years; for example those outlined by Clark (1946) and Simon (1966). The method is used by paleontologists to reveal morphological characteristics of fossils (Rolfe, 1965). However, the application of this technique to the study of sedimentary structures is very recent. Rhoads and Stanley (1966) and Stanley and Rhoads (1967) examined unconsolidated sandy samples collected from dunes, tidal channels, and mud and sand deposits in intertidal environments. Consolidated sediments such as Cambro-Ordovician quartzose sandstones of the Ottowa region have been examined by means of transmitted infrared radiation. Stanley and Rhoads indicate that this technique is not a substitute for radiography. Infrared and X-ray sensitivity to organic matter differ. The primary advantage of infrared photography is the simplicity of the technique. Only an ordinary camera, commercially available infrared film and filter, and an easily constructed, portable light-box are required. This method can readily be used in the field and other areas where X-ray equipment is not available.

4.4A Sample Preparation

The sample to be photographed should be cut into a thin slab, bounded by two parallel faces. Better results are obtained when the slice is less than 6 mm thick. Consolidated samples should be cut to less than 5 mm, and saw marks should be removed. Rhoads and Stanley (1966) place a 100-watt bulb beneath the thin slice to test the thickness of the specimen. If a small amount of transmitted visible light passes through the sample, proper thickness has been achieved and the slice should be transparent to infrared radiation.

Unconsolidated samples can be impregnated prior to slicing, and the pore spaces can be slowly saturated with water, after which the sample is frozen. Once frozen, the specimen can be cut. Rhoads and Stanley did not observe any disturbing influence due to this freezing procedure.

After cutting, the ice binder is allowed to melt and the sectioned surface, still wet, is coated with a water-soluble glueing agent such as Elmer's Glue-All (see Section 1.5B of Chapter One). A glass plate is cemented to this glue-covered surface, and, when dry, the sample can be trimmed to the desired thickness (this can be done with sandpaper). A light is placed beneath the slice to make sure that it is thin enough for transparency to infrared radiation.

4.4B Photographic Equipment

The method can be carried out best in a darkroom. In this way no false illumination takes place, and an easy check can be obtained on light leaks coming from the light source around the specimen.

A light-box with two 100-watt bulbs (tungsten light sources) has proven to be sufficient as a light source. A glass plate on top acts as sample holder. All light leaks around the sample must be eliminated (Fig 4.13).

Almost all types of cameras can be used. Because of time exposure the apparatus should be mounted either on a tripod or to a camera support (Fig. 4.13).

Depending on the type of infrared film used, a special filter must be placed in front of the lens

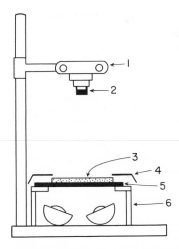

Fig. 4.13. Schematic darkroom setup for transmitted infrared photography: (1) camera; (2) Kodak No. 15 (G) series 6 filter; (3) sample slice; (4) opaque paper; (5) glass plate; (6) light box with two 100-watt tungsten filament lights. (Courtesy D. J. Stanley, Smithsonian Inst., Washington, D. C.; redrawn after Rhoads and Stanley, 1966.)

to eliminate the blue portion of the light, thus allowing the photograph to be made entirely by infrared radiation. Rhoads and Stanley (1966) recommend a Kodak Wratten filter 25 (A), but filters no. 29 (F), 70, 87, 88 A, 98 B, and 15 (G) can also be used.

Kodak IR 135 black and white film (35mm) and 4 x 5 in. infrared sheet film can be used. These films can be developed in the same way as ordinary black and white film. The infrared Ektachrome Aero film, type 8443 (35mm), requires focusing corrections and a special developing technique (for information see Kodak, 1963).

4.4C Procedure

The exposure time depends on the thickness and density of the sample slice, the type of film, and the type of filter used (Table IV.1). The exposure time increases with increase of the content of organic matter. Coarse-grained samples are more transparent to infrared radiation than are fine-grained samples. The exposure time may vary between 2 and 100 sec when a focal distance of 50 cm is used (Fig. 4.14).

4.5 IMBEDDING OF SAMPLES IN A CLEAR MEDIUM

Unconsolidated sediments will crack and fall apart with time when exposed to normal atmospheric conditions. Consequently, the sedimentary structures are easily disrupted. Under atmospheric conditions fossils and nodules containing marcasite decompose to ferrous sulfate and sulfate and sulfuric acid. Certain fossils may be too fragile to be handled.

As far as sediments are concerned, methods to preserve their nature and structures have been discussed in the first two chapters and in earlier parts of this section. From these discussions it is clear that wet and moist pelitic samples are difficult to dry in a short time without cracks resulting.

When unconsolidated samples, slices of sediments, or split cores have to be preserved for comparative studies, or as representative samples from certain environments or areas, it will be much easier to imbed the samples in a clear medium than to impregnate the material or to make a peel. The imbedment technique is well known in biology and in paleontology (Shrock, 1940). The advantage of imbedding in a clear medium is that the sediment is well preserved and remains unchanged in this condition, which is particularly important for moist, fine-grained sediments.

The imbedding technique has many advantages over Sawdon's (1948) use of plastic tubes. Plastic tubes can be used as core liners, or even as core barrels, or cores can be placed in them later. In all instances, smearing effects will appear along the wall of the tube, resulting in obscured sedimentary characteristics.

Table IV.1.   Exposure Data for Infrared Photography. Focal distance 50 cm

Type of Sediment	Thickness of Section (mm)	Type Film	Type Filter	Diaphragm	Exposure Time (sec)
Tidal channel sand (m)	7	35 mm IR 135	15 (G)	f 5.6	5
Tidal channel bottom (m)	7	4 × 5 in. IR	25 (A)	f 32	100
Sand dune (f-m)	6	4 × 5 in. IR	25 (A)	f 32	100
Sand dune (f-m)	7	35 mm IR 135	15 (G)	f 5.6	15
Sand dune (f)	6	35 mm IR 135	25 (A)	f 5.6	20
Sand dune (f-m)	6	35 mm IR 135	15 (G)	f 5.6	15
Sand dune (f-m)	6	35 mm IR 135	15 (G)	f 5.6	4
Sand dune (f-m)	6	35 mm IR 135	15 (G)	f 5.6	5
Sand dune (f-m)	4.5	35 mm IR 135	15 (G)	f 5.6	15
Sand dune (f-m)	4.5	35 mm IR 135	15 (G)	f 5.6	4
Intertidal flat (f-m)	6	35 mm IR 135	25 (A)	f 5.6	15
Intertidal flat (f)	1	35 mm IR 135	25 (A)	f 5.6	10
Nepean Sandstone (f)	4	35 mm IR 135	25 (A)	f 5.6	5

Data collected from Rhoads and Stanley (1966) and Stanley and Rhoads (1967).
f = fine; m = medium.

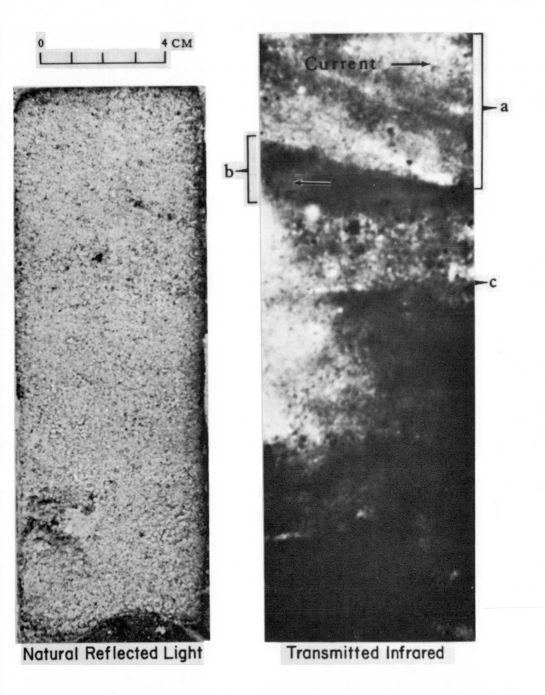

Natural Reflected Light   **Transmitted Infrared**

Fig. 4.14. Oriented tidal channel sand photographed in natural reflected light and by transmitted infrared radiation. Both photographs are oriented with their right margin to the mouth of the tidal channel (Sippewissett Marsh, West Falmouth, Cape Cod, Massachusetts). Core collected in center of tidal channel in 4 ft of water during ebb flow. The infrared photograph reveals dipping laminae produced by ebb current (zone a), truncated laminae dipping to the left, produced by a previous flood current (zone b), and a former change in current direction (zone c). The core shows a herringbone stratified unit, which is typical of tidal influenced regimes in coastal deposits. Slice thickness 7 mm, IR 135 film, f 5.6, 5 sec, focal distance 50 cm. (Photograph and text courtesy D. J. Stanley, Smithsonian Inst., Washington, D.C.)

Imbedded samples can be used for many types of investigations, for example, lithological research and other surface studies. The artificial product does not interfere with the making of radiographs. If necessary, the imbedded sample can be cut open and used for all techniques requiring fresh material such as granulometry, mineralogy, petrology, paleontology, many chemical analyses; even peeling and impregnation techniques can be applied.

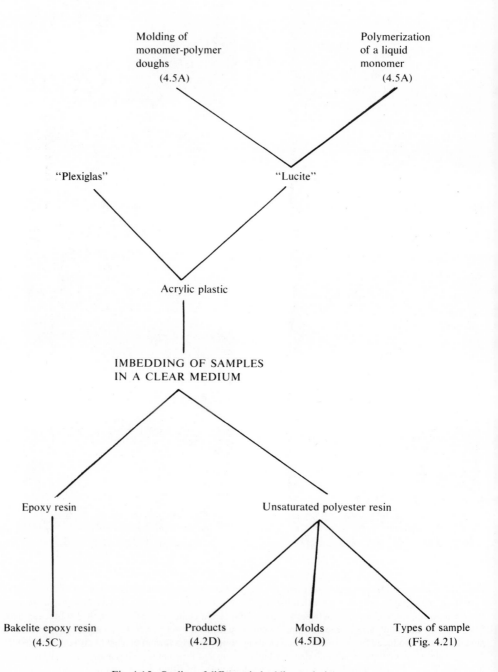

Fig. 4.15. Outline of different imbedding techniques.

Four different types of product (outlined in Fig. 4.15) will be mentioned and their proper techniques will be discussed: acrylic plastic "Plexiglas," acrylic plastic "Lucite," epoxy resin, and unsaturate polyester resin. The acrylic plastics have the advantage of being colorless and as transparent as the finest optical glass. However, they require more equipment or more laboratory operations than the other two. The epoxy also produces very good results when completely dry samples are used. Most polyester resins contain a slight color, which becomes more pronounced when cobalt accelerators are used (see Table II.2). The advantage of polyesters is the simple and inexpensive procedure, and the fact that moist samples often can be imbedded very satisfactorily. Unsaturated polyesters also have the advantage over the other three products in that the costly polishing can be replaced by a simple "covering" method, which produces practically the same effect.

4.5A Imbedding in Plexiglas

The only laboratory from which information about this technique has been obtained is the Coastal Studies Institute of Louisiana State University (J. M. Coleman).

Equipment

The plexiglas mixture must be cured in an autoclave at a temperature of 60°C (140°F). The temperature may rise as high as 77°C (170°F) with no adverse effect on the final product (J. M. Coleman, personal communication, February, 1963). A pressure of 120 lb/in.² (11.4 kg/cm²), built up by nitrogen, is required to prevent combustion and cooking of the samples. Autoclaves are discussed in Chapter Five.

A large container, equipped with a propeller which rotates at a low speed (1000 rpm) using a standard drill press, is the most simple setup for mixing the plexiglass mixture ingredients. Mixing should be carried out under a hood or in a well-ventilated room since two volatile components must be used.

Molds can be made of glass, metal, or porcelain. The nonglass trays should have tapered sides to facilitate later removal of the specimen. At the Coastal Institute small trays made of pieces of plate glass taped together are used. The sizes of the small trays are 66 x 8.3 x 6.3 cm (26 x 3¼ x 2½ in.). By using nontapered glass molds of this size, three sides of the final product

are smooth enough to eliminate polishing. Only the top of each cast must be milled and polished. Milling can be carried out on a large buffing wheel, planing machine, or a belt sander (see Chapter Five). Grids number 100, 200, and 320, respectively, should be applied. Final polishing can be carried out on a felt disk using a pure tin oxide or aluminum oxide powder as polishing powder.

Chemicals

The plexiglas mixture consists of plexiglas powder type Y-100 from Rohm & Haas Co. and the liquids methyl methacrylate (hydroquinone 0.006%) and ethyl methacrylate (hydroquinone 0.01%), both available from the same company (Rohm & Haas, not dated). The mixing ratio by weight is 65% plexiglas powder, 26% methyl methacrylate, and 9% ethyl methacrylate. As a base layer one can use slabs of ¼ to ½ in. thick plexiglas instead of preparing it from the given mixture.

The molds have to be covered with a mold release agent to facilitate removal of the cured cast. Krylon Crystal Clear No. 1301 or 1303 is sprayed lightly onto the sample to insure adhesion to the base layer and to prevent air adhesion.

Procedure (for figures see section 4.6)

1. Molds are made ready for use. If they consist of pieces of plate glass, they should be taped with pressure-sensitive masking tape. The molds are sprayed or brushed lightly with a mold-release agent.
2. The autoclave is turned on and set so that it will reach a temperature of 60°C (140°F).
3. Plexiglas bottoms are cut to a size of 64.8 x 7.6 x 0.6 cm (25½ x 3 x ¼ in.).
4. The moist core is carefully cut in half with a cheese cutter, piano wire, or knife, applying the electro-osmosis method (see Chapter Six).
5. The half-core is transferred carefully onto a plexiglas base and aligned properly. A frame is placed around the core half (see Chapter Six). The sides of the frame are about 6 to 12 mm (¼ to ½ in) high. By using this frame as a guide, the protruding part of the half-core is cut and removed, leaving an uniform, thick slice behind. Only when dealing with large objects in the sediment — shells, wood fragments, etc. — the previously mentioned slicing methods are nonapplicable. A knife or a spatula is more suitable to

remove excess material and to free protruding elements. Sediment particles should be removed with a brush.

6. The sample surface is studied and freed from smearing effects. The frame is removed and loose particles brushed from the base.

7. Sample data are written on ordinary paper as a label and placed underneath the sample. The plexiglas base with sample and label is now placed in the mold.

8. The sediment slice and label are coated with Krylon several times to insure adhesion to the base and to prevent floating. This also prevents the adhesion of air bubbles to the sample surface and, to a certain extent, "sweating" of salts onto the sediment surface. Each spray coat applied should be very thin (see also point 4 of Lucite procedure).

9. The plexiglas mixture is made in a large container. Approximately 20 min at a temperature of 22°C (72°F) is required to obtain complete dissolving of the powder and to reach the proper viscosity. Lower temperatures require longer mixing times. Mixing should be carried out under a hood or in a well-ventilated room. For 9 molds (67.3 x 9.5 x 5.1 cm = 26½ x 3¾ x 2 in.) the following amounts are necessary: 16.3 lb (7400 g) Plexiglas molding powder, 6.52 lb (2960 g) methyl methacrylate, and 2.25 lb (1025 g) ethyl methacrylate.

10. The obtained plexiglass mixture should be poured as soon as possible. It is important not to pour it onto the sediment itself but next to it. About 12 to 18 mm (½ to ⅜ in.) of mixture should stand above the sediment slab.

11. The filled molds are then placed in the autoclave, and when closed, enough nitrogen is allowed to flow through to flush out the oxygen. Next all valves are closed and a nitrogen pressure of 120 psi (11.4 kg/cm²) is built up.

12. The samples are "cooked" for at least 8 hours (overnight). During this time the mixture forms a transparent solid and bonds to the plexiglas base.

13. After curing, the nitrogen pressure is released, the heat supply is shut off, and the autoclave door is opened slightly. The specimens should be allowed to cool slowly for a period of 3 to 4 hours.

14. The molds can then be taken out and the casts removed.

15. Each cast is cut to its proper size, after which the top is milled and finally polished.

Comments.

1. Clayey and silty samples can be cut into slabs as thin as 3mm (⅛ in.); sand and gravel slices require thicknesses of 6 to 12 mm (¼ to ½ in.).

2. As soon as the pressure of the autoclave is released, the casts should be examined. If they are not well-cured (not completely transparent), the curing can be continued for another 4 hours. If this inspection is carried out quickly, it is still possible to obtain good results.

3. The final polishing is difficult and often expensive since it sometimes is necessary to have this polishing done by a special firm.

4. Arthur (1949) suggests dipping moist, fragile, geological specimens in a mixture of 20% molding powder Y-100 and 40% ethyl methacrylate, or spraying them with Krylon Crystal Clear.

4.5B Imbedding in Lucite

The Plastics Department of DuPont de Nemours & Co. manufactures acrylic resins under the name Lucite. Two different procedures can be used: (a) molding of monomer-polymer doughs and (2) the polymerization of a liquid (DuPont, 1966). Esso Production Research Company, Houston, Texas, has applied Lucite doughs successfully for more than 15 years to encase unconsolidated sediment slices in a permanent, durable mount (Shannon and Lord, 1967).

Showalter (1950) described a technique for imbedding ancient core specimens in Lucite to carry out relative permeability studies. His method was improved by Barret and Fatt (1953).

Molding of Monomer-Polymer Doughs

The polymer Lucite 4 or 41 is a methyl methacrylate polymer supplied as fine granules. The monomer, methyl methacrylate, is a thin, waterwhite liquid. When mixed together the two components form a doughlike mass (DuPont, not dated, 1966).

When methyl methacrylate monomer undergoes polymerization, a shrinkage of about 20% takes place. The process is strongly exothermic. These two factors account for difficulty sometimes encountered in bulk polymerization of this monomer. The application of these doughs is of value when low temperature, low-pressure moldings, low mold costs, and simplicity of operation are required. The temperature ranges from 43 to 100°C (100 to 212°F) and the pressures from 2.5 to 10.5 kg/cm² (35 to 15 psi).

Procedure (Esso Production Research Company).

1. The cores are cut in half lengthwise with a cheese cutter, a knife, or the electro-osmosis method. "Cores composed of large shell fragments and sand are difficult to slice with a cheese cutter; cores of this type should be frozen and cut into slabs on a band saw" (Shannon and Lord, 1967).
2. Half of the core is transferred carefully in a close-fitting wooden trough or directly onto a ½ in. thick, precut Lucite base. Next, a slice is cut, using the sides of the box as a cutting guide. The slab is then moved onto the precut Lucite base. When the half-core is placed upside down directly onto the Lucite slab, a frame should be placed around it (see point 5 of the procedure for imbedding in plexiglas).
3. The sample is aligned (Fig. 4.16) and trimmed. A label with data can be placed underneath the sediment slab. Several Lucite slabs with sample slices can be arranged in a shallow metal or porcelain tray. The tray should have tapered sides to facilitate later removal of cured mounts. Clay and silt samples can be cut as thin as ⅛ in.; sand and gravel normally require thicknesses of ¼ to ½ in.
4. Samples of clean, uncontaminated material are sprayed with a thin coat of Krylon (see point 8 of the procedure for imbedding in Plexiglas). Core material that harbors anaerobic bacteria should *not* be sprayed with Krylon but instead with a solution of pentachlorophenol (C_6Cl_5OH) and methyl methacrylate monomer in equal parts. A light spray is sufficient to inhibit bacterial growth (Shannon and Lord, 1967).
5. The monomer-polymer dough is prepared by mixing 65% by weight Lucite bead polymer 4F-NC-99 powder, 26% methyl methacrylate (hydroquinone 0.006%), and 9% ethyl methacrylate (hydroquinone 0.01%). The mixing temperature should be 72°F, and the mixing procedure should be carried out under a hood or in a well-ventilated room. Stirring by hand intermittently for about 5 min with a spatula is usually sufficient for complete dissolution of the powder. When the temperature is lower than indicated, a longer stirring time is necessary.
6. The mix is poured immediately over the specimens, taking care not to pour directly onto the slice surfaces (Fig. 4.17). About 1 in. of the mixture should stay above the slabs.
7. The tray is now placed in an autoclave and "cooked" at 150°F under a pressure of 120 lb/in.² using nitrogen. Overnight cooking for about 15 hours normally is sufficient. The inert gas pressure is released slowly and the mounts are left in the autoclave for 3 to 4 hours to cool slowly to room temperature.
8. The block of mounts is removed from the tray (Fig. 4.18) and cut into appropriate segments on a cutting machine. Each mount is then milled and polished.

Comments. According to the manufacturer, it is necessary to filter the monomer through a hard Whatman filter paper. The monomer is stirred vigorously while the polymer is added slowly in a fine stream to prevent lumps from forming. Agitation is continued until the mixture no longer shows a tendency to settle.

Thin casts (⅜ in. or less) can be made in hand presses. They require a curing cycle of 1 hour at 100°C (212°F) to insure complete polymerization. The cycle can be reduced by adding 0.02% benzoyl peroxide (based on the monomer content) before adding the polymer (DuPont, not dated, 1966).

Polymerization of a Liquid Monomer

The liquid methyl methacrylate monomer polymerizes into a solid polymer under the influ-

Fig. 4.16. Aligning and trimming of the sediment slice on a precut Lucite base slab. (Courtesy Esso Production Research Company.)

Fig. 4.17. Pouring the Lucite dough onto the prepared core slabs, taking care not to pour directly on the sediment face. (Courtesy Esso Production Research Company; Shannon and Lord, 1967.)

Fig. 4.18. Removal of the cured mount from its tray. (Courtesy Esso Production Research Company; Shannon and Lord, 1967.)

ence of heat, light, or a catalyst. The monomer is also available from DuPont (1955, 1966).

Casting Procedure. The monomer contains a trace of inhibitor, which must be removed before polymerization can take place. This inhibitor prevents polymerization from acting during shipment and storage. A method of removing the inhibitor is to vacuum-distill the inhibited monomer (boiling point 212°F at 760 mm; 142°F at 200 mm) through a short packed column. In the DuPont information bulletin a figure of a typical distilling setup is given. Another method based on washing and filtration is also discussed in this bulletin (see also Bell, 1939). The inhibitor-free monomer should be stored at a temperature about 4°C (40°F).

For mass casting operations it is advisable to use a syrup, that is, a partially polymerized monomer. This will shorten the total polymerization time. The monomer is partially polymerized by heating in a water-bath under a reflux condensor at a temperature of 90 to 100°C (194 to 212°F) until a fairly viscous mixture is obtained. The syrup should still be pourable. If the monomer does not thicken within 4 hours, a trace of inhibitor may still be left. The addition of 0.02% benzoyl peroxide (Luperco AC from Lucidol) by weight will usually cause it to thicken.

The casting should be carried out in closed molds. Glass or metal molds are very suitable. Gelatin capsules are satisfactory for small castings. Glass bottles make very inexpensive molds but often are not completely round.

Water spoils the appearance of the finished cast by interfering with the polymerization and clouding the resin. "Dehydration at temperatures below freezing has been effective in some cases. The procedure of dehydrating by immersion in a succession of mixtures of alcohol and water, of increasing strength, and finally in absolute alcohol, will be effective with some objects" (DuPont, 1955).

First a bottom should be made. When the supporting layer is ready the sample can be placed on top. Spraying a few thin coats of Krylon Crystal Clear 1301 or 1303 will improve the final results.

The monomer (partly polymerized or not) is poured, and polymerization can be started by heat, light, or a catalyst. To avoid vaporization of the monomer and, consequently, the forming of gas bubbles, do not use extreme temperatures.

Since we are dealing with large volumes it is practical to follow the layer-on-layer method (DuPont, 1966). Successive layers of monomer (as thin syrup), no thicker than 12 mm (0.5 in.), are poured and polymerized. At a temperature of 71°C (160°F), less than 24 hours is required for each layer. A shrinkage of 20% should be taken into account.

After the complete casting is finished, it should be cooled slowly to room temperature to prevent differential shrinkage.

It may be desirable to give the cast an aftertreatment to insure completion of polymerization. This can be carried out at a temperature of 90 to 100°C (194 to 212°F) for a few hours.

4.5C Imbedding in Bakelite Epoxy Resin

Gendron (1958, 1959) developed a technique for mounting metallographic specimens in a clear, cold-setting, epoxy resin. The following properties are required:

1. The mounting material should be transparent.
2. Molds of a variety of sizes and shapes should be readily available at low cost.
3. Heat and pressure which might damage delicate specimens should be avoided.
4. The method should be simple and time-saving.

Procedures

Bakelite ERL 2795 epoxy resin can be obtained from the Union Carbide Plastics Company. As hardener Gendron prefers Hysol 3416 from the Hysol Corporation. The best mixture proved to be 100 g resin with 10 g hardener. Mixing is carried out in a waxed-paper cup using an electric mixer. The time required varies between 1 and 1½ min. The mixture is then poured into the molds which contain the previously set metal samples. The molds are placed immediately in front of an air conditioner, set at its coldest temperature, for approximately 5 hours, after which the molds are removed and allowed to reach room temperature. After approximately 15 to 20 min at 150°F, the mounts can be removed from the molds. The surface exposed to air is the only one that needs polishing. When it is desirable to examine the core at high magnification, a cut face can be polished to reveal a cross-sectional view of the contents. This material is medium hard and, consequently, easy to grind and polish.

Molds up to 7 in. in diameter can be made

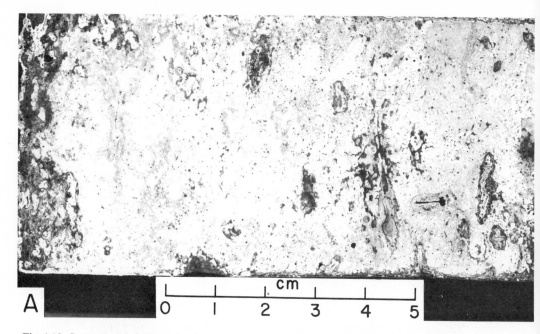

A

Fig. 4.19. Impregnated air-dry carbonate core part from the Campeche Shelf, Gulf of Mexico. Impregnated with Bakelite ERL 2795 epoxy resin and then cut and polished. (A) Core part, oblique illumination. (B, C) Details of the core, bright field illumination. (Courtesy N. J. Gendron, Materials and Processes Lab., General Electric Co., Schenectady, N.Y.)

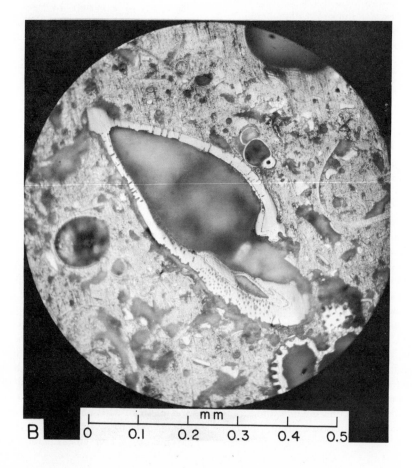

B

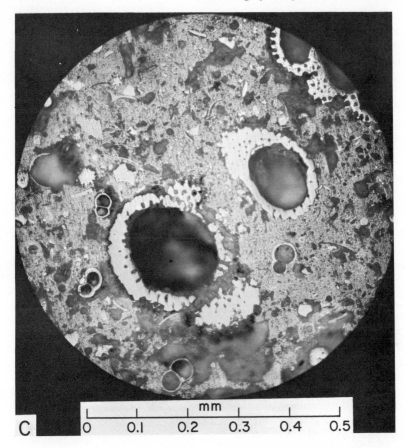

C
mm
0   0.1   0.2   0.3   0.4   0.5

without complication. Only 9 g of hardener should be applied instead of 10 g per 100 g resin, and the molds should be placed in a pan with water set before the air conditioner.

Use of a mold release agent is recommended if the mold is made of rubber. Tin cans and polyethylene cups do not require a mold release agent.

Brush Carbons

Pincus and Gendron (1959, 1960) also apply this mixture to vacuum impregnation and imbedding of brush carbons and metal samples. The impregnation equipment "consists of a vacuum desiccator, into which a glass funnel with a stop cock has been placed, and which is connected to a vacuum pump and a McLead gauge. A rubber feed hose within the desiccator leads from the funnel to the [specimen container]. The sample is then degassed at about 200 u for 1 hour." The mixture is prepared and poured very slowly into the sample containers until a liquid level of about ¼ in. is reached in the funnel. The stop cock is then closed and evacuation is continued for another 10 min. Initially, much degassing

and boiling occurs in the resin. After about 3 min, the bubbling slows down to 3 to 4 large bubbles per second.

The vacuum is then released slowly, the mold is removed, it is filled with the additional resin and cured as described above. The mount can be ground until the required level of research is reached. After grinding with different grades, polishing can be carried out. After final polishing, the surface has to be washed and brushed with fine sterile cotton under running water and then blown dry with clean compressed air.

Carbon specimens can thus be prepared for research. The advantage of this method is that the sample is not only imbedded but also impregnated.

Comments

According to Gendron (personal communication, February, 1967) over 18,000 metal samples have been mounted and polished each year at the Materials and Processes Laboratory of General Electric Company. Besides metals, brush carbons, and cokes, a variety of subjects have been potted, ground, and polished.

Since epoxy and water do not mix, this method is not suitable for moist cores. Water causes the epoxy to turn milk-white and the setup is not good. An air-dry core slice of fine-grained calcareous material from the southwest part of the Gulf of Mexico proved to be dry enough for imbedding (Fig. 4.19).

4.5D Imbedding in Unsaturated Polyester Resins

The imbedding of moist as well as dry geological samples is easier to carry out in an unsaturated polyester resin than in an acrylic resin, and the results are more or less equal. The main difficulty is to obtain clear, colorless plastics and colorless accelerators. Besides the chemicals only an oven, molds, plate glass, and a thick slab of marmor or other material are required. A vacuum unit is convenient.

The Resins, Initiators, and Accelerators

A few colorless plastics have been mentioned previously (see Section Two). The number of unsaturated polyesters suitable for imbedding is large. Before starting imbedding, the operator should make a trial in order to find possible inhibiting influences of the sediment, influences of the sediment, influence of the temperature, and the properties of the cured resin. Moreover, each new vessel with resin may differ slightly from another order.

Bio-plastic from Ward's Natural Science Establishment, Inc., is a pre-accelerated colorless resin which can be obtained in kits. Some kits also contain molds and polishing tools, a light bulb oven, forceps, needles, spatula and stirrers, and special liquids for the preparation of biological specimens. Ward (not dated a, b) provides a curriculum aid in which the most commonly encountered imbedding situations (mainly biological) are described.

The Norsodyne 50 of the Norsodyne polyester series of the North French Coal district is an excellent resin for imbedding, especially when their colorless initiator C 6 and colorless accelerator LM/5 are used (H.B.N.P.C., 1959, 1962, not dated). The resin viscosity is about 3.2 poises at 25°C (77°F).

For clear imbeddings, the company suggests for the base and top layers a combination of 100 parts by volume of resin, 2 parts of initiator C 6, and 1.5 parts of accelerator LM/5. For the sustaining layers (Fig. 4.20) the amounts should be 100, 1 to 1.5, and 0.5 to 1.5 cc, respectively.

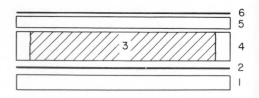

Fig. 4.20. Schematic diagram of the layer-on-layer procedure: (1) base layer or supporting layer; (2) glueing layer; (3) sample; (4) sample-retaining layer; (5) covering layer (may enclose part of the sample); (6) final top layer.

The polyester resins Lamellon 230 and Scadopol S 811 and S 813 are manufactured by Scado-Archer-Daniels N.V. (Scado, 1965, not dated a, b; Billiton, 1963 a, b, c). Lamellon 230 has a viscosity of 17 to 20 poises at 20°C (68°F), the two Scadopol types have a viscosity of 300 to 350 cp at the same temperature. Hobilon C (a peroxide) is the initiator and Hobilon Z (containing cobalt) the accelerator. Clear results can be obtained by following the data presented in Table IV.2.

Polyester 5132 of Frencken (1962) with initiator FF 1694 (a peroxide) and accelerator FF 1697 (a cobalt-containing liquid) produces good results. The surface may be sticky. Postcuring in an oven is recommended. The manufacturer prescribes the mixture as given in Table IV.3. The manufacturer suggests adding 1 drop of methyl violet to each 50 cc of mixture to prevent coloration due to the accelerator.

An excellent product manufactured by Reichhold Chemicals Inc. is 32-032 Polylite (HU-332), which has a viscosity of 300 to 400 cP at 25°C (77°F), a specific gravity of 1.11 to 1.13, and a gel time at 77°F of 14 to 20 min when 1% MEK is used (see Table IV.4). The resin is pre-accelerated. The gel time can be varied by varying the amount of initiator (Table IV.4; Reichhold, 1965). The advantage of this resin is that the surface of the cast is already so smooth that grinding and polishing are not absolutely necessary.

In Chapter Two the accelerator lauryl mercaptan (NL-70 of the Noury Van der Lande) is mentioned. The advantage of this product is that no more pronounced discoloration due to the influence of ultraviolet light will result with the application of this accelerator than ordinarily appears in the resin without the addition of this chemical. The composition is

Table IV.2.   Mixing Ratios for Scado-Archer-Daniels Unsaturated
Polyester Resin Lamellon 230

Lamellon 230	100 g	74.3 cc
Hobilon C	4–8 g	3.56–7.04 cc
Hobilon Z	2–10 g	0.04–0.2 cc

Billiton (1963c).
Note. Curing time at room temperature varies from 8 to 24 hours.

Table IV.3.   Mixing Ratios for Frencken Polyester 5132 Unsaturated
Polyester Resin at 18 to 20°C

Polyester resin 5132	100 g	77 cc	100 cc
Initiator FF 1694	0.8 g	20 drops	26 drops
Accelerator FF 1697	0.3 g	10 drops	13 drops

Frencken (1962).
Note: For large volumes the amount of initiator can be lowered to 0.6 g or less. At higher temperatures the amount of initiator should be lowered.

Table IV.4.   Variations in Gel Time of Reichhold's 32-032 Polylite Unsaturated
Polyester Resin with Different Amounts of Initiator

32-032 Polylite (cc)	100	100	100	100
MEK peroxide (cc)	1.5	1.0	0.75	0.50
Gel times in minutes	10–15	14–20	14–23	20–30

Reichhold (1965).

5% lauryl mercaptan derivative in 95% dimethyl phthalate. It can be used with all peroxide initator types mentioned at an amount of 0.5 to 3% (Noury and Van der Lande, 1961a, 1962, not dated b; Table II.2). This product is very active at room temperature, but the resin rapidly gels into a soft and flexible solid state. Complete cure can only be obtained by heating at temperatures above 60°C (140°F).

Molds and Mold Release Agents

Molds can be made from various materials such as glass, wood, plaster of Paris, silicone rubbers, and metal. For an occasional imbedding, molds of plate glass are the simplest ones. The glass strips are taped together on the inside as well as on the outside with cellotape. Wood and plaster of Paris are also inexpensive, but they require a coating of paraffin, petroleum jelly, silicone rubber, or other such products to seal the pores.

For permanent work, the silicone rubber materials or steel molds are more convenient. When dealing with rigid materials, the mold must have a tapered shape.

A convenient size is the tapered stainless steel mold used by the author. The bottom inside dimensions are 10 x 55 cm (4 x 21⅝ in.), the top dimensions are 11 x 56 cm (4⁵/₁₆ x 22 in.), and the inside height is 4 cm (1⅝ in.). The material thickness is about 1.2 mm (³/₆₄ in.).

Each type of mold release agent mentioned in Chapter Two, section 2.1D can be used. It is very important to use a type that produces a uniform, smooth film since irregularities in the release agent film remain formed in the final cast. If irregularities are not important since sides of the cast have to be cut, or grinding is not a problem, one can use overlapping strips of cellotape (Dollé, personal communication, June 1964).

For the final cover on a glass plate (discussed later) a coat of Hobilon 30, Ortholeum, or another similar viscous mold release agent may be followed by a coat of a 4% solution of polyvinyl alcohol (filtered).

Procedure

The layer-on-layer procedure is the best method for this type of imbedding (Fig. 4.20). By keeping the layers thin there will be no risk of cracking.

Under the following points 2 to 6 different

materials, and consequently different handling steps, are discussed (Fig. 4.21). For convenience, a core is used under points 2 to 5. The same procedure is valid for other shapes of samples, if the whole sample has to be imbedded or if a slice of it does. For the same reason a metal mold is used.

1. The Preparation of a Base Layer (Supporting Layer)

a. Thoroughly clean the mold with acetone, taking care that all release agent film or plastic is removed. The bottom should be especially clean since the smoothness of the back of the cast depends on this.

b. Apply a not too thin coat of mold release agent to the inner sides of the mold using a soft brush, a piece of flannel, or a tissue. Special attention should be given to the corners. Place the mold upside down, inclined on a piece of filter paper to allow excess fluid to drain before hardening.

c. The plastic mixture now can be made. Be careful not to trap too much air during mixing. A

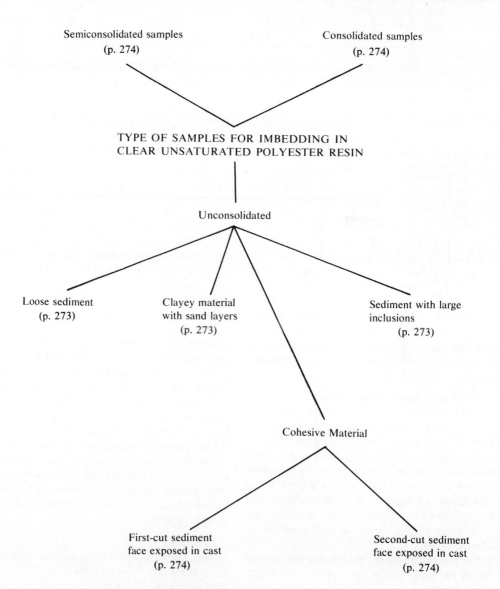

Fig. 4.21. Outline of the imbedding of different types of sediment in a colorless, unsaturated polyester resin. All are discussed in Section 4.5D.

good principle is to fill the mixing beaker with the amount of resin needed and then cover the beaker. Allow the plastic to stand for ½ to 1 hour to let all the air, trapped during pouring, escape. Then the other component(s) can be added. Vigorous mixing should be avoided.

The mix can be placed in a vacuum chamber, if one is available, to facilitate removal of air. It often is necessary to apply vacuum several times for short periods, releasing the vacuum between periods rather vigorously to break the surface air bubbles.

d. Pour a layer of about 6 mm (¼ in.) into the mold. The best pouring method is to pour steadily in one central strip and not in separate strips since air may become trapped when the strips flow together.

Examine the poured layer for air inclusions. If a vacuum chamber is available in which the mold fits, it is convenient to apply vacuum for about 5 min. Otherwise, bubbles on and near the surface can be burst with a dissecting needle. At the plastics laboratory of the North French Coal mines, cigarette smoke is blown hard over the surface. (Inhaling is necessary in order to blow hard enough.) The smoke particles burst the surface bubbles, thereby allowing the deeper ones to move upward.

Place a cover on top of the mold to prevent dust from falling in.

2. Sample: Unconsolidated, Loose Sediment or Clayey Material with Sand Layers. First-cut sediment face becomes exposed as upper face in the cast.

a. Take the base layer out of the mold. If it is curved it must be placed in an oven at 60°C (140°F) for about 20 min and then laid on a flat table with a long weight placed on it. Allow the slab to cool off.

b. Clean the mold and apply the mold release agent again. If polyvinyl alcohol was not used, and no visible damages can be observed in the film, a coat of polyvinyl alcohol can be applied instead of cleaning.

c. Apply release agent to the bottom and sides of the base layer as well.

d. Slice the sediment core into two halves. For sandy cores a cheese cutter can be applied; for clayey parts it is better to apply the electro-osmosis method. One also can use a cheese cutter and use the "electro-osmosis knife" to remove the smearing effects. Place a piece of Saran Wrap, same width as the core but longer,

on top of the cut face. Next a plexiglas tray can be placed upside down over the wrap, thus enclosing part of the core. Turn the entire sample upside down and remove the protruding core part. Remove loose particles that fall off.

e. Make a mixture of a few cubic centimeters for use as glue. Brush the mix to the base plate over a surface similar to the size of the sediment slab.

f. Take the bottom of the base plate with the glueing layer, turn it upside down and place it on the sediment slab. Next carefully turn over the entire sediment and remove the tray. The Saran Wrap can be taken off next. Remove loose particles and examine the surface for artificial marks. Place the base layer back in the mold without damaging the mold release agent film.

g. Type all data required on paper, trim it to the correct size, spray it lightly at least twice with Krylon Crystal Clear, and glue it with a little mixture underneath the sample.

h. Spray the sample as well as the label several times with a thin coat of Krylon Crystal Clear 1301 or 1303. This is to seal the surface and reduce the formation of cracks and escape of moisture. Sealing prevents water from coming in contact with the plastic, which would result in a milky appearance of the plastic. Lack of water movement also prevents formation of salt crust along the sediment surface.

3. Sample: Sediment with Large Inclusions Such As Shells, Wood Fractions, and Other "Hard" Objects.

a. See points 2a, b, and c.

b. Slice the core in two halves. The operator may try to use a cheese cutter or the electro-osmosis knife but will often encounter difficulties. In both cases, place the core in a half-cylindrical core holder and cover it with a similar holder. Place the seams vertically and use a knife for slicing. If this is unsuccessful, remove one half-core piece by piece and fill the holes of all objects removed to obtain a smooth surface. This automatically excludes the possibility of using the first-cut face as a later exposed face.

c. Apply a layer of glue to the base plate (see point 2e).

d. Assuming the first-cut face will not be the exposed face (if it is, follow point 1), the base plate is placed upside down on the sediment face, the entire apparatus is turned over carefully, and the core holder is removed. Fallen particles should be removed. Place a frame around the half-core. The better this frame fits, the less disturbance can be expected during removal of the protruding sediment. The walls of the frame should be 6 to 18 mm high, depending on the coarseness of the sediment. Removal of excess sample can be done with a small spatula or a knife. Clayey parts can be trimmed with the electro-osmosis knife to prevent disturbance and smearing effects. Protruding inclusions can be freed with a needle and/or small hand brush.

e. Remove the frame and all particles that have fallen beside the slice, make a label and place it in position. Apply several thin coats of Krylon (points 2g, h).

4. Sample: Unconsolidated Cohesive Material. (I) First-cut sediment face is exposed as the upper face in the cast (see point 5).

a. Cut core in two halves (see point 2d).

b. Remove all traces of cutting and loose particles.

c. Place a piece of Saran Wrap of the same size on the sediment face.

d. Place a strong plate on top, smaller than the size of the base plate. Turn the entire apparatus upside down.

e. Remove the core holder. This may be difficult if the sediment is sticky. A spatula or a long, sturdy rod (knitting needle) may be of help, especially when applying electro-osmosis.

f. Place a tight-fitting frame around the core. The height of the frame should not be more than 6 to 12 mm (¼ to ½ in.).

g. Remove the protruding part of the core, using the electro-osmosis method with the cutting knife.

h. Remove the frame and all loose particles.

i. Place a "glue" layer of plastic mixture onto the base layer, which still is in the mold and/or apply this glue layer to the top of the core slab.

j. Take the plate with the sample in one hand and the mold with the bottom layer in the other. Turn the mold upside down and place it over the sample. After contact is made, turn the entire piece upside down, and remove the loose plate and the Saran Wrap. This wrapping is necessary to prevent the plate from sticking to the sediment.

k. Align the sediment slab, remove artificial scratches, etc., and place it on the sediment. Spray the slab and label again (see points 2g, h).

5. Sample: Unconsolidated Cohesive Material. (II) Second-cut sediment face is exposed as

the upper face in the cast (see point 4). If the slab slides during steps 4d to h, damage to the slice surface should be expected, and it is therefore better to follow the rules given here. These rules also may be easier to apply because of the small amount of handling involved.

a. Cut the core and place a piece of Saran Wrap on top (see point 2d). Place a sturdy plate on this core face, turn the entire piece upside down, remove the core holder, and place a frame around the half-core (see points 4d to f). Remove the protruding sediment part and take the frame away.

b. Examine the sediment surface and remove damage marks. Place a piece of Saran Wrap, of the same size, on top, followed by a sturdy plate (see point 4d). Turn the entire piece upside down.

c. Place a "glue" layer onto the slab and take the plate with sediment slice in one hand and the mold with bottom layer in the other. Turn mold upside down and place it over the slab. Turn the entire piece back over, remove the sturdy plate, wrapping, and loose particles. Examine the surface. Make a label, place it on the slab and spray several thin coats of Krylon on it (see points 4 to 1).

6. Sample: Consolidated and Sturdy Semi-consolidated Sediment Slabs That Can Be Handled without Support (for porous-sediments, see also comments, point 4). (Under comments, point 6, another method is discussed).

a. As soon as the bottom layer starts gelling and is stiff enough to support the sample slice (see point 1d), the sample and the sprayed label can be applied. The surface of the plastic should be sticky enough to act as a glueing surface.

b. Spray the slab and label with Krylon.

7. Sample Retaining Layer. When the gluing layer is more or less hardened and, consequently, the temperature is far below the peak exotherm (Fig. 2.2), a new layer can be poured. The mixture is poured around the sample slice and *never* onto the sediment, thus allowing air present in the sample to escape. Large cracks should be filled.

Pour the mixture until it just covers the surface of the sample, unless the slice is thicker than 1 cm (⅖ in.). If the slice is thicker, more successive sustaining layers have to be poured. It is better, with respect to air bubbles, to enclose the sediment surface, which will prevent cracks being formed in the sample during the curing of the plastic (time element).

Before pouring, be sure that all air is removed from the mixture (see point 1c). During pouring new bubbles may become trapped. With a dissecting needle or smoke and/or vacuum application these bubbles can be removed (see point 1d). Cover the tray to prevent dust from falling in.

8. Covering Layer. As soon as the sample retaining layer is cured, a covering layer can be poured. Its thickness should not exceed 8 mm (⅓ in.). The method of pouring is discussed under point 1d. Cover the mold again and allow the mixture to cure thoroughly.

9. Postcuring. The free plastic surface often is tacky. This tacky material can be removed with acetone, but it is better to cure the cast first in an oven for about 1 to 4 hours at 60°C (140° F). If tacky areas are still present they can be removed with acetone.

10. Removing the Mold. Sometimes the cast is easily removed from the mold. However, if they stick together, turn it upside down and slap it with the hand or place the mold upside down on a table, with some soft material on top to prevent breakage of the mount, and slap it hard.

Glass molds can be taken apart by removing or cutting the tape. Molds made of plaster of Paris normally have to be broken before the cast can be removed.

11. Trimming. Due to shrinkage of the plastic, a small rim is formed along the sides. For tapered molds this is not serious since the sides have to be trimmed anyway to obtain straight sides.

The easiest method in all cases is to cut off strips along the sides on a diamond-impregnated cutting machine. If not enough plastic is available along the sediment slab, a planing machine, buffing wheel, or belt sander must be used.

12. Reflattening of the Cast. It often happens that the mold curves slightly during the curing of the different layers. Since it is necessary to have a completely flat upper surface for the preparation of the final top layer, it is a good practice to flatten all casts made.

a. Place the cast in an oven for about 1 hour at 60°C (140°F). This weakens the plastic.

b. Remove the cast from the oven and place it upside down on a thick glass plate or other smooth surface.

c. Place a strip of heavy material (steel or stone bar) directly on top. This bar should be a little smaller than the size of the cast (see point 13f). Allow the cast to cool.

13. Final Top Layer. The top of practically each unsaturated polyester resin mixture becomes irregular during curing due to irregularities in shrinkage. Milling and polishing the surface will produce the desired results; however, this process is quite expensive.

At the Plastics laboratory of the North French Coal mines another method is applied, which is easy and inexpensive. Their procedure can be compared to placing a cover glass on a thin section.

a. Clean the surface of a thick glass plate thoroughly with acetone. Place the glass plate on a horizontal table top.

b. Apply a uniform thin coat of mold release agent to the entire glass surface. The agent should be a semiviscous liquid such as Hobilon 30 or Ortholeum. Place the glass plate nearly vertical for drainage and allow the film to dry. Pour a 4% aqueous solution of polyvinyl alcohol (filtered) over this film; let it drain and dry.

c. Make a mixture of resin, initiator, and accelerator. The quantity needed is about equal to a 1 to 2 mm-high section of the cast. Be sure that no air bubbles are left.

d. Examine the surface of the cast and clean it with excess acetone. Be sure that no parts of the rags used are left. The acetone evaporates quickly.

If holes from air bubbles are present or if air is trapped, open the holes, remove most of the thin hole covers, and fill the holes with mixture. A dissecting needle often is the best instrument to open or enlarge bubble holes.

e. Pour a thick row of mixture onto the glass plate over a length a little larger than the length of the trimmed cast. Remove all air bubbles.

f. Take the cast, without touching the upper surface, and place it on one long side on the glass plate near the plastic mix row. Tilt it over at an angle of about 60° and move the cast slowly forward. As soon as the lower side is completely in contact with the mixture and a high roll is built up in front, the slab is turned down slowly. By doing this slowly air bubbles along the mixture surface are forced out. When the cast is flat it is important to make sure that the fluid mixture is all around. Moving the cast slightly may help to squeeze out mixture and distribute it evenly, especially in the corners. Place a heavy steel bar or stone slab on top (see point 12c).

g. Due to the differences in thickness of the mixture along the cast, gelling will not start at each point at the same time. As soon as gelling begins at a point, the operator should go along the cast with a dissecting or spear needle to loosen excess mixture from the cast. This has to be repeated until the cast is completely loose. Allow the plastic to cure overnight.

h. Remove all excess plastic and remove the cast from the glass plate. Removal of the cast may be difficult. Insert a knife or a spatula between the glass plate and a corner of the cast. Be very careful not to break the cast while lifting.

It is sometimes necessary to use a knife at many points along the cast, which may cause small damage marks. To prevent this two coats of mold release agent of different types (see point 13b) may be needed.

The polyvinyl alcohol film may stick to the cast. Often it can be torn off easily; otherwise it can be removed with water.

14. Final Polishing. In some instances, the operator may not be pleased with the gloss of the final top layer. The luster can be improved by hand polishing the surface (Fig. 4.22). One starts with waterproof silicone paper (for example, grades 320 and 600 of the 3M Company), followed by final polishing on a felt polishing disk applying abrasive liquid or aluminum or magnesium oxide. The final polishing can also be carried out using a polishing wax, Brasso, copper polish, or another commercial product, and a flannel rag.

Comments

1. It sometimes is nice to color the supporting layer, especially when dealing with biological objects which are imbedded for decorative purposes. Several types of dyes can be applied (see Chapter Two). Intensity of color depends on the amount of dye used. Excess dye may prevent curing of the plastic.

2. When dealing with objects which have protruding parts (e.g., the echinoid in Fig. 4.23a), it is necessary to keep all parts in the right position until enough sustaining layers are poured to prevent the parts from falling.

3. If there is a long delay between two pourings, dust and grease will collect on the surface and it is sometimes impossible to clean the surface. Even when a surface looks clean, enough dirt can be collected to prevent proper binding with the next plastic layer. This is demonstrated in Fig. 4.23. (After the cast was finished as far as indicated in Fig. 4.23a, about 3 months passed before the work was continued.) Figure 4.23b demonstrates clearly that patches of dirt will show up distinctly. Furthermore, color differences are present above and below the contact. The reason for this is not known; presumably different amounts of accelerator have been used.

4. Porous materials and corals are difficult to imbed without causing many air bubbles, especially if a vacuum is applied. The best method is

Fig. 4.22. Imbedded cores. These cores were dried out partly and cracked before time was available to imbed them on board ship. Continental shelf off Surinam, S. America.

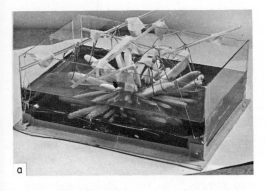

Fig. 4.23. (a) Setup with rope and tape to imbed an echinoid in a clear unsaturated polyester resin. (b) The cast, as set up in (a), is finished. Several months went by before the upper half of the echinoid could be imbedded. Cleaning of the old surface was not carried out properly, resulting in a greasy film between the two pouring periods. Slight differences in products and mixtures account for the color differences.

to soak the specimens under vacuum in a pre-accelerated polyester resin (no initiator) for 2 to 5 days. During that time air is replaced by plastic. It is a good idea to turn the specimen each day to allow the air to escape from lower positions. At the end of the soaking period, leave the specimen at least 1 day in the position in which it will be inbedded. The specimen is then ready to be placed on the supporting layer.

The present author had bad experiences with imbedding pieces of coal. Applying the method discussed under point 6, air bubbles became visible during gelling of the mixture. Application of vacuum resulted in a forest of vertical rows of air bubbles (Fig. 4.24).

Soaking in pre-accelerated resin under vacuum was the best way to remove the bubbles. It is easy to make a test on the degree of soaking by releasing the vacuum and applying it again. If no air is visible after an hour, the soaking is completed.

Slightly porous objects can be sealed with a number of Krylon sprays. However, there is no guarantee the plastic will not dissolve part of the lacquer.

5. Shells and other concave forms often are difficult to imbed without trapping air. The easiest way to prevent such entrapment is to reverse the building order of the plastic layers. Start with the covering layer, then a glueing layer, sustaining layer(s), and finally the (un) colored base layer. If the bottom of the mold is very smooth, it may not be necessary to apply a final top layer and/or polish it.

6. The process of reversal imbedding, as discussed in Comments, point 5, has been tried successfully with consolidated and semirigid samples. Weak unconsolidated sediments' slabs are more difficult to imbed in this way since sliding of the slab may result in smearing. The last-exposed side of the sediment slab was sprayed several times with Krylon before being placed on the covering layer.

4.6 REPLICA OF SURFACE CHARACTERISTICS

It sometimes is very useful for study and instruction purposes to have, besides photographs, a series of replicas of surface characteristics of structures. Replicas from the large variety of ripple marks present on the beach and adjoining areas are made by many investigators.

The investigator is not restricted to making casts of the surface of an area; he can also combine the cast with vertical samples collected with the small sampling boxes (Senckenberg box) used for making peels. These combinations, as made by H. E. Reineck in Wilhelmshaven, are the most instructive models ever encountered by the author.

The most common material used for making casts in the field is plaster of Paris (gypsum). These field casts are used in the laboratory as molds for making positive copies of the original structure. For the positive copies one can use ordinary plaster of Paris, dentistry gypsum,

Fig. 4.24. Imbedded piece of Belgium anthracite. Imbedding was carried out in a vacuum unit; the coal was not soaked with polyester before, resulting in an abundance of rows of air bubbles.

which is harder and does not break off as easily, or silicone rubbers. The author prefers dentistry gypsum or normal plaster of Paris since they can be painted with water paint to obtain more natural color effects.

The operator should be careful when working near the water line. Casts, with or without boxes, should be removed when the area is still dry to avoid disturbance due to escaping water.

4.6A Surface Models

These are merely models from surface characteristics such as the morphology of ripples. They are easy to make and require little equipment and time (see also method of McMullen and Allen, Chapter One, Section 1.3B).

Equipment

Plaster of Paris, a mixing pan which is large enough for about half of the required plaster, a mixing rod, a container with water, a frame or material to make one, and a compass.

Field Procedure

1. Select the area and place the frame on it. The sides of the frame should be pushed into the sediment to secure its position and to prevent leakage of the plaster of Paris.

2. Make a thin mixture of water and plaster of Paris. Viscous mixtures have the disadvantage that they do not flow evenly over the sediment surface during pouring and create flow marks on the cast.

3. Start the pouring near a corner. Never pour directly onto the sediment, but onto a spoon or in a hand to avoid disturbing the sediment. Move with the flowing motion of the mixture to produce a smooth covering.

Save a little of the mixutre and pour it in a hole. This can be used as a test spot to check the progress of the hardening.

4. Make a thick mixture of plaster and water and pour this onto the first-poured layer. The total cast should be about 2 cm (¾ in.) at the thinnest spots when the cast is not over 15 x 30 cm (6 x 12 in.). Larger casts should be thicker to avoid breakage during removal and transport.

5. Let the mix harden. Since the first mixture was very fluid it is necessary to check the "check hole," because this thin mixture needs more time to harden than the viscous mix. Just before the surface is hard, the operator can engrave all data needed, such as locality, orientation.

6. When the plaster is hard, loosen the frame by inserting a trowel or a knife between the cast and frame. The frame can then be removed.

7. Allow the bottom of the cast to dry and remove all loose material.

Laboratory Procedure

1. Brush the lower surface of the cast carefully to be sure that no sediment is left.

2. Apply a mold release agent to the surface. Many products can be used. The author prefers normal nitrocellulose lacquers because they produce a thin film when sprayed and are not greasy. Petroleum jelly and similar products are not recommended since they may cause greasy spots to which paint will not adhere well. Since the plaster is very porous, one has to be sure that enough coats are applied, but not so many that the relief becomes smooth.

3. Place a frame around the cast. The frame should be high enough to allow the preparation of a thick replica.

4. Pour the material (plaster of Paris, dentistry gypsum, silicone rubber, plastics, etc.) into the mold, and allow it to harden thoroughly. Copy all information engraved on the cast, remove the frame, and collect the replica. For large-sized replicas one can enclose the frame to give it more strength.

5. Examine the replica and restore it if necessary. Use ordinary water paints to obtain more natural color effects.

4.6B Three-Dimensional Models

Three-dimensional models are a combination of surface models and high-relief peels. The same equipment that is used for the surface models in addition to a number of Senckenberg boxes and some coarse sand are necessary. If not enough boxes are available to make a frame, several strips of metal or wood can be used.

Field Procedure

1. Select the area and push the boxes into the sediment (Fig. 4.25). The open side of the base part of each box should face the cast so that the relief of the peel protrudes. The base part should not be pushed in completely, thus saving the ripple surface and allowing the walls to act as a frame. The cover of each box should be pushed in carefully to prevent damage of the surface of the sediment.

The operator should realize that the thickness of the box covers makes it necessary to have a little space between two base parts. The author has often made models using 10 or 12 boxes, producing sides 2 or 3 boxes long, or all sides consisting of 3 boxes. The advantage of using boxes all around is that the boxes along two sides are kept in reserve if something goes wrong with the other boxes. For the final product it is not necessary to have high relief peels along more than two sides.

The investigator is not obligated to make square or rectangular shapes with sides parallel and at right angles to the ripple crests. However, it sometimes is useful to have at least one other direction.

2. Once the peels are framed, make and pour a plaster of Paris mixture (see Field Procedure, points 2 to 5) (Fig. 4.26).

The boxes should be numbered, with the corresponding numbers engraved in the cast to prevent later difficulties.

Fig. 4.25. Making a frame from boxes for replica purposes in the field.

3. When the cast is ready to be removed, the operator first collects the boxes (see Chapter Six). If possible, he should fill the open space in the box above the sample with another, preferably coarser sand to prevent damage to the sample during transport and to prevent difficulties when separating the high-relief peel from the filling material.

4. The cast is sampled as described in Field Procedure, points 6 and 7.

Fig. 4.26. The space inside the frame from boxes is filled with a mixture of plaster of Paris and water.

Laboratory Procedure

1. High-relief peels can be made of all boxes or of the boxes belonging to two adjacent sides, as described previously. Do not remove too much material from the exposed vertical surface, and be sure to pour the resin mixture onto the original surface line to preserve that surface. If coarse material is used as boxfilling, no difficulties are encountered when removing the impregnated part of that coarse material from the rest of the peel.

2. The replica is made as described in Field Procedure, points 8 to 12. It is recommended, however, that one make the replica the same thickness as the peels to give the peels sufficient support.

3. When the replicas and high-relief peels are ready, mount the peels in place, taking care that the upper surface of each peel fits well with the surface shapes of the replica. It is therefore important that little or no material is removed from the exposed vertical surface after the boxes are opened.

4. The seams between the peels can be colored to make them less distinct.

5. Reineck (1961a; personal communication, February, 1962) filled his replica with water at different levels to obtain contour lines in intervals of 1 cm. By applying different colors for each interval, a distinct relief picture of the surface resulted.

These three-dimensional models are very instructive (Fig. 4.27), for they combine surface characteristics with current indications and facilitate the understanding of structures formation (e.g., Reineck, 1960a,b).

4.7 CEMENTING OF BROKEN SAMPLES

It is not unusual that samples break and, consequently, lose part of their value as demonstration or study material. It may not be difficult to put the pieces together again, but much patience is required in the glueing process. A number of examples will be treated here.

4.7A Cementing of Broken Sedimentary Peels

Whether a sedimentary peel can be cemented depends completely on the type of break, the type of peel, the thickness of the peel, whether or not the peel is to be mounted, and how well or how visible the cementing will be.

Unmounted peels can be glued best with the same binding material that is used for making the peel. Mounting the peel is recommended to give it extra strength. The front side can be restored better if loose material is available. This may be time consuming, and therefore the amount of effort should depend on the main use of the peel.

High-relief peels are very difficult to cement because of restoration of the protruding laminae.

4.7B Cementing of Impregnated Sediments

Impregnated blocks which are broken should be cemented with the same artificial product used to fill the sediment pores. The operator seldom will be able to remove all internal traces of the break due to missing grains. On the outside of the block, however, this should not be too difficult.

Thin sections which are broken, including the glass, should be removed from the object glass when the cover glass is removed. When Canada balsam is used, heating will be sufficient to bring the balsam to its fluid state. Apply some new balsam for mounting. When a resin is used for mounting it becomes difficult. One can try to clean the thin section with a little acetone as soon as the cover glass is removed, then mounting that side to a new object glass. The old object glass may have to be removed by grinding.

4.7C Cementing of Broken Hard-Rock Samples

Several products can be used to cement consolidated sediments. The unsaturated polyester resins, and some epoxy resins that cure at room temperature (discussed previously) are very suitable. A dye also can be added to obscure the cementing material.

Another good product is Glyptal 1276 from General Electric (discussed previously). The author has used this product more than any other artificial resin.

Stonemasons and marble workers normally use cold-curing resins that are especially developed for their work. Pigments or dyes can be added to these products. Every marble worker is familiar with his own special products. The only products used by the author are Carsolith Normale from Carlos Desmedt, stone-glue from Rhodius, and a glue from Akemi.

All of these products consist of two parts: the resin and a hardener. The amount of hardener is critical since it influences the hardening time.

Fig. 4.27. Positive cast made from the field mold. Paraffin peels, collected from the boxes, are mounted against the cast to obtain a three-dimensional picture of the ripples. Photograph shows unfinished stage. Exposed river bank with washover structures. Brazos River near College Station, Texas.

After cementing, one should remove excess cement before it hardens since hardened cement is difficult to get off.

4.7D Repairing Imbedded Sediments

The author has repaired imbedded sediments with unsaturated polyester resins only. First, one examines how the pieces fit together and how much restoration must be carried out on the sample itself. All sides of the break are cleaned and a similar resin mixture is used as cement. All excess mixture cannot always be removed since the mixture acts as a strengthening agent. Since it is difficult to keep the upper surface smooth, it is better to repeat the imbedding step as far as the fabrication of the final top layer is concerned.

CHAPTER FIVE

SOME LABORATORY INSTRUMENTS

Most procedures used to study sedimentary properties have to be carried out in the laboratory (Fig. 5.1). It is necessary to utilize certain technical instruments not only for saving man-hours and for series production but also for obtaining good results. Some techniques are so simple that common chemicals and household tools form the required equipment. However, other methods require more and specific purchases, especially when dense samples have to be lithified and transferred into thin or thick sections.

Thin sections of normal size, 32 x 24 mm, can be made from hard rock with a minimum inventory on apparatus. When large-sized thin sections are required, especially from unconsolidated samples, it is necessary to use special vacuum-impregnation, sawing, and grinding equipment. At the same time, the presence of such equipment enables the operator to carry out part of the work in series. Since it is impossible to discuss all commercial equipment and institution-invented apparatus related to techniques described in this book, the author has confined himself to the description of his own experience.

In studying sedimentary properties, low-viscosity resin mixtures are used because they easily penetrate the sample when only vacuum is applied. If, in addition, the temperature is to be raised, an oven will be suitable. For some impregnations with epoxy resins, however, it is desirable to have an instrument to which vacuum as well as pressure can be applied. For imbedding sample slices in plexiglas a combination of heat and pressure is required.

Since the devices for vacuum-only are different from those to which vacuum as well as pressure can be applied (Sections 5.1 and 5.2, respectively), schemes of both systems are given in Fig. 5.2.

5.1 VACUUM-ONLY IMPREGNATION APPARATUS

We shall discuss three such apparatuses. Jongerius and Heintzberger (1962, 1963) deal with large-sized samples (15 x 8 x 5 cm) and therefore use larger equipment than Altemüller (1956, 1962).

5.1A. Equipment according to Jongerius and Heintzberger

This impregnation equipment consists of a steel cylinder (internal height 20 cm, internal diameter 30 cm) with a connection pipe to a vacuum pump, and a flat round metal cover. A ring is welded to the top of the cylinder, which has a groove for an O-ring. Both corresponding edges are greased with vacuum or stopcock grease to make the seal airtight. In the center of the cover a hole is drilled (diameter 10 cm) in which a thick-walled spherical observation glass is mounted. In addition, a small, round ring (diameter 32 mm) with raised edge is welded in the cover; this holds a glass vacuum sphere with a capacity of about 2 liters. The connection between cylinder and sphere can be closed by a hollow stop, which is connected to a vacuum pump. In both the vacuum tubes, running from

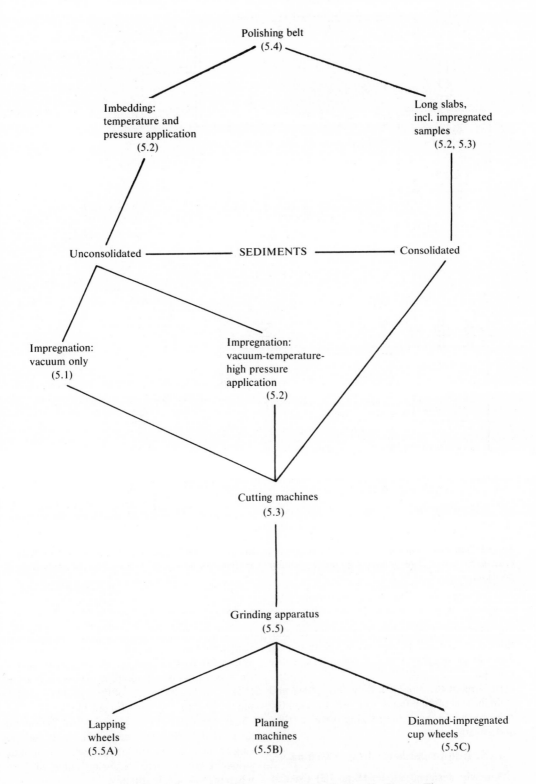

Fig. 5.1. General outline for the different apparatuses discussed in Chapter Five.

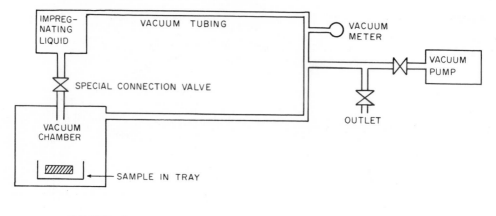

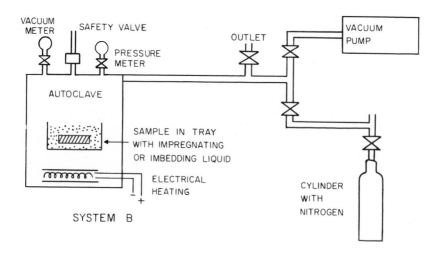

Fig. 5.2. Schematic impregnation outfit. *System A:* impregnation with the application of vacuum, with or without heating. *System B:* impregnation with the application of vacuum and heat, later followed by the application of pressure.

the cylinder and the sphere to the vacuum unit, three-way control valves are placed; one of these is an open connection for releasing the vacuum.

In Utrecht the stopcock between sphere and cylinder was often difficult to clean. Forcing the stopcock could result in breaking the glass tube underneath.

5.1B. Equipment according to Bouma

A newly designed model (Fig. 5.3) used at Utrecht has a larger cylinder (1). All parts are made of stainless steel. The inner dimensions

are: height 30 cm, diameter 35 cm. The cylinder is mounted to a base plate (2), with two support angles (3) welded underneath to provide space for a vacuum connection pipe (4) on the bottom (Fig. 5.4). To the top (5) of the cylinder a ring is mounted with a groove for an O-ring (7). The cover consists of clear plexiglas (6). To secure an airtight sealing, the edge of the cover and the ring of the cylinder are greased with vacuum or stopcock grease since the O-ring will let air pass when the vacuum is still low.

In the cover (6) three holes (8) are drilled to give alternative entry points for the impregnating liquid. One hole is in the center, one at 8 cm and one at 15 cm from the center. The holes can be closed by heavy metal mushroom-shaped stops, also fitted with an O-ring groove.

A metal funnel-shaped cylinder (9) replaces the glass sphere type of Jongerius and Heintzberger. Its cover (10) is made of plexiglas, sealed to the funnel by an O-ring (11), with a conical hole that allows connection with a vacuum tube. The stopcock (12) is made of stainless steel, thus overcoming the difficulties encountered with the glass stopcock. A cylindrical chamber (13) contains a cylindrical center (15), through which a hole (17) (diameter 4mm) has been drilled. Two O-rings (16) prevent air and resin leakage. The outer sides of the chamber are closed by metal covers attached by two screws each (14) and can be removed easily. A steel plate (18) with O-ring groove acts as an airtight connection when the funnel is placed in one of the holes of the cylinder cover (6).

A ring (10) is welded to the cylindrical part of the funnel (9), acting as its support. The support consists of a vertical rod screwed to the base plate, and a horizontal rod with a ring at its end. The horizontal rod can move up and down and around the vertical rod and at the same time slide through a block tightened to the vertical rod. In this way the funnel can be placed in any position the operator desires (Fig. 5.3). The whole apparatus is easy to dismantle for cleaning and operational purposes.

Vacuum Circuit

Vacuum is obtained by a vacuum pump, sometimes in conjunction with a filter pump. Dr. Slager (Department of Soil Science, Agricultural University in The Netherlands) found that if the vacuum pump alone is not capable of producing a vacuum of about 10-15 mm Hg, a filter can be inserted in the circuit. In Fig. 5.5 a scheme of the vacuum circuit is given. From the pump (1) a vacuum tube with stopcock (2) goes to a series of 10-liter bottles (3) which serve as a vacuum reservoir. From the last bottle, a tube runs through vacuum meter (4) to a T-piece (7).

Fig. 5.3. Vacuum impregnation apparatus (right-hand side) consisting of a stainless steel cylinder with plexiglas cover and a stainless steel funnel with steel vacuum stopcock and plexiglas cover. The funnel is movable around its support. (For details see text and Fig. 5.4). On the left-hand side is the model according to Jongerius and Heintzberger.

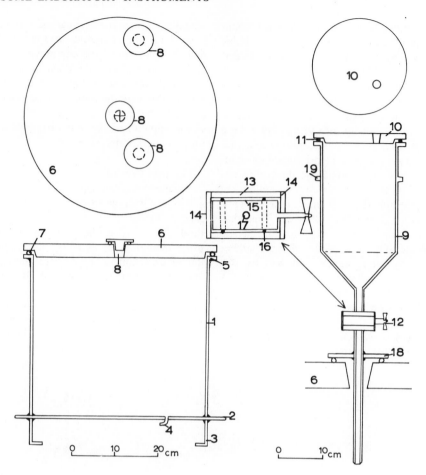

Fig. 5.4. Schematic drawing of the vacuum impregnation apparatus: (1) cylinder; (2) base plate; (3) sled; (4) connection to vacuum line; (5) ring welded to cylinder; (6) cover; (7) O-ring; (8) hole with stop and O-ring; (9) funnel; (10) plexiglas cover with hole for connection to vacuum; (11) O-ring; (12) stainless steel stopcock; (13) stopcock house; (14) closing plates; (15) cylindrical inner part; (16) O-ring; (17) hole, 5 mm diameter; (18) steel plate with O-ring; (19) ring on which funnel hangs in support.

In the middle of this tube another T-piece is inserted, of which the center circuit is connected through a stopcock to a filter pump. The stopcock (6) separates the impregnation apparatus from the vacuum supply tube. From the T-piece (7) two equivalent tube circuits run to the vacuum apparatus (one type according to Jongerius and Heintzberger and one modified type). Each of these small circuits can be closed by a stopcock (8). Each contains three in/out lets: one for release vacuum (9), one connection (10) to the impregnation cylinder, and one connection (11) with stopcock to the funnel.

For each vacuum unit the procedure is as follows. Close stopcock (6) and the stopcock in tube (5). Start the vacuum pump (1) and open the stopcock (2). Place the samples in the cylinder, close the cylinder, close the stopcocks in tubes (9) and (11) and open the stopcock (8). Fill the placed funnel with impregnating liquid, close funnel and open the stopcock in tube (11). Open the stopcock (6). Start filter pump, open stopcock in tube (5), close stopcock (2) when the electrical pump no longer creates vacuum; shut off the pump. Open the steel stopcock of the funnel. After impregnation the stopcock in tube (5) is closed before the filter pump is shut off. The stopcock (9) can be used if vacuum is still present when the cylinder must be opened. When working continuously, the operator should close the stopcock (6) before opening the tube (9) to preserve the vacuum in bottles (3).

Between the filter pump and the stopcock in tube (5), a cylinder with water-absorbing chemicals is placed, for water prevents good curing of the plastic. The samples should not be allowed to absorb water. The styrene monomer vapors are bad for the electrical vacuum pump. A barrel with oil inserted in the circuit or a freeze unit prevents these vapors from reaching the pump (Jongerius, written communication, December, 1966).

It is also useful to place a double-neck bottle between the filter pump and the stopcock in tube (5) to prevent water from running into the circuit if something adversely affects the water pressure.

5.1C Equipment according to Buehler

Vacuum impregnation equipment is also commercially available at Buehler Ltd. Examples are given in the AB Metal Digest, Volume 12, No. 1 (1966).

5.2 VACUUM-HIGH PRESSURE-TEMPERATURE IMPREGNATION APPARATUS

For the imbedding of samples in plexiglas as well as for the impregnation of sediments with certain epoxy resins it may be necessary to apply vacuum, pressure, and temperature. The vacuum serves to evacuate air out of the sediment pores. Temperatures between 80 and 180°C are necessary to lower the viscosity of the impregnating or imbedding liquids, or to cure these synthetic products. Pressure can be utilized to force the resin into the sample, and it is also required to cure the plexiglas mixture (see Chapter Four).

For imbedding with plexiglass, temperatures up to 57 to 60°C (135 to 140°F) and pressures up to 8.44 to 9.14 kg/cm^2 (8.1 to 8.8 atm, 120 to 130 lb/in.^{2}) are required. To impregnate unconsolidated samples with the epoxy resin D.E.R. 332 from Dow Chemical, vacuum as well as high pressure (1400 lb/in.^{2}, 98.42 kg/cm^2, or 95.2 atm) is necessary.

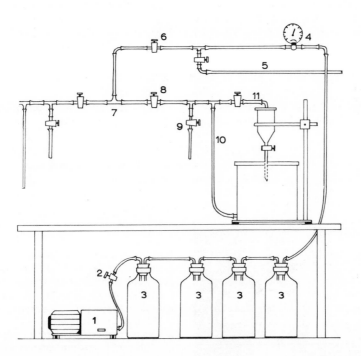

Fig. 5.5. Schematic outline of the vacuum circuit (for two apparatus): (1) vacuum pump; (2) stopcock; (3) 10-liter bottles; (4) vacuum meter; (5) tube running to filter pump; (6) stopcock; (7) T-piece; (8) stopcock; (9) tube for release vacuum; (10) tube running to cylinder; (11) tube running to funnel.

At the Coastal Studies Institute of the Louisiana State University, two apparatus (autoclave and high-pressure impregnation device) are in use; in Utrecht a combination of these has been developed.

5.2A. Autoclave of Coastal Studies Institute

The Research Center of Humble Oil in Houston, Texas, works with a commercial type of autoclave, while the Coastal Studies Institute group has developed its own atuoclave to imbed samples in plexiglas (Fig. 5.6). It consists of a horizontally positioned steel cylinder in which a tray, for the imbedding of nine core slices simultaneously, can be placed. The inner dimensions are 68.6 cm (27 in.) and 35.6 cm (14 in.). An insulation cover of fiberglass assists in holding a constant temperature. The steel cylinder is double-walled. The inner wall is strong enough to hold a pressure of at least 150 lb/in.^{2} (10.55 kg/cm^2 or 10.2 atm). Water can circulate between both mantles, and the cylinder is therefore connected to a hot water supply and equipped with a drain valve at its base and an overflow valve on top. Temperature is controlled by a thermometer.

Pressure is obtained from high-pressure nitrogen cylinders, because oxygen reacts adversely on the curing of the plexiglas. The nitrogen cylinders are connected to the autoclave by an ordinary set of pressure gauges. A pressure relief valve on top of the autoclave makes a decrease and total release of pressure possible while a "pop off" pressure safety valve must be mounted to prevent the autoclave from bursting. The autoclave can be closed by a hinged, thick-walled door. Twenty large bolts with coarse thread allow quick and easy tightening of the instrument. To prevent leaking, a set of two different-sized high-pressure gaskets are placed in polished grooves between door and autoclave.

5.2B. High-Pressure apparatus of Coastal Studies Institute

The high-pressure impregnation device for impregnating unconsolidated samples with a synthetic resin, as developed at the Coastal Studies Institute, is a very simple instrument (Fig. 5.7). It is made of a high-tensile material to withstand a pressure of 1500 lb/in.^{2} (105.5 kg/cm^2 or 102 atm). The device consists of two parts: a lower and an upper part. The lower part is a vertically placed tube (1) with an inner diameter of about 6.3 cm (2½ in.) and a length of about 63.5 cm (25 in.). Three stands have been welded to the bottom to facilitate mounting the device to the floor. A vacuum-pressure pipe (3, 6) is connected to the bottom end. Just below the top, a metal collar (7) has been welded with eight holes through which the tightening bolts (10) run. The upper part is made of a thick metal plate (9) to which a small piece of the main cylinder has been welded. Eight holes have also been drilled through the metal plate. On top of the upper part a hand-operated high-pressure release valve (11) and a "pop off" pressure safety valve (12) are mounted to a protruding tube. The parts fit together with a high-pressure gasket (8) between; the eight coarse threaded bolts secure the tightening.

The vacuum-pressure pipe line ends in a T-piece. From this T-piece a pipe with a high-pressure valve runs to a vacuum pump, while the other side goes to high-pressure cylinders with nitrogen through a normal set of pressure gauges.

Fig. 5.6. Autoclave for imbedding cores in plexiglas as used at the Coastal Studies Institute of the Louisiana State University: (1) cylinder with a fiberglass insulation cover; (2) water tube connected between autoclave and hot water supplier; (3) thermometer; (4) chain for holding nitrogen cylinders; (5) pressure gauges for nitrogen cylinders; (6) pressure relief valve; (7) "pop off" pressure safety valve; (8) autoclave door; (9) coarse thread bolts for quick tightening; (10) bucket for mixing plexiglas mixture. (Courtesy J. M. Coleman, Coastal Studies Institute, L.S.U.)

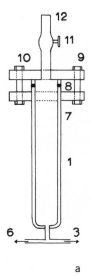

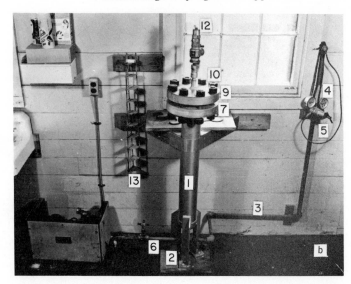

Fig. 5.7. (*a*) Schematic drawing of the high-pressure impregnation device as developed at the Coastal Studies Institute of the Louisiana State University. (*b*) Photograph of the high-pressure impregnation device: (1) lower cylindrical tube; (2) stands for connecting device to the floor and giving space for pipe line; (3) high-pressure pipe with high-pressure valve running to nitrogen cylinders; (4) gauges for nitrogen cylinders; (5) high-pressure valve; (6) pipe running to vacuum pump; (7) collar welded to cylinder, (8) high-pressure gasket; (9) upper collar; (10) coarse thread bolts; (11) pressure relief valve; (12) "pop off" pressure safety valve; (13) sample holder. (Courtesy J. M. Coleman, Coastal Studies Institute, L.S.U.)

A sample holder (13) for 10 samples, made of three metal rods with metal plates welded to them at regular intervals, fits the vertical tube of the impregnation device. Waxed cups, containing sample and synthetic resin, are placed in the holder before this is lowered in the device.

5.2C. Autoclave according to Bouma

The author developed an instrument that is a combination of both apparatus of the L.S.U. Coastal Studies Institute. The requirements were as follows (Fig. 5.8). Minimum workable diameters: a cylindrical shape with an inner diameter of 16 cm and an inner length of 60 cm. It must be used at a nitrogen pressure of 102 atm (1500 lb/in.^{2} or 105.5 kg/cm^2) at 20°C (68°F), with vacuum at a temperature up to 200°C (392°F), or with a nitrogen pressure of 10.5 atm (154 lb/in.^{2} or 10.8 kg/cm^2) at 80°C (176°F). It should be equipped with valved connections for vacuum, pressure, and release, a variable thermostat, pressure safety relief, thermometer, vacuum and pressure meters. The door must be hinged and easy and quick to tighten.

A calibrated high-pressure relief valve (2) is attached to the upper side of the cylindrical ap-paratus, a thermometer (3), a manometer (4) with a valve in the connecting pipe to the auto-clave, and a vacuum meter (5), also equipped with a valve. At the front side is a thermostat (8), a high-pressure valve (10) with a pipe-end for nitrogen cylinder connections, a high-pressure valve (11) with a pipe-end for vacuum connection, and a high-pressure valve (9) for release of pressure or vacuum.

The closing part consists of a thick collar (12), which is welded to the open end of the auto-clave, and a hinged door (13, 14) made from a thick plate. To secure tightening, a high-pressure gasket (16) (asbestos compound packing) is placed in a polished groove made in the collar, while a ring which fits the gasket sticks out of the door. Eight bolts (15), with coarse thread, keep the door closed.

Three electrical tubular heaters with metal covers (18), each 1000 watts, are mounted inside along the basal part of the autoclave cylinder. On top of these rods a loose metal tray (19) with holes can be pushed in, on which sample containers will be placed.

Electrical connections for the tubular heaters run through the closed end of the autoclave cylinder and through porcelain insulators where an

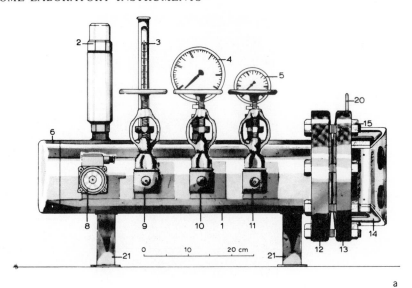

a

Fig. 5.8. (a) Schematic drawing of the high pressure-vacuum-temperature apparatus as used at the Geological Institute at Utrecht, The Netherlands. (b) Photograph presenting the apparatus with the hinged door open: (1) cylinder of the autoclave; (2) high-pressure relief valve; (3) thermometer; (4) manometer with valve; (5) vacuum meter with valve; (6) loose cap protecting electrical parts; (7) relay; (8) thermostat; (9) high-pressure valve with pipe-end for connection to nitrogen cylinders; (10) high-pressure valve with pipe-end for connection to vacuum pump; (11) high-pressure relief valve; (12) thick collar welded to cylinder; (13) door; (14) door hinge; (15) bolt with coarse thread; (16) high-pressure gasket; (17) protruding ring fitting gasket; (18) electrical tubular heaters; (19) metal tray for sample containers; (20) door grip; (21) triangular legs with ground plates.

electrical circuit runs to a separate relay (7) to the thermostat (8) and to the main tension (300 V, 3 phase).

The operator should be provided with instructions concerning the handling of an apparatus to which temperature and/or pressure and/or vacuum can be applied to prevent accidents. Since one may easily forget to close the correct valves each time, and especially when high pressures are applied, safety precautions should be strictly observed. Ignoring safety precautions may have

dangerous results. Electrical safety contacts can be built in.

5.2D. Mount Press according to Buehler

Buehler, Ltd., manufactures a specimen mount press for 2.54 cm (1 in.) or 3.18 cm (1¼ in.) mounts (AB Metal Digest, Vol. 10, No. 2, 1964). The maximum pressure which can be applied is 10,000 psi (4,540 kg/cm²; Buehler, 1962, 1966).

5.3 CUTTING MACHINES AND SPECIAL TOOLS

In the field of cutting machines a large variety both in capacity and performance is available. Geological institutes generally require a large-size machine, which enables the operator to cut fairly large rock specimens and also fist-sized samples. The thickness of the sawblade should be limited, especially when many parallel slices are required.

The spinning blade produces a centrifugal effect on the water, and consequently the closer the water is injected to the point where the specimen touches the blade, the more water runs into the sawed groove, and the more effective cleaning and cooling action is obtained. The machine should be very sturdy, easy to clean and grease, and the blades should be easy to change. Very essential is the table on which the sample rests. This table should be easily movable and movement should be so precise that no sideway movements are possible during the sawing operation. It should be large enough to hold large samples and to carry additional tools for keeping a specimen in place.

5.3A. Cutting Wheels

Diamant Boart (1964) published a very valuable booklet in which several types of sawblades and their properties are given. Mounting and technical difficulties are also discussed. Some topics will be treated briefly here. Cutting a wheel with an uninterrupted abrasive edge which produces very accurate cuttings are manufactured by Diamant Boart for diameters ranging from 30 to 400 mm. Blades with diameters from 20 to 250 cm normally have a segmented rim, which allows better cooling during cutting. It is very important that the wheel is firmly clamped in the cutting machine, that the rotation direction corresponds with the arrow on the wheel, and that there is no visible wobble in the blade. The depth of cutting depends on the hardness of the material.

Most continuous-rim diamond wheels may be run in either direction equally well. They provide a much smoother finish and thinner cut than the rimlock-type wheels (G. W. Graves, Buehler Ltd., written communication, October 1966).

When the operator observes that the wheel loses its cutting ability because of low specific cutting pressure, the depth of cutting should be decreased. The specific cutting pressure is a very important part of the operation. When this pressure is too high, diamond grains are torn off and early wear results. If the specific cutting pressure is too low, not enough cutting force will be available to move the blade freely through its groove and the diamond grains will wear. Another important point is the peripheral speed which should not be less than 30 to 35 m/sec. Before continuing cutting the operator must wait until the wheel has obtained its normal speed again. If the cutting operation is still slow and the wheel slows down again and again, the blade can be sharpened by cutting a few centimeters into an abrasive material such as a sandstone or a limy sandstone brick composed of round, well-sorted sand grains.

The peripheral speed and the specific cutting pressure will result in the cutting speed, which normally is given in cm^2/min. This speed can be checked regularly by measuring the time necessary for cutting a certain surface; for example, 0.5 m^2 in 10 min gives a cutting speed of 500 cm^2/min (see Table V.1).

Ample cooling water should be provided since the water not only serves for cooling but also as a cleaner for the groove cut and as a lubricator. The water for cooling should be injected into the blade by a tube shaped like a dinner fork. Wheels with diameters ranging from 200 to 350 mm need at least 5 to 8 liters/min, those of 400 mm need 12 liters/min, and those of 500 mm diameter need 15 liters/min.

The peripheral speed of the cutting blade depends on the hardness and abrasive texture of the specimen. Diamant Boart (1964) suggests a speed of 20 to 30 m/sec for hard material with a dense texture (porfirite, quartz), for less hard and dense samples (limestone, granite, marmor) a speed of 35 to 55 m/sec (granite: 35 m/sec; limestone: 55 m/sec), and for abrasive specimens (tuff, soft sandstone, etc.) a peripheral speed of 55 to 65 cm/sec. Figure 5.9 presents the number of revolutions for different diameters corresponding to peripheral speeds.

5.3B Large Cutting Machines

Excellent results have been obtained with type A of Diamant Boart. The machine is capable of cutting long slabs with a maximum height of 150 mm. Blades with diameters of 200, 250, 300, and 350 mm can be used (Diamant Boart, 1961).

Type B_2 of Diamant Boart is a machine comparable to type A; however, the suspension of the spindle of the cutting wheel has a swan's neck shape, which allows the operator to cut samples that overlap the sides of the table (Diamant Boart, 1961).

At the Shell Development Company Exploration and Production Research Division, Houston, Texas, a standard cutting machine (Stone saw type SS 16 from the Stone Machinery Company, Mantius, New York) is used. It is mounted on rails (Ginsburg et al., 1966). The rails are fixed to a thick table deck, which rests on concrete foundations. The machine is rigged for automatic feed by the addition of a cement block with a weight of 100 kg. A simple boat trailer winch is used to pull the saw and weight back.

Cutrock Engineering Company Ltd. manufactures a core and rock cutting machine, model VCC 500, which allows cutting of samples up to 152 cm (5 ft) long and 15.2 cm (6 in.) in diameter with a 45.7 cm (18 in.) wheel, or 20.8 cm (8 in.) in diameter with a 61 cm (24 in.) wheel. The

depth of the cut can be adjusted. Without reclamping the sample, parallel slices can be cut (Cutrock, not dated a).

Cutrock's refractory brick cutting machine, model BCM 200, can be fitted with a square angle rest and adjustable stop to allow repeated cuts to be taken from long lengths (Cutrock, not dated b).

Cutrock auto-feed rock cutting machines, models ARC 100 P and H, can handle a cutting wheel with a diameter of 35.6 cm (14 in.) (Cutrock, not dated c).

Buehler, Ltd. manufactures two large models. The 11-1152 AB DI-MET cutter 120 is equipped for a 35.6 cm (14 in.) diamond wheel with a plain rolling table, whereas no. 11-1150 has a 30.5 cm (12 in.) rotary table with graduated lateral, transverse, and tilting adjustments. The 11-1173 AB cutter HIPA J-3 features a heavy welded steel saw box and accommodates a wheel of 40.6 cm (16 in.) diameter, which will readily section a 15.2 cm (6 in.) rock.

Table V.1.   Cutting Speeds for Different Types of Material. The Velocity Obtained Depends on the Cutting Wheel Used. The Manufacturer Should Provide Similar Data for His Products

Material	Average Cutting Speed (cm²/min)	Remarks
Slate	600– 900	
Granite	70– 100	Cutting depths ranging from 2 to 5 mm
Granite	100– 150	Cutting depths ranging from 5 to 10 mm
Soft sandstone	500– 700	
Hard sandstone	200– 500	
Very abrasive soft sandstone	700–1000	
Basalt	400– 600	
Marble	600– 800	
Quartz	4– 8	Cutting depths ranging from 0.3 to 1 mm
Quartzite	50– 80	Cutting depths ranging from 0.3 to 1 mm

After Diamant Boart (1964).

Note: The Cutrock Engineering Co. Ltd. (written communication, July 1966), uses deeper cuts. They cut at least 25 mm deep with no serious difficulty.

As a general rule it can be stated that the maximum cutting depth is ⅓ of the diameter of the cutting wheel. In the laboratory, where the operator deals with small samples, the cutting depth can be more (J. Gramberg, Mining Lab. Delft, written communication, February, 1967). Further, the peripheral speeds, as given in Fig. 5.9, apply more to industry than to laboratories.

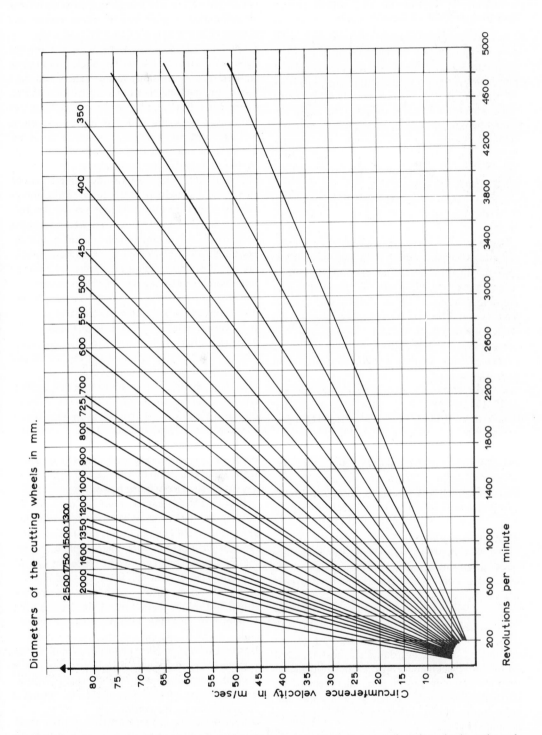

Fig. 5.9. Relation among peripheral speeds (circumference velocities), diameter of cutting wheels, and revolutions of the wheels. (Redrawn after Diamant Boart, 1964, p. 10.)

293

5.3C. Cutting Machines for Small Samples and for Mounted Slices

It is very convenient and time-saving to have a smaller additional cutting machine for cutting small samples and for taking off a slice from rock pieces or from specimens mounted on glass. The thickness of the sawblade is less than that of the large wheels, and, consequently, the force obtained by placing the blade against the object is smaller. There are many such instruments. Some of these cutting machines are for cutting only, whereas others are combined cutting-grinding machines.

Type D_2 from Diamant Boart is similar to their B_2 type, only smaller, but it is still very sturdy.

Wirtz manufactures cutting machines DIMA capable of cutting samples ranging from 35 to 200 mm in thickness. The axis of the cutting wheel is placed underneath the table. It is also possible to mount a device for thin-slicing samples mounted on object glasses of 28 x 48 cm (Wirtz, not dated a, b, c).

Jongerius and Heintzberger (1962, 1963) use a small Winter cutting machine for preparing their thin sections.

Buehler produces two types. Number 11-1160 AB DI-MET cutter 11R holds a 20.3 cm (8 in.) wheel. The 11-1175 AB cutter HIPA E-2 is a precision trim saw (Buehler, 1966).

Cutrock manufactures a series of cutting-grinding machines. Model OAC 200 is an over-arm cutoff machine (Cutrock, not dated d). The Mark II Unicutta is similar to normal technical grinding machines. It can cut sections up to 1¼ x 1 in. and grind sections mounted on 3 x 1 in. microscope slides (Cutrock, not dated e).

Diamant Boart (1961, 1964) also manufactures a cutting-grinding machine, type LM, which can hold a 200 mm diameter diamond wheel.

Special core-cutting machines are the core drilling and cutting machine model CD 200, which can handle cores up to 10.2 cm (4 in.) diameter (Cutrock, not dated f), the VCC 200 core-cutting machine for longitudinal and lateral sections from cores up to 6 in. diameter and 16 in. long (Cutrock, not dated g), and core- and cube-cutting machine model GSP 250 (Cutrock, not dated h).

5.3D. Clamping Devices

Different types of clamps are available; the one to be used depends upon size and shape of the object. Very irregularly shaped samples can be imbedded in plaster of Paris. A simple, but strong clamp consists of two vertical rods with screw thread, which can be fastened on the table. A horizontal bar with a thick rubber strip underneath moves along these vertical rods and can be pressed on the sample by using nuts that fit the rods. For samples with parallel flat upper and lower sides this is ideal (Cutrock, not dated b, c, f). Instead of a clamp which rests on the sample, the operator can use vertical rods with thread mounted on the table along which fingers can slide until they rest on the sample, which then can be tightened by nuts. This allows clamping of different-sized samples. If the bottom is not flat enough, wooden wedges can be used to obtain a nonmovable mount (Cutrock, not dated a). Clamps that work as bench-vices are not ideal, even when the claws are covered with wood, since irregular objects turn easily or move under the pressure of the cutting blade. Very convenient are clamps that can be screwed to a raised edge of a cutting machine table. These are often straight strips to which the sample can be held during hand cutting. Normally the combination of a raised edge such as a clamp and, if necessary, one or more wooden wedges enables the operator to place the sample firmly on the table and prevent it from moving away. Jongerius and Heintzberger (1962, 1963) deal with impregnated samples of approximately brick size. They have a few clamps (Fig. 5.10), which hold the sample in place and prevent the sawed-off slice from being hurled away.

Special clamps can be made or obtained for holding samples or slices mounted on an object glass. These clamps can move, pressed against a straight strip clamp, with the sample along the cutting blade. They can also be placed in a grinding machine or can be used for holding the mounted slice during polishing operations. Buehler (1962) manufactures Thin Section Slide Holder No. 30-8000. This device holds a glass slide on which a sample chip has been cemented. Boron carbide blocks protrude slightly. The sample is approximately 50 μ thick when it is ground down flush with the blocks. Jongerius and Heintzberger (1962, 1963) developed a large wooden jig for cutting large-sized mounted soil sections (Fig. 5.11). At the Geological Institute at Utrecht a brass clamp is used; this is primarily for cutting samples or slices mounted on object glasses (Fig. 5.12). Dollé (1959) used pieces of glass (60 x 30 mm and 6 mm thick) against which the object glasses can be mounted

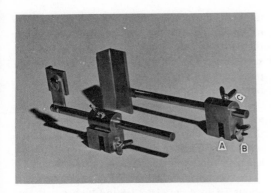

Fig. 5.10. Metal clamps as used on the Diamant Boart cutting machine, type A. The grooves (A) are pushed on the raised edge of the table and fastened by the butterfly nuts (B). The length of the arm is adjustable by means of butterfly nut (C). The retaining piece of the right clamp is much longer than the left one. Apart from clamping, the right clamp has a bearing position to prevent the sawed-off plate from being hurled away. (Courtesy of the Soil Survey Institute, Wageningen, The Netherlands; Jongerius and Heintzberger, 1962, 1963.)

Fig. 5.11. Wooden clamp for cutting mammoth-sized sections. The object glass is clamped to the jig by thin bakelite strips. (Courtesy of the Soil Survey Institute, Wageningen, The Netherlands; Jongerius and Heintzberger, 1962, 1963.)

by means of a temporary cement (e.g., Lakeside No. 30L), which melts at a temperature lower than Canada balsam or Lakeside thermoplastic cement No. 70 C. These glass pieces are used during cutting, grinding, and polishing.

5.3E Comments

Gramberg (Techn. Univ., Mining Lab., Delft, The Netherlands, written communication, February 1967) carried out many experiments with a cutting machine from Conrad, Model WOCO

200. The cutting blade spindle is placed underneath the table (Conrad, not dated a, b, c), as in the models from Buehler, Winter, and Wirtz. The lubricating proved to be completely successful, and none of the experiments failed. Rock slabs from quartzite, gneiss, diabase, dense limestone, and poorly consolidated sandstone of a size 9 x 12 cm were mounted on glass with Araldite epoxy resin from Ciba. The cutting blade vibrated so little, because of a balance which applies the pressure equally, that without any difficulty enough could be cut off that 0.5 mm rock was left on the object glass. The object glass itself is held in position by means of vacuum, created by a filter pump.

5.4 POLISHING BELT

Grinding and polishing activities of flat surfaces can be carried out on grinding or polishing wheels of different sizes depending on the size of the object. For long samples such as imbedded sediment slices in plexiglas or in plastic, it is advisable to use a long sanding belt such as the one used at the Coastal Studies Institute. The advantage of such a belt is that the sanding is carried out semiautomatically, which is important since this process is very time consuming.

The major part of the belt has been obtained from the glass industry. The sandpaper belt runs

Fig. 5.12. Thin-section slide holder, made of brass. The object glass is fixed tight by means of two screws. Geological Institute, Utrecht, The Netherlands.

around two drums, of which one has a fixed position while the other drum is movable to allow adjustment of the belt tension and removal of the belt. An electric motor produces the force through a belt to the fixed drum. Both belt ends are protected by caps. At the side of the variable drum a water pipe with spray nozzle provides the water necessary for the sanding. (Fig. 5.13a).

The object to be polished is placed on top of the belt (Fig. 5.13b). An adjustable platform with a glass-covered top is mounted underneath the upper side of the sandbelt to secure a good and flat contact with the object. Since the object must be held in position, a stop can be placed in a holder fixed to the adjustable platform.

Weights can be placed on top of the object to obtain more pressure at the contact with the sanding belt. A series of belts with different grades are required to carry out the grinding-polishing work in steps, similar to the manufacture of thin sections.

5.5 GRINDING APPARATUS

Grinding and polishing are the last steps to be taken in finishing a thin section or a cut section, or before an acetate peel can be made from a hard-rock specimen.

The best-known apparatus is the polishing wheel, made of a circular, balanced steel plate. The specimen is pressed to the disk, while different grades of carborundum powder (from coarse to fine), with water as lubricant, are used to obtain the effect wanted. It is very important to regularly check the surface of the wheel for the right flatness. To protect the disk, it is necessary to use the same number of wheels as applied carborundum grades to avoid contamination. The coarsest grade also ruins the surface of the disk, and therefore polishing is senseless since grains may stick to the surface and scratch the sample. The wheel must be cleaned regularly with water because small pieces, broken off the sample, will ruin the ground sample surface.

It is important that the sides of the box are about 2½ to 4 cm (1 to 1½ in.) higher than the top of the wheel to prevent the water from wetting the whole surrounding area because of the centrifugal force. The space between the outer side of the wheel and the box sides should be wider than the size of the small specimens or thin section object glasses. This is to avoid breakage when the object slips out of the operator's hand. When not in use a cover should be placed on top of the cleaned unit to prevent dirt from falling in.

5.5A. Lapping Wheels

Buehler (1962) manufactures several lapping wheels such as the Nos. 38-1442 and 38-1440 AB thin section grinders and several petrographic polishers for small specimens and thin sections. The types 38-1420 and 38-1422 grinders HIPA K18 and K12 are heavy models.

Diamond compounds for polishing are available in disposable plastic syringes, for example, the Buehler Metadi Diamond Pastes.

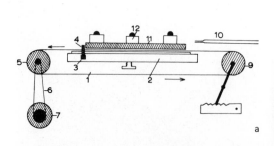

Fig. 5.13. (*a*) Sanding belt as used at the Coastal Studies Institute of the Louisiana State University, Baton Rouge, Louisiana. (*b*) Scheme of the sanding belt: (1) sandpaper belt; (2) adjustable platform with glass covered top; (3) object stop holder; (4) object stop; (5) fixed drum; (6) belt; (7) electric motor; (8) switch; (9) movable drum to adjust belt tension and to remove belt; (10) water pipe with spray nozzle on head; (11) object; (12) weights. (Courtesy J. M. Coleman, Coastal Studies Institute, L.S.U.)

Cutrock (not dated g, j, k, l) also manufactures different types and sizes of lapping machines with wheel diameters varying between 15 and 62 cm (6 and 24 in.).

Wirtz (not dated d) manufactures extensive series of grinding and polishing machines. They all are equipped with changeable wheel speeds. The Wirtz Company recently came out with an automatic grinding-polishing apparatus under the name Vibropol. A maximum of 10 samples can be treated simultaneously (Wirtz, not dated e, f).

The market also provides polishing wheels to which abrasive paper can be clamped using water under centrifugal force. For small specimens they can be used successfully, but the author ordinarily prefers metal wheels and carborundum powder for routine operations.

For large surfaces such as stonemasonries large movable polishing disks are used. These disks may be equipped with carborundum or diamond-impregnated segments of different grades. For laboratory use hand-operated grinding-polishing disks can be obtained since the masonry models are too large and too expensive for occasional use (Diamant Boart, 1964).

5.5B Planing Machines

For large thin sections and for series of small-sized thin sections a planing machine can be used. A table is movable along three perpendicular guides operated by levers equipped with micrometers. The table can make a transverse horizontal movement. Over this table a carborundum or diamond-impregnated wheel rotates on a horizontal spindle. The sections to be thinned down are placed on top of this table. Because of the continuous horizontal tranverse movement in longitudinal direction, together with a slow transverse traveling of the table, a thin slab is taken off from the rock sections. After each complete sweep the wheel must be lowered 1 to 75 μ, depending on the type of rock. For good results it is necessary to have similar rocks in a series.

Cutrock (not dated g, i) manufactures the Universal Rock Cutting and thin section machine, model GH 3. It has a table surface of 43 x 12.6 cm (17 x 5 in.) and is able to grind 6 mounted 3 x 1 in. thin sections simultaneously.

Following Jongerius and Heintzberger (1962, 1963), a combined horizontal and vertical surface grinding machine, model HV IIa from Klaiber, is employed at the Geological Institute in

Utrecht (Klaiber, 1964a, b, 1964c, 1966). The clamping area of the table is 500 x 150 mm (19^{11}/16 x 5^{29}/32 in.), the electromagnetic area is 300 x 150 mm (11^{13}/16 x 5^{29}/32 in.) or 380 x 150 mm. The table can be leveled in all directions. After each grinding sweep the abrasive disk is slightly lowered, initially about 50 μ every time for sandy samples and about 75 μ for clay samples. This must be reduced to a few microns when the section becomes thin (Jongerius and Heintzberger, 1962, 1963).

Jongerius and Heintzberger of The Netherlands Soil Survey Institute at Wageningen, The Netherlands, use at present another grinding machine with a larger electromagnetic clamping table. It is model MF 80/40 from the Matra-Werke (Matra, not dated a, b, c). The apparatus has a clamping table which enables the operator to use a planing length up to 80 cm and a planing width up to 40 cm (Fig. 5.14). To prevent clays, especially montmorillonite, from swelling, it is necessary to use oil instead of water for cooling. The Shell S 4919 oil has proven to be the best product found by Jongerius and Heintzberger.

At the Koninklijke Shell Exploratie en Produktie Laboratorium in The Netherlands, a Müller precision grinding machine, model MPS 3H-131 is applied. The (diamond-impregnated) disk turns on a vertical spindle and, depending on its size, is able to cover the complete table area in one longitudinal movement of the table. The surface of the area which can be ground at once is 540 x 170 mm. The apparatus can be equipped for complete automatic operation. It switches off automatically when the preset condition is reached (Müller, not dated).

Fig. 5.14. Electromagnetic table and grinding disk of the Matra planing machine. (Courtesy Matra-Werke and Miller Holding Maatschappij N.V.; Matra, not dated c.)

Another fully automatic grinding machine is the Blohm-HF 6 with a table area of 60 x 30 cm used in Germany for the production of thin sections of soils (Geyger, 1962). The Blohm-HFS 15 has a table measuring as much as 162 x 30 cm.

Clamping Samples to the Table of a Planing Machine

There are several ways to clamp the sample(s) to the table. It is absolutely necessary that the objects do not move, since a slight movement may result in breakage. When a series of mounted samples are thinned down simultaneously, the operator is required to use object glasses of the same thickness. This should be checked by a micrometer.

Metal strips can be placed on the table and fixed to it by electromagnetic forces. The object glasses or sample specimens should be clamped between the metal strips. This means that when a maximum number of samples must be ground at the same time, their surfaces should be the same size. In practice it often occurs that during the grinding operation one or more object glasses jump out of their position and consequently become damaged by the abrasion wheel.

Instead of metal strips a metal mold can be made in which the object glasses fit. Since the object glasses are not always exactly the same size, there is the risk that some glasses may jump from their mold. The best way is to make a large mold, which has a large area for object glasses (Fig. 5.15) to be placed in rows, or of which the outer sides of the molds have an adjustable screw (Fig. 5.12). For large-sized thin sections, as made by Jongerius and Heintzberger, of which only one object is placed upon the table at one time, the operator can apply broad metal strips to enclose the object glass.

A large metal plate with a polished upper surface is applied in Utrecht with some success. The plate has a thickness of 8 mm and a size of 250 x 120 mm. The object glasses, having similar thicknesses, were mounted on this plate by means of a temporary cement (Lakeside no. 30L). The object glasses were mounted tightly next to one another, allowing a large series to be ground at once. The only disadvantage was that the lubricating oil dissolved the Lakeside slowly, which sometimes resulted in loosening of the glasses and, consequently, breaking them.

At the Koninklijke Shell Exploratie en Produktie Laboratorium in The Netherlands a nice sample holder has been invented (Fig. 5.15). In a flat, stainless steel plate rectangular areas of object-glass size are raised to a depth a little less than the thickness of an object glass. The bottoms of these depressions are well polished. Four shallow holes are made in the corners and one in the center. The corner holes are connected to the central one by grooves. The cen-

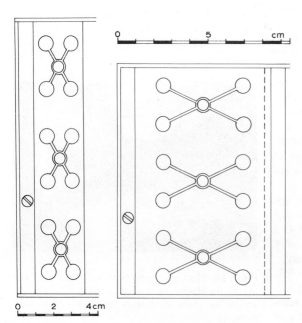

Fig. 5.15. Drawings of parts of metal object glass holders for different sizes of object glasses as developed at the Koninklijke Shell Exploratie en Produktie Laboratorium at Rijswijk (Z. H.), The Netherlands. Each rectangular area is a little less deep than the thickness of the object glass. Its bottom is well polished. The four shallow holes near the corners of each rectangular area are connected by grooves to a central hole. Each central hole is in turn connected internally to a vacuum system. The object glasses are held in position by vacuum. (Redrawn after construction drawings obtained from the Koninklijke Shell Exploratie en Produktie Laboratorium, The Netherlands.)

tral holes is connected to other central holes through drilled hole inside the plate. The whole system is connected to a vacuum unit. The plate is wet with water or slightly greased with acid-free vaseline. The object glasses with samples mounted on them are placed on the plate and vacuum is applied. The metal plate is fixed to the table by electromagnetic forces.

The size of the sample holder depends on the size of the table of the planing machine. In a series it is necessary to use object glasses of the same thickness. The object glass holder plate should be cleaned thoroughly before use to avoid leakage in the vacuum circuit, which would cause improper fixing of object glasses. It is better to connect the vacuum holes of one short row in the plate to one another and to construct for each row an outer connection opening to which a tube can be fitted than to make one internal connection circuit. Outside the metal plate these tubes are connected to one tube running to the vacuum. In this way certain small tubes can be closed if the operator is not able to fill the entire metal plate with objects. He also decreases the risk of making his plate useless if one of the rectangular areas should not work well. As can be seen in Fig. 5.15 plates can be made for different sizes of object glasses.

thick-thin section. The last step of the procedure

In operating planing machines it is not only necessary that the object glasses have the same thickness and are well mounted to the machine table, but also that all samples in one series have more or less the same hardness in order to obtain the same grinding rate for all objects after each sweep. Only in such cases can the operator thin his sections so far down that it nearly has the thickness required. Care should be taken to use lubricants that do not dissolve the mounting media. Epoxy resins are very strong, inert bonding agents (AB Metal Digest, Vol. 10, No. 2, 1964; Vol. 12, No. 1, 1966; Vol. 13, No. 1, 1967).

5.5C. Diamond-Impregnated Cup Wheels

Some of the smaller cutting machines are equipped also with a grinding attachment. A diamond-impregnated cup wheel rotates on a horizontal spindle. The object glass, to which the sample is mounted, is placed in a special clamp. The object holder is moved back and forth by hand along the cup wheel, while a micrometer attachment enables the operator to move the object toward the cup. In this way a mounted sample is thinned down quickly into a

Fig. 5.16. Grinding machine for grinding samples mounted upon object glasses, as built at the Geological Institute at Utrecht, The Netherlands.

should be carried out on a grinding disk or on a glass or plexiglas plate with a fine-grade carborundum powder (see Chapter Two). The use of such an instrument is time saving.

Diamant Boart (1961, Fig. 216) manufactures type LM, which has a cutting disk with a diameter of 200 mm and a diamond-impregnated cup wheel with a diameter of 150 mm. The number of revolutions is 2950/min. The largest size of object glasses which can be clamped are 70 x 27 mm.

Cutrock (not dated e, g) manufactures a similar apparatus under the name Mark II Unicutta. It can cut 32 x 25 mm sections, and it can grind sections mounted on 76 x 25 mm object glasses.

Dollé (1959) uses a similar instrument. His object glasses are mounted with a temporary cement (Lakeside No. 30L) to glass pieces (60 x 40 x 6 mm) which fit the special clamp of the apparatus.

At the Geological Institute in Utrecht such a grinding instrument was made of angle-iron, an electric motor, a diamond-impregnated cup wheel, and some additional parts (Fig. 5.16). It

is necessary to obtain a certain narrow range of revolutions per minute since diamond grains are rubbed off when the circumference velocity is too low or damaged when the circumference velocity is too high.

5.6 DISCUSSION

When samples have to be impregnated or imbedded or when thin sections have to be made of certain sizes, the investigator must decide which laboratory apparatus is necessary. The decision depends completely on the number of samples, their types and sizes, the amount of time an operator has, and the funds available. Completely automatic instruments save an enormous amount of time, and when handled properly they will give many years of trouble-free usage. However, when only small series of samples have to be treated, the investigator does not want such expenses.

For all instruments discussed in this chapter it is necessary to follow the local regulations concerning safety precautions.

CHAPTER SIX

SAMPLING

A very important aspect of the study of sedimentary structures is sampling techniques and handling. All disturbances introduced before the actual study begins influence the observations and may result in misinterpretations. Undisturbed, unconsolidated cores actually are still a dream, but a number of newly developed devices do allow the collection of nearly undisturbed samples. Very little is known about the influence of pressure relief, especially when the sampler comes above water. Moreover, the compaction introduced by sampling is not known for all devices. When working from a ship in deep water, no direct control of lowered instrumentation is possible. In very shallow water more control can be obtained if rods instead of wire can be used or when SCUBA divers can control the operation. Closed-circuit television systems are still in initial development stages, but they may be used more frequently in the future for making better evaluations of the procedures. However, reliable evaluations can be obtained only if *in-situ* measurements and observations by SCUBA-diving geologists and from submersibles (Terry, 1966) can be carried out simultaneously with lowerings from surface research vessels. Closed-circuit TV, bottom photography, observations by SCUBA divers or from submersibles are all necessary together with sampling to obtain an insight into the bottom topography and characteristics, the water-sediment interface characteristics, and the changes in these properties within an area of study. Both types of diving mentioned are really necessary to get a better feeling for how it looks on the bottom (see also Dill and Shumway, 1954; Shepard et al., 1964).

Several sedimentary structures are indicative for current directions and/or sediment movement (particle or mass). Since it is seldom known beforehand if such indicators will be collected, it is good practice to take "oriented" samples as much as possible.

The author does not intend to go into such geophysical investigations as gravity, magnetometry, seismic, continuous seismic profiling (Sparker, Sonarprobe, etc.), in spite of the important information they provide on the general build-up that forms a necessary base for studying sedimentary processes and basin filling. It is similar to carrying out field geology and geophysics when working on fossil sediments on the continents. It is also beyond the scope of this book to discuss drilling operations as carried out by industry.

This book is restricted to "small" devices with which the author has become familiar, although this information cannot include all new devices that are being developed and built daily (see also Section 6.7). Sampling of unconsolidated as well as consolidated sediments will be discussed; however, very little about the consolidated sediments is incorporated. Boundary-layer condition measuring devices, discussed by Sternberg and Creager (1965), will not be included in this book.

Not only is the collecting of a sample important but the manner in which an unconsolidated sample is extruded, dried, sliced, and stored is also. Sometimes the noncoring operations introduce more disturbances than actual coring.

Figure 6.1 outlines the main parts of this Chapter.

As previously mentioned, it is very important

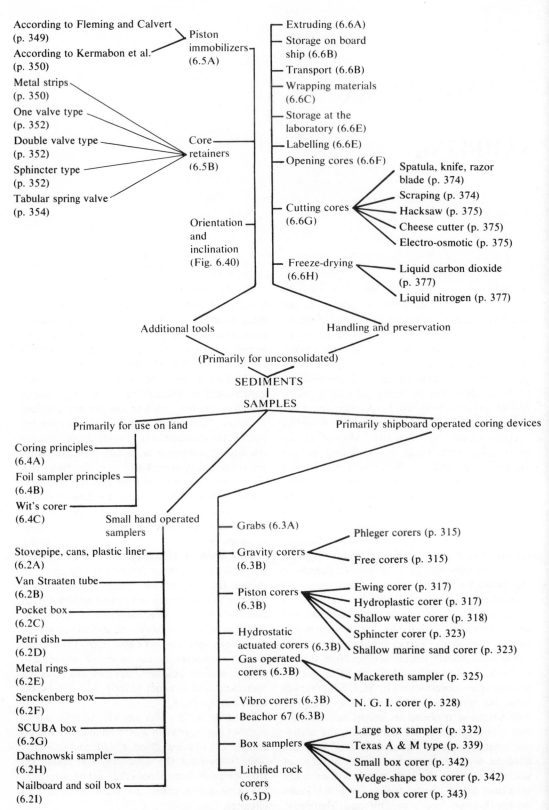

According to Fleming and Calvert (p. 349)

According to Kermabon et al. (p. 350)

Metal strips (p. 350)

One valve type (p. 352)

Double valve type (p. 352)

Sphincter type (p. 352)

Tabular spring valve (p. 354)

Piston immobilizers (6.5A)

Core retainers (6.5B)

Orientation and inclination (Fig. 6.40)

Extruding (6.6A)

Storage on board ship (6.6B)

Transport (6.6B)

Wrapping materials (6.6C)

Storage at the laboratory (6.6E)

Labelling (6.6E)

Opening cores (6.6F)

Cutting cores (6.6G)

Spatula, knife, razor blade (p. 374)

Scraping (p. 374)

Hacksaw (p. 375)

Cheese cutter (p. 375)

Electro-osmotic (p. 375)

Freeze-drying (6.6H)

Liquid carbon dioxide (p. 377)

Liquid nitrogen (p. 377)

Additional tools

Handling and preservation

(Primarily for unconsolidated)

SEDIMENTS

SAMPLES

Primarily for use on land

Primarily shipboard operated coring devices

Coring principles (6.4A)

Foil sampler principles (6.4B)

Wit's corer (6.4C)

Small hand operated samplers

Stovepipe, cans, plastic liner (6.2A)

Van Straaten tube (6.2B)

Pocket box (6.2C)

Petri dish (6.2D)

Metal rings (6.2E)

Senckenberg box (6.2F)

SCUBA box (6.2G)

Dachnowski sampler (6.2H)

Nailboard and soil box (6.2I)

Grabs (6.3A)

Gravity corers (6.3B)

Phleger corers (p. 315)

Free corers (p. 315)

Piston corers (6.3B)

Ewing corer (p. 317)

Hydroplastic corer (p. 317)

Shallow water corer (p. 318)

Sphincter corer (p. 323)

Shallow marine sand corer (p. 323)

Hydrostatic actuated corers (6.3B)

Gas operated corers (6.3B)

Mackereth sampler (p. 325)

Vibro corers (6.3B)

N. G. I. corer (p. 328)

Beachor 67 (6.3B)

Box samplers

Large box sampler (p. 332)

Texas A & M type (p. 339)

Small box corer (p. 342)

Wedge-shape box corer (p. 342)

Long box corer (p. 343)

Lithified rock corers (6.3D)

Fig. 6.1. General outline of the contents of Chapter Six.

to have a photograph of the area around the sampling locality. For ancient sediments which are exposed in outcrops, it is common practice to take photographs. It is equally important to do this underwater (Figs. 6.2, 6.3). Several underwater cameras or housings have been developed for divers. There also are several shallow and deep-sea cameras on the market, such as the Edgerton camera (E. G. & G.), Ewing camera (G. M. Mfg. & Instr. Corp.), Shipek camera (single and stereo; Hydro Products-Oceanographic Engineering Corp.), (Buffington and Shipek, 1963), and the Alpine underwater camera (Alpine Geophysical Assoc., Inc.). Excellent results obtained with these instruments are published by Ewing et al. (1946), Northrop (1951), Menard (1952), Carsola (1954), Zenkevitch and Petelin (1956), Shipek (1960, 1962, 1966), Hamilton (1963), Laughton (1963), Heezen and Hollister (1964), and others.

Many contributions on sampling devices, techniques, and sample handling have been published. A number of these contributions will be mentioned in the text. Very few books deal with this general subject; the only ones the author is aware of are Terry (1961), Richter (1961), and Cochran (1959: *Instruction Manual for Oceanographic Observations*). A number of devices are discussed in Pratje (1933, 1952), Shepard (1963), Hill (1962–1963), Hopkins (1964), and Fairbridge (1966); a selected bibliography of oceanography books published between 1959 and 1966 has been compiled by Sinha and Strauss (1967). Beside these contributions, surveys on sampling devices and/or bibliographies can be found in Turnbull (1953), Züllig (1956), Vargas (1957), Holme (1964), Hopkins (1964), and Rosfelder and Marshall (1967). Cochran (1965) discusses the statistics of sampling.

6.1 SAMPLING CONSOLIDATED AND SEMICONSOLIDATED SEDIMENTS

Collecting samples of consolidated or semiconsolidated sediments is such a normal procedure that hardly and words need be spent on this. Especially when dealing with massive (not broken) deposits, only experience is necessary to collect the desired samples. Normally a hammer and one or more chisels are the only tools necessary. It is important to note the position and orientation (see Section 6.5) before sampling. If cracks are present, the investigator should take advantage of them to avoid breaking in undesired directions. The sample can be trimmed to size once it is removed from the outcrop. The investigator should also try to collect material that is as unweathered as possible. This may mean that he has to remove a lot of rock before actual sampling starts.

If field-operated cutting or drilling units are available, care should be taken to use enough water or other cooling liquids to prevent the breakdown of such instruments from development of heat.

6.1A Cracked and Fissile Deposits

When the formation to be sampled is cracked or fissile and breaks easily along lamination planes, decent samples are hard to obtain in the manner just discussed. If possible, pieces can be removed and taped or glued together and then packed carefully for transport. However, this procedure is seldom easy or satisfactory. The operator should first remove weathered material. If it is possible to cut openings along the sides of the desired sample, the operator can cover the sides as well as the front of the sample with plaster of Paris to sustain the parts prior to removal.

Where it is possible to remove material above the desired sample location, a good technique is to fill the cracks with plastic or with one of the products mentioned in Section 4.7. Once those materials are hardened, a good sample can be collected. Additional glueing may be desired before shipment of the samples. Good results have also been obtained by placing the sample or sample parts on a wet layer of plaster of Paris inside a cardboard box and filling the box with plaster. Before pouring the plaster of Paris, a few remarks and directions should be written on the box to indicate orientation and inclination as well as directions for cutting the sample.

6.2 SMALL HAND-OPERATED SAMPLERS FOR UNCONSOLIDATED SEDIMENTS

Sampling unconsolidated sediments is rather easy and inexpensive when the deposits are found above water or in very shallow water. The only restriction is on the size of the sample the operator can handle. Simple instruments allow the investigator to collect only the upper part of a series when no vertical outcrops such as excavations are present.

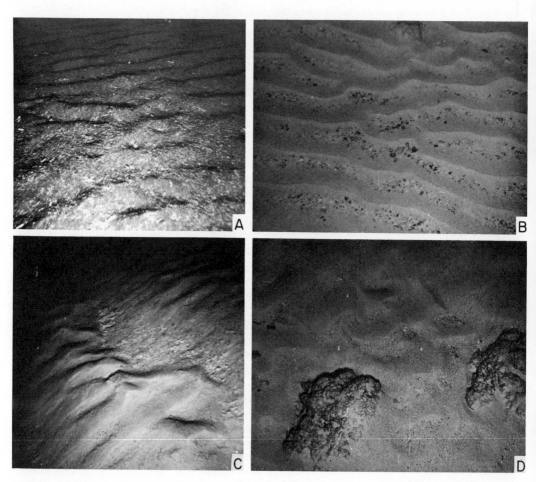

Fig. 6.2. Submarine bottom photographs, mainly sandy material. (A) Ripple marks in brown to black coarse volcanic sand and coral material on marine shelf at St. Paul Island, Indian Ocean. Depth: 448 m (245 fathoms). (B) Medium sand with pronounced concentration of sorted coarse particles in troughs of symmetrical ripple marks which have very sharp crests. Wave length approximately 6 in. (15 cm). Coarse particles in troughs are broken fragments of manganese crusts and small nodules. Southwest flank of Eniwetok Atoll near outer slope break. Depth: 2013 m (1100 fathoms). (C) Cross-rippling in sorted calcareous sand on a topographic high, Rift Mountains, Indian Ocean. Depth: 2940 m (1607 fathoms). (D) Medium sand with sorted concentrations of coarse particles in ripple troughs formed as interference ripples around the exposed boulders of bedrock. Largest exposure on left side is approximately 18 in. (45.5 cm) long and is partially covered by medium sand. Southwest flank of Eniwetok Atoll at outer slope break. Depth: 2013 m (1100 fathoms). (A) and (C) Photo area 60 cm wide and approximately 4 m deep. Type II camera, 2.4 m target distance. (B) and (D) Photo area approximately 18 ft² (16,200 cm²). (Official Navy Photos. Photos by Carl J. Shipek, Naval Undersea Warfare Center; B and D: Shipek, 1962; C: Shipek, 1966.)

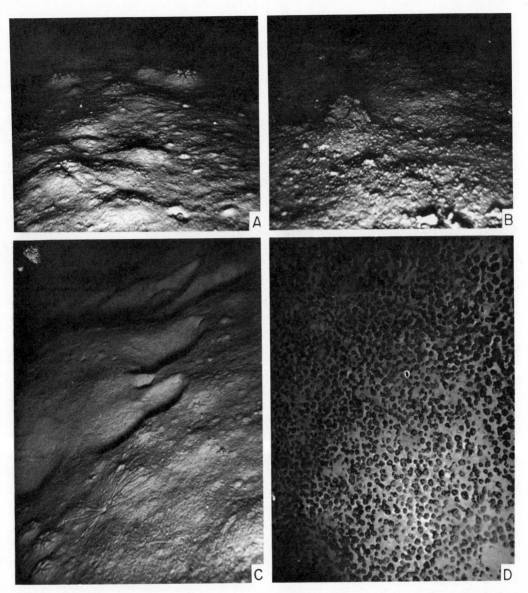

Fig. 6.3. Submarine bottom photographs with secondary phenomena. (A) Sutured animal mounds in churned red clay. The mounds are 5 cm high and probably created or altered by crustaceans. Intermountain Valley, Rift Mountains, Indian Ocean. Depth: 1760 m (962 fathoms). (B) Worm and excrement on mottled clay. Mounds are 3 cm high. Near Cocos Island, Indian Ocean. Depth: 5100 m (2731 fathoms). (C) Slumping in calcareous clay on slope of Intermountain Valley, Rift Mountains, Indian Ocean. Depth: 1760 m (962 fathoms). Note sutured mounds and radial feeding pattern. (D) A high concentration of evenly spaced spherical manganese nodules, 3.48 cm in diameter, resting on Globigerina ooze. Animal churning has caused the partial covering of some nodules. The freshness of the nodule surfaces indicates a low rate of sedimentation in this area. South central Pacific Ocean, northwest of Tahiti. Depth: 2695 m (1473 fathoms). All photographs taken with type II camera, 2.4 m target distance [except (D): 2.6 m]. Photo area for (A), (B), and (C): 60 x 400 cm; for (D): 120 x 120 cm. (Official Navy Photos. Photographs A, B, and C by Carl J. Shipek. Photograph D by Stephan Calvert. Naval Undersea Warfare Center; A, C, and D: Shipek, 1966).

The samples obtained can be brought to the laboratory for further analysis, and since they are easy to orient before collecting, rather complete investigations can be carried out.

Only a few sampling devices will be described here. However, they cover most of the basic elements of these types of tools.

6.2A Stovepipes, Cans, Plastic Liner

These sampling devices have already been mentioned in Chapter Two. They work well in wet and moist material and often produce good results in shallow water.

Both stovepipes and open cans have a seam, which is ideal for orientation. It is important that the lower end (cutting edge) is as thin as possible to facilitate penetration and to prevent disturbance.

Clear plastic tubes or PVC pipe can also be used, but their wall thickness makes it necessary to sharpen the cutting edge (Section 6.2B).

Normally one loses the sample during pull-out unless precautions are taken as discussed in the next section, or unless the operator digs a hole next to the pipe and gets his hand or a plate underneath before pull-out.

6.2B Van Straaten Tube

Van Straaten (1954) modified the preceding principle by adding a handgrip and a valve to the pipe to facilitate penetration and removal without losing the core. These modifications also allow the collection of longer samples. Van Straaten uses brass tubes of 1 and 2 m length, with an internal diameter of 6 cm and a wall thickness of 1.2 mm. Compaction proved to be negligible. Sometimes it is necessary to apply a little rotation to and fro at a small angle to prevent the sediment sticking to the wall, thus preventing compaction.

Equipment

Two variations on this type of coring device have been published by Van Straaten (1954). Figure 6.4A shows a sampler for use on land and in shallow water up to wading depth. Two rings are welded to the tube; between them fit a pair of iron bars; the bars, in turn, enclose the pipe and stick out to the sides as handles. Winged nuts keep the bars together. The valve on top consists of a metal lid with a thick rubber plate underneath. A closing mechanism enables the operator to open the valve completely or close it tightly.

Fig. 6.4B presents the corer as used from a small boat. Welded to two metal rings is a rod with holes. A pipe (gas tube) fits over this rod as an extension piece. The self-closing valve on top is similar, but the closing bracket is replaced by two vertical rods, along which the lid can move up and down. The lid has to be heavy enough to slide down easily and to close well. One side of the bar that fits around the vertical rods has a slit instead of a hole to enable turning the lid completely away. By using extension rods instead of cables, it is possible to keep the corer oriented.

Procedure

The corer is pushed into the sediment. Hammering sometimes is useful, but once the tube stops in sand during pushing, hammering is useless. A rapid succession of alternating short pulls and pushes often is the best method to promote penetration. However, all these movements may disturb the core.

Before pulling the corer out, the remaining space above the sample is filled with water and the valve is closed. The water acts as a piston and allows the core to be brought up without any loss.

The valve can be removed or opened completely and an extruding rod with tight-fitting disk or piston can be inserted. The core is next extruded in a half-cylindrical core holder or a V-shaped holder (see Figs. 4.2 and 4.3).

Comments.

Coring sand, as well as extruding it, is usually very difficult, if not impossible. Extruding it without tapping the sides often results in the pressing of sand grains sideways against the walls, so that the core becomes quite inextricably fixed in the tube and has to be washed out.

In dry or compact deposits above the waterline, it frequently is impossible to obtain cores more than a few tenths of a centimeter long. In such cases the operator extrudes the short part before collecting a deeper part until he reaches the water table.

The geological oceanography section of Texas A & M University has collected good samples by using stainless steel pipes with a beveled bottom side. Instead of having rings welded to the pipe, a set of pipe coupling flanges are used. These fit around the pipe with a strip of inner tube between to prevent sliding. A 2 x 4 (a piece of lumber which is 5 x 10 cm in section) about 30 in. long is placed on top to press the corer in. When this is too difficult, a sledge hammer is used.

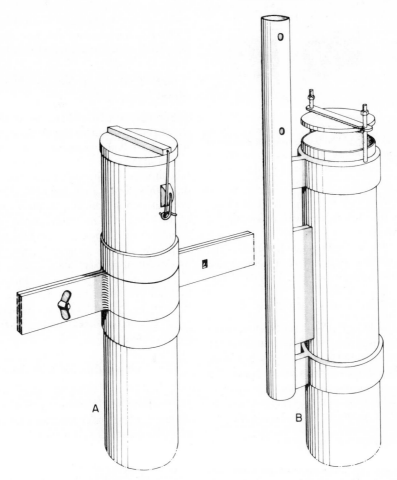

Fig. 6.4. Two executions of the Van Straaten tube. (A) For use on land and in water up to wading depth. (B) For use from a small boat. For explanation see text.

Before pull-out, the upper part is filled with water and a rubber stopper is placed as a valve. A rope around the pipe underneath the flange helps remove the pipe from the sediment. The samples were not extruded in the field. In this case two men can walk through marshes and collect two to three cores per trip.

6.2C Pocket Box

Small boxes for obtaining oriented, undisturbed samples are illustrated in Fig. 6.5. The sample obtained is 6 x 3.5 x 2.8 cm. Sampling pure sands and keeping them undisturbed is nearly impossible; for soils and clayey sands they are good. The idea is to impregnate the sample with a resin while it is still in the box.

The box consists of two parts like a match box. The sampling frame is rectangular (6 x 3.5 cm and 2.8 cm high), made of stainless steel. Two sides come together like a hinge with pin to facilitate later removal of the sample (Fig. 6.5). The other part is the cover, in which the sampling frame fits like a match box. One of the sides is slightly shorter to allow for the hinge, thus securely closing the frame. Both parts have the same number to prevent errors.

Procedure

A horizontal or vertical sediment face is made smooth, after which the frame is pushed in. Having all lower edges of the frame beveled facilitates insertion. A trowel or spatula can be used to free the frame, or the cover can be pushed

Fig. 6.5. Pocket box. (a) The box in closed position. (b) The box taken apart. Note the pin in the hinge of the frame being taken out halfway, and the shorter side of the cover to give place to the hinge of the frame.

over it directly. Before removal, orientation is written on the cover.

Once the sample is impregnated, the pin of the hinge can be removed, and, with a knife, the impregnated sample can be freed from the walls.

6.2D Sampling with a Petri Dish

Howard and Henry (1966) successfully used plastic Petri dishes for sampling as well as for storage. Plastic is used to apply X-ray radiography while the sample is still in its container. No problems exist in photographing the sample, making sedimentary peels, impregnating, or drying it for size and mineralogical analysis.

Procedure

The sediment surface, vertical or horizontal, is made as smooth as possible and the Petri dish is pressed into it evenly. Dishes over 150 mm in diameter are not inserted in this manner. In those instances it is necessary to dig a shallow groove with a diameter approximately ½ in. greater than that of the dish to allow for slight expansion during pushing in. Next a knife is pushed behind the dish at a distance of 1 to 2 in. and the sample is cut loose. After the Petri dish is placed horizontally, excess material can be removed and the dish can be covered.

Semiconsolidated samples can be sampled in the same way with the only difference being that a groove is made while pressing the dish into the sediment. A slight undercutting is necessary to allow the dish to move in.

Subaqueous samples were also collected by Howard and Henry. A 1-gallon can, without bottom and with a hole in the top, is pressed in the sediment and the whole is lifted out on a shovel. The can with sediment is placed in a cardboard box and put in a refrigerator or cold room for several hours to firm up. Next the can is laid on its side and cut open, after which the

sediment can be sampled according to the previously described method. For vertical exposures it is easier to use a rectangular can.

6.2E. Cylindrical Samples Collected with Metal Rings

Borchert (1961b, 1962) collects samples from the vertical wall of an excavation. A sheet-iron lining, 6 cm high, is bent into a cylindrical shape with a diameter of 10 cm. One end of this strip has three cuts in it, each 1 cm deep. The pieces on both sides of a cut are bent slightly inward and outward alternately, to allow the other side of the lining to fit in, thus securing the circle. A metal cylindrical box fits over the just-described ring. Ring and box are pressed into the sediment as described in Section 6.2D. Collecting the sample from the deposit and smoothing the surface is identical to the Petri dish method. When back in the laboratory, the cylindrical box is removed and the sheet-iron ring opened prior to sample removal.

6.2F. Sample Box for Use Mainly above Water

Unconsolidated, oriented, and undisturbed samples can be collected above the water table easily be means of rectangular boxes, known as Senckenberg boxes (Reineck, 1957, 1961a, 1963a; Bouma, 1963, 1964b; Hydro Products: box sampler model 610, not dated a; McMullen and Allen, 1964).

Equipment

The box sampler consists of two parts: an open-ended base and a cover (Fig. 6.6f). The sizes used by the present author are 8 x 6 x 3 in. To avoid rusting or corrosion, stainless steel is preferred. The outside dimensions of the base are: height 8 in., width 6 in., depth 3 in. Material thickness is important: too much thickness makes the box too heavy and difficult to push into the sediment; too thin walls makes the box too weak and disturbs the sample during insertion. The preferred wall thickness is $\frac{3}{64}$ in. The author prefers having all sides welded together to give the box more strength.

The inside dimensions of the cover are: height 8 in., width 6$\frac{1}{16}$ in.; the top is 3 in. deep, but the sides are only 1 in. The cover should slide easily over the base.

The lower ends of both parts are beveled (Fig. 6.6f) to facilitate the forcing of the pieces into the sediment.

In addition to the boxes, the following items are needed: a plate or trowel which covers an area of 6 x 3 in., adhesive tape, old rags for cleaning, marker, compass, spade, piece of lumber (2 x 2 in. and about 8 in. long), and a hammer.

Sampling Procedure

Samples can be collected either vertically or horizontally, or in any inclined direction. When dealing with dense or stiff sediments, it is recommended to use a spatula or knife to precut the sediment while pushing in the box parts. Sometimes it is easier to dig away material around the sampling site, leaving at least 4 to 5 in. material around the 6 x 3 in. sampling surface. During pushing in, the outside sediment falls aside slightly, thus decreasing friction.

It is necessary to have the sediments moist, especially when dealing with sands, in order to prevent disturbance. Dry and loosely packed sediments such as dune deposits have to be moistened thoroughly before sampling.

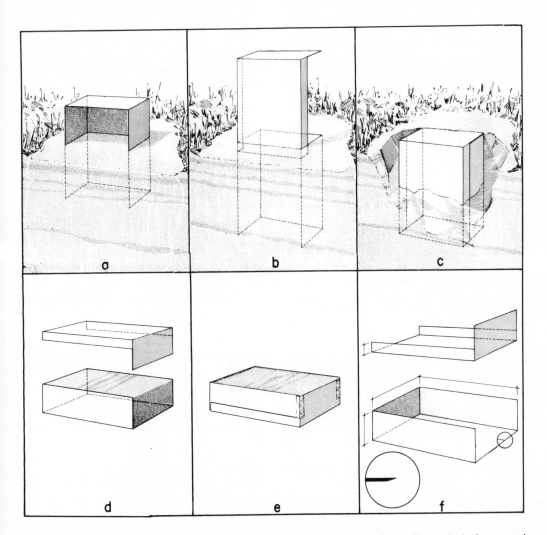

Fig. 6.6. The Senckenberg box and its operation procedure. (a) After a section is flattened, the base part is pushed in. (b) The cover is pushed over the base part, thus enclosing the sample on all sides but the bottom. (c) When both parts are pushed in, sediment is removed from the front side. (d) When the sample is removed, the cover is taken off and turned 180°, after which it is placed, thus enclosing the complete sample. (e) In the laboratory the cover is removed, turned 180°, and placed so that only one original vertical side of the sample is exposed. (f) The parts of the sampling box.

The actual sampling procedure consists of a number of simple steps:

1. When a vertical sample is required, the first step is to flatten an area larger than the size of the box. A flat top ensures proper filling of the box.

2. Take the base part of the box, place it on top of the flat area, and push it in (Fig. 6.6a). Apply an even force on both sides of the top; otherwise, distortion will result. It is the author's experience that by placing a piece of lumber (2 x 2 in. for example) on top of the base near the back side, the box can be hammered in without noticeable distortion.

3. Once the base is completely in, thus in firm contact with the flattened sediment surface, the orientation and other information should be written on the top side of the box.

4. The top part is next pushed over the base part, thus enclosing the sample on all sides but the bottom (Fig. 6.6b, c).

5. Dig away the sediment at the operator's side and on both sides of the box (Fig. 6.6c). Insert the plate or trowel under the open end of the box. It is advisable to insert the plate ½ in. lower than the box reaches. With the hand underneath the plate, the box can be removed from the sediment.

6. The box is turned over and laid down gently on the back side of the base part. With rather loose sediment it is better to keep the open side a little higher than the other end to prevent sediment from breaking off.

7. The plate is removed by sliding it off to avoid sediment breaking off while it sticks to the plate. Excess material is cut off and the box is freed from the sediment.

8. The cover is removed next by sliding it off, it is cleaned inside and turned 180° about a vertical axis and then placed on the base part (Fig. 6.6d), thus enclosing the complete sample. Openings are cleaned and taped.

9. The box is turned to its original vertical position for transport. Any compaction that occurs during transport will not damage the sample.

10. Back in the laboratory, the tape can be removed, the box laid on its back, and the cover part slid off. However, before sliding it off, insert a plate inside the top of the cover part to keep all material in place when the cover is removed. The cover is cleaned inside, turned 180° about a horizontal axis, and slid under the base part, thus enclosing the sample while leaving one original vertical side of 6 x 8 in. exposed (Fig. 6.6e). Tape the openings and remove a little of the sediment in order to obtain a fresh and undisturbed sample face. Fill any space formed during sampling or transport with another type of sand to avoid disturbances during drying.

6.2G. Sampling Box for Use Under Water

The Senckenberg box, as just described, does not work too well under water, especially in loose sands. Considerable distortion results during pull-out. Reineck (1963a) and the present author independently developed a box sampler for use under water. It is a combination of the Senckenberg box and the large box sampler. This device can be used while wading, as well as from a boat, using extension rods.

Equipment

The underwater box sampler consists of a box holder, a handgrip, a closing arm, and a sampling box (Fig. 6.7). The box holder is made of ³⁄₆₄ in.-thick stainless steel and has the shape of an open box. Inside fits the actual sampling box, which has dimensions of 3 x 6 in. Two pair of corresponding holes in the box holder and in the box allow the insertion of two pins to keep the box in place. On top of the box is welded a cylindrical holder to which a rod or pipe fits, acting as a handgrip. On each side of this holder two valves are placed on top of the box holder. The holes in the original design [Fig. 6.7 (4)] are too small. They are 1 in. in diameter and are closed by a rubber disk pressed to the hole by a keeper spring. To the short sides of the box holder are welded two pins about which the closing arm can turn.

The handgrip can be of different design. It can be short, for example, 3 ft, with a handgrip on top and a side piece sticking out near the bottom. This enables one or two SCUBA divers to insert the sampler into the sediment. Another possibility is to attach extension tubes to it so it can be handled from a small boat.

The closing arm [Fig. 6.7(2)] turns about two pins, which are welded to the box housing. It is made of two strips of metal, bent into the proper shape and welded together over the length they meet. At the very end a hole is made through which a rope can be fastened when working from a boat. Each lower end has two threaded holes to which a box cover can be screwed.

The sampling box consists of two parts, which

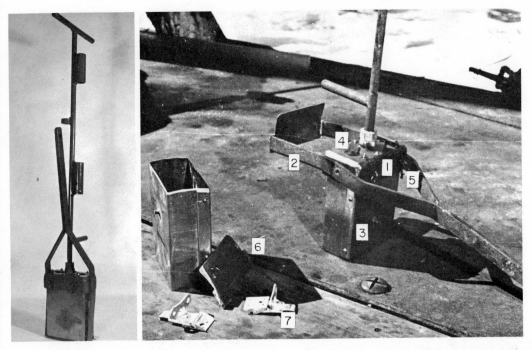

Fig. 6.7. Sampling box for use under water. Left-hand side: the instrument in closed position. Right-hand side: the sampler in open position: (1) box holder; (2) closing arm with cover screwed to it; (3) sampling box; (4) valves; (5) pins with chains to attach box to box holder; (6) loose cover; (7) Link Locker.

are screwed together. The box is 3 x 6 in. in section and 8 in. high. Removing the 6 x 8 in. side exposes a vertical section of the sample. The lower parts of the 3 in. sides are rounded as is the box cover to insure proper closing. To the box are welded two small knobs to which Link Lockers[1] can be attached to keep the box cover in place.

Sampling Procedure

The procedure is rather simple. A box is attached to the sampler, and a bottom plate is attached to the end of the closing arm. It now is ready for operation. (SCUBA divers may have a hard time pushing the sampler into the sediment due to their very small negative buoyancy.) When working from a boat, the operator needs enough extension rods to reach bottom. A rope should be attached to the end of the closing arm.

When the sampler is closed and above water, the Link Lockers are placed, thus securing the bottom to the box. The bottom plate can then be loosened from the closing arm and the box can be removed from the sampler.

[1]See footnote 2, Chapter Six.

6.2H. Dachnowski Sampler

This sampler gives satisfactory results in unconsolidated clays, sandy clays, and peat. The Geological Survey in The Netherlands, as well as the present author, have collected "undisturbed" samples up to depths of 10 m. Bouma (1963) collected samples from marsh deposits with this instrument. Lammers (1965) describes the ones used by the Geological Survey cited.

Equipment

The sampler consists of a stainless steel tube in which a spindle with guide strips can move up and down. The upper part of the tube is closed, leaving a hole for the spindle and both guides. The lower part of the tube is beveled. To the lower end of the spindle a cone-shaped point is connected (Fig. 6.8).

To the upper part of the spindle extension rods can be connected, each 1 m long. A handle on top facilitates the operation.

Two sizes proved to be convenient. Normally samples will be collected of a size 30 x 3 cm, but for peat a sample size of 35 x 4.5 cm works well. Larger sizes become too difficult to handle.

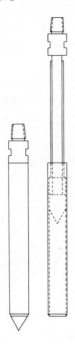

Fig. 6.8. Dachnowski sampler in open and closed position. For explanation see text. (Redrawn after Lammers, 1965.)

Procedure

The spindle is pulled up and turned 90° in order to let the guide strips rest on the upper side of the sampling tube. The sampler is then pushed into the sediment over the length of the tube. After pulling up, the spindle is turned again 90° and the sample is pushed out.

The next sample can be collected by lowering the sampler in a locked position. The point sticking out below the tube assures that no sediment penetrates. When the sampler is in position, the spindle is pulled up and turned 90°, after which the sampler can be pushed in.

It was found that very good samples can be obtained in this way. It is important that there be enough free space between the spindle and the top of the tube so that sand grains cannot lock both halves.

6.2I Nailboard and Soil Box

Jager (1959) sampled wet and heavy soils with a nailboard consisting of a heavy wooden board through which nails are driven at regular intervals. The board is hammered into a vertical section.

An improvement on this technique was pub-

lished later by Jager and Schellekens (1963) in which the authors replaced the board with a heavy sampling box (Chapter One).

6.3 PRIMARILY SHIPBOARD-OPERATED SAMPLING DEVICES

Bottom samples have to be collected from a boat or vessel when the water is too deep for wading or diving or when the equipment is too heavy for hand operation. Although most samples will be collected from unconsolidated deposits, a short description of underwater rock coring will be given.

No distinction will be made between sampling at sea and in a lake. Since the number of coring devices is very extensive, mainly due to modifications of a few basically similar corers, and since industry as well as marine geologists, limnologists, and others develop new devices regularly, only a restricted number of devices can be described. Also, those samplers which do not collect "undisturbed" material that can be used for the study of sedimentary structures will not be considered. Theoretically no sampler is able to collect a completely undisturbed sample, but many cores contain a large or small portion to which the label "undisturbed" can be attached.

As stated in the introduction to this chapter, it would be ideal to obtain in just one lowering information such as orientation of the core, a picture of the bottom, current measurements, amount of suspended material, temperature, etc. Part of this combination has materialized and is described in the section on box samplers.

A few devices described in Section 6.2 can be used from a small boat, for example, the Van Straaten tube and the underwater sampling box.

Unless the weight of the sampling device is great enough so that its release gives a clear indication that the bottom has been reached, it is advisable to attach a pinger or a ball-breaker (Cochran, 1959) to the sampler to prevent letting out too much wire, which may tangle around the sampler. A pinger is the best solution since it tells the operator the depth of lowering at any given time. Care should be taken to use a strong abyssal pinger when working in deep water, since too often pinger signals fade out before the bottom is reached.

6.3A. Grabs

As far as geological application is concerned,

grab samplers are primarily developed to collect bottom samples for lithological and petrographical analyses (Cochran, 1959; LaFond and Dietz, 1948; Kahlisco, not dated; G.M.M.C., not dated b, c; Alpine, not dated a). Most grabs are easy to handle, but even with the heavier ones little success is obtained in real deep water. Presumably due to their high surface area/weight ratio and their shape when they are open, grab samplers start to swing and do not hit bottom in the proper position.

When filled grabs are opened properly in a tray that fits the sampler, relatively little disturbance of the center part will result. Smaller samples can then be collected from it.

The variety of grab samplers developed is very extensive. The oldest type may be the Van Veen grab, which is still used. The modified version, having two hinged valves on top, gives better access to the sample (Van Veen, 1936; Richter, 1961; G.M.M.C., not dated a).

A combination of the Van Veen grab and a camera, which records the appearance of the surface before the grab disturbs it, is especially useful for studying sediment transport and bioturbation (Menzies et al., 1963; Smith, 1964).

The Shipek sediment sampler is composed of two concentric half-cylinders. The inner one is rotated at high torque by two helically wound external springs. Upon contact with the bottom, triggering is automatically accomplished, and at the end of its 180° travel the sample bucket is stopped and held in closed position (Hydro Products, not dated b, c).

The mud snapper and clamshell snapper also use a spring-actuated closing mechanism to prevent malfunction or loss of sample when proper operation is impossible due to hard pieces. The present author does not know how "disturbed" the collected sample will be.

A combination of a grab sampler and the box corer closing mechanism was developed by Auerbach (1934). A rectangular box is closed by two lever arms, each closing half of the box. Jonasson and Olausson (1966) developed a similar device and added a compass to it.

6.3B. Coring Devices with Cylindrical Core Barrels

The largest group of surface coring devices use a cylindrical pipe as a core barrel. This group can be subdivided, based on the principles of penetrating the sediment and/or collecting the sample. First to be developed were the open-barrel gravity or punch corers, which depend only on gravity to cause penetration of the sea or lake bottom. The term "gravity corer" is not correct since other types of devices are also based on gravity as far as penetration is concerned. To this punch corer group also belong the free corers, which operate without a winch cable.

Piston corers constitute another large group of coring devices. These have a piston to facilitate penetration of the core into the barrel (Swedish Committee on Piston Sampling, 1961).

These two groups of surface coring devices are very common. Many oceanographic instrument manufacturers offer them, for example, Alpine Geophysical Associates, Inc. (not dated b), Benthos Inc., CM^2 (not dated), G. M. Mfg. & Instrument Corp. (not dated d), Kahl Scient. Instr. Corp. (not dated), and Hydro Products (not dated d).

Other types of coring devices have been developed, and it is expected that some of them will become important samplers in the near future. Examples are the vibrocorer, the compressed air corer, and the hydrostatic corer. For further references, consult the introduction to Chapter Six.

Kögler (1963) compares different coring devices with regard to inside and outside diameters of core barrel and core bit. Table VI.1 clearly demonstrates that most devices have wall thicknesses which are much higher than those recommended by Hvorslev (1949), as reported by Richards (1961).

Piston corers may not always have gross recovery ratios of 100%, while the gross recovery ratios of gravity corers may be much less than 100% unless they are properly designed (Richards, 1961; Richards and Parker, in press).

Gravity or Punch Corers

Most gravity corers are equipped with free-fall devices to give them the necessary terminal velocity upon impact (Fig. 6.9) (Hvorslev and Stetson, 1946). Burns (1966) made measurements on free-fall velocities on three corers of different lengths, different weights, and different barrel diameters. He reached the following conclusions:

"Although only three specific corers were tested, conclusions may be reached which appear to be applicable to a general class of small corers which use a tripping device to give measured free-fall, and are not so heavy as to require elaborate winch and rigging facilities.

Table VI.1   Some Technical Data on Inside Clearance Ratio (C_i), Outside Clearance Ratio (C_o), and Area Ratio (C_a) of Some Submarine Sediment Coring Devices

	Hvorlev's Recommendations for Cohesive Sediments and Long Corers	Corer after Pratje	TPR 51	Corer Geol. Inst. Kiel	Corer Alpine-Bandy	Corer Wilhelms-haven	Long Box Corer Kiel	Hvorslev-Stetson Free Corer
Type corer		—	—	—	—	—	Box	Gravity
Barrel diameter, outside (mm)	D_t	54	62	70	70	114	154	—
Barrel diameter, inside (mm)	D_s	45	52	61	63	102.5	151	—
Core nose diameter, outside (mm)	D_w	70	82.5	75	83	127	166–170	—
Core nose diameter, inside (mm)	D_e	43	50	59	60	101	150	—
$C_i = \dfrac{D_s - D_e}{D_e}\,\%$	0.75–1.5	46	40	3.3	5	1.4	0.6	1.6
$C_o = \dfrac{D_w - D_t}{D_t}\,\%$	<3	29.6	33	7.1	18.5	10.2	7.7–10.3	1
$C_a = \dfrac{D_w^2 - D_e^2}{D_e^2}$	<10	165	172.2	61.5	91.3	58.1	22.4–28.4	35
Core diameter (cm²)		14.52	19.64	27.34	28.27	80.12	225	—
Displaced sediment (cm²)		23.96	34.47	16.84	25.84	46.58	64	—
References	Richards, 1961; Kögler, 1963; Hvorslev, 1949	Kögler, 1963; Pratje, 1933, 1952	Kögler, 1963; Udinzew et al., 1956	Kögler, 1963	Kögler, 1963	Kögler, 1963	Kögler, 1963	Richards, 1961; Hvorslev and Stetson, 1946

After Richards (1961); Kögler (1963).

"(1) The optimum free-fall distance is 2 to 3 m, since the terminal velocity for the corer will be reached by this distance.

"(2) The efficiency of settings less than 2 m is questionable, since the corers have distinctive and unsteady accelerations during their initial fall prior to reaching terminal velocity.

"(3) Settings of greater than 3 m should be avoided, particularly if the corer is not fitted with stabilizing fins"(Burns, 1966).

Emery and Dietz (1941) and Emery and Hülsemann (1964) collected data concerning the mechanics of sediment coring with open-barrel gravity corers. For standard pipe with an inner diameter of 40 to 63 mm, the conclusion was reached that cores of silty clay have a shortening of about 50% compared to the total depth of penetration. For sand-silt-clay cores the same shortening may result, but for globigerina ooze a shortening of about 33% was observed. Other authors (Hvorslev, 1949; Hamilton, 1960) state that no shortening results in the upper 40 to 60 cm if the corers are properly designed. Richards (1961) reports an average shortening of 26% for 20 open barrel gravity corers, for 7 of which no shortening at all was claimed (Richards and Parker, in press).

Marine geologists at the Navy Electronics Laboratory, San Diego, California, "have used only the 2-inch outer diameter and 2¾-inch inner diameter acetate-butyrate clear plastic liners to take cores in clayey silt up to two feet in length with 100% recovery. The liner is sharpened at the bottom end, a thin plastic core retainer is cemented inside the liner at the lower end, and a flapper valve is taped to the upper end. These corers are used by SCUBA divers, or are mounted on deep-diving submersibles. In some cases a stainless-steel core nose, designed according to the specifications of Kallstenius (1958), is taped to the bottom end of the liner. The flapper valve must allow free flow of water

Table VI.1 (Continued)

USNHO Hydroplastic Corer	USNHO Phleger Corer	USNHO Ewing Corer	Lamont Geological Observatory Ewing corer	USNHO Kullenberg Corer	USNEL Standard Corer	Kullenberg Corer	N.G.I. Gas-Operated Corer
Gravity-piston	Gravity	Piston	Piston	Gravity-piston	Gravity-piston	Piston	—
—	—	—	—	—	—	—	—
—	—	—	—	—	—	—	—
—	—	—	—	—	—	—	—
—	—	—	—	—	—	—	—
1.3	10.2	0.6	5.3	1.8	2.4	0	0.9
13	0	23	18	10	22	25	0
57	62	84	87	105	130	373	12
—	—	—	—	—	—	—	—
—	—	—	—	—	—	—	—
Richards, 1961; Richards and Keller, 1961	Richards, 1961; Cochran, 1959	Richards, 1961; Cochran, 1959	Richards, 1961; Heezen, 1952	Richards, 1961; Cochran, 1959	Richards, 1961; Richards and Keller, 1961	Richards, 1961; Kullenberg, 1955	Andresen et al., 1965

out of the corer as it enters the sediment; otherwise back pressure will inhibit recovery. Keller was able to take 10-foot long cores with 100% recovery in St. Andrews Bay when he removed the valve from the top of the hydroplastic corer designed by Richards" (E. L. Hamilton, personal communication, January, 1968).

Ross and Riedel (1967) found that the upper sections of piston cores are shortened relative to simultaneously collected open-barrel cores. The inconsistency of results may be due to slight differences in the coring devices, technique used, and types of sediments. Some results are also published by Bezrukov and Petelin (1951).

Phleger Corers. Most manufacturers of oceanographic instruments sell a small-diameter (1.5 in.) gravity corer under the name Phleger corer. They all consist of an upper weight mass to which a core barrel can be attached. A plastic liner is inserted inside. Some models have a fixed weight, others have changeable weights. A one-way valve is mounted on top; this closes as soon as the corer is pulled up.

Free Corer. The free corer is a gravity corer which can be dropped from a ship or helicopter without any attachment. "The free corer consists of two basic assemblies: (1) a recoverable core barrel, check valve, buoyant chamber assembly, and (2) an expendable weight and casing assembly. When these two assemblies are combined, the core barrel fits loosely inside the casing" (Moore, 1961). In the prototype used by Moore, a delay timer (Van Dorn, 1953) releases the unit from its weight-casing assembly. The barrel is lifted up out of its casing and floats to the surface (Fig. 6.10).

The prototype has been changed by other investigators as well as by industry. Bowen and Sachs (1964) and Sachs and Raymond (1965) published a modified version, which is built now by Benthos, Inc. (not dated). This design weighs 82 kg, and the recoverable part has a positive

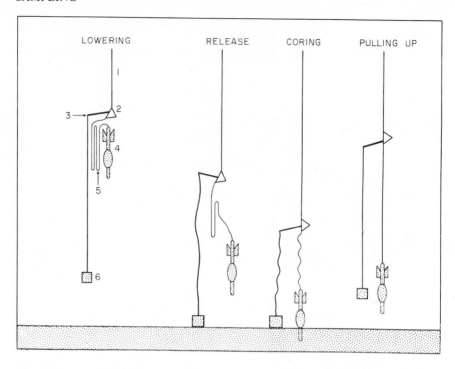

Fig. 6.9. Principle of free-fall in coring devices that are lowered on a cable: (1) lowering cable; (2) release mechanism; (3) release lever arm; (4) corer; (5) loops of wire of equal length as the free fall; (6) trigger weight. (Redrawn from Cochran, 1959.)

buoyancy of 8 kg. The float unit consists of two hollow borosilicate glass spheres with an outside diameter of 25 cm. One sphere is empty, the other contains an electronic flash unit. The spheres have been pressure tested to a depth of about 6400 m. A bag of knotted nylon netting holds both spheres together. The weight is an iron casting.

The Hydro Products Moore free corer model 880 has a coring tube with a length of 48 in. and a 3 in. outside diameter. A battery-operated flashing light beacon or a citizen's band radio beacon is enclosed in one of the glass spheres. The descent velocity is 12 ft/sec, the recovery velocity is 15 ft/sec. A hydrodynamic release mechanism unlatches the floats when the unit comes to rest.

Piston Corers

The main difference between the gravity and the piston corers is the addition of a closely fitting piston attached to the end of the lowering cable. The piston is placed inside the core barrel or the liner, just above the core catcher. The piston should not move relative to the sea bottom while the barrel penetrates the sediment. The hydrostatic pressure will not allow a vacuum to be created between the top of the core and the piston as long as the frictional forces do not exceed the hydrostatic pressure. The slightly lower pressure underneath the piston helps to move the sample into the barrel (Fig. 6.11). Kullenberg (1944, 1947, 1955) designed a corer utilizing this principle and no basic changes have been made since. An extensive report on standard piston sampling is published by the Swedish Committee on Piston Sampling (1961).

Burns (1963) described some possible mistakes in using information from cores obtained by piston-type coring devices. A simple experimental setup resulted in some interesting data on different sediment sequences (Fig. 6.12). In each case the corer was pulled up by the piston cable. Fewer mistakes in sediment sequence occur at full than at partial penetration. Correct length of the free-fall portion of the piston cable is very important. Partial penetration combined with delayed stopping of the downward movement of the piston results in the greatest number of misinterpretations. Normally, it is not too difficult to detect the "flow in" part of a core. "Flow in" occurs when the core barrel penetrates partially into the sediment. At the beginning of pullout the piston must travel the remaining distance to the top of the core barrel, during which time additional sediment will be drawn into the barrel (see Fig. 6.52) (Bouma and Boerma, 1968).

Ewing Piston Corer. Several companies manufacture a variety of the piston corer developed by Ewing. Different sizes and weights are available, depending on core length requirements, core diameter, and ship facilities. Basically, the corer consists of a tripping mechanism, stabilizing fins

on top of the weights, core barrel with liner, cutting edge with core retainer, and a tripping line with weight (Fig. 6.13). The piston often has a one-way valve. To make the piston tight-fitting several leather collars usually are used. A small gravity corer is often used as a trigger weight. Heezen (1954) presented a slightly different release mechanism (Fig. 6.14), which is used also by Texas A&M University. Recently the geological oceanography section of this University changed the weight to an egg shape in order to streamline it. This group does not use core liners but extrudes the core directly on board ship.

Hydroplastic Corer. Richards and Keller (1961) were not very successful in using the

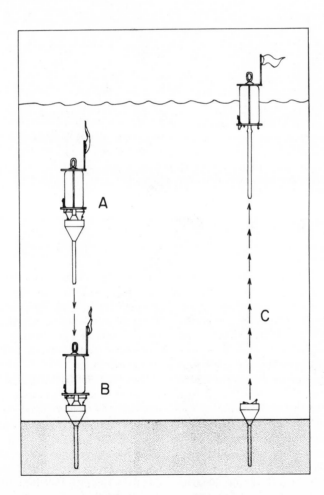

Fig. 6.10. Three operational stages of the free corer. The assembled unit is thrown overboard (A), hits bottom and penetrates (B). After the release-delay timer releases the core barrel-check valve-buoyant chamber assembly, this assembly floats back up to the surface, leaving the weight-casing assembly behind (C). (Redrawn from Moore, 1961.)

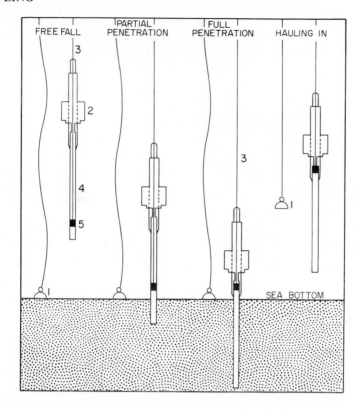

Fig. 6.11. Principle of operation of piston cores. As soon as the trigger weight (1) hits bottom, the corer (2) is released and free fall starts. The lowering cable (3) has reserve length, similar in length to the free fall, between the release mechanism and the top of the corer (see Fig. 6.9). As soon as the core barrel (4) hits bottom, the free loop in the lowering cable is gone and the piston (5) stays stable above the sea bottom, while the corer penetrates. (Redrawn from Cochran, 1959.)

conventional gravity and piston corers for their investigations on engineering and mass physical properties of marine sediments, since metal barrels rust and are difficult to cut to short lenths for vane shear strength tests. Cellulose acetate butyrate core liners readily lose water if not protected from desiccation (Keller et al., 1961). A coating with Victory Brown-155 wax from the Bareco Wax Company of Barnsdall, Oklahoma, proved to be better than a paraffin coating. Richards and Keller (1961) replaced the core barrel with plastic liner by high-impact grade PVC pipe (Fig. 6.15A). Neoprene gaskets under PVC attachment bolts and a sponge-rubber gasket at the top end of the barrel provide an adequate water seal to allow a one-way valve to be effective. Each weight is 22.7 kg (50 lb). The core nose (Fig. 6.15B) is mounted to "the barrel using hex-socket set screws that have been sharp-

ened on the cutting end to be self tapping. The cutting edge of the nose has a double bevel: a 30° edge angle for the first 6 mm (0.25 in.) and a 5° angle for the remaining 5 cm (2 in.)" (Richards and Keller, 1961). A brief description of a newer model is given by Richards and Parker (in press).

Reineck (1967) developed a plastic-barrel piston corer which allows collection of cores with a diameter of 4 in. and a length up to 4 m.

Shallow Water Piston Corer. Several piston corers for use in shallow water have been described by Silverman and Whaley (1952), Ginsburg and Lloyd (1956), Reish and Green (1958), and Byrne and Kulm (1962). Wright et al. (1965) describe different coring devices used for collecting lake sediments. The Mackereth corer may even collect better samples than the devices discussed here. Byrne and Kulm (1962) had objections to both the heavy and the complex designs, and therefore designed a lightweight, inexpensive corer with a 3 in. outside diameter barrel which has a length of 6 ft.

The barrel is made from aluminum irrigation tubing. A 3-ft long rod fits two opposing holes in the barrel and acts as a handle. The piston is made from a rubber stopper sandwiched between two large steel washers and fastened by wing nuts around an eye bolt that runs through the stopper. The stopper has to be milled down slightly to fit the barrel, which has an internal diameter of 2⅞ in.

The position of the piston within the barrel is controlled by a lightweight chain which is attached to a monopod.

Another lightweight piston corer was discussed by Livingstone (1955). The piston contains two leather washers and a spring pin near its bottom to hold the piston in place within the barrel during lowering. Vallentyne (1955) modified the Livingstone design to reduce costs, to avoid distortion of the uppermost part of the sample, to prevent the piston from rising prematurely, and to allow the use of a falling weight to force the sampler when dealing with hard sediment. The main disadvantage is that two wire lines are necessary instead of the single line as used in the Livingstone corer.

The piston rod (Fig. 6.16A, B) is about 5 ft long and made from brass rod. A spring catch (4) is mounted to its lower end. A sliding collar (2) with screw can be fastened at any level on the piston rod. The head piece (Fig. 6.16C) is attached to the sampling tube by three screws. This head piece has three holes, one (6) for the piston rod, one (7) for the piston line, and one (8) for the lowering line. The piston is presented in Fig. 6.16D. The upper part (D_1) has a beveled cup (9) in which the lower end of the piston rod fits. The L-shaped hole (10) is for the piston line. It is widened at its end (11) to house a knot tied in the piston line. A leather collar (D_2) is held in place in between the pieces D_1 and D_3 (Fig.

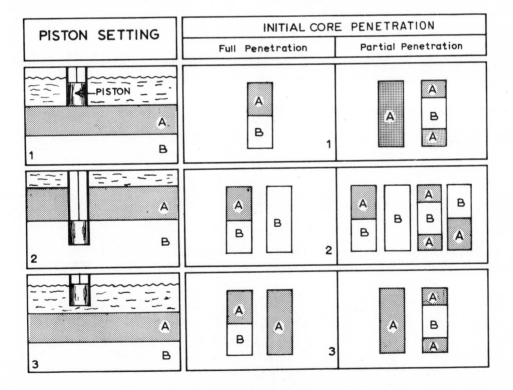

Fig: 6.12. Sequence of sediment obtained with different combinations of piston setting and initial core penetration as collected from an experimental setup. (Courtesy R. E. Burns, ESSA, Joint Oceanographic Research Group, Seattle; Redrawn from Burns, 1963.)

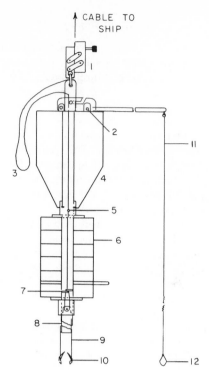

CABLE TO
SHIP

Fig. 6.13. Schematic drawing of the Ewing piston corer: (1) cable clamp; (2) release mechanism; (3) loop of the piston cable; (4) stabilizing fins; (5) piston cable; (6) weights; (7) connection of piston to piston cable; (8) piston; (9) core barrel with liner; (10) core nose with core retainer; (11) trigger line; (12) trigger weight. (Redrawn from G. M. Mfg. & Instr. Corp.; Kahl, not dated.)

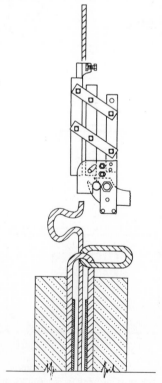

Fig. 6.14. Tripping mechanism for heavy piston coring devices. (Redrawn from Heezen, SYMPOSIUM ON OCEANOGRAPHIC INSTRUMENTATION, Publication 309, Division of Physical Sciences, National Academy of Sciences — National Research Council, Washington, D.C., 1954.)

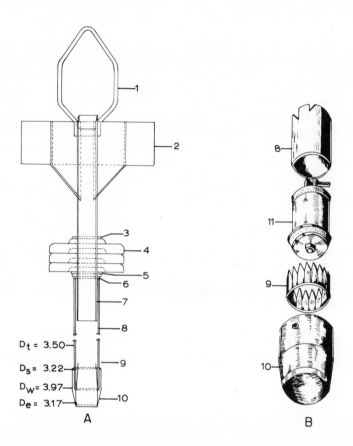

Fig. 6.15. (A) Schematic drawing of the Hydroplastic corer. Dimensions in inches. (B) Detail of the core nose, spring-leaf core catcher, piston and end of PVC barrel. For (A) and (B): (1) bail; (2) shroud; (3) lock ring; (4) weight; (5) weight support; (6) gasket; (7) weight stand; (8) PVC barrel; (9) core catcher; (10) core nose; (11) piston; (D_t) barrel diameter, outside; (D_s) barrel diameter, inside; (D_w) core nose diameter, outside; (D_e) core nose diameter, inside. (Redrawn from Richards and Keller, 1961.)

6.16D) by the piston bolt (12), which screws into the upper part. The complete corer is shown in Fig. 6.16E.

"The sampler is operated in two different ways depending on whether a loose surface sediment or a compact subsurface sediment is being sampled. In either case, the sampler is first assembled. A wire piston line extends from the piston to about five inches beyond the top of the sample tube. A loop is knotted on the upper end of the piston line and any length of wire may be attached to this by means of a catch. The piston rod is lowered into the bevelled cup of the uppermost piston piece. The bevelled nose piece of the piston protrudes from the sample tube and the brass collar on the piston rod is tightened in position. The sampler is then ready for operation and need not be taken apart again" (Vallentyne, 1955).

When the operator is sampling loose sediments, the bottom of the piston rod is locked to the head piece with the piston at the bottom of the core barrel. The sampler is lowered by holding on to the extension rods or the lowering line. At the desired depth, the piston line is clamped to the boat or raft, and the rods pushed down until the sample has been collected. The device is then pulled up and the core is extruded.

For coring more solid subsurface sediments, the bottom of the piston rod is rested in the beveled cup of the piston, and the sampler is lowered by the piston line. "With hard sediments, it is necessary to push on the connecting rods until the desired level has been reached. Since this force is directed against the piston, there is no danger of the piston rising prematurely in the sample tube. When the desired level has been reached, the piston line is secured to

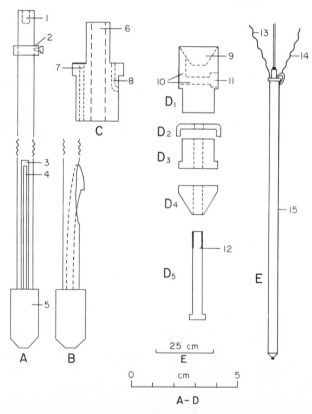

Fig. 6.16. Schematic drawing of the modified Livingstone piston corer for lake deposits. (A) Piston rod with threaded hole (1), brass collar with screw (2), slit (3) for catch (4), and bottom piece (5). (B) Bottom part of piston rod as shown in (A) but turned 90°. (C) Head piece with hole (6) for piston rod, hole (7) for piston line, and hole (8) for lowering line. (D) Parts of the piston. (D₁) Upper piece with beveled cup (9), L-shaped hole (10) for piston line with widened end (11). (D₂) Leather collar. (D₃) and (D₄) Lower parts of piston. (D₅) Piston bolt with threaded top. (E) Assembled corer with piston line (13), lowering line (14), and core barrel (15). (Redrawn from Vallentyne, 1955.)

the raft. The extension rods are pulled up slowly until the catch is heard to lock to the head piece. The friction of the sample tube against the surrounding sediment is sufficient to keep the sample tube in place while the rods are being raised. The sample is collected by pushing down on the extension rods" (Vallentyne, 1955). Sometimes a tripod car jack, bolted to the raft, is used for pulling the barrel out of clay. Each retrieved sample is 1 m long. Successive samples can be collected below each other. Sometimes a casing is necessary. If a release pin is attached to the piston line, the lowering line becomes unnecessary.

Deevey (1965) also described the Livingstone sampler and introduced some modifications. The maximum depth of water in which it has been operated successfully is 30 m. For greater depths another raft will be necessary. Cores have been collected through sediments as thick as 18 m, including 1 m or more of gravel or till.

The Sphincter Corer. Kermabon et al. (1966) described a wide-diameter piston corer. The core catcher and the nose piece are discussed in a later part of this Chapter. It is a modified Kullenberg corer with a steel barrel and a plastic liner, capable of collecting cores with a diameter of 120 mm. An electrical release system, which is not subject to premature operation, has been developed. A special "split" piston was designed, as discussed elsewhere. The gross recovery rate is close to 100% throughout the full length.

A stainless steel power container with electrolytic capacitors and a mercury switch replace the trigger weight. A spike is connected by a short chain to the bottom of this unit. As soon as the spike reaches bottom, the power container falls over from vertical to horizontal, and the mercury switch closes as soon as an angle of 60° off the vertical is reached. The capacitors discharge into the solenoid, which is attached to the tripping mechanism. The operation of the solenoid causes a linkage system to release a tripping arm which permits the corer to fall.

Shallow-Marine Sand Corer after Van der Bussche and Houbolt. A streamlined corer, needing a free fall from a height exceeding 7 m to penetrate up to 1.5 m into shallow marine sands, has been developed at the Koninklijke/Shell Exploratie en Produktie Laboratorium at Rijswijk (Z. H.), The Netherlands (Van der Bussche and Houbolt, 1964). The device has been devel-

oped to collect oriented, undisturbed cores from clean sands on the North Sea bottom.

Experiments proved that a weight of 3500 kg was needed to force a tube (outer diameter 70 mm, wall thickness 1 mm) hydraulically into a water-covered beach to a depth of 1 m. Kinematically this force can be obtained by using a streamlined body weighing about 300 kg when it has a free fall through the water column of about 7 m.

The corer (Fig. 6.17A) consists of a streamlined body (1) whose lower end is filled with lead (2). The body is made from steel plates welded around a central tube (3). The tail is provided with vanes (4) to insure stability. The core barrel (5) is 1.5 m long, has an outside diameter of 70 mm, and can be fastened to the lower end of the central tube by means of a nut (6). The corer can be released by a trigger, to which it hangs by ring (7). A nylon rope [Fig. 6.17C (16)], wound on a very light and easily running spool, is also connected to ring (7). This nylon rope (breaking strength 4500 kg) serves to recover the device. A nonreturn valve (8) is inserted within the lower part of the central tube. The impact was so high that a core catcher could not withstand it, which led to the development of turning the corer about a spindle (9) as soon as it was pulled out of the sediment. This spindle is fixed in the weight just below the center of gravity ("x" in Fig. 6.17C). The nylon hoisting rope is connected to the spindle by two steel wires [Fig. 6.17C (17)] which are 2 m long.

A rubber piston employed was unsuccessful, since shortening commonly resulted. The authors therefore developed an aluminum piston with shear disk (Fig. 6.17B). It consists of an aluminum body (10) with 0-ring (11) that moves easily through the core barrel (5). To prevent it from moving up the barrel before the bottom is reached, a shear disk (12), made from 0.3 mm-thick soft aluminum with a diameter slightly larger than the outside diameter of the barrel is connected to the piston by means of bolt (13), cover plate (14), and washer (15). The cover plate is only slightly smaller than the inside diameter of the barrel. Upon impact with the sediment surface, the protruding part of the shear disk is sheared off and the sample can enter the barrel easily. Fig. 6.17 C explains the operation of this corer. As soon as the device is pulled out of the sediment, the corer turns over about the spindle allowing no sediment to fall out since it rests on the piston.

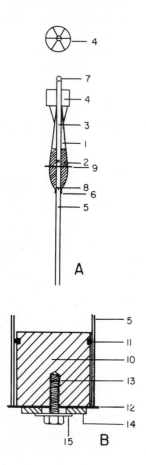

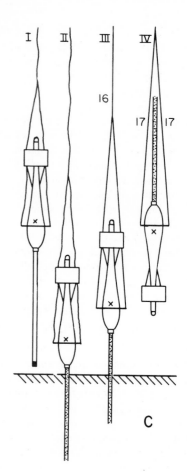

Fig. 6.17. Shallow-marine sand corer after Van den Bussche and Houbolt. (A) Schematic drawing of the corer. (B) Aluminum piston. (C) Operation cycle of the corer. (A-C): (1) streamlined body; (2) lead filling; (3) central tube; (4) stabilizing vanes; (5) core barrel; (6) coupling nut; (7) connecting ring; (8) nonreturn valve; (9) spindle; (10) aluminum piston body; (11) O ring; (12) shear disk; (13) bolt; (14) cover plate; (15) washer; (16) nylon rope; (17) steel wire. (CI) Corer falls freely; piston is held in position by the shear disk. (CII) Upon impact with the sand, the shear disk is sheared off and the piston is pushed up by the penetrating sand; this causes some water loss from the sandy core, which increases the adhesion of the sample to the inside of the barrel. (CIII) During the time the barrel is pulled out of the sediment, the core remains in the barrel due to the increased adhesion. (CIV) As soon as the bottom of the core barrel is out of the sediment, the corer turns over. (Redrawn from Van den Bussche and Houbolt, 1964.)

Excellent results were obtained in water depths up to 30 m. At greater depths the impact proved to be too heavy for the core barrel. However, this was overcome by using core barrels with thicker walls. A nonmagnetic variety of this corer was used on the North Sea to attach a compass-clinometer.

Hydrostatic Actuated Corers

"Hydrostatic pressure can be used to actuate (1) implosion systems, (2) mechanical drive systems, (3) hydraulic drive systems, (4) Magde-

burg systems, which involve an evacuating pump instead of an atmospheric chamber, and (5) systems using differential compressibility of fluids or solids" (Rosfelder, 1966c). Hydrostatic pressure has been used sporadically until now, but may become important in the near future (Rosfelder, 1966b, 1966c).

The first attempts for an implosion corer were made by Varney and Redvine (1937) and Petterson and Kullenberg (1940). The coring tube was separated from an atmospheric chamber by a seal, which was broken by a trigger upon bottom contact. It was expected that the hydrostatic pressure should drive the barrel into the sediment, but the results were not satisfactory.

Kermabon (1964) used the implosion to actuate a coring device. In his tube accelerator, the coring tube is closed at the bottom by a tripping line. It rises with upward acceleration, knocks off an upper cap, and by reaction causes a downward acceleration of the core barrel. Fig. 6.18A presents the operation principle of his first prototype. Kermabon (written communication, February 1966) writes that "the present prototype is quite different from the one described in the report. It now consists of just a tube closed at one end by a glass diaphragm and a removable cap at the other; no more piston, the mass of water involved acts as a piston. The results are quite encouraging but the acceleration is so great that I have a few problems to solve regarding strength of materials." The glass plate is ruptured by an explosive charge.

Kermabon (written communication, February 1968) made more improvements on the device presented in Fig. 6.18A. A schematic drawing, which is close to the real model, is given in Fig. 6.18B. Equality of the lengths L_1 and L_2 is important for obtaining full implosion in the water.

When the lead shot in its bag (j) reaches the sea floor, the rubber band (k) flips the expandable oil-mercury switch (1) in reverse position, short-circuiting the electrodes. The high-pressure explosive cap (m) is actuated and the glass diaphragm (n) ruptured. The piston is pushed upward while the upper cap (o) and tube are accelerated downward after the weak link (p) in the retrieving cable is ruptured. The lead weight battery and switch are expandable.

However, this has the defect inherent to all heavy piston corers: the piston does not remain stationary after release. Kermabon and Cortis (in press) describe a system for monitoring the piston relative to a fixed platform lying on the sea floor.

Gas-Operated Corers

Since it is difficult to keep the piston rigidly fixed in position relative to the sediment surface, to build a corer with a large area ratio, and to collect a long sample without exceeding the maximum safe length-to-diameter ratio of 20 proposed by Hvorslev (1949), another penetration principle is used by some authors. The Mackereth (1958; Smith, 1959) corer cannot be modified easily to collect successive short samples (Andresen et al., 1965). At the Norwegian

Geotechnical Institute a corer has been developed to overcome these difficulties.

Mackereth Sampler. This is a portable corer which operates at depths up to 90 m (300 feet). Smith (1959) reported that cores up to 20 ft in length and 1½ in. in diameter had been collected. The corer operates as a piston which is pushed into the sediments by compressed air. The coring device is held in place near the water-sediment interface by the hydrostatic pressure during the coring operation. The sampler is raised to the surface by using compressed air.

The corer consists of two metal tubes, one tube inside the other, each about 20 ft long, and a cylindrical anchor chamber which is 4 ft long and 1½ ft in diameter [Fig. 6.19A (1-10)]. The outer chamber (1) is connected to the lower part of the outer tube (2). During the coring operation an air pressure hose (3) is attached to the top of the outer tube, and another hose (4) runs from the anchor chamber to a pump. A short pipe (5) leads from the anchor chamber to the lower part of the outer tube. An air release valve (6) is mounted to the top of the anchor chamber. A piston (7) closes the inner tube (8) inside the outer tube (2). Core extrusion is carried out by pumping water in the space between piston (7) and the fixed piston (9) through the tube (10).

The operation of this corer can be explained in a number of steps (Fig. 6.19B):

a. The coring device is lowered on a nylon rope and both hoses are payed out at the same time. When the anchor chamber (1) is in contact with the sediment surface and the corer supported by the nylon rope, one hose is connected to a pump and the other hose to compressed air.

b. The water in the anchor chamber is pumped out through the hose (4) and the hydrostatic pressure forces the chamber into the sediment. The chamber has to be pumped out completely. The hose is then sealed off.

c. Compressed air is forced through the hose (3) and the core tube is pushed into the sediment. A pressure which is 100 lb/in² greater than the hydrostatic pressure at the sediment surface usually is sufficient.

d. When the piston (7) passes the small pipe (5), air escapes from the outer tube into the top of the anchor chamber. The core is taken by this time. As soon as the pressure in the anchor chamber exceeds the hydrostatic pressure, the chamber is lifted out of the sediment.

e. The corer with core is raised to the water

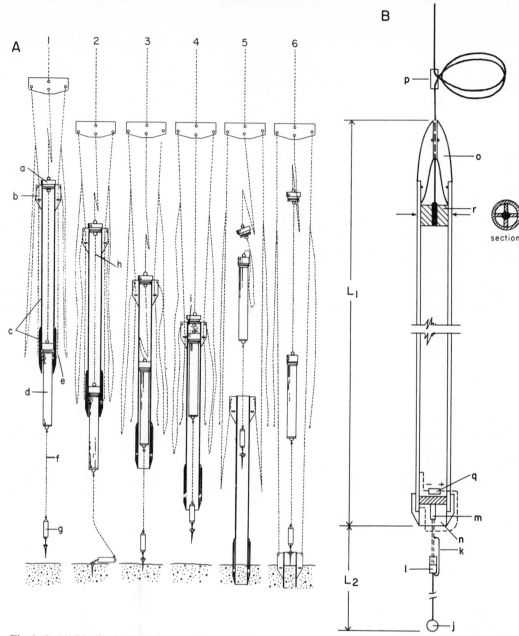

Fig. 6.18. (A) Idealized schematic operation principle of the first prototype of Kermabon's hydrostatic corer. (1) The apparatus approaches the bottom; the mercury switch has not yet made contact. (2) The switch hits bottom and falls over; this fires an explosive charge, which sets the piston releasing device in motion freeing the piston from the tube. (3) The liberated piston starts to move up and the coring tube begins to accelerate downward; the two traction cables are released when the piston begins to move; as the tube moves downward these cables pay out through two guiding rings in the tube fins. (4) The absolute watertightness is no longer insured when the piston ascends; the piston reaches the end of its course; since the air in the tube is compressed to a pressure equal to the external pressure, the movement of the heavy piston separates the lightweight cap from the relatively heavy tube. (5) The tube is moving at high speed and starts to penetrate the sediment. (6) The tube penetrates the sediment; its downward motion is stopped by arresting cables, which are used also to pull out the corer; the cap, piston, and switch are attached in series to a cable and are retrieved with the corer: (a) waterproof cap; (b) fins; (c) tube nose assembly; (d) assembly piston; (e) traction cable before release; (f) electrical cable; (g) battery and switch; (h) cable for piston recovery; (i) water penetrates slowly. (Redrawn from Kermabon, 1964.) (B) Schematic drawing of the hydrodynamic accelerator: (j) lead shot in bag; (k) rubber band; (l) expandable oil-mercury switch (in this position the oil is on top of the mercury); (m) high-pressure explosive cap; (n) glass diaphragm; (o) upper cap; (p) weak link in the retrieving cable; (q) battery; (r) cross-plate. (Courtesy A. Kermabon, Compagnie Maritime d'Expertises, Marseille, France; redrawn from Kermabon, written communication, February, 1968.)

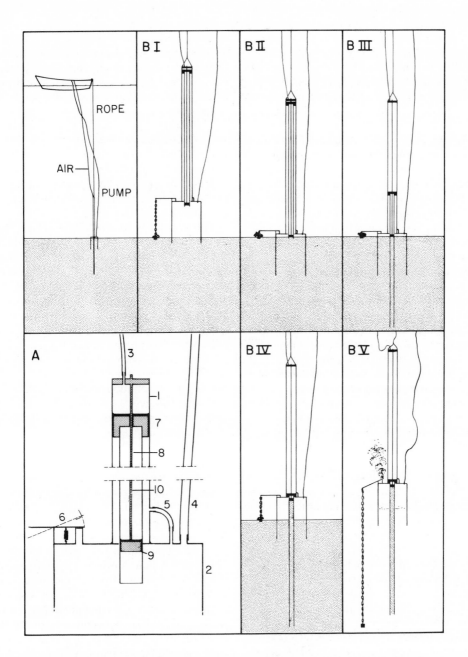

Fig. 6.19. Diagram of the main features (A) and operational sequence [B(I-V)] of the Mackereth core sampler. The usual working position is indicated in the upper left-hand drawing. (A): (1) outer chamber or anchor chamber; (2) outer tube; (3) air pressure hose; (4) hose running to pump; (5) short pipe for air escape from the outer tube into the anchor chamber; (6) air release valve; (7) piston; (8) inner tube; (9) fixed piston; (10) tube. The operation sequence I-V is explained in the text. (Courtesy A. J. Smith, Univ. College, Univ. of London, England; redrawn from Smith, 1959.)

surface by means of the buoyancy of the trapped air. The air release valve (6) acts during the ascent to avoid an excessive rate of ascension. The valve is held closed by a spring when the corer rests on the sediment surface but is pulled open by a weight on a chain when the corer is higher above the bottom than the chain is long.

When the corer is back on board it is taken to the shore to extrude the core. Water is pumped into the space between the pistons (7) and (9) through the tube (10). The water forces the piston (7) to move up, thus retracting the core tube and simultaneously extruding the core.

The advantage of the Mackereth sampler is that it can be operated from a small boat and does not require a special rig. The one described is not suitable for collecting material coarser than sand since water-saturated sand falls out during ascent if a core retainer is not used.

The Mackereth sampler has been modified recently. Plastic core tubes are now used instead of metal ones (A. J. Smith, personal communication, January 1968). The pump has been changed and the pump outlet from the chamber is no longer at the surface of the water.

Walker (1967) describes another modification: a diver-operated pneumatic sampler. The outer tube, core tube, and air-bottle mounting are manufactured from high-impact, rigid PVC. The air-release valve [Fig. 6.19A (6)] is replaced by a rubber plug. A diver pushes the anchor chamber into the sediment and then closes the rubber plug. The coring tube is pushed in using compressed air from a cylinder with a capacity of 750 liters at 126 kg/cm². Two valves are mounted on the outer tube, one for forcing the corer into the sediment and one for forcing the anchor chamber out of the sediment.

N. G. I. Corer (Torpedo). In the introduction of the section on gas-operated corers, we mention that engineers at the Norwegian Geotechnical Institute developed a corer to overcome a number of difficulties (Andresen et al., 1965).

The streamlined device with its internally contained sample tube is 3 m long, 22 cm in diameter at its widest part, and weighs 510 kg in air. The overall length is about 4.7 m when the sample tube is fully extended. The torpedo-shape was designed to increase stability in both water and soft sediment, and to reduce friction between the corer and the sediment when penetrating the desired depth into the bottom, as well as during the pull-out.

Three principal sections form the body of the corer: a cap to allow easy access to the gas generator; the propulsion section to which the hollow piston rod with the fixed piston is screwed; and the housing for the movable piston and its protective steel jacket with a poured-lead weight between.

The coring tube is 180 cm long, which enables the collecting of samples 165 cm long. The fast, constant-drive speed, fixed piston, small area ratio of 12%, outside clearance ratio of 0%, inside clearance ratio of 0.9%, and very sharp cutting edge permit sampling with very little disturbance (Table VI.1).

The sampler is lowered on a cable. Shortly after the device becomes emplaced in the bottom, the time switch closes, initiating a chain of reactions: the electrical circuit fires the fuse in the ignitor; this produces the required temperature and pressure to start the main charge, which burns like a cigarette and develops gas; the gas moves at a controlled rate through the orifice and enters the cylinder behind the movable piston; the piston pushes forward at a speed of about 20 cm/sec. About 20 sec may be needed for the total penetration operation.

During successive lowerings the corer could be positioned sufficiently deeper to obtain samples that overlap one another.

Vibrocorers

The vibropiston core sampler was recently adapted to underwater use by Kudinov (1957) to overcome difficulties encountered in collecting samples with other corers. A vibrating device on top of the barrel and a stationary piston mounted on a rigid frame form the basic new elements. This combination "disturbs the marine unconsolidated formations through thixotropic changes and water migration in clays and muds and through individual vibration of particles in loose sands" (Gumenskii and Komarov, 1961). However, due to its penetration power, vibrodrilling can be placed between gravity coring and rotary coring and is suitable for semiconsolidated formations such as hard clays, shales, recent calcareous sandstones, etc. (Rosfelder, 1966c).

The Kudinov vibrocorer is activated by an electromagnetic vibrator and has a one-way permeable piston inside the sampling tube, which allows only an upward flow of water. The weight of the device and the up and down vibration of the sampling tube cause penetration. Water from the sediment passes upward through channels in the piston. As the sample enters, the piston rises. The piston is not connected to the winch cable but to a counterweight housed inside one of the two vertical supporting tubes. The sampling tube pivots to a horizontal position on board ship when the sample is removed (Sanders, 1960; Zenkovitch, 1962).

Sand cores, 4 to 6 m long, have been collected from the shallow shelf areas of the Black Sea. The coring tube has a diameter of 53.5 mm with a wall thickness of 5 mm. The maximum operation depth is 150 m of water.

Alpine Geophysical Assoc., Inc. (not dated b) constructed an air powered "Vibracore" to retrieve 12 to 40 ft long samples in depths up to 150 ft. The operational water depth is limited only by the type and size of the air compressor. Electronic signals indicate the degree of bottom penetration. About 16 ft can be drilled in 120 sec. The required air supply is 300 ft³/min at 100 to 125 psi. The frame is an aluminum "H" beam with four side legs to give it the proper stability. A one-way check valve at the top of the coring pipe allows water to escape during sample collection. The inside and outside diameters of the core pipe are 4 and 4½ in., respectively. Usually a plastic liner is used, reducing the core diameter to 3½ in.

The Compagnie Maritime d'Expertises developed a vibrocorer that operates on only one cable. Two vibrators are fed from a tank with compressed air. However, divers are necessary to direct the operations. Normally a depth of about 50 m will not be surpassed. The air pressure most commonly used is 80 kg/cm² (COMEX, not dated).

Beachcor 67 Sand Corer

Several attempts have been made to reduce the friction between core and barrel by using a type of sleeve that envelops the core during sampling and moves up in the barrel with the sample. The Swedish foil sampler discussed previously, has proven to be very successful. Instead of foil, one can use types of plastic sleeve or other material to completely enclose the core. Chmelik (personal communication, November, 1965; *et al.,* 1968) developed two prototypes at Texas A&M University. The sleeve can be stored just above the core nose or it can be stored higher on the barrel and fed into it by an 180° turn at its base. In the second case the high impact due to the free fall often broke the sleeve prematurely. Rosfelder (1966c) discusses the possibility of using strips of plastic material, which can be stored as a roll and

zipped together longitudinally to build a hose. Future developments will certainly solve the technical problems. Chmelik demonstrated the value by comparing his result with a piston core collected from the same station. It was striking that the top of his core was practically undisturbed, since nearly no bending of laminae along the sides could be observed by X-ray radiography analysis.

At American Undersea, extensive tests were carried out to determine the mechanisms affecting the motion of sand moving as a column through a pipe. Measurements of the force required to insert a pipe with loose piston into sand showed a rapid increase in the resistance of the sand. At a certain point, extremely high forces were required to obtain any further movement (Coffee, 1968). The results were not markedly different for wet or dry sand. The experiments resulted in the development of their Beachcor 67.

In the Beachor 67 the sand friction is limited to a pipe section only 5 in. long, which is not long enough to obtain enough friction to influence coring action (Fig. 6.20). The flexible core container is made of extruded polyethylene tubing, 0.004 in. thick and 3⅛ in. in diameter, sealed at the unattached end. This tubing is stored around the 5 in. long container placed on top of the core nose.

Once the core passes the sleeve container it enters the larger diameter core barrel with negligible friction. The inside diameter of the sleeve container is 2⅞ in., the inside diameter of the barrel 3⅛ in. This results in a core shortening of ¼ of the total length (Fig. 6.20). No cumulative errors were observed due to this diameter increase. (However, it is obvious that grain movement takes place and destroys the original fabric. When examining the layering of the sediment and making correlations between holes, this should not interfere.) Drilling is accomplished by pumping water down between the double wall of the barrel and allowing it to flow back up along the outside. This facilitates penetration, and little force is needed to lower the corer when no outside friction is present. The water movement has no effect upon the core, and even dry cores can be collected since the core nose is below the outflow of the water jet. It was found that there is a limit to the length of the input pipe with regard to the total core length. A 6 in. long input pipe allows the collection of a 6 to 7 ft core, a 5 in. long one takes cores well over 10 ft in length before the sand packs. Surface

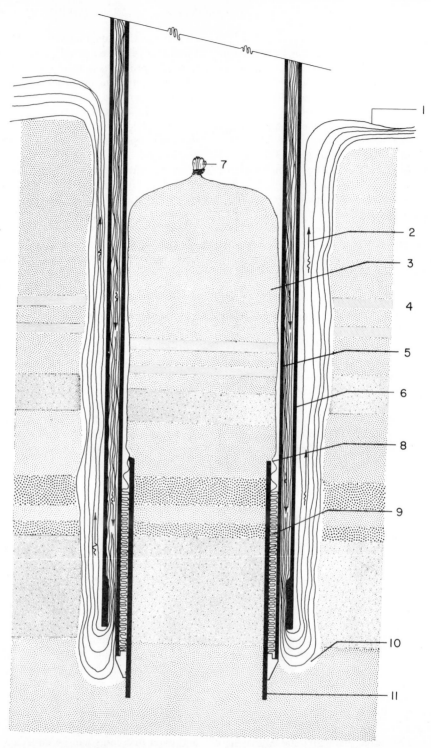

Fig. 6.20. Schematic drawing of the lower part of the Beachcor 67 in operation. Note the sample shortening due to widening of the internal diameter once the core passes the core sleeve container: (1) surface of the beach; (2) direction of water motion; (3) core in plastic sleeve; (4) beach laminations; (5) core tube; (6) water jacket; (7) closed top of core sleeve; (8) widening of the inside diameter on top of the core casing cartridge (9); (10) water-cutting area; (11) corer nose. (Courtesy C. E. Coffee, President, American Undersea, San Francisco, U.S.A.; redrawn from Coffee, 1968.)

330

SECTION A-A

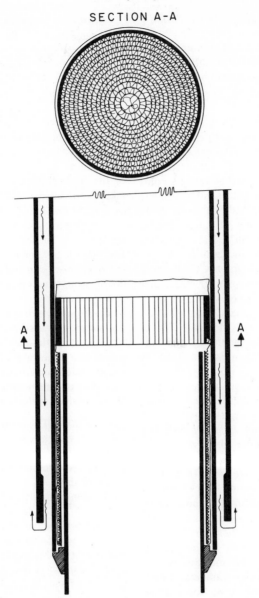

Fig. 6.21. Shallcor adaptor for use of the Beachcor 67 under water. The adaptor is placed inside the core sleeve and lies on top of the sample. It prevents movement of solid particles while the core liner expands during penetration. Section A-A shows the 1 in. wide strips of corrugated and flat brass rolled together and inserted in a stainless steel ring. See Fig. 6.20 for names of the different parts. (Courtesy C. E. Coffee, President, American Undersea, San Francisco.)

finish of the tool also proved to be important. At the present time American Undersea is working on a system for very long cores that has no diameter change at all (Coffee, personal communication, October 1967).

A modification of this device for underwater use is the "shallcor adaptor" (Fig. 6.21), which is fitted into the end of the core sleeve. Water flows through the adaptor with little or no tendency to move the core liner. Sand, or even mud, finds considerable impedence.

Another extension of the patented design of the Beachcor 67 is the use of a "plastic zipper" to assemble the liner at the top of the corer during sampling, which will reduce the length of the input tube to about 1 in.

The corer can be carried and operated easily on the beach by two men with only approximately 5 min required from start to finish.

6.3C Box Samplers

Reineck (1958, 1961, 1963a, b) developed a box sampler for collecting rectangular samples in tidal flat areas. The present author adapted the sampler to deep water and attached a compass to it, simultaneously with similar changes made by Reineck (Bouma and Marshall, 1964; Bouma and Shepard, 1964; Bouma, 1964a; Füchtbauer and Reineck, 1963). The instrument has been operated very successfully at the Scripps Institute of Oceanography. Cores have been collected with the box sampler where piston and gravity corers failed (Bouma and Shepard, 1964). Rosfelder modified and enlarged the sampler (Rosfelder and Marshall, 1967). The latest known modification and enlargement was made by the present author in 1967 at Texas A&M University.

The box sampler collects rectangular samples of the upper part of the sediment column. It is an ideal instrument for collecting the sediment at the water-sediment interface. The sample boxes fit in the sampler and serve also for storage. Some advantages of the box sampler over a pipe corer are the ability to remove a side of the box, thus exposing a vertical section of the sample, as well as the large cross section of the box, which allows a wide variety of investigations to be carried out on the same sediment horizon.

A box corer with a different closing mechanism was made by Klovan and Imbrie and modified at Hudson Laboratories (Sanders, 1966). The apparatus is primarily used for shallow water.

Kögler (1963) described a long box corer used by the marine geology group at the University of Kiel, Germany. The cross section of the sample is less than that of the preceding box sampler, but its length is many times longer. This type is also used successfully at the Woods Hole Oceanographic Institution (D. A. Ross, personal communication, November 1967). At the Scripps Institution of Oceanography three long box corers are used. Their results are not always successful, which may be due to differences in type of sediment (N. F. Marshall, personal communication, February 1968) (see also Section 6.7).

Large Box Sampler

The model first built at the Scripps Institution of Oceanography will be discussed extensively to explain the principles and the coring operation. Modifications made are described after this discussion.

The box corer consists of a number of basic parts: gimbaled frame, central stem with box holder, closing mechanism, tripping mechanism, sampling box, and orientation unit (Fig. 6.22). The central stem slides through the gimbaled top of the frame in such a way that samples are taken vertically. The German sampler, as well as the first one built at the Scripps Institution of Oceanography, had boxes which were 20 x 30 cm (8 x 12 in.) in section and 45 cm (18 in.) high. Rosfelder lengthened the box to 24 in. The closing mechanism consists of a blade at the end of a double arm, which pivots about the box holder. A release mechanism on top of the central stem makes it possible to use only one wire for lowering, sampling, closing, and returning to the surface. Very little free fall is possible, which means that penetration into the sediment is primarily based on gravity. Very little or no disturbance of the sediment can be discovered along the sides of the box.

The Frame. The frame not only insures vertical penetration at slopes up to about 18° but also prevents the sampler from falling over. This preventative facilitates extracting the filled box on board ship.

The frame consists of hollow pipe (Figs. 6.22 and 6.23). The base is 5 ft on a side. Four legs run up from the corners of the base to a square, which consists of angle iron. These in turn are bolted to a gimbal mount (Fig. 6.24). The frame is so constructed that it can be separated into two parts for easier storage and transport (Fig. 6.12). A number of holes are drilled in all frame parts for drainage and to avoid collapse of the pipes under hydrostatic pressure. A few guides are also welded to the base of the frame to prevent the closing arm from swinging around (Fig. 6.23).

Central Stem. The central stem is made of tubing (65 in. long, 2½ in. in diameter) and filled with lead. The stem slides through the inner ring of the gimbal. At 35 in. from the top, a 1 in. diameter hole is drilled through the stem, through which a safety pin fits (Fig. 6.22). This pin prevents the central stem from sliding through the gimbal, thus enabling attachment or removal of the sampling box. An ad-

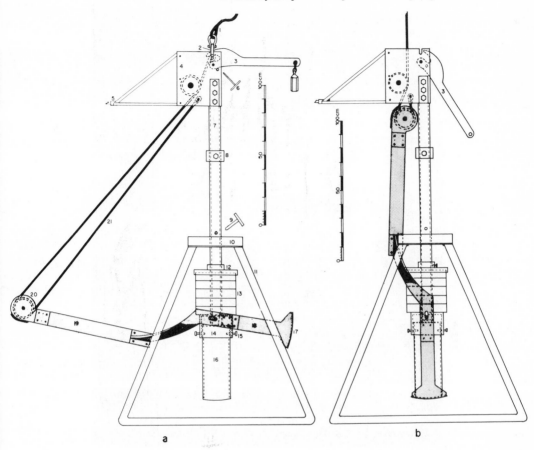

Fig. 6.22. Schematic drawing of the large box sampler built at Scripps Institution of Oceanography, La Jolla, California, in 1962. (A) Open position for lowering, (B) Closed position with sample collected. (A and B): (1) main wire; (2) oblong ring in the main wire; (3) tripping arm with weight and hook which fits through ring (2); (4) plate of the tripping mechanism; (5) brackets for compass attachment; (6) safety pin to prevent arm (3) from falling down (pin is removed when sampler is over the side); (7) central stem; (8) stopper; (9) heavy pin inserted to prevent central stem sliding through the gimbal (10) when sampler is on deck; (11) leg of the frame; (12) protection cap for the weights (13); (14) sample box holder; (15) pins to attach box (16) to box holder (14); (17) curved base of the spade of the closing mechanism consisting of lower lever part (18), upper lever part (19) with pulley (20); (21) lower part of the wire serving as closure wire.

justable ring can be secured to the stem to stop the box from penetrating too far. It is one of the safeguards against sample disturbance, especially compression.

A frame housing for the sample box is welded to the bottom of the stem (Fig. 6.25). A plate is welded inside the housing and acts as top for the sample box, while 26 holes (½ in. in diameter) are drilled through this plate to allow water to escape from the box during penetration. Above this plate, holes are drilled in the sides of the box housing for the same reason. Two holes are drilled on each of the long sides of the housing at a distance of 1 in. from the bottom. Matching holes exist in the sampling box.

Short pins are used to attach the box to the housing, while electrical tape is wound around them (Fig. 6.23) to prevent loss.

The top of the housing is flat and serves as base for the weights. Each weight is made from lead and has a self-locking key design for easy removal, which holds them in place. A thick metal sliding plate is attached rigidly to the stem by set screws to keep the weights in place and to prevent them from damage by the legs of the frame.

At the top of the central stem an 8 x 1½ in. slot is cut in which the plates of the tripping mechanism fit. The plates are connected to the stem by 3 bolts.

Fig. 6.23. Box sampler. Note that both safety pins are inserted: (1) compass; (2) place where the frame can be taken apart; (3) guides to prevent sampler from turning around in the frame. See Fig. 6.22 for discussion on the parts of the sampler. (After Bouma and Marshall, 1964.)

Tripping Mechanism. The main wire is divided in two parts by an oval ring. The part below the ring serves to close the sampler. The tripping mechanism disconnects the closing wire from the central stem so that a separate lowering cable and a closing-raising cable are not needed.

The tripping mechanism consists of two plates and an arm (Fig. 6.22). The closing arm is 30 in. long and turns inside these plates. At the outer end a lead weight is attached. The other end of the arm has a hook that curves slightly downward to fit through the oval ring in the cable.

The wire goes through the plates, guided by a pulley, down to the pulley of the closing mechanism and then back to the plates, where it is attached permanently.

Closing mechanism. The closing mechanism turns under the box to insure closure before the unit is pulled out of the bottom (Fig. 6.26). On one end of this mechanism a pulley is placed between the plates to guide the closing wire. At 36 in. from this end the arm forms a yoke, which separates the plates to a width of 18 in. The closing mechanism turns about two heavy rods (1½ in. in diameter), which extend 4 in. from the housing. The holes in the yoke are oval (2½ x 1⅝ in.), which is necessary for the operation (discussed later). A curved knife (14¾ x 10½ in.) is welded at right angles to the other end of the yoke. This plate knife is curved identically to the short side of the box and sharpened on one side.

Sample Box. The box is made of ⅛ in. thick stainless steel and consists of two parts. One part forms three sides, the other part fits over it and can be secured by six countersunk screws on each side (Fig. 6.27).

The top and bottom of the box are open. The shortest sides of the bottom curve in the same way the closing mechanism is curved. All lower edges of the box are beveled to obtain better penetration.

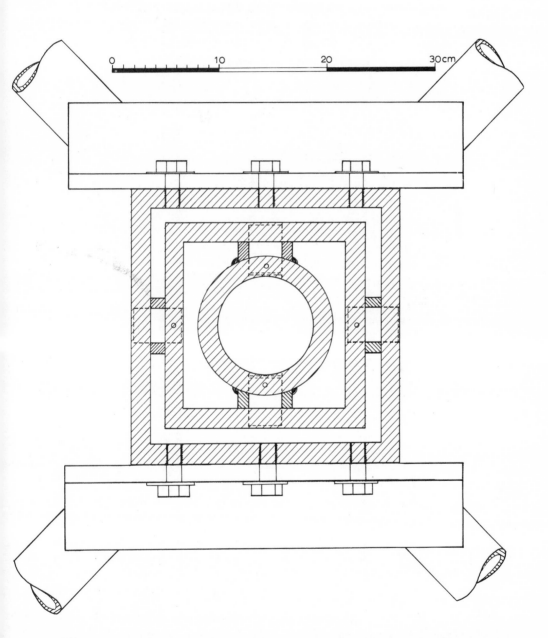

Fig. 6.24. Gimbal unit forming the top of the box sampler frame. The angle iron parts are welded to the legs and the gimbal unit is bolted to them. The unit consists of two square parts and one round part, both inner ones turning inside their outer one about heavy axles. Greasing openings are indicated.

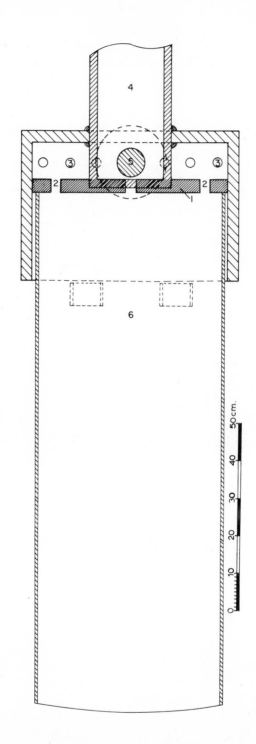

Fig. 6.25. Box holder with box. Inside the holder a plate (1) with holes (2) is mounted. The holes allow water to flow out. Above this plate holes (3) are drilled in the side of the box holder. The central stem (4) is welded to the box holder. Rod (5) is the axle about which the closing mechanism turns. When box (6) is inserted it closes against plate (1).

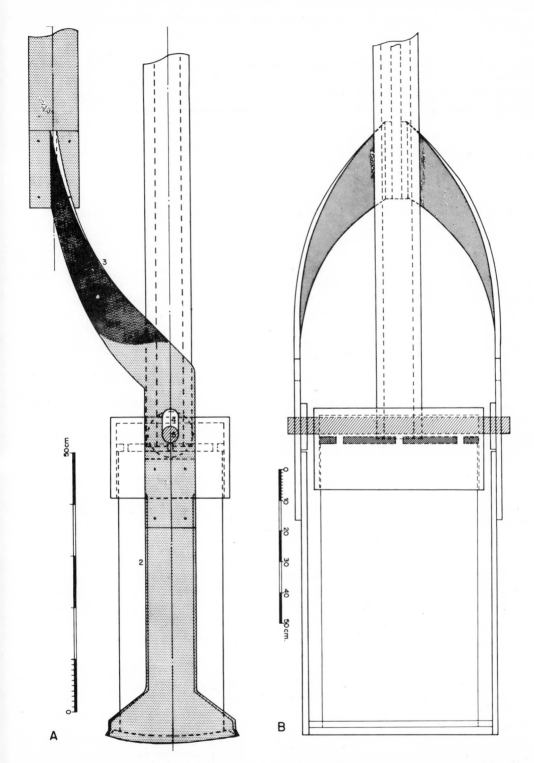

Fig. 6.26. Lower part of the closing mechanism showing the spade (1), lower lever part (2), bent part of the yoke (3) with the oval hole (4) and axle (5) about which the closing mechanism turns. (A) Mechanism seen from the side. (B) The view when mechanism is turned 90°.

In the middle of each long side, 2.69 in. from the bottom, a metal button is welded to which Simmon's No. 2 Link Lockers[2] can be attached to secure a bottom plate to the box. On each short side, 3.75 in. from the top, two U-shaped metal brackets are welded, into which handles are fitted (Fig. 6.27).

The bottom plate is made of the same material and thickness as that of the box. Its size is 12½ x 9 in. and it is curved like the knife of the closing mechanism. Along the 9 in. sides small guides are welded at right angles to the plate. In the middle of the long sides, ¼ in. from the sides, grooves are cut in which the bent lower sides of the Link Lockers fit. The bottom plate is beveled at one of its long sides.

Another hole is drilled in the center upper part of the plate of the Link Locker so that it fits over the button welded to the sampling box. The locks are fastened by turning the wing, thus shortening the locker.

The loose handles are made of ⅛ in. thick ⊔-shaped strip (Fig. 6.27) to which a round ⊔-shaped metal rod is welded.

Lever. It often is difficult to slip the bottom plate easily between the lower side of the sampling box and the knife of the closing mechanism. Using a hammer to insert the plate may ruin the sample. Therefore a lever was constructed of ⅜ in. thick strap metal 1½ in. wide. A long handle (55 in. long and 10½ in. wide) is curved at one end for hanging it at the rods about which the closing mechanism turns. A curved-bottom plate holder is welded to the handle near its curved end (Bouma and Marshall, 1964).

Sampling Procedure. On deck, the sampler hangs in the frame with the safety pin through the central stem to keep it off the deck (Fig.

6.22). The box is attached to its housing, the operator taking care that the cover side of the box faces the knife of the closing mechanism. If this is not done, the bottom plate guides cannot pass when this plate has to be inserted. The four short pins, used to attach the box, work a little easier than two long ones, but they have to be taped to prevent loss (Fig. 6.23).

The oval ring is placed in the slot of the tripping plates, and the closing arm is turned up, the hook of this arm going through the ring. A safety pin (Figs. 6.22, 6.23) is inserted through a hole in the tripping plates underneath the arm to prevent it from falling down. The compass is set and the sampler is ready for use.

A few lines should be placed around the legs of the frame to prevent the sampler from swinging when it is picked up by the winch. As soon as the sampler is free from the deck, the lower safety pin is removed. When the sampler hangs outboard the safety pin underneath the tripping arm is removed.

As soon as the frame comes to rest on the sediment surface, the sampler turns to a vertical position and then pushes the box into the deposit due to the weights. If the sediment does not stop the penetration, the ring around the central stem will. Only through trial and error will it be known for each area at what height this ring should be placed, as well as the amount of weight. The tripping arm falls due to slack in the main wire.

By pulling up the wire the closing mechanism is turned underneath the box, and the compass is locked. This all occurs before the box is pulled out.

When the sampler comes back on board ship it is necessary to keep the main wire tight, even when the safety pin is placed through the central stem. The lever is then used to insert a bottom

---

[2] The Link Lockers often rust due to drainage of salt water during storage and break off. H. E. Reineck uses springs between rods to attach the cover to the handgrips of the sample box. D. S. Gorsline applies large hose clamps, which can be tightened with a screwdriver. Nielsen Hardware Corp. (not dated a,b,c) manufactures turn catches similar to the Link Lockers as well as compression spring catches. Their SC-D-20648 and SC-D-20649 models are open spring catches which have the advantage that sediment can readily be removed from the springs, which is difficult when the springs are pocketed. The smaller catch can take a strain of about 500 lb. The spring loading will equal about 50 lb. if the springs are deflected ⅛ in. For the larger catches these amounts will be twice as high. These catches are made of stainless steel. The spring loading feature does permit some flexibility in the application of the catches and the proximity to the linking with the strike. At Texas A & M University the larger compression spring catches have been tried successfully on the large boxes. Part of the strike (J-shaped) was welded to the lower rod of the catch to fit the groove of the box cover. The upper part of the catch was broadened in order to get a little more material around the hole that fits the knob of the box.

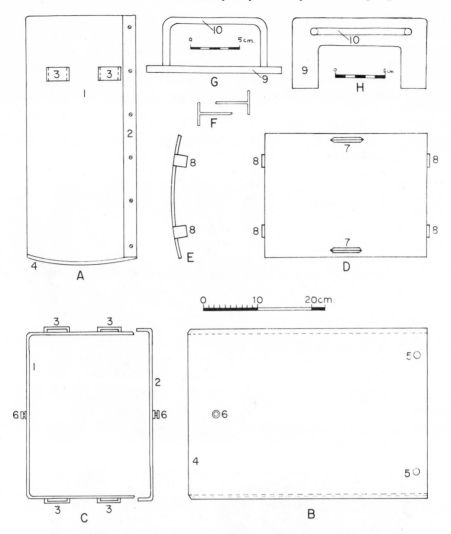

Fig. 6.27. Sample box with cover, pins, and handles. (A) Box seen from the 8-in. side showing the main part (1), side cover, (2), flat metal brackets (3) into which the handles fit, and the beveled lower side (4). (B) Side cover with beveled cutting side (4), holes (5) for the pins (F) to attach box to the sampler, and button (6) to attach Link Lockers to. (C) Cross section through the box with main part (1), side cover (2), flat metal brackets (3), and buttons (6). (D) Box cover seen from below with grooves (7) to which base parts of Link Lockers (see Fig. 6.7) fit, and guide pieces (8). (E) Box cover seen from the side with guide pieces (8). (F) Pins to attach box to sampler. (G, H) Handle consisting of a flat ⊔-shaped piece (9) whose legs fit into brackets (3), and round handle (10).

plate beneath the box. When slack is given to the main wire, the closing knife lowers due to the oval hole in the yoke of the closing mechanism. The bottom plate is pushed gently in place and attached to the box by means of the Link Lockers. A strong bar or screwdriver often is needed to lift the bottom plate to enable the Link Lockers to hook on.

The closing mechanism is next turned to a horizontal position, the tape is removed, and the handles are placed on the box. The pins can be removed and the box taken out. This should be done carefully to avoid disturbing the sample top.

The Texas A & M Box Sampler. The drawings of the improved model, designed at the Scripps Institution of Oceanography by Rosfelder and built under the name NEL Spade Corer at the Navy Electronics Laboratory in San Diego, California, served as base for the A & M box sampler. The present author made a few changes such as size, location of the compass tripper, placing of penetration stoppers on the central stem, variability in weight, and reducing the length of the closing wire.

The frame base is not square but has the shape of a trapezium, while three legs extend to the gimbal (Figs. 6.28, 6.29). The closing mechanism has a rectangular shape, which is easier to

Fig. 6.28. Photograph of the A&M box sampler on board R/V ALAMINOS: (1) frame with three legs; (2) gimbal unit; (3) central stem of which one side can be removed to insert or remove wafers of lead; (4) stopper; (5) pins; (6) box housing with rubber valves with iron weight attached to them; (7) gusset to support the box, made from angle iron and welded to the box housing; (8) rectangular closing mechanism with spade on one side and pulleys on the other; (9) tripping mechanism (hook partly visible); (10) part of the oblong ring of the wire is visible; (11) black painted compass housing with compass; (12) compass release wire.

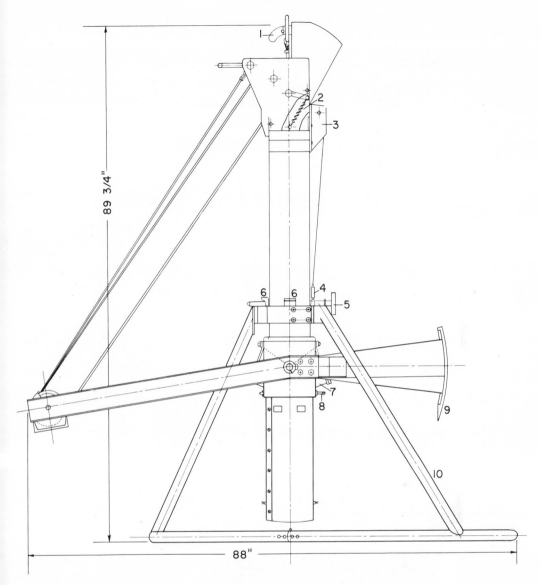

Fig. 6.29. Overall drawing of the box sampler modified by A. M. Rosfelder and built at U.S. Navy Electronics Laboratory in San Diego, California. Most of the parts are given in Fig. 6.28: (1) hook that fits through the oblong ring in the wire (a safety pin must be placed through the hook in front of the ring); (2) elastic band, which tries to pull the hook back; (3) compass housing; (4) compass wire with release mechanism; (5) long pin to prevent central stem moving through the gimbal; (6) sway preventors; (7) valve housing with valve; (8) long pin to attach box to sampler; (9) spade of the closing mechanism with lead plate inside; (10) frame.

make than the yoke. The central stem is square, and part of one side can be removed to place or take out wafers of lead. These wafers are about ¾ in. thick. The box housing does not have the perforated inside plate, but four ½ x ½ in. strips are welded inside the housing.

Two black rubber pieces with iron plates on top act as valves to let water out during penetration. At two corners of the housing strong angle iron strips are welded to give additional support to the sampling box when the closing mechanism operates. The design of the tripping mechanism has been changed and the tripping arm replaced by elastic surgical tubing. (This tubing is similar to the tubing used in underwater spearguns). A free-floating compass is used instead of the self-locking disk design; this is placed to the central stem in an aluminum housing. The sampler is made from stainless steel; the upper part is nonmagnetic. The frame is not made from stainless steel. The Texas A & M box sampler was built by Del Industries, the boxes by Ambox, both in Houston, Texas.

The sampling box is 12 x 12 in. in section and 36 in. high, the total sampler as seen in Fig. 6.28 is 290 cm (114¼ in.) high. The wall thickness of the boxes is 3/16 in. The top part of the closing knife is covered with lead to provide a tighter seal. Since difficulties are still encountered in getting the bottom plate in place, the present author made a lever (Fig. 6.30). This lever differs slightly from the other one since less space ıs available to attach it to the sampler. The bottom plate does not fit in a holder but is only pushed in place by this lever. A screwdriver is needed to lift the bottom plate to hook the Link Locker onto it. The filled boxes weigh over 400 lb. Storage is discussed in Section 6.7.

Rosfelder used two pullies in the tripping mechanism and two at the end of the closing arm to give the closing wire a double loop. The author replaced it by a single wire and single pullies.

Since it was nearly impossible to handle the Texas-sized boxes, a special dolly was developed to carry the boxes in any position between vertical and horizontal, retaining the bottom plate in place (Fig. 6.31).

Small Box Corer. Reineck (1963) dexcribed his ¾ ton box sampler as well as a small one whose weight is ⅕ of the large one. The principle still is the same. The base of the frame is round, three legs extend to the gimbal, and the sampling box is much smaller. Two wires are used, one for lowering and one for closing and pulling up.

Wedge-Shaped Box Corer

The Klovan-Imbrie box corer has been modified at the Hudson Laboratories of Columbia University (Sanders, 1966). The sampler consists of three parts: sample box, handle, and closing plate.

The sample box is wedge-shaped. Its dimensions are 14 in. high, 11 ¹⁵/₁₆ in. wide, and 6 in. deep at the top. It is closed on three sides and on top, leaving one side open. The horizontal top plate is ⅜ in. thick, so that it can resist any pounding which may be necessary to move the box into the sediment. This top plate bolts to the small angles that support the edges where the vertical sides join the top of the box. A number of 3 in. diameter holes are drilled in the top plate to allow pouring of plaster of Paris or other ma-

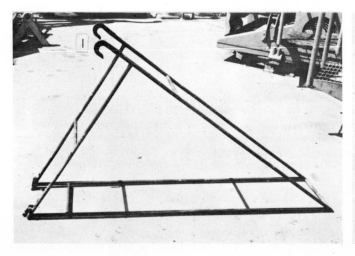

Fig. 6.30. Lightweight lever made from 1 in. tubing. The flat hooks (1) just fit along the box housing. The small hooks (2) at the base (see also detail) can turn around freely. The end side of the box cover fits in the little slits. The cover is not supported by the lever but serves only to facilitate its positioning.

Fig. 6.31. Four-wheel dolly for handling the large box sampler boxes. The wheels can turn 360° but can be locked when in line with the dolly, and each wheel has a brake. A hydraulic jack pushes the short side of the lever down. To the other side of the lever an L-shaped frame of angle iron is attached in which the box fits. The upper side of the box is clamped to it with a vise-grip.

terial on top of the sample to stabilize the surface before moving the sample. A short 1¼ in. pipe is welded to the top plate to which a removable handle can be attached. A ⅜ in. hole through the handle and the 1¼ in. pipe will accommodate a bolt.

The two vertical triangular sides as well as the other box side and the closing plate are made from No. 16 gauge stainless steel plate. The closing plate makes an angle of 67° with the top plate. The closing plate is 15¾ in. long, reinforced along its length by two strips of angle iron spot-welded to it, and strengthened on its top by welding on two end strips ¼ x ½ x 13⅛ in. A handle of ⅜ in. rod is welded to the top. A curved guide strip is spot-welded onto each edge of the closing plate. Each guide fits around a rod welded to the inclined side of each vertical triangular box side.

Long Box Corer (Kastenlot)

Kögler (1963) demonstrates (see Table VI.1) the advantage of the box corer over the pipe corers with regard to sample size, amount of sediment to be pushed away during penetration, and total weight required.

The long box corer consists of three main parts (Fig. 6.32):

1. Weightstand. A square tubing with a valve inside and a hook on top has a weightstand welded to its bottom part. Up to 14 lead weights, each about 50 kg, can be placed on this stand.

2. Boxes. Four different lengths are applied: 1, 1.5, 2, and 4 m. The boxes consist of two halves and are held together by countersunk screws. The distance between the screws is 5 cm. To prevent leakage a rubber strip (1 x 0.3 cm) is placed in between (Fig. 6.32a). The inner side of the box is coated with PVC. The wall thickness of the box is 1.5 mm.

3. Core nose with catcher (Fig. 6.32b). The nose-catcher combination is square in section and operates on the principle given in Section 6.5B (Udinzew et al., 1956). Two plates are held against the inner wall of the core catcher during lowering. As soon as pulling up starts they are released from the wall and pushed inside by springs. When they are free from the wall the sediment pushes them down until they close (Fig. 6.32c, d). Both levers which hold the plates in open position have a tendency to turn outward but are held in place by the spring action that tries to close the plates. During penetration, the sediment pressure from below keeps the plates open by applying pressure against the levers. As soon as the corer is pulled up the pressure on the levers reverses, thus releasing the plates. The core nose with catcher is attached to the box by four screws (M 10).

The box is also coupled to the weightstand by four screws (M 12). An outboard rack allows detaching the weight from the box before the box is taken on deck. This is particularly necessary when dealing with the longer boxes.

The Hydrowerkstätten Company in Kiel manufactures a model which has been improved by Kögler. Up to 17 leadweights, each 50 kg, can be placed on the weightstand. The weights

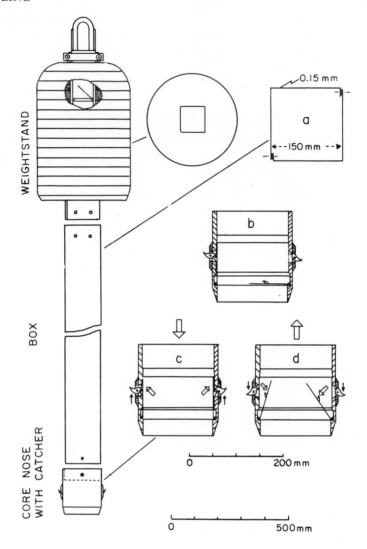

Fig. 6.32. Schematic drawing of the long box corer and details on the closing mechanism (see also Fig. 6.37c). For explanation see text. (Redrawn from Kögler, 1963.)

have in addition to their central hole a hole for housing an orientation-inclination device. The total weightstand is nonmagnetic. A one-way valve is mounted on top. The boxes are 15 x 15 cm, 2.4 or 6 m long. The wall thickness is 0.175 cm. Instead of these boxes, the operator can substitute a steel pipe (OD 10.2 cm, ID 9.5 cm) with PVC liner (OD 9.4 cm, ID 9.0 cm). The pipe may be up to 10 m long. A tripping mechanism is necessary to give the device sufficient free fall (Kögler, written communication, February 1968).

6.3D Coring Lithified Rock on the Sea Floor

I am aware of only two existing devices to collect a core of consolidated rock from the sea bottom. Rosfelder (1967) discusses the possible use of several other types (such as dart corers, bottom-standing drills, rotary corer with hydrostatic feed, and vibrating mole corer). The present author had little success with an instrument (manufactured by Lane Wells Company in the United States) he worked with on the Florida Escarpment and Campeche Shelf in the eastern Gulf of Mexico; no results were obtained in either deep water or shallow water during the trials. The corer was invented in Sweden and has been operated successfully in the Gulf of Bothnia.

The corer consists of two main parts: a flat-topped platform with a strong handle welded to the upper side for cable attachment. A threaded heavy cylinder, in which the firing cap is placed, fits in the center of this platform and the firing mechanism is attached eccentrically. To the lower part a heavy barrel is welded in which the second part—the actual core barrel—fits. This barrel is closed at its upper end. When it is inserted into the heavy barrel, an O-ring maintains a tight fit. A triggering rod, longer than the core barrel, hits bottom first and sets off an explosion which propels the core barrel into the rock. Pull-out is established by two cables attached to the heavy barrel and the core barrel.

Under the direction of Dr. N. Nasu, Professor of Oceanography at Tokyo University, the Tsurumi Precision Instruments Company of Yokohama and the Asahi Chemical Company developed the Free Fall Rocket Corer, which is available in the United States through CM², Inc.

The corer free-falls to the ocean bottom; upon impact a set of thrusters ignite and push the core barrel into the floor. After penetration stops, a second set of thrusters ignites and lifts the coring device out of the bottom. At this point, a gas generator is activated and gas is discharged into a folded balloon, which provides enough positive buoyancy to return the coring device and core to the sea surface.

The corer is available in a variety of sizes and special order units can be made to any size. One standard model is 4 m long, weighs 130 kg, and will take a 2-m long, 64-mm ID core. Additional 1-m sections of core barrel can be added as necessary.

6.4 SAMPLERS FOR USE PRIMARILY ON LAND IN UNCONSOLIDATED SEDIMENTS

Numerous coring devices for collecting "undisturbed" cores in unconsolidated deposits on land have been developed, especially by Quaternary geologists, hydrologists, and soil engineers. New developments are made regularly, as are improvements on existing tools.

The small hand tools which allow coring to depths of a few tenths of meters have been discussed in Section 6.1. An attempt is made here to restrict the discussion to a few devices that can be handled without a trained crew and without the application of rotary drilling or hydraulic pressure. A number of such drilling rigs have been used successfully for coring unconsolidated sediments. The interested reader should consult contributions such as those of Kjellman et al. (1950), Begemann (1961), Van der Sluis and Schaafsma (1963), and Hageman (1963). Core drilling in frozen ground is described well by Hvorslev and Goode (1960). Extensive references are given by all these authors. The soil sampler with metal foils developed at the Royal Swedish Geotechnical Institute (Kjellman et al., 1950) will be mentioned.

6.4A Coring Principles

Kjellman et al. (1950) described some disadvantages of existing soil samplers and some attempts to improve them.

Vertical changes in the properties of soils normally are more rapid than those in any horizontal direction. It is therefore important to try to attain a continuous sampling program without missing parts of the vertical section between two lowerings. The authors distinguish the following types of sampling:

1. Intermittent sampling. "This means that representative samples are taken from the main layers of the ground according to the results of a previous sounding."

2. Nearly continuous sampling. Theoretically, each following sample can be collected immediately below the previous one. Hvorslev called this method continuous sampling, which is not correct since top and bottom of each sample are disturbed. Bottom parts may be missing due to fallout, and the top of each sample normally consists of some material that fell down from the wall of the hole or from the previous sample. In order to get optimum results, Hvorslev (1948) defines the "safe length of sample." This length (L_s) depends upon the diameter (D_s) of the sample, the character of the soil, the depth below the surface, and the design and operation of the sampler. For cohesionless soils $L_s = (5$ to $10)D_s$, and for cohesive soils $L_s = (10$ to $20)D_s$. Swelling of soil in the upper and lower ends of a core as well as the torsion or tension involved

in separating the sample from the subsoil may cause severe disturbance.

3. Nearly continuous undisturbed sampling. The preceding procedure can be improved by drilling two holes as close together as possible without risking mutual disturbance. Samples are taken alternately in one hole and the other so that the undisturbed parts of the core from both holes form a more-or-less continuous core.

4. Continuous undisturbed sampling. When the two holes, as mentioned above, can be very close together and the undisturbed parts overlap each other, a "continuous" coring result may be obtained.

With increase in length of sample entering the sampler, the sliding resistance increases, causing an increasing excess pressure in the mouth of the coring device. When the "safe length of sample" is reached, the excess pressure is so high that part of the sediment under the nose bit of the sampler is pushed aside instead of being caught by the sampler. The sliding resistance can be decreased considerably by using a stationary, tight-fitting piston (see 6.3B discussion of Piston Corers).

6.4B Principle of the Foil Sampler

To eliminate the sliding resistance between the core and the inside of the barrel, the core can be insulated from the barrel by means of a number of thin axial metal strips or foils (Fig. 6.33A). One end of each foil strip is attached to a piston, which is placed inside the barrel immediately above the core head (Kjellman et al., 1950). The piston fits easily inside the core pipe and is kept on a constant level while the barrel penetrates. There is no sliding between the sediment core and foil strips. If, for some reason, the core should tend to move up or down, this movement is immediately prevented by friction and adhesion between the foils and the core. The recovery ratio thus is 100%. (The present author has compared cores collected by the Swedish foil sampler with cores obtained by a piston corer. Radiographs from both types clearly revealed the difference in disturbance along the sides. The Swedish foil sampler cores showed very little down-bending of sediment near the sides of the barrel. The other core exhibited considerable marginal disturbance and provided much less material that could be used for analyses.)

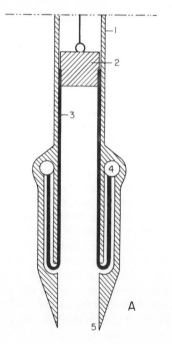

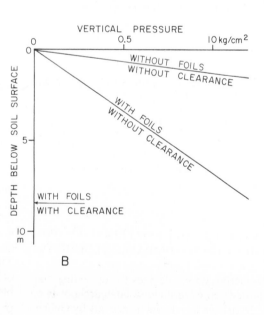

Fig. 6.33. Sketch showing the principle (A) of the Swedish foil sampler, and the influence (B) of foils and clearance on the vertical pressure in a clay core (diameter 6 cm, unit weight 1.5 ton/m³, adhesion 0.01 kg/cm²): (1) core barrel; (2) piston; (3) foil; (4) foil storage; (5) cutting edge of core nose. (Redrawn from Kjellman et al., 1950.)

Initially the foil is stored in coils (similar to steel tape measures) mounted outside the core barrel. At the bottom of each magazine a slot is made in the inner wall of the barrel through which the foil passes. Kjellman et al. (1950) discuss several different foil storage and feeding mechanisms.

If a tube is inserted into clay, the downward shearing stresses are transmitted by adhesion from the barrel to the core. The vertical pressure in the core increases and on each level it will be greater than the vertical pressure in the soil outside the barrel. This excess pressure in the core increases from zero at the top of the core to a maximum at its lower end (Fig. 6.33B). This high excess pressure would cause the core to push the underlying sediment aside, and the core itself would move downward.

The core is protected from shearing stresses if foil strips are used. The vertical pressure will be equal inside the core and outside the barrel (Fig. 6.33B). The core cannot push the underlying sediment aside and move downward.

Horizontal pressure is exerted by the core, which causes friction between the barrel and the foil strips. This friction produces a pulling force in the strips and the risk of breaking them is great when dealing with long cores. To eliminate this risk the corer entrance is made slightly smaller than the internal diameter of the barrel. When entering the barrel, the core tends to simultaneously expand horizontally, contract vertically, and to move downward. The core thus hangs on the foils by adhesion, and both the vertical and the horizontal pressure will be smaller inside the barrel than outside. If the clearance in the barrel is great enough, the core will completely hang on the foils, and the vertical pressure in the core will be zero (Fig. 6.33B). Consequently, the horizontal pressure in the core and the friction between the barrel and core will also become zero.

The pulling force in the foil strips can be reduced by lubricating the outer side of the strips (Kjellman et al., 1950).

6.4C Wit's Coring Apparatus

Wit (1960, 1961, 1962) developed a sampler for collecting cores in shallow and moderately deep bore holes. The need for undisturbed samples was felt particularly in the hydrogeological studies carried out in the Delta region in the SW Netherlands. Most Quaternary and Tertiary formations are sands, gravel-bearing sands, and clayey sands. The coring apparatus discussed here can be used for coarse sand and fine gravel when core bit A is used, as well as for fine sand, clayey sand, clay, and peat when core bit B is taken. Good results also have been published by De Ridder (1960).

Description of the Coring Device

The apparatus is illustrated in Fig. 6.34. In the left-hand side of this figure, the device is shown with cutting shoe A, which can be closed off from the inside. The right-hand drawing presents the corer with cutting shoe B. In this part of the figure the top of the core is turned 90° and the tubes in the sampler head are left out.

A rubber air hose (1) runs from a vacuum-pressure pump to the sampler head. The air line (7) is connected at one side to the air hose and at the other side to cutting shoe A. At the point indicated by (7) in the figure, the air tube can be disconnected and closed when cutting shoe A is exchanged for cutting shoe B. A nylon cloth (2) filters the air before it enters the air line (7).

The sampler is driven into the sediment by the sliding steel weight (3), which is attached to the cable. During penetration, the displaced water flows off through two outflow openings (5) into a rubber tire, which is placed in a small chamber (4) with a perforated bottom. When the rubber tire is inflated, it closes the outflow openings. The tire has the advantage that it closes well even when sand and silt enter the chamber.

The outer barrel (8) is connected to the sampler head by means of a swivel with a rubber washer for airtight closure. The total length is 60 cm, its lower 312 mm is reamed out to give space for inserting a sampling tube. Each sampling tube (9) is made out of zinc, 1 mm thick. The length is 30 cm and the internal diameter is 62 mm.

Cutting shoe B (10) has a bayonet catch and a bolt to secure it to the core barrel. When the outflow holes in the sampler head are closed, the sample adheres to the sampling tube. Whenever coring is done above the water table, the lowest part of the bore hole must be filled with water before the sampling device is brought in to prevent air coming above the sample, which would destroy the effect of closing the sampler head and possibly cause the sample to fall out.

Cutting shoe A (12) is connected to the air line (7). Before coring is started, air is removed by a vacuum pump to force the rubber ring (11) flat against the inside of the cutting shoe. Underpressure is maintained during coring. As soon as the desired coring depth is reached, air is pumped into the line. The outflow openings be-

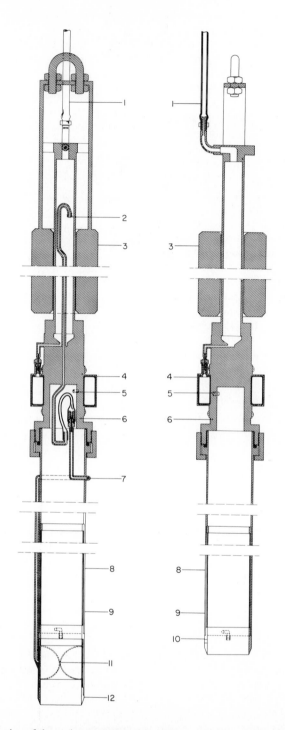

Fig. 6.34. Schematic drawing of the coring apparatus according to Wit: (1) rubber air hose; (2) nylon cloth; (3) steel weight; (4) rubber tire; (5) outflow opening; (6) sampler head; (7) air tube; (8) outer barrel; (9) sampling tube; (10) cutting shoe B; (11) rubber ring in cutting shoe A (12). (Courtesy Institute for Land and Water Management Res., Wageningen, The Netherlands; redrawn from Wit, 1960, 1961, 1962.)

come closed and the rubber ring (11) completely closes the core barrel. The maximum permissible pressure of the rubber tire in the housing (4) is 3 atm over and above the atmospheric pressure plus the pressure of the water column in the bore hole. Two atmospheres of overpressure is allowed in the rubber ring (11). An overpressure of about 1 atm will completely close the outflow openings as well as the lower side of the core barrel.

The coring device has been used successfully to depths of 50 m. Deeper holes have not been tried. There are some disadvantages of this coring apparatus.

"Clean bailing of the borehole produces some difficulties when strongly flowing sands, among other things, are present. As a result of this there is a possibility that the apparatus will remain stuck in the borehole. Most satisfactory results were achieved by taking cores in combination with cased flush borings. Before the sample is taken, the borehole should be flushed out until clear water appears at the surface" (Wit, 1962).

6.5 SOME ADDITIONAL TOOLS FOR SAMPLING

It is difficult to separate certain tools or parts from the samplers or sampling methods just discussed. However, certain items, especially those which are not restricted to a particular instrument or method, will be treated separately here. Core catchers, piston immobilizers, and orientation tools are the main subdivisions. The last group includes instruments to be attached to shipboard-operated samplers, or for collecting unconsolidated samples on land and in very shallow water, as well as those employed for sampling ancient sediments.

6.5A Piston Immobilizers

Fleming and Calvert (not dated) describe a small instrument that can be mounted to the head of the tripping arm of a piston corer. It releases the piston cable after the corer has penetrated the sediment. Immobilization of the piston prevents additional disturbance and flowing in of sediment during pull-out (Bouma and Boerma, 1968). This instrument does not have any waterdepth limitation, as may be the case with the method of piston immobilization described by Emery and Broussard (1954). These authors

release the piston wire from the tripping arm by a heavy messenger, which is very time consuming in deep water and impractical due to the flexibility and slack on the pay-out wire. Kermabon et al. (1966) describe a "split" piston, which has the same effect as the piston immobilizer. PVC tubes are used as core liners. Kermabon and Cortis (in press) developed a simple system of cables and pulleys for controlling the position of the piston relative to a fixed platform lying on the sediment.

Piston Immobilizer of Fleming and Calvert

The piston immobilizer consists of a cylinder, a spring-loaded piston, a piston rod, and a metered orifice (Fig. 6.35). The piston (2) is provided with an O-ring (6) to eliminate leakage. The cylinder (1) is open on only one end. The closed end has a metered orifice (5), the diameter of which depends on the required travel time of the piston.

The piston (2) and the cylinder (1) have a clearance of 0.003 in. The spring (3) is a valve spring from a G.M.C. 71 Series diesel engine. It has a ¾-in. travel, of which ⅝ is used. The clearance of the piston rod (4) in the guide blocks is 0.062 in. The piston pin is a ³/₁₆-in. stainless steel "roll pin."

The piston immobilizer is bolted to the head of the tripping arm. The loose end of the piston wire, with a thimble attached, is inserted between the guide blocks. "The cocking bolt is inserted through the thimble into the threaded screw (8), and the piston is cocked by taking up on the bolt until the lock-pin (7) can be inserted into the piston rod. The cocking bolt is then slackened until the lock-pin rests against the plate; the cocking bolt is then removed completely. The slack from the piston wire and the winch wire is taped along the barrel and weight-stand. A further length of wire is attached to the lock-pin and the bail of the weight-stand. This wire must be shorter than the amount of desired free-fall. ... This wire triggers the piston by pulling the lock-pin from the piston rod when the core trips and free-falls. The piston then travels through the guide blocks and releases the piston wire; by this time the corer will have penetrated fully. Piston travel-time can be regulated by changing the size of the metered orifice, and it is felt that 10 to 15 seconds is sufficient time for the corer to have penetrated fully, from the time of tripping, using a 20-foot barrel and 300

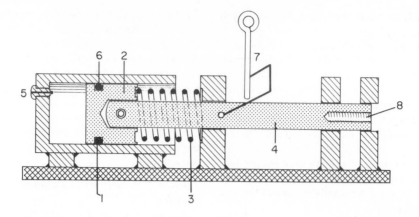

Fig. 6.35. Scheme of the piston immobilizer according to Fleming and Calvert: (1) cylinder; (2) piston; (3) valve spring; (4) piston rod; (5) metered orifice; (6) O-ring; (7) lock pin; (8) threaded hole for cocking screw. (Redrawn from Fleming and Calvert, not dated.)

pounds weight" (Fleming and Calvert, not dated).

Fleming and Calvert claim that this prototype needs some modifications:

"1. Larger diameter cylinder to give a better piston travel time range.

"2. A modified spring retention, to shorten the overall length of the unit.

"3. A piston-wire release of the lever type, which requires no guide blocks. On occasions, the piston wire was bound in loops on the guide-blocks.

"4. A more efficient method of cocking the piston."

Piston Immobilizer According to Kermabon, Blavier, Cortis and Delauze

Kermabon et al. (1966) describe a "split" piston, which becomes freed from the pulling line after a precalculated time (Fig. 6.36).

A hollow piston (1) has a cylindrical hole in its upper part in which another piston (2) can slide. This inner piston is connected to the pulling line. It has a small calibrated hole (3) with a diameter of 2 mm, through which water can move from chamber (4) to chamber (5). Chamber (4) is in open connection to the surrounding water. During penetration of the corer, a downward force acts on piston (1), while piston (2) is motionless. The pins (6) are then sheared and piston (2) starts to move up. This movement is slow during the short time of the penetration.

Pull-out should start very carefully, the operator applying 100 to 200 kg pulling force to allow piston (2) to become disengaged completely from the hollow piston.

6.5B Core Catchers

Normally the operator will insert in the cutting head of his piston, punch, vibration of jet corer, a device to prevent loss of sediment during pull-out and emergence from the water. The most common core retainers are bent metal strips, but the one-valve, double-valve, "sphincter" core catchers and tubular spring valve are better.

Rosfelder (1966a) lists five conditions required for a good core-retainer: "(1) the core-retainer should act efficiently against the extraction suction and be closed as soon as the pull-up begins; (2) the retainer should be kept tight all the way up to the deck to protect the core from any washing; (3) it should be as thin-walled as possible, avoiding any change of the optimum area ratio of the core nose; (4) it should not protrude inside the barrel and scratch, disturb, or exert excessive pressure on the sediment during penetration; and finally, (5) it should preferably be built into the core nose in order to be easily removable and adapted to any barrel length."

Metal Strips as Core Catcher

Copper or brass strips usually form the core catcher. The strips are cut from a single piece of metal, with all strips remaining attached at the lower part of the metal piece (Fig. 6.37). This piece is placed inside a metal ring and fastened, and the strips are bent inward to give them a domelike shape.

The material used must be flexible enough to be pushed aside when sediment penetrates the barrel, and their domelike shape should bring the strips together as soon as pull-out begins.

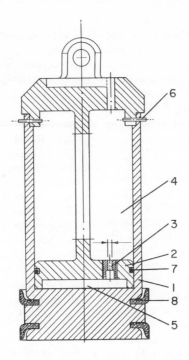

Fig. 6.36. The "split" piston according to Kermabon, Blavier, Cortis, and Delauze: (1) hollow piston; (2) inner piston; (3) calibrated hole; (4) outflow chamber; (5) inflow chamber; (6) pins; (7) O-ring; (8) leather collar. (Redrawn from Kermabon et al., 1966.)

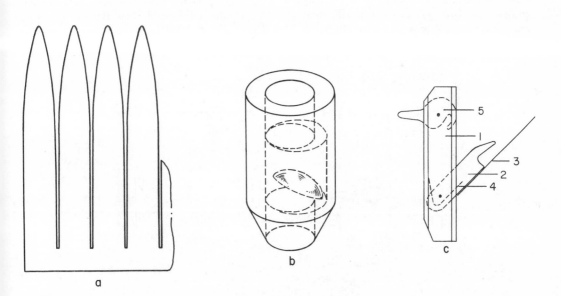

Fig. 6.37. Some types of core retainers. (a) Layout of the finger springs core retainer, made from thin copper, brass, or other metal. The strip with fingers is soldered inside a metal ring, which fits the core bit. (b) One-valve core catcher, consisting of a convex valve which fits inside the wall of the core bit during penetration. A spring pushes the valve out of its housing as soon as penetration stops. (c) Double-valve core catcher (one half): (1) wall of the core bit; (2) blade strengthening; (3) phosphor-bronze blade; (4) spring; (5) trigger (Courtesy E. Seibold, Geol. Inst. Kiel, Germany.)

This type of core catcher, however, is not ideal. The strips are often too strong to be completely pushed aside when the upper soft sediment penetrates, and they often are too weak to close the barrel when the lower sediment is stiff. As occasionally happens, the strips are pushed through the closing position and either hang down or form a "buttoned" closure.

One-Valve Core Catcher

This type of core catcher fits inside the core nose to allow the sediment to penetrate normally. Its lower part is attached to the core nose by a hinge. A spring gives it a tendency to be pushed out of its housing, and as soon as the upper part is freed the sediment above it pushes the valve down. Since it is hinged it turns and closes the barrel. In a closed position it has an angle of about 30° with its housing (Fig. 6.37b).

This valve type works well when it is cleaned of sediment particles and properly oiled after each use. The main disadvantage is that due to its thickness (about 5 mm), a rather thick total wall thickness of the core nose is required.

Double-Valve Core Catcher

The Geological Institute of the University of Kiel (Dr. E. Seibold) made an improved version of a Russian core catcher (Fig. 6.37c), which consists of two half-cylindrical, thin-walled, phosphorbronze plates which close the barrel.

Two opposing slits are drilled in the wall of the core nose to house a trigger and a narrow blade support. The half-cylindrical blade is 0.4 mm thick and is attached to the support by screws. The blade is flexible enough to fit snugly to the side of core nose. A spring pushes the support with blade around its hinge inward, thus bringing the blade into a horizontal position. A rim keeps the blades from being pushed down further. The trigger keeps the blade against the core nose wall by keeping the support inside its housing. A small trigger arm protrudes outside the barrel. During penetration of the sediment, this trigger arm is pushed up, releasing the blade.

The advantage is that the blades are not released as soon as the soft upper sediment is penetrated due to the small surface of the trigger arm. The disadvantage is the total wall thickness of the core nose.

The Nylon Sleeve "Sphincter" Core Catcher

Kermabon et al. (1966) describe a core catcher which closes the barrel like a diaphragm of a photo-apparatus. It is a nylon sleeve core catcher developed by Delauze for core diameters of 200 mm, and it has been adapted by Kermabon et al. for 120-mm diameter cores.

A sleeve of thin nylon fabric is held at each end by the solid metal rings R1 and R2 (Fig. 6.38a). The height (H) of the sleeve is equal to

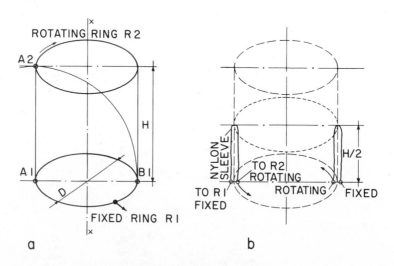

Fig. 6.38. Principle of the nylon sleeve "Sphincter" core catcher. For explanation see text. (a) Original construction. (b) Modification. Both rings are mounted adjacently. (c and d) Explanation of the operation of the core catcher. (c) Driving mechanism of the "Sphincter." (d) Nose bit of the core catcher. Only the three lower parts are presented here: (1) 2-mm cable to close the nylon sleeve; (2) sliding support in the driving mechanism to which the 2-mm cable is attached; (3) piston line; (4) rotative ring; (5) support; (6) nylon sleeve; (7) fixed ring; (8) teeth; (9) catcher. (Courtesy A. Kermabon, North Atlantic Treaty Organization, Saclant ASW Research Centre; redrawn from Kermabon et al., 1966.)

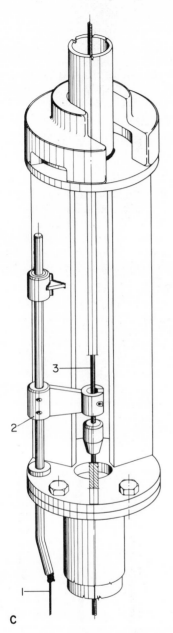

c

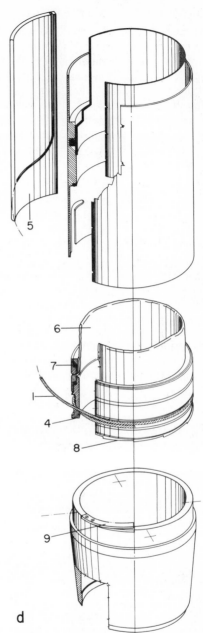

d

the internal diameter (D). The lower ring is fixed in position, the upper ring (R2) can rotate about axis xx' and be moved downward in such a way that point A2 of the generatix A1A2 will remain on the imaginary sphere with the center A1 and radius A1A2. A half-revolution of ring R2 will bring point A1 to B1 and all the points of R2 on the circumference of ring R1. Consequently, the internal area of the rings will then be completely closed by a twisted diaphragm of nylon.

For technical reasons a modification has been made (Fig. 6.38b). Both rings are mounted with no separation; the nylon sleeve is turned back as a cuff over half the distance H. It takes less space than the original but provides the same result.

In Figs. 6.38c and d the working of the core catcher is explained. The force necessary to drive the rotative ring R2 is provided by the final stroke of the piston which pulls on a 2-mm cable (1). This cable runs inside a tube mounted along the core barrel. The sliding support (2) to which this cable is attached moves with the piston line (3) over the last 35 cm of its total upward movement. The force obtained is transmitted to the rotative ring (4) through the support (5). The end of the cable (1) is wound around the ring (4) and then clamped to it. The nylon sleeve (6) and both rings (4, 7) are mounted in the core nose above the nose bit. The fixed ring (7) is fastened by a screw and a waterproof seal. Due to teeth (8) on the catcher (9) the rotating ring cannot slide backward once it is closed.

The Tubular Spring Valve

Rosfelder (1966a) made the cloth valve more readily applicable by simplifying it. The closing wire as well as any external protruding sources of energy could be eliminated by putting the cloth valve inside a torsion spring. The valve is open when the torsion spring is under tension. Tests indicated that the closing action is not a quick motion but a progressive "strangulation" of the sediment column.

"The spring valve core-retainer[Fig. 6.39a,b] consists basically of a torsion spring with two rings at its extremities and a nylon tube placed inside the spring and attached to the rings. This cloth cylinder is made of nylon 0.1 mm thick and is firmly held by the springs, which are built in two inserting parts for that purpose. The spring is designed in such a way that its inner diameter, when expanded under tension, equals the inner diameter of the barrel plus the thickness of the cloth. One ring is permanently attached to the core nose, the other, which can rotate freely is secured by a pin. The pin is merely a screw with a high angle thread that is attached to a flat leaf on the outside of the core nose, which releases the free ring when it is turned down by the pressure of the sediment during pull-out. The core-retainer is cocked when the spring is held under tension by the pin, and when released, the cloth cylinder is twisted, securing the sample" (Rosfelder, 1966a).

A slightly different design was made for free corers; here the core catcher is attached directly to the inner liner. No external latch is present. The catcher is released and closed when the inner liner is disconnected from the expendable outer barrel and its ascent begins (Fig. 6.39c, d).

According to Kermabon (personal communi-

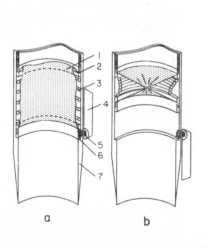
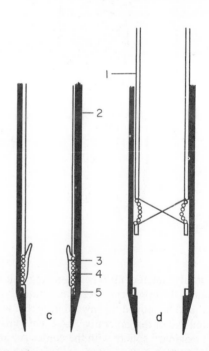

Fig. 6.39. Two types of tubular spring valve core retainers. (a) Open position. (b) Closed position. This type can be used in any coring device which is lowered on a cable: (1) cloth; (2) fixed ring; (3) torsion spring; (4) lever; (5) latch; (6) rotating ring; (7) cutter. (c) Open position. (d) Closed position. This type has been designed for free-fall corers: (1) liner; (2) barrel; (3) cloth, (4) spring; (5) teeth. (Courtesy A. M. Rosfelder; redrawn from Rosfelder, 1966a.)

cation, January 1968) sphincter core catchers and similar designs are difficult to close in sand layers. This is why he had to use the retrieving pull of the corer as a source of energy instead of the rubber bands or springs used initially.

6.5C Orientation and Inclination Instruments

Since some sedimentary structures can be used to reveal former current directions, it is important to collect oriented samples from consolidated as well as from unconsolidated deposits. It further is important to measure dip directions from mega and macro foresets.

Ancient sediments often are moved from their original position by tectonics and the amount of rotation and tilting must be compensated when pretectonic current directions are sought. When extraneous local magnetic fields are present a solar compass can be used.

Figure 6.40 presents an outline for the different instruments discussed in this section.

Tilt-Compensating Instruments

When an investigator wants to obtain paleocurrent directions from sedimentary structures present in a folded series, he can measure strike and dip of the bed, the angle between strike and the direction revealed by sedimentary structures, and the axial pitch of the structure. With the help of a stereographic projection the pretectonic current direction can be established. However, many measurements are necessary. It saves time when the rotating of the tilted direction back to its pretectonic position can be carried out in the field and when the desired reading can be obtained directly.

The most simple, but not the most accurate method is to use a sturdy notebook, a pencil, and a geological compass. The book is placed on the bedding plane with its back parallel to the strike. A pencil is placed on the book and turned parallel to the sedimentary structure. The combination book-pencil is then turned back about the strike to a horizontal position. The direction of the pencil can be measured with the compass.

When an oriented sample is to be marked, for example, with a North direction, book and pencil are placed in a horizontal position with the back of the book along the strike and the pencil oriented to North. Next the book-pencil combination is turned to the bedding plane and the pencil direction noted on the rock after which a sample can be collected.

These measurements are very crude since three loose items are involved. When done carefully, decent results can be obtained. Two rubber bands around the fieldbook to keep the pencil in place will increase the accuracy. Influence of dip and axial pitch are discussed later.

These measurements can be carried out on a lower, upper, as well as inner bedding plane if they are parallel to the main bedding of the series, or directions within a layer can be moved parallel to a bedding plane and then rotated.

Ten Haaf (1959) developed a tilt-compensation compass and from this instrument the author derived a tilt compensator (Bouma, 1962), and later another tilt-compensating compass (Bouma et al., 1965). Armstrong (1967) described an instrument for measuring planes and vectors in space. Parizak (1967) modified an instrument for measuring strike and dip.

Tilt Compensator. This instrument replaces the previously mentioned fieldbook and pencil. However, a geological compass is still needed. The tilt compensator is like a pair of compasses, made of nonmagnetic metal. Each leg is 150 x 12 x 8 mm. Two spirit levels are mounted in one leg facing in opposite directions, thus enabling the investigator to use either side of this leg to face upward (Bouma, 1962). The spirit level-containing leg is placed along the strike of the bedding plane, and the instrument is turned about this leg until it is parallel to the bedding plane. The other leg is turned parallel to the sedimentary structure, leaving the first leg along the strike (Fig. 6.41a). The instrument is then turned back to a horizontal position (Fig. 6.41b) and the direction is measured with a geological compass.

To collect oriented samples, the position of Fig. 6.41b is started with the solid leg pointing North. The instrument is then turned about the strike and the reading is taken on the bedding plane.

Tilt-Compensation Compass after Ten Haaf. The instrument is made out of brass and consists of a base plate, straight on one side and round on the other side (Ten Haaf, 1959, Fig. 51; Bouma, 1962, Fig. 4d). A raised edge is perpendicular to the straight side and parallel to the edge. The angle thus formed contains a housing with a spirit level. To the round side a small nautical compass is attached so that it can turn freely over 360°. To the upper side of the compass housing, a hinged sighting-frame (alidade) is mounted; to its base is a pointer parallel to the

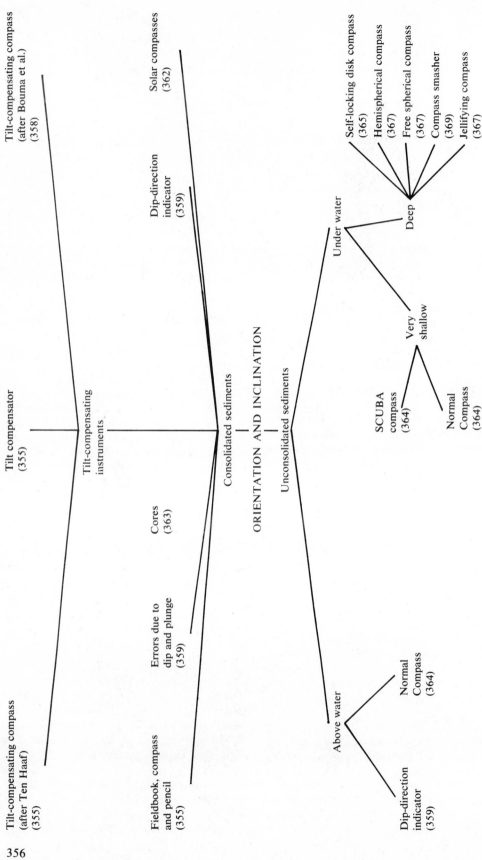

Fig. 6.40. Outline for the different types of instruments (all discussed in Section 6.5C) to obtain orientation and/or inclination of samples from consolidated as well as from unconsolidated sediments.

356

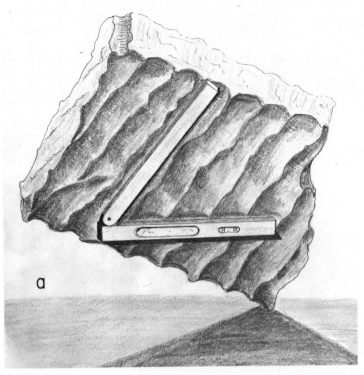

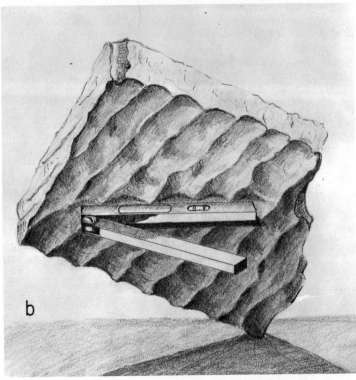

Fig. 6.41. The tilt-compensator and its application. (a) The spirit level-containing leg is parallel to the strike of the bedding plane, the other leg is parallel to the desired structure. (b) the instrument is turned back about the strike to horizontal position. The solid leg now indicates the pre-tectonic direction of the troughs of the convex wave ripples.

alidade. A silvered, central portion of the compass cover glass forms a mirror for the alidade and allows one to carry out measurements on bedding planes which are too high above the investigator for *in situ* work. The instrument can also be used as a normal geological compass.

The raised edge is placed along the strike and the instrument is tilted until it lies against the bedding plane. The pointer is then turned parallel to the desired direction. After turning the instrument back to a horizontal position, the reading is given directly by the compass. Only one instrument is needed, which facilitates the work and avoids mistakes that may occur when different loose parts, which easily move out of position during tilting, are used.

Tilt-Compensating Compass after Bouma, Mutti, and Maarschalkerweerd. The authors were not able to duplicate the invention of Ten Haaf since no similar nautical compass could be found. Successful efforts were obtained by using a Finnish type of geological compass manufactured by Suunto Oy, which consists of an aluminum frame (Bouma et al., 1965). The lower part of the frame is rectangular with dimensions of 52 x 55 mm and the upper part is round with a diameter of 50 mm. The edges of the rectangular frame and of its lower protection window were rounded until a diameter of 60 mm had been obtained (Fig. 6.42a). A ring of plastic with nylon reinforcement (11) was made with an outside diameter of 70 mm, an inside diameter of 50 mm and a height of 11 mm. The diameter of the

lower 7 mm of the ring was enlarged to 60 mm, so that it fits over the rounded lower part of the compass. A brass ring (10) of 0.3 mm thickness and inside and outside diameters of 40 and 60 mm, respectively, is inserted between the base plate (see below) of the instrument and the lower protection plate of the compass, after which the plastic ring could be mounted with four screws to the base plate.

The base plate and raised edge are made of plexiglas. The spirit level is imbedded with clear plastic in a plexiglas bar, through which a long hole (22 x 7 mm) is drilled and polished. The procedure of imbedding the elongated, slightly curved spirit level has to be carried out very precisely. The bar with polished hole is placed on an absolutely horizontal table, which is given a thin oil cover (or mold release agent). A 2-mm base is poured, and as soon as the mixture starts gelling the spirit level is placed in the proper position and temporarily held there by modeling clay. When the base layer has hardened, the spirit level is mounted at the same time and the hole can be filled in one pouring. After hardening, the top of the bar is polished.

Base plate (1), raised edge (2), and bar (3) are mounted together with plastic or with a special glue for plexiglas.

In the upper part of the frame (6) a threaded hole is drilled in which the pointer (12) can be screwed. The pointer is 55 mm long and 5 mm in diameter. In the ring (11) a mark (13) is made parallel to the raised edge to facilitate a turn of 0 to 180° parallel to this raised edge. In this way

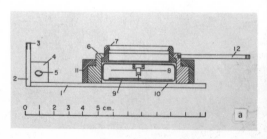 

Fig. 6.42. Section (a) through and photograph (b) of the tilt-compensating compass: (1) plexiglas bottom plate; (2) raised edge made from plexiglas; (3) protecting bar; (4) spirit level housing; (5) spirit level; (6) compass frame; (7) aluminum direction ring; (8) liquid filled chamber containing the magnetic needle; (9) clinometer; (10) brass ring; (11) ring to mount compass to base plate; (12) pointer; (13) mark. (Redrawn from Bouma et al., 1965.)

the instrument can be used as a normal compass, and the clinometer (9) inside can also be used.

This tilt-compensating compass is similar to the one of Ten Haaf; it lacks only the mirror, and it is much lighter in weight.

Since the upper rim of the raised edge is placed along the strike, when dealing with lower bedding planes, it is protected by a U-shaped brass bar (3). Mutti (written communication, March 1966) replaced this metal bar by a hard rubber bar to create more friction in order to hold its position when the instrument is turned about the strike. The pointer can be turned 360° (Fig. 6.42b). The pointer facilitates lining up with the direction obtained from sedimentary structures.

Errors due to Dip and Plunge

Ten Haaf (1959) described the errors obtained by projecting a direction vertically instead of swinging it back around a horizontal line and by neglecting the axial pitch of a sediment series. His two graphs are given in Fig. 6.43 (see also Potter and Pettijohn, 1963, p. 261).

When current directions are obtained from sole markings, which never are perfectly aligned, dips up to 25° can often be disregarded since they cause a maximum error of less than 3°. As long as the strike makes a very small angle with the alignment of the current-indicating structures, even steeper dips can be ignored.

When a tilt-compensating compass is used, the plane of the instrument needs to be kept only roughly parallel to the bedding when the pointer is set, or horizontal when the needle position is read. With steeply dipping beds, however, a considerable error can be introduced in this way when the edge of the instrument does not coincide with the strike (Fig. 6.43a).

As can be seen easily in a stereographic diagram, an error in the reading is introduced when dip has been compensated for but axial pitch has not (Fig. 6.43b). This graph clearly shows that for strata with a dip of less than 45° the error from any amount of unrecognized axial pitch cannot exceed 10°. However, this error is equal to the inclination of the true fold axis for vertical strata. The error may run up "to 180° in the inverted limbs of plunging folds. Even with the local axial pitch as slight as 10° — unlikely to be detected in an isolated outcrop — the error becomes considerable if beds are tilted more than 120°" (Ten Haaf, 1959).

"It is now possible to formulate the following rules to be observed in direction mapping:

"1. Dips up to 25° can be neglected altogether.

"2. Where strata are highly inclined or vertical, allowance should be made for any considerable axial pitch.

"3. Measurements on overturned beds are inadmissible, unless the tectonic axis at that particular locality has been ascertained exactly" (Ten Haaf, 1959).

"If a known axial plunge must be reckoned with, it is easiest to measure the current direction in the ordinary way and then apply a correction. The magnitude of this correction only depends on the angle of dip of the bed and the estimated angle of inclination of the tectonic axis, and can be read immediately from the graph of [Fig. 6.43b or else constructed on a stereographic net.] As it is only required for steeply dipping strata, where the azimuth of the axis can make only a small angle with the strike direction, a simple rule gives the sense in which the correction must be applied:

"Positive or clock-wise, when the sole of the bed faces to the right for an observer looking in the down-plunge direction; negative when the sole faces left" (Ten Haaf, 1959, p. 76).

An advantage of using a tilt-compensating compass, together with the graphs of Fig. 6.43, is that many readings can be obtained or oriented samples collected in part of a flank of a series where the dip does not vary, and afterward the same correction can be applied to the readings. The use of these instruments facilitates and speeds the work considerably, while mistakes can be avoided since no separate readings need be made or noted for strike, dip, and angle between strike and direction of the indicative sedimentary structure.

Dip-Direction Indicator

The direction of the dip of cross-bedded foreset beds can be obtained in many ways. If a foreset plane is exposed, one can measure the direction of the strike and the maximum dip. However, there is seldom enough surface exposed to carry out these measurements.

In unconsolidated sediments the investigator often can remove enough material to expose a foreset face. Strike and amount of dip can be measured by placing a field book at the sedimentary face to prevent errors due to damaging the

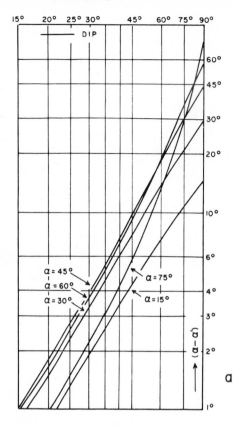

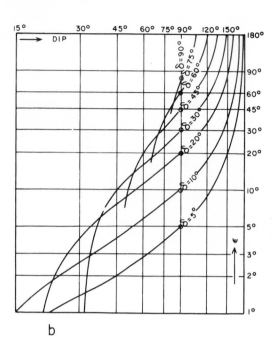

Fig. 6.43. (a) Graph presenting the angular error $\alpha - \alpha'$ which results when the dip of the layer is neglected. α is the true angle between strike and the direction to be measured, α' is the projection of α on a horizontal plane. (b) Graph presenting the angular error ϵ due to neglecting the axial pitch δ. Logarithmic scales. (Redrawn from Ten Haaf, 1959.)

sediment. Glashoff (1935), as well as Berry (1960), describe instruments for measuring cross-bedding and determining planar orientation (Potter and Pettijohn, 1963). Burachek (1933, p. 432) cuts the loose sand with a knife so that the foreset beds trace as horizontal lines on a vertical face, which then represents the strike of the foresets.

Pryor (1958) constructed a simple device (Fig. 6.44A; see also Potter and Pettijohn, 1963, p. 78), called a dip-direction indicator. It consists of a round disk of nonmagnetic metal to which a level is mounted, and a line grooved at right angles to the level. The investigator digs a V-shaped opening into the sediment and inserts the indicator into the sand along two foreset bed lines, which come together at the end of the V-shaped opening. These lines indicate the plane of the foreset bed. The indicator is then turned until the level indicates a horizontal position.

The direction of the level housing represents the strike, the grooved line the dip.

Foreset bedding directions in consolidated sandstones cannot be measured in this way. The investigator must measure apparent dip in two intersecting faces. The azimuth of each face is recorded along with the apparent dip of the line which presents the exposure of the foreset bed. The maximum dip and direction can be determined with a stereographic net (Bucher, 1944, pp. 195, 196). A nomograph has been constructed by Steinlein (1950) to determine the maximum dip direction from two apparent dips. The use of a stereographic net can be avoided by holding a field notebook parallel with the two apparent dips of a foreset and determining the maximum dip of the fieldbook (Potter and Pettijohn, 1963).

The author designed an instrument which is a combination of the dip-direction indicator and the tilt-compensator (Fig. 6.44B). To a rectangular, nonmagnetic base plate with a round level is mounted a housing with a locking swivel joint, as used in tripods for cameras. A screw can fix the position of the locking swivel joint. A pipe is mounted to this joint. In this pipe fits a telescopic leg of a camera tripod. (A rim assures that the tripod leg fits only in one way and cannot turn.) At the other end of this leg another locking swivel joint with fastening screw is mounted. On top of the ball a short, nonmagnetic cylindrical rod is mounted through which a rectangular hole is made. Through this hole a rectangular, nonmagnetic strip can move. Its position can be secured by a fastening screw. The strip is about 30 cm (12 in.) long. On top of this cylindrical piece another, similar short rod is placed. This rod is fastened to the lower one by a locking screw to allow it to turn 360° and to

fix a desired position. A rectangular hole is made through this upper block in an off-centered position so as not to interfere with the locking screw. Through this hole moves another strip, whose position also can be secured. At one end of this strip a dip-direction indicator can be mounted.

The operation is rather simple but not always easy to perform by one person. The base plate is placed in a horizontal position with one side parallel to the strike and then turned parallel to the bedding. The telescopic leg is extended and turned in such a direction that both strips are close to a point where two apparent dip lines cross. The strips are placed parallel to these apparent dip lines. The unit is then turned back about the strike until the base plate obtains a horizontal position. The dip-direction indicator can then be turned until the level indicates a horizontal line which is the pretectonic strike of the

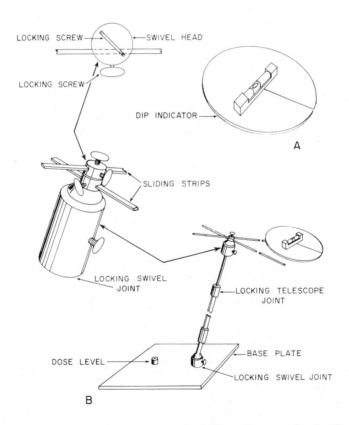

Fig. 6.44. Dip-direction indicators. (A) Pryor's dip-direction indicator for measuring the direction of the dip of foresets in unconsolidated sediments. The round disk is made of nonmagnetic metal. A level is mounted to it and a groove is made at right angles to the level housing. (B) Bouma's dip-direction indicator for measuring the direction of the dip of foresets in consolidated deposits. Two locking swivel joints and one locking telescope joint allow the investigator all movements necessary. The sliding strips make it possible to measure on apparent dips crossing toward the observer as well as away from the observer.

foreset bed plane. The strike direction can be measured, as can the dip.

The movement of the strips through their housings is necessary to allow measurements on rocks where the apparent dip lines cross toward or away from the observer.

Geological Solar Compasses

The magnetic properties of some rocks may cause large and unpredictable aberrations in the readings made by the usual geological compass. This was the reason that led Ten Haaf (Geological Institute at Utrecht, The Netherlands) to develop instruments based on sundials (Ten Haaf and Wensink, 1962). Both types (estival and perennial) will be discussed briefly.

Equatorial solar compasses do not involve any calculation or reduction, for the directions can be read immediately. They do have to be used in combination with a timepiece. One degree on the firmament is equal to 4 minutes of time, and since the accuracy in the field is within a few degrees, a normal watch, set on solar time, will be good enough.

The length of the solar day varies during the year, and a seasonal difference exists between true solar time and mean local solar time. A simple graph (Fig. 6.45) in the compass box can replace astronomical tables for our purpose.

". . . The projection upon the celestial equator of the sun's apparent motion is always the same. The principle is embodied in the well-known garden sun-dial, where the shadow of the axis falls on an equatorial circle with equally-spaced hour divisions. If the gnomon is mounted correctly on the pedestal, with its hour-circle parallel to the equator (that is, if the axis has the azimuth N and an elevation equal to the geographical latitude) it always marks true solar time. But the converse of this relation can also be used: if, maintaining the right elevation, the gnomon is rotated on the pedestal till it marks the correct solar time, it is properly oriented — which means that the axis points due north.

"Thus, the magnet needle of a geological compass can be replaced by a small revolving equatorial sun-dial."

Figure 6.46 presents two types of equatorial solar compass. "For use in a given region, the gnomon's elevation is fixed to equal the geographical latitude. Then all measurements are made by simply rotating it till the shadow marks the correct solar time, and reading at the pointer N. The estival variant is intended for regions where no field work can be done in winter anyway (e.g., Iceland). Its flat time scale will function only as long as the sun is above the equator, but it has the advantage of being more easily made and less bulky than the perennial.

"Working with these equatorial solar compasses is as fast and as accurate as with an ordinary geologist's compass, at least in high and temperate latitudes. Towards the tropics the accuracy diminishes: our 2° maximum error

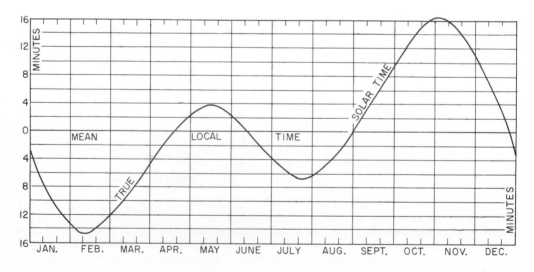

Fig. 6.45. Graph indicating the seasonal difference between solar and mean local time. (Redrawn from Ten Haaf and Wensink, 1962.)

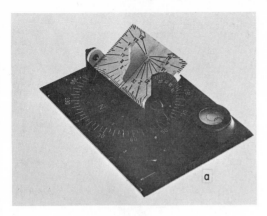
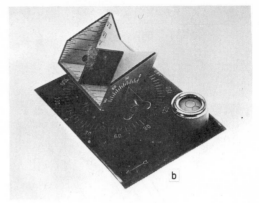

Fig. 6.46. Two types of equatorial solar compasses. (a) Equatorial solar compass (estival). It consists of a rectangular base plate with spirit level and compass scale. Mounted to it is a revolving gnomon with solar time scale. The elevation (latitude) can be adjusted by means of a graduated arc and a locking screw. (b) Equatorial solar compass (perennial). The difference between this type and the estival type is the shape of the time scale which functions as well when the sun stands below the equator. (Photographs courtesy E. Ten Haaf, Geol. Inst. Utrecht, The Netherlands; Ten Haaf and Wensink, 1962.)

would grow to 3° in latitude 40°, and ca. 4° in latitude 30°. Beyond only the horizontal solar compass is practicable" (Ten Haaf and Wensink, 1962).

A horizontal solar compass is very simple: it consists of a metal plate with straight edge, a spirit level, and an engraved, graduated circle with a thin perpendicular rod in its center. A number of reductions are necessary to obtain the desired azimuth.

The estival equatorial solar compass has a straight edged base plate with spirit level and a compass scale around whose center is mounted a revolving gnomom, with solar time scale, which can be adjusted to the proper elevation (latitude) (Fig. 6.46a). The perennial type is a little less bulky. Its only difference is the shape of the time scale which functions also when the sun stands below the equator (Fig. 6.46b).

Orientation of Cores from Consolidated Sediments

It is not only important that cores and other samples from consolidated sediments are oriented, but also that cores from consolidated deposits have an orientation mark. In reservoir studies, orientation of sedimentary structures may be related to fluid flow properties as well as grain orientation and inclination, and without orientation markings, no accurate basin analyses can be given.

Johnson and Kiel (1961) describe "a simple orienting device built for use with a light shot-hole type drilling machine. The tool is simpler and more dependable in operation than commercial orienting tools designed for greater depth."

The orienting tool (Fig. 6.47a) "consists of a 20 feet length of heavy-gauge steel pipe (1) tipped with a 1¾ inch insert-type drilling bit (2). Around the pipe is a 3-foot sleeve (3), slotted at its lower end. On the sleeve is mounted a lever-and-arm mechanism (4). A small projection (5) on the pipe fits into the slot (6), preventing the relative movement of the pipe and sleeve.

"Before coring, the orienter is run into the hole and the sleeve rides on the projection, which is at the top of slot (6). Each piece of drill pipe is marked so that the orientation of the tool is known after it has been lowered into the hole. On reaching the bottom of the hole, the lever-and-arm mechanism, which has been collapsed next to the sleeve, jams the sleeve and pipe to one side and the projection moves to the bottom of the slot (6). The entire drill string is raised 1 inch with the hoist and is then turned slightly by hand until the pipe projection reaches a second slot (7) on the sleeve, allowing the bit to be lowered to the bottom of the hole. The orientation hole is then drilled, the orienter is withdrawn, and the hole is cored with a conventional core barrel. Part of the small hole appears as a groove parallel to the length of the core [Fig. 6.47b].

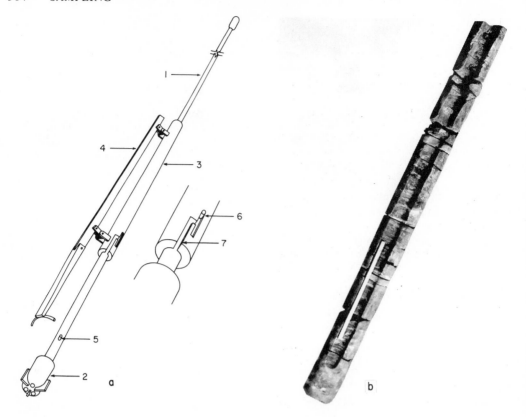

Fig. 6.47. Orienting tool for cores of consolidated sediments. (a) Diagram: (1) steel pipe; (2) drilling bit; (3) sleeve; (4) lever-and-arm mechanism; (5) pin; (6) slot; (7) slot. (b) Offset hole of known orientation visible as a groove. Late Pennsylvanian sandstone from a depth of 250 ft, Oklahoma, U.S.A. (Courtesy C. R. Johnson, Jersey Prod. Res. Co., Tulsa, Oklahoma; Johnson and Kiel, 1961.)

Because the orientation of the tool is known; the direction in which the grooved side of the core was oriented is known. This provides a positive means of orienting all of the core pieces" (Johnson and Kiel, 1961).

Figure 6.47b shows the offset hole in a friable late Pennsylvanian sandstone in central Oklahoma.

Underwater Compasses

At rather shallow depths, when wading or working from a small boat, the operator can orient his samples before sampling as long as he can direct the device. A normal compass then is sufficient.

Inclination and orientation in place by SCUBA divers can be obtained using a simple white plexiglas plate to which an inclinometer and a small compass in a watertight housing filled with oil are mounted (Dill and Shumway, 1954).

The first category of underwater compass is compasses enclosed in a housing to protect them from the outside pressure. They mostly are derived from the housed clinometers currently used in the oil industry (see below). They are very numerous; the first patents on this type of housed compass for high-pressure environment were applied for before 1900 (Rosfelder, written communication, November 1967).

H. E. Reineck of the Senckenberg Institute in Wilhelmshaven, W. Germany, uses a special Leutert clinometer (Leutert, 1954, 1964), an instrument which is used in deep-bored holes. Orientation as well as inclination are recorded photographically on a disk.

Harrison et al. (1967) describe a photographic compass inclinometer. Their device uses a compass photographed by a single shot camera shortly after the sampling device hits the bottom. All components are housed in a deep sea pressure case. The firing is done through a magnetic switch.

A continuously recording compass-clinometer for piston corers for depths up to 1000 m has been developed at the Koninklijke/Shell Exploratie en Produktie Laboratorium in The Netherlands (Plankeel and Van der Sluis, 1960).

"A magnetically oriented opaque cap with a vertical slit to the north balances in an opaque cylindrical box provided with a circumferential horizontal slit. Light from a lamp within the cap passes through the two slits to form a point on photographic paper surrounding the box. The box moves upwards past the photographic paper at a speed of 1 mm per minute and the line traced by the light spot shows the corer's orientation. The cap has a toothed rim. When the instrument is vertical, light can just pass between all the teeth; when it inclines, more light passes between the teeth on one side and none at the other. Spots of light then become line segments. The angle of inclination is estimated from the length of the longest segment, its direction from that segment's position with respect to the north."

Free-Flooding Compasses. Rosfelder and Marshall (1966) list a number of "devices that work at ambient pressure without protective casings: (i) a locking disk having a circumferential line of pins that drop down and lock the compass needle (Aime, 1845) or the compass disk (Bouma, 1964a); (ii) small balls that fall into a partitioned compass disk (Ekman, 1905); (iii) a pointed pendulum that punches a hole in the compass disk (Ekman, 1901; Nansen, 1906); (iv) a compass disk with raised numbers that stamps an aluminum foil (Dahl and Fjeldstadt, 1949); (v) a compass needle with a raised point that punches "north" on an acetate film (Felsher, 1964); (vi) a sliding pivot that pushes and jams the compass against the cover glass (Meyer, 1877, quoted from Richard, 1908); and (vii) a compass secured in position by means of a jellifying compound (Carruthers, 1958)."

Self-Locking Disk Compass. A self-locking disk compass has been developed by the author (Bouma, 1964a) to determine the orientation of samples collected with a box sampler. Since this sampler is assumed to penetrate more or less vertically, no inclination has to be measured. It can be attached to any other device where orientation is the only requested information. The accuracy is 5°.

This compass consists of the following parts: (1) housing; (2) top part; (3) orientation disk; (4) locking disk; (5) holder; and (6) tripping line (Fig. 6.48).

Housing (Fig. 6.48). The housing (C) is made from PVC tubing with a diameter of 15.2 cm (6 in.). The height is 8.9 cm (3.5 in.). The bottom plate (A) consists of the same material and is screwed to the housing with an O-ring placed between. On the center of the bottom plate a tungsten pin is mounted.

Top part (Fig. 6.48E). This part is made from clear plexiglas, has the same diameter as the bottom plate, and is attached to the housing in the same way. Sealed to the top part (E_1a) is a thick plexiglas plate (E_1b) in which there are two holes that correspond to two holes in the housing. Pressure-equalizing tubes (H) fit in these holes. They are secured to the housing by their own O-ring at their open end. The caps (I) fit over them. A hole is drilled through the caps to obtain an open connection from the pressure-equalizing tubes to the surrounding sea water.

A hole is drilled through the center of the top part (E_1a), through which the central pin of the locking disk moves. To prevent leaking, an O-ring is placed around this central pin and secured to the lower side of the top plate by a disk. Another hole is drilled off center through the top part to allow the compass to be filled with oil. A screw with O-ring seals this hole.

A plexiglas ring (E_1c, E_2) is glued to the part E_1b. Three vertical grooves are made in this ring. A pointer is glued to the base of part E_1b.

Orientation disk (Fig. 6.48B). The disk is made of ⅛-in.-thick plate. On the upper side (B_1) an orientation rose is inscribed with intervals of 5°, where two wedge-shaped nonmagnetic pieces are also mounted. To the lower side (B_2) two magnetic needles are attached. At the center of the lower side a sapphire-bearing assembly is mounted. This allows a free rotation on the tungsten pin of the bottom plate. This orientation disk can turn freely within an inclination of 15°.

Locking disk (Fig. 6.48D). The disk is made of an epoxy resin or polyester resin. It contains pins arranged in a circle with 5° intervals. These pins fit over the wedge-shape pieces on top of the orientation disk and lock it in position. Three linen bakelite pieces are mounted to the side of the locking disk. They fit the grooves of ring E_1c. A pin is mounted through the center of the disk.

Assembly procedure. A bottom plate with pin

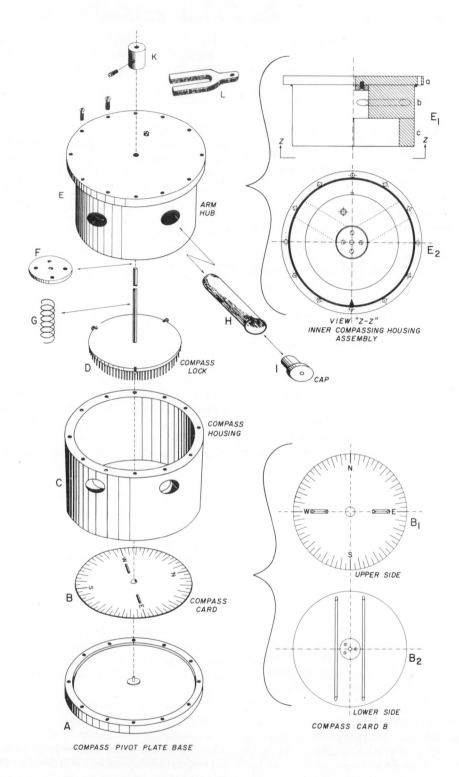

Fig. 6.48. Scheme of the self-locking disk compass. (A) Bottom plate with tungsten pin. (B) Compass card. (C) Compass housing. (D) Compass lock. (E) Top part: (a) top part; (b) plexiglas ring; (c) small plexiglas ring. (F) Guidance plate. (G) Spring. (H) Rubber pressure-equalizing tubes. (I) Cap. (K) Top cap. (L) Wedge. (Redrawn from Bouma, 1964a.)

is attached to the housing. Some clear oil is poured into the housing and the orientation disk is placed. Plate (F) is mounted to the top part, with an O-ring between them. A spring (G) is placed on the locking disk and its pin is moved through the center part of the top plate and then secured by cap (K). The thus assembled top part-locking unit is mounted to the housing and then tubes (H) with caps (I) are placed. The compass is filled with clear oil, which has low compressibility and can withstand the low bottom temperatures.

Mounting to the sampler. Depending on the metals used, a certain distance has to be maintained between sampler and compass to avoid magnetic influences of the sampler. A brass housing with or without protective top strips fits around the assembled compass. Set-screws secure the compass position (Figs. 6.22, 6.23).

Cap (K) is placed so that wedge (L) fits underneath. To this wedge a tripping line is attached. The other end of the tripping line is connected to the arm of the closing mechanism of the sampler.

When wedge (L) is in place, the compass disk can turn freely. As soon as this wedge is removed, the locking disk is pushed downwards by spring (G) and secures the wedges on the orientation disk without touching the disk itself.

The compass proved to operate very well. Since the box sampler at the Scripps Institution of Oceanography was used frequently in submarine canyons, a ball head was placed in the bracket connecting the compass to the sampler and a safety chain was used as additional security.

The self-locking disk compass is now manufactured by N. V. Observator in The Netherlands. A few improvements have been made without changing the basic idea.

Jellifying Compound Containing Compass. The instrument is designed to measure direction and velocity of currents, but it can be used directly to obtain orientation and inclination of lowered instruments. Carruthers (1958) designed the apparatus, which is now manufactured by G. M. Mfg. & Instrument Corp. (not dated e, 1967; J. Kahl, personal communication, November 2, 1967).

The basic design consists of a glass (for depths less than 65 m = 200 ft) or a polycarbonate bottle which is filled with one or two special gelatin solutions setting at either 33°C or 21° C. A powerful ceramic disk magnet, marked at the cardinal compass points, is suspended in such a way that it does not touch the bottle walls (Fig. 6.49). For deep-water work, the plastic bottles are completely filled, having an oil above the gelatin solution, and the sealing cap has a rubber diaphragm to be sure that internal and external pressures are in equilibrium.

The bottle is filled hot and lowered into the sea with the jelly in solution. For depths up to 330 m (1000 ft) the plastic holder provides sufficient insulation to permit a lowering time of about 10 min. When the gelatin solidifies it also freezes the disk compass. The compass is tilted as well as the jelly and the direction is observed by noting the highest point of the compass nearest to the surface of the gelatin. With a protractor the angle of the frozen jelly is measured.

The main disadvantage of this design is the estimation of the time necessary before gelling takes place.

The abyssal Pisa No. 231 WA 607 has a jelly and oil filling of the plastic bottle. Hot water is poured into the tube to prevent freezing. The tube is sealed at the bottom while a rubber cap with elastic cord closes the top. When the instrument hits bottom a spring release whips off the rubber cap to allow the bottom water to replace the hot water.

Hemispherical and Free Spherical Compasses. Rosfelder and Marshall (1966) tried many types of compasses and came to the conclusion that it was best "to seek a mechanical compass that will give both orientation and inclination, will be reliable, accurate, small, compact and easy to set, and will require a minimum of handling."

Hemispherical compass (Fig. 6.50A). In this design a standard fluid navigational compass is used. It allows a high angle of inclination. A lightweight hemispherical dome of permeable synthetic foam is mounted to the compass disk. This dome can be locked by a double pin, which is attached to the glass dome. The spring-restrained pin is actuated by the release of an eccentric pin, which is loaded with a torsion spring.

A later model employs two matched pins, while the foam is replaced by a light stainless steel sieving cloth (1 mm mesh), which is easier to shape into a hemisphere. Foam requires machining while frozen.

Free spherical compass (Fig. 6.50B). Two magnets and an orienting balance-weight are placed in a polyethylene ball with a diameter of

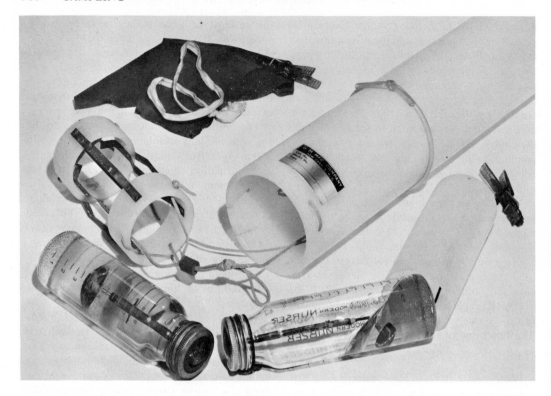

Fig. 6.49. Components of the Carruthers jellifying compound containing compass. (Courtesy G. M. Mfg. & Instrument Corp., New York.)

3.8 cm. This ball fits a spheroidal chamber in which it is immersed in a fluid of nearly the same density as the ball. The fluid is a mixture of water and glycerine. The ball has a slight positive buoyancy. Three tiny sapphire bearings on the upper wall of the chamber keep the compass ball free and in stable motion. The compass chamber is pressure-stabilized through a second connecting chamber, which contains a diaphragm. This compass is locked by a spring-loaded plunger, which is cocked in an open position by the hook of a disconnecting mechanism (Fig. 6.50D). The hook stays in its housing due to tension on the release wire. The present author used this model very successfully on the large box corer.

Rosfelder and Marshall (1966) improved this compass model (Fig. 6.50C) to reduce costs. They use a cylindrical chamber and a rolling diaphragm to eliminate the lower stabilizing chamber and the plunger seal. The ball is made from polypropylene, whose compressibility matches more closely the fluid compressibility than does polyethylene.

An improved version of the free spherical compass is now manufactured by Rextant Inc., La Jolla, California.

The problem of triggering the compass at bottom impact is rather important. The disconnect mechanism (Fig. 6.50D) described here is a solution that can only be implemented when one of the parts of the sampling device is moving, thus allowing for slack in the release wire. "The various ways for triggering an underwater compass at bottom impact could be listed as follows:

1. Use of the slack from the main cable or from a tripping line, with winched equipment. This slack allows for the displacement of a trigger, of a locking pin or of a magnet actuating a magnetic switch, under the restoring force of a spring.

2. This slack can also be obtained by fastening the tripping line at the bottom of the core barrel, on a releasing lever tilted by the sediment on impact, or on a shearing pin sheared by the sediment. We found this way of actuating the compass far more reliable than with a weight

hanging on the taut line. With devices having two disconnecting parts (such as components of a free corer), this type of triggering is easily implemented" (A. M. Rosfelder, written communication, November, 1967).

Compass Smasher. Rosfelder and Marshall (1966) built a cheap device to take along as reserve if the spherical compass was lost, damaged, or stopped working.

The compass smasher is a very simple device for which inexpensive tool compasses are used. A cylinder serves as a housing in which the compass can be inserted. A weight is held in place by a tripping lever, while a spring on top of the weight pushes the weight down as soon as the tripping lever releases it (Fig. 6.51).

The inexpensive compasses used in this device have cardboard parts which swell in water.

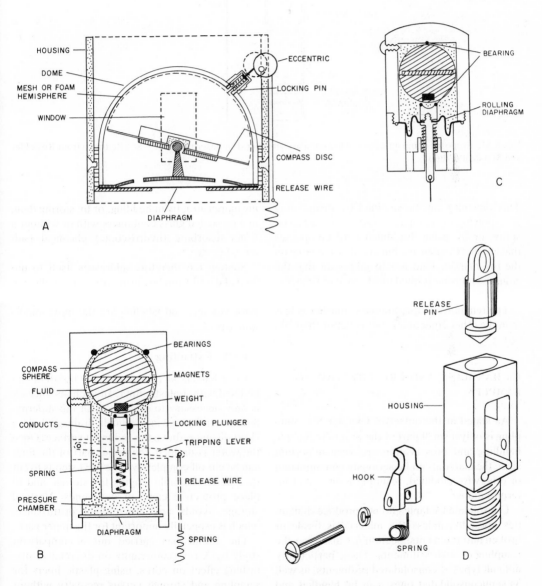

Fig. 6.50. Hemispherical and free spherical compasses in schematic presentation. (A) Hemispherical compass. (B) Free spherical compass. (C) Proposed simplified design of the free spherical compass. (D) Disconnect mechanism acting as trigger for the tripping lever of the compass. (A, B, and C redrawn from Rosfelder and Marshall, 1966; D drawn from disconnector obtained through courtesy of A. M. Rosfelder, Scripps Institution of Oceanography, La Jolla, California.)

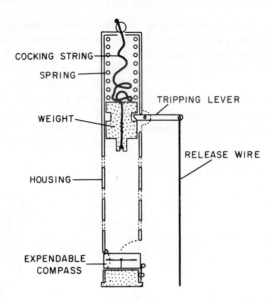

Fig. 6.51. Proposed compact design of the compass smasher. For description see text. (Redrawn from Rosfelder and Marshall, 1966.)

This difficulty can be avoided by dismantling and spraying the compass housing with a plastic spray, or by using fluid-filled card compasses that are more expensive but are also available on the toy market. One has to take care that the compass housing is filled inside to avoid squeezing due to hydrostatic pressure.

The reliability of the compass smasher is less than the other compasses, but is better than 20°.

6.6 HANDLING AND PRESERVATION OF SAMPLES

As stated at the outset of Chapter Six, sampling is only a small part of the program. Sample handling and preservation are very important since the material can be seriously contaminated or otherwise modified, resulting in incorrect interpretations.

Consolidated samples do not produce difficulties normally, unless the material is fissile or broken into fragments. Section 6.1 discusses the sampling of such sediments. These particularly difficult types of consolidated sediments, as well as semiconsolidated ones, can be handled and preserved using the same materials as needed for sampling generally. Exceptions are those sediments that may fall apart due to humidity, dryness, or oxidation. Often such specimens can be preserved in kerosene (e.g., samples contain-

ing marcasite), by imbedding, or by storing them in well-sealed glass containers with or without a water-absorbing (hydroscopic) chemical (salt cores).

Section 6.6 therefore addresses itself to unconsolidated samples, primarily cores collected underwater. Extruding, slicing, drying, transport, storage, and labeling are the most important aspects.

6.6A. Extruding

Core barrels of piston and punch corers can be lined with a plastic liner. In such instances it is only necessary to drain most of the supernatant water by sawing a small drainage hole a short distance above the sediment surface. Once the water is drained, the upper part of the liner can be cut off completely. The next step is to cut the liner to suitable lengths for storage and to place protective caps on both ends. Vertical storage avoids any longitudinal disturbance, which is especially important for the upper part.

The author has carried out a comparison study by X-ray radiography on the relative disturbing effect on cores, using plastic liners for sampling and storage versus recovery without plastic liners and direct extruding on board ship. No apparent difference could be observed. Moreover, the increase in core barrel wall-thickness due to a liner—which is cut lengthwise later—is nearly as objectional as direct extruding because it reduces core recovery length. Extruding should be done in the direction in which the core penetrates the barrel, thus preventing any force with a piston on the top of the core. The advantage of direct extruding is that "flow-in" can be detected on board ship (see Fig. 6.52) and that many analyses can be carried out almost immediately on board such as radiography, vane shear, sampling for water content, density, granulometry, pH, Eh, and conductivity (Bouma and Boerma, 1968). Shipboard extrud-

ing is more economical because flow-in parts can be thrown away directly, reducing the volume to be transported.

The Geological Oceanography Department of Texas A&M University recently developed a satisfactory system for extruding piston cores. The barrel used normally is 12 m (40 ft) long with an internal diameter of 7.5 cm (3 in.). The barrel consists of pipe cut into 10-ft lengths to facilitate handling. Starting with the upper barrel, the disassembled core is extruded by using a

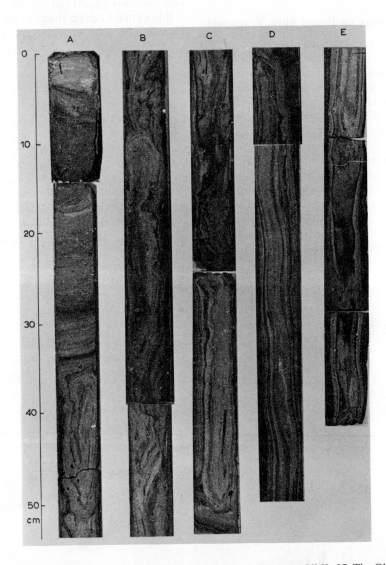

Fig. 6.52. Black and white photograph of successive parts (A-E) of piston core VML-27. The Globigerina core was cut lengthwise. "Flow-in" starts at a depth of 36 cm. The part above 36 cm has slightly curved laminae due to the coring operation. The vertical line pattern of the flow-in is somewhat irregular and has intervals where the pattern is rather disturbed, presumably due to jerks in pulling up the piston. NAVADO project, Station K2, Mid-Atlantic Ridge, eastern middle step, depth 3368 m, 43°04′ .9 N and 23°12′ .9 W, length of core 253 cm. (After Bouma and Boerma, 1968).

loose piston which fits onto a long, sturdy pipe. The lower end of this extruding pipe fits a re-straining holder on deck. A line is fastened around the barrel and, with the help of an electri-cally driven capstan, the barrel is pulled over the piston. The barrel is extruded into a half-cy-lindrical PVC pipe (internal diameter ³⁄₁₆ length 3 ft 10 in.), which fits in a wooden ⊔-shaped trough. Several PVC trays are required per barrel section. Identifying data are written on the side of each PVC unit.

6.6B. Storage on Board Ship, Transport

Several different methods of storing core ma-terial on board ship are known. Deciding factors depend primarily on the facilities on board ship and further on the type of core. Piston and punch cores in capped plastic liners can be stored vertically and/or horizontally. If a refrig-erator is on board for this purpose, such storage is preferable.

Extruded cores can be cut lengthwise, apply-ing the effect of electro-osmosis when they are clayey, and left in two half-cylindrical PVC tubes. The next step is to place the PVC holder, containing the cores, in plastic hose which can be heat-sealed or taped. Cores can also be placed in "high-impact" polystyrene "D-tubes,"

an ideal storage technique on board ship as well as in the core locker (see Fig. 6.53). The tech-nique was developed at the Scripps Institution of Oceanography; the tubes can be obtained from Jet Specialities.

The Geological Oceanography Section of Texas A&M University places four half-cy-lindrical PVC tubes in a wooden tray. Two trays fit one box (Fig. 6.54). The advantage of this system is that the filled boxes can be conven-iently carried by two persons. The boxes are very easy to transport and to store.

Box samples can be stored in the refrigerator. On board the R/V ALAMINOS of the Depart-ment of Oceanography, Texas A&M Universi-ty, an inconvenience is encountered since the refrigerator is below decks and lacks easy ac-cess. Another disadvantage is that water drains from the boxes, which makes the samples less valuable for carrying out analyses on geotechni-cal properties. The solution may be to place each box in a watertight box of marine plywood covered with several coats of epoxy paint. The box can then be filled with sea water to the top of the sample, thus keeping the water content of the sample at its original concentration.

Small samples, such as extracted from the cores, can be wrapped in protective materials and then stored in a cool place.

Fig. 6.53. "High-impact" polystyrene "D-tubes" used for storing half-cores. A polyethylene envelope is placed around each half-core to seal in moisture. The envelope covered section is then placed on top of a sheet of for-malin soaked blotter paper and sealed in the D-tube. (Courtesy G. A. Rusnak, U.S. Geol. Survey, Marine Geol-ogy and Hydrology, Menlo Park, California; Rusnak and Luft, 1963.)

Fig. 6.54. Wooden core boxes with trays for storing and transporting piston cores, as is in use by the Geological Oceanography Section of Texas A&M University. Cover and bottom of the box are made from marine plywood. The handles are attached to the box by long bolts. The laths visible on the turned-over cover prevent the cover from slipping off and help close the box better and tighter. The outside length of the box is 4 ft. Two trays fit the box. The bottom also is made of marine plywood. Each tray is divided into four spaces by laths, which prevent the core holders (half-cylindrical PVC tubes, 3 ft 10 in. long) from rolling. Two rope handles on each tray facilitate handling. All wood is sprayed with epoxy paint. The cores in their PVC tray are placed in nylon sleeve whose ends are taped around the PVC.

6.6C. Wrapping Materials

Short cores collected by hand can best be wrapped in aluminum foil and then dipped a few times in molten wax. This combination gives the core more strength and prevents loss of water.

Plastic foil and Saran Wrap can also be used, especially when the sample is stored in a container.

Samples left in their metal coring pipe can be capped with plastic caps or covered by plastic foil. Wet rags or cotton waste should be placed on top of the sample before capping. Sealing the caps with tape or dipping them in molten wax insures proper closing.

Box samples stored in a refrigerator can be covered by a gel. The author successfully used the AM-9 grout from Cyanamid. On top of the gel, soaked rags and/or papers are placed, after which a sheet of heavy plastic foil is placed over the box and taped to it. Wet material on top of the gel prevents the gel from drying out and cracking. When no gel is present it can be replaced by Saran Wrap with aluminum foil on top, which is covered again by soaked papers and plastic foil.

Broken or fissile consolidated samples and semiconsolidated samples can be wrapped with tape or covered with plaster of Paris (see Section 6.1).

6.6D. Transport

Transport of samples especially road transport, presumably is the worst part of the sampling-transport-storage system. The samples should be transported as much as possible vertically, as they were sampled (see also Araldite Peels, Section 1.6A) to avoid disturbance. Long cores usually cannot be transported vertically and therefore present special problems when driving over bad roads. Small Senckenberg boxes can be placed on foam plastic or rubber to decrease the effects of shocks.

6.6E. Storage at the Laboratory, Labeling

Very few research institutions are equipped with good storage facilities. A large core-locker with temperature and humidity control saves the wet unconsolidated samples from drying out and cracking, which is essential for studying sedimentary structures. Cores can be stored for a reasonable length of time without serious water loss when they are wrapped properly. The wooden boxes mentioned earlier (Fig. 6.54) close rather well, but less than the D-tubes, which seal better (Fig. 6.53). Tar paper is rather impermeable but has to be stapled or taped, which makes access to the material difficult.

It is important to keep the temperature low (not below freezing point) in the storage area and to have an additional walk-in refrigerator for storing samples for geochemical work.

For reasons of easy access, the cores should be placed on racks to avoid moving the material any more than necessary.

Short cores, still in their coring tube, should be stored vertically, as is the case with all box cores. Small samples, wrapped in aluminum foil, plastic, or other material should be taped to sustain their shape. Thin core slices of small size as well as thick ones from large box samples should be placed in strong trays (plexiglass, plastic, sturdy cardboard, or wood lined with plastic sheet) and should be stored with a certain inclination to prevent them from falling apart when shrinkage takes place.

Labeling Unconsolidated Cut Cores

Different techniques are known for indicating depth intervals on cut surfaces of unconsolidated cores. Labeling is very important since core shortening, due to drying out and the forming of cracks, makes it impossible afterward to relocate the proper depths.

Thumbtacks or pieces of waxed paper may be the most common labeling materials. However, both have many disadvantages. Pyle and Hall of the geological section of the Department of Oceanography, Texas A & M University, developed a better system while trying to label cores with wet and muddy hands. They used a commercial label-maker and printed the required depth intervals on the tape (10 cm, 20 cm, 30 cm, etc.) leaving about ½ in. between the depth indications. The tape is then cut directly after each depth indication. The ½-in. excess in front of each mark is bent downward till it makes a 90° angle with the depth indication. The down-bent end is cut diagonally, leaving a pointed part which is placed into the edge of the core.

These tapes are not too expensive and have the advantages that they do not corrode or cause contamination. The numbers do not rub off or smear when wet, and they provide more labeling space than thumbtacks.

6.6F Opening of Stored Cores

Once the cores are stored in a liner or in their sampling pipe for an appreciable amount of time, it becomes impossible to extrude them without serious difficulties and disturbances. The core material settles, which means that enormous

friction must be overcome by extruding, while metal pipes often corrode and form a binding between sample and barrel.

In order to save stored samples that still are in their sampling tubes, the operator has to cut the tubing. Rusnak and Luft (1963) use a hand power saw to cut the plastic liner. It is possible to cut just through the liner and use tape to prevent the halves from falling apart. It is also possible to make the saw cut so that a sturdy knife or an electrical tile cutter is needed for the final opening. The best procedure for cutting is to make a tray in which the liner fits and to "ride" the electrical saw over it along a guide to insure a straight cut. The present author has used successfully a rock cutting machine with diamond-impregnated blade to cut zinc pipes.

6.6G Cutting Unconsolidated Sediments

The cutting of unconsolidated sediments often gives unsatisfactory results due to the pushing of material or particles, generally larger than the average grain size of the sample, ahead of the cutting device and the consequent disturbing and/or smearing of the surface. There are, however, at least five methods of cutting an unconsolidated sample and obtaining a smooth, undisturbed sediment face. Frequently it is advantageous to use a combination of two or more of these methods.

Spatula, Knife, and Razor Blade

The spatula, knife, and razor blade are possibly the oldest tools used for cutting unconsolidated samples. Except when dealing with clays, these tools are easy to use and give good results. The tools are most effective when they are kept clean and wet. The razor blade can be applied for the final touch (Van Straaten, 1954). When the sample contains significant amounts of clay, smearing effects usually result.

Scraping

Samples containing sand-size material occasionally are very difficult to cut. It may be necessary to use a spatula or a knife to remove material by careful scraping until a satisfactory face results. In addition, scraping has to be done when obstructive particles such as pebbles and shells are present. A dissecting pin and a brush often facilitate the removal of these particles from the surface. It is particularly important that the surface be clean when the sample is to be imbedded.

Hacksaw

Unconsolidated sediments, especially carbonates, can become very hard when dry and consequently cause difficulties during cutting. Coarse particles form obstructions, which when forced forward by the cutting edge cause the sample to crack. In such cases, where a knife or a wire does not work, the use of a hacksaw may help. For long slices the blade has to be turned 90° in the frame and the sample sustained on all sides to prevent longitudinal or vertical movement. The following procedure proved useful when dealing with carbonate cores collected from the Florida Shelf and Scarp. The core was enclosed in a PVC tube which had been cut in half longitudinally. The lower end was blocked by a stopper. By slow sawing movements with light forward pressure, the hacksaw blade was advanced through the core. The longitudinally cut edges of the PVC tube acted as a guide for the blade. Little damage occurred to such well-supported samples. Additional cleaning by scraping and local cutting was usually necessary.

Cheese Cutter

A "cheese cutter," consisting of a coping saw in which the blade is replaced by a thin wire, is commonly used to cut unconsolidated cores. The wire often leaves cutting marks behind and causes some smearing of the clayey parts. The greatest disadvantage in using the cheese cutter lies in the frequent difficulty of separating the parts of the cut sample. Sandy samples, however, when not excessively wet, often can be successfully cut by this method. A spatula may be needed to lift away the upper portion of the sample.

Electro-Osmotic Core Cutting

Chmelik (1967) applied the principle of electro-osmosis to cut unconsolidated pelitic cores in order to reduce distortion and cutting marks on the sedimentary face. This technique has been used successfully on board ship on piston cores and in the laboratory on cores and on small and large box samples.

The electro-osmosis effect only works when the sample is clayey since the principle "involves (1) negative electrical charge on clay particles, (2) existence of an electrical double layer, (3) small diameter interconnecting pores which act as capillary tubes throughout the clay mass and (4) electrical potential applied across the sample. The reader is referred to Van Olphen (1965) for an excellent review of electro-osmosis. In a clay mass, the interstitial fluid is forced to migrate from the positive to the negative electrodes. This effect can be used to continuously lubricate the blade of a cutting knife ..." (Chmelik, 1967).

Nearly dry clayey samples can be cut better when this principle is applied. Wetting the surface of the sediment facilitates cutting.

Chmelik built his instrument from surplus parts to avoid expenses in the testing stage. A variac, a transformer, two rectifiers, two fuses, and an ampere meter are all the parts needed. Gilmore (Texas A&M University) recently made a modification, upon the author's request, to make the unit less bulky (Fig. 6.55). The selenium-type rectifiers and the transformer were replaced by four silicon diodes. The use of four diodes instead of the two rectifiers also changed the output from a half-wave rectified sine wave to a full-wave rectified sine wave. The scheme is given in Fig. 6.56.

Many types of cutting device can be used. The already discussed cheese cutter operates even better when the electro-osmosis principle is applied. However, as soon as the wire has passed, the two halves of the core may stick together again. A chrome-vanadium steel butcher knife, 30 cm long (12 in.), keeps the two halves apart when the cut is made vertically. Horizontal cuts can be made by using a hacksaw frame in which a blade, with the teeth ground off, is mounted. A strip of aluminum foil is folded along the leading edge of this blade.

All these cutting devices work only when they are connected to the negative output of the DC source (see Fig. 6.56).

The positive electrode can be a strip of aluminum foil laid on top or bottom of the sample, a copper rod, or a spatula stuck into the sample. The positive electrode will be eroded slowly and metal ions will move into the sediment. High operating amperage relative to the conducting area can cause excessive heating and drying of the sample at the positive electrode. The operator should therefore be careful in his selection of contact locations.

The distance between cutting knife and positive electrode, amount and type of clay, and its moisture content determine how much voltage is needed to provide sufficient current. The pulsating direct current creates a vibrating effect, which facilitates the cutting operation. One to four amperes and a voltage varying between 20 and 120 volts is sufficient. Excessive heating at

Fig. 6.55. Parts used for electro-osmotic core cutting. The box contains a variac (black dial), transformer, two rectifiers, two fuses, an ampere, and a switch. In front of the box lies a ⊔-shaped cutter of which the sides move over the rims of the half-cylindrical PVC pipe. The cutting edge of the central part can be raised or lowered to give it the desired cutting depth. Black electrical tape is mounted to the sides as insulation. The butcher knife is connected to the negative output, copper rod to the positive output of the box.

the positive electrode results when the amperage becomes too high.

The advantages of applying the principle of electro-osmosis to cutting of clayey cores are that (1) an undisturbed surface will be obtained on which no cutting marks are left, (2) very thin slices, suitable for microradiography, can be made without having the disadvantage of shortening the sample, and (3) fine sedimentary structures remain undisturbed. Cutting has to be carried out slowly to avoid drag due to lack of "lubrication." The water moves with a certain velocity toward the cutting blades, depending mainly on the current and moisture content of the sample. The knife may not move faster through the sediment than the rate at which elec-

tro-osmosis lubricates the blade. The electrical circuit must remain complete during withdrawal of the knife. A slow-blow fuse (Fig. 6.56) is necessary to prevent blowing of the fuse behind the ampere meter when an accidental contact between both probes occurs. Extreme caution should be used to avoid placing the operator in the circuit at these potentials because electrical shock can cause death.

When larger particles are encountered in a core, the operator can apply a sawing movement to the knife, when vertical, to prevent pushing the particle in front of the knife. Chmelik (1967) attached a spring-loaded, on-off electrical button to his knife, which has many advantages such as having more control on the current when some-

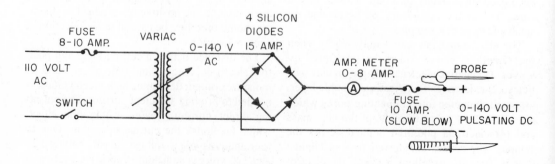

Fig. 6.56. Electrical scheme of the power supply for electro-osmotic core cutting.

thing goes wrong. At the same time it is the fastest way to break the electrical circuit if the operator gets an electrical shock.

6.6H Freeze-Drying

Waiting time often is a nuisance to the investigator once he has collected his samples. Clayey samples may require 6 months to 2 years to dry, especially when the samples are bulky and when cracks have to be avoided as much as possible.

Freeze-drying is a quicker method to remove the water from the pores of the samples. Some investigators still are not certain what type of distortions will result by this method, which is the reason that different techniques exist. Two methods will be discussed.

Freeze-Drying with Liquid Carbon Dioxide.

Whitehouse and McCarter (1957) modified the method described by Anderson (1951) for their electron microscopy investigations of clay minerals. Previous studies in freeze-drying techniques have been carried out by Kistler (1932), Wyckoff (1946), and Williams (1953). Their techniques suffer from the effects of the movement of two different phase boundaries through the material during freezing and sublimation (ice-liquid water boundaries and ice-water vapor boundaries) and random piling of clay material after sublimation of the ice often was induced.

The principle followed by Whitehouse and McCarter as indicated by Anderson is that a liquid can be changed completely and imperceptibly into the vapor phase without the formation of liquid-gas phase boundaries if the liquid is heated under pressure to its critical temperature.

Procedure. The specimen first passes through a series of saline water-ethyl alcohol mixtures. The saline water employed has the same composition and ionic strength as that of the water which forms the environment in which the clay material was deposited. The series of saline water-ethyl alcohol mixtures increase in concentration from 5% by volume alcohol to 100% alcohol.

The samples then pass through a similar series of mixtures consisting of ethyl alcohol-amyl acetate starting at 5% by volume of amyl acetate to 100% amyl acetate.

The samples are than placed in a pressure chamber and flushed with liquid CO_2 at 25°C. Next the chamber is closed and filled with liquid CO_2, after which it is warmed to 34°C. At the pressure employed, the CO_2 passes through its

critical point at 31°C. After this, the chamber is opened a little to release the compressed CO_2 gas. The sample can be removed and placed in a vacuum desiccator to await examination.

Freeze-Drying with Liquid Nitrogen

Werner (1966) developed an easy method of freeze-drying water-saturated, fine-grained sediments. It has the advantage over impregnating moist sediments with Carbowax 6000 or with the Arigal C-epoxy resin combination since less time is involved when a hard, impregnated sediment sample is required. Werner states that freeze-drying has the following advantages:

"1. No shrinkage,

"2. Quick withdrawal of interstitial water,

"3. Possibility to choose between different plastics for hardening,

"4. Good sample condition for impregnation (high porosity of the dried sample),

"5. Minimal fabric deformation during the procedure.

"A disadvantage is the formation of cracks when sediments with high amounts of expanding clay minerals are used."

During freeze-drying one eliminates the capillary forces of the water, which are responsible for the volume decrease during normal drying. Drying via sublimation takes relatively little time, and does not limit the operator's choice of resins for impregnation. Freeze-drying consists of the following four steps.

1. *Sampling.* Partial samples have to be collected from the original sample. Werner uses galvanized metal trays with a base size of 40 x 60 mm and sides of 16 mm. The wall thickness is ¾ mm. The sides are tapered to facilitate later removal. Aluminum foil is used as lining (see Fig. 2.5) (Werner, written communication, April 1967). Werner collects slices of only 4 to 5 mm thick by pressing the tray no further than 5 mm into the sample. A label with sample number and orientation is pasted onto the tray, and then the slice is cut loose with a perlon thread. The slice hangs at the top side of the tapered tray, and therefore a band of 2 mm along the sides must be removed to allow the slice to fit the bottom. The space above the slice is needed for desalting and later impregnation.

A sturdy, thin-walled, metal frame with an internal size of 60 x 40 mm and a height of 5 mm is pressed into the sample. By applying the electro-osmosis method one can collect the slice much easier. The electrical charge can also be used on a knife to cut the slice loose. To prevent damage, one can place a plate on top of the frame, after Saran Wrap is placed between sediment and plate. The tray is placed upside down over the frame and the whole turned back over. Remove the plate and the wrap and apply the effect of electro-osmosis allowing the slice to slide out of the tray.

2. *Desalting.* When dealing with marine sediments it is better to remove as much salt from the slice as possible since salt may ruin the sample due to its hygroscopical action or by its crystal growth.

The operator fills the upper part of the tray with distilled water and decants this after about 2 hours. This procedure should be repeated about 6 to 10 times. The metal trays are not used for the following steps. The aluminum foil is sturdy enough to keep its shape.

3. *Quick freezing.* Slow freezing leads to the formation and growth of ice crystals, which leads to the forming of cracks and complete disturbance of microscopic-sized particle arrangements. On the other hand, quick freezing transfers pore water into ice in such a short period of time that no ice crystal growth is possible. This quick freezing can be carried out by using liquid nitrogen. It is important that this freezing does not take place from all sides toward the center which would result in high pressures inside due to volume increase of the frozen parts. This can be prevented by having contact with the liquid nitrogen on only one side of the sample. The cold front then moves through the slice (see Fig. 2.5). The aluminum foil trays with samples actually float in the liquid nitrogen. The time required to quick freeze 4 to 5 mm-thick slices is about 20 sec. The operator should take care not to touch the trays with bare hands. Small cracks may be present in the corners of the aluminum foil trays. Liquid nitrogen passing through such cracks does not disturb the sample.

4. *Dry freezing.* The samples must now be dried. In principle a desiccator with a cool aggregate to create a strong temperature drop under vacuum can be used. Werner uses a unit which is sold on the market (Beta apparatus from Christ Company). He is able to dry 36 samples at the same time in a desiccator which is 28 cm high and 18 cm in diameter. The time required is about 24 hours when a vacuum of 1.2 lb/in.$^2 \times 10^3$ is applied. By sticking a thin needle into the center of a sample one can find out if the slice is dry. If no resistance is observed, the operator can be sure that no ice is present.

The samples are powdery-voluminous, but they still have enough cohesion to prevent collapsing. To avoid any risk it is better to impregnate them directly since the samples are very sensitive to the absorption of humidity (forming of cracks). There is no restriction as far as impregnating resins are concerned. Werner, however, prefers Pleximon 808 (see earlier discussion of Method of Werner).

Comments

A high montmorillonite content in the sample easily leads to the forming of cracks during dry-freezing. It is likely that water is removed from the interlaminar position of the crystal lattice. Freeze-drying under a lower vacuum prevents this cracking; however, more time is required to reach the dry state.

Sometimes the cracks form along inhomogeneities and, consequently, makes them stand out more clearly. Fecal pellets, when present, show up nicely in such instances.

6.7 DISCUSSION

The field of sampling devices and related tools and instruments is extensive and very active in its development. In this chapter only a few coring devices, selected from the group the author has worked with or is familiar with, are presented. All instruments discussed may be used to collect samples to which the methods discussed in the preceding chapters can be applied. No attempt has been made to be complete or to present a thorough literature survey. Several books and papers, which have either discussions of sampling devices or extensive bibliographies, have been mentioned in the introduction to this chapter.

Some devices are described rather extensively since they are relatively simple and do not require very detailed drawings. The author is not in the position to try out all devices in order to give a critical review of what can and cannot be done by coring in shallow water, in deep water, in bore holes, etc. In addition, the reader should keep in mind that any device is developed for a more or less specific use and often for a certain type of sediment. It is well known that successfully tested samplers sometimes fail when conditions—especially type of sediment—are changed. Very little is published concerning bad experiences.

APPENDIX ONE

RADIOGRAPHY ON BOARD SHIP

The application of X-ray radiography on board ship on fresh cores has many advantages with only a small amount of time involved. Some of the advantages are the elimination of one packing-unpacking sequence in core handling, the absence of disturbing influences due to drying, the direct evaluation of the collected samples, expecially when the radiographs are combined with lithological descriptions, and the detection of flow-in, which can be thrown away.

X-ray radiography can be applied only if proper shielding as well as darkroom facilities are available. The present author successfully used the radiography technique on board H. M. S. Snellius (Hydrographic Office, The Netherlands) while investigating the shelf off Surinam, and on the R. V. ALAMINOS (Dept. of Oceanography, Texas A&M University) during cruises in the Gulf of Mexico. During the Surinam cruise, the Philips-Müller Macrotank B X-ray apparatus was placed in the drafting room and the radiation beam was directed upward. Special shielding eliminated all radiation danger. On the R. V. ALAMINOS the Picker "Hotshot" X-ray apparatus was placed in one of the below-deck laboratories with the X-ray beam directed downward. No special shielding was necessary since the laboratory walls were thick enough. Special wooden developing tanks were made since no permanent installations could be mounted on board.

PROVISIONS ON BOARD H. M. S. SNELLIUS

The little available space and the large crew made it impossible to find a special place for the X-ray apparatus which could be used as exposure room only.

Exposure Room

Since the exposures were made in an active part of the ship, special precautions were necessary to avoid radiation and scattering outside the small area in which the X-ray apparatus was placed. For this purpose an antiscatter cone was made from 1-mm-thick plate-steel (Fig. I.1). To support the 2-mm-thick lead cover, a frame of angle iron was necessary; this held the steel and lead plates in place. All bolts were covered on the inside with pieces of lead (see Fig. 3.37g). The corners of the frame had an additional cover of lead since radiation leaked through the openings between the sheets of lead.

The antiscatter cone was connected to the X-ray apparatus, as well as to the frame (see Fig. 3.16), by metal strips (Fig. I.1). The frame was attached to the ship's deck, with hard rubber blocks placed between. It was observed that the unit was stable enough to make radiographs while at anchor or underway.

Plexiglas trays were made to contain the

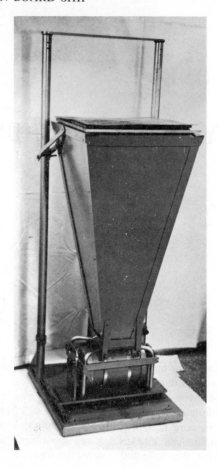

Fig. I.1. Antiscatter cone used as protection against radiation when exposures are not made in a special exposure room. Exposure setup: The antiscatter cone is mounted on top of the Philips-Müller X-ray apparatus. The
X-ray head is attached to the X-ray support. The upper side of the cone is also attached to the support. On top of
the cone is placed a plexiglas plate on which the sample is laid, overlain by the filled exposure holder. A lead
plate is placed on top. On board, a special frame was placed over the sides of the plexiglas and the upper sides of
the angle iron frame, leaving free a window of plexiglas large enough for the samples.

sample slab to be radiographed. Two sizes were
used: 27½ x 40 cm inside and 1 cm high internally (10⅞ x 15¾ x $^{13}/_{32}$ in., respectively), and
7 x 40 cm (2¾ x 15¾ in.) with the same height.
The thickness of the plexiglas is 3 mm (⅛ in.).
The parts were glued together with Tensol Cement No. 6 from Imperial Chemical Industries
Ltd. The large trays were used for box samples,
the small ones for cores. The cores were first
split lengthwise, then cut to length to fit the tray,
and then inserted with the flat face down. The
protruding part was removed.

The sample boxes were opened, a little material removed from the exposed face, and the tray
inserted after some slits were cut to fit the sides
of the tray. A heavy stainless steel plate was
next inserted about 5 to 7.5 cm (2 to 3 in.) be-

hind the tray and a thick slice of sediment was
removed. Protruding material was removed
piece by piece, often with a knife, and, finally,
the sides of the tray were used to obtain a plane-
parallel sediment slice. The trays are very good
for storage, as well as for transport, and the
plexiglas does not interfere with the X-rays.

From fibrous wood a cabinet was made with
12 shelves. Each shelf could contain one large
tray. The door of the cabinet closed well due to a
foam plastic strip, which was necessary to keep
the moisture in. The fibrous wood was lacquered
several times with an outdoor varnish. A wet
sponge was placed inside the cabinet to keep the
moisture content high. A vertical space for
moisture circulation was left open behind the
shelves.

Developing Room

A darkroom with large sink was present on board H. M. S. SNELLIUS and the hard rubber developing tanks could be placed easily. Since there was no air-conditioning in the darkroom the operator had to follow the tropical processing method discussed previously.

PROVISIONS ON BOARD R. V. ALAMINOS

During the geological oceanography cruises, at least two laboratories are not in use and can be transformed into an exposure and a developing room. All parts of the radiography outfit may be installed and removed in a short time.

The application of X-ray radiography on half-cylindrical cores has been accepted as a near-routine procedure on board ship. Slices from a box core or a piston core are radiographed on board ship occasionally. For this the plexiglas trays are used.

Exposure Room

The lower deck bulkheads all are thick enough to act as protection from scattered radiation, and, consequently, no special lead housing had to be made. The X-ray head is mounted in one laboratory, while the control panel is placed in another laboratory. The connecting cable runs through a hole in the bulkhead.

The "Hotshot" X-ray head is attached to the special core rack by means of four long legs in order to give it a focal distance of 39 in., which is enough to cover the special mold (see Fig. 3.27) that fits in the middle of the rack (Fig. I.2). Two aluminum plates are screwed to the legs to make it more stable. The rack can be fastened to the deck.

Fig. I.2. X-ray radiography setup for radiographing lengthwise cut piston cores on board ship. The X-ray head is mounted to the core rack by four long legs giving it a focal distance of 39 in. Two plates are screwed to the legs to make the mount rigid. The core rack is made from wood (2 x 8 in.) with a false bottom 2 ⅛ in. below top to give space to the hinged door in the middle on which the exposure cassettes are placed. Above this door a sand mold is placed to eliminate the half-cylindrical shape of the PVC core containers. Four wooden core holders are mounted in the rack to guide the PVC containers and to keep them horizontal. Exposure holders for 17 x 14 in. film can be handled. The PVC containers are 3 ft. 10 in. long and four exposures are needed to cover them completely.

The core rack is made from 2 x 8 in. wood with a ¾ in.-thick bottom at 2⅛ in. below the top. The internal length is 78 in., the internal width 18 in. On each end of the special sand mold, as well as midway on each side of the remaining rack, wooden holders are mounted to guide the PVC trays in which the half-cylindrical cores lie. Underneath the sand mold, the bottom of the tray is replaced by a hinged wooden plate on top of which the exposure holder and the lead letters and numbers can be placed.

The unit is easy to transport and to install and proves to be stable enough to make exposures while underway.

Developing Room

One of the laboratories could be adapted to function as a darkroom. Beside special developing tanks, an electrical dryer, a rack for developing hangers, and the special X-ray darkroom lamps could be placed and/or mounted. One laboratory table, with drawers underneath, served as dry bench.

The developing tanks are made from marine plywood, ¾ in. thick. Two units, each with three tanks, were built, since one unit became too heavy and too large to take below decks. Each unit has the following outside dimensions: 25 in. high, 17¾ in. wide, and 24 in. long. A four-legged frame with strengthening beams, all made from 2 x 4 in.-thick wood, is attached to each developing tank to bring it to a convenient height (Fig. I.3). The upper side of each tank is 40 in. above the floor.

The tank for developer-stop bath-fixer is divided into three parts, each of 10 gallons. Each tank is 7 in. wide (Fig. I.4). At 3 in. below the top of each tank a frame is made on which the tank cover or the developing hanger can rest. The 3 in. proved to be sufficient to prevent spilling over when the tanks are open. Along the short sides of each part (7 in. long) the frame parts are made from ¾ x ¾ laths; on the 16¼ sides they are ¾ in. high and ¼ in. wide. The other tank unit has one part for hypo-removing solution of 10 gallon volume (7 in. wide), followed by a 15 gallon washing part (10½ in. wide) and a 5 gallon Photo-flo tank (3½ in. wide). A tank with hypo-remover is inserted to reduce the amount of water necessary for washing. The covers serve to keep the solutions free of dust and to prevent spilling in rough seas. Each cover is made of plywood. In its middle a 7 x ¾ x ¾ in.

Fig. I.3. Developing tanks made from marine plywood and coated with epoxy paint. For sizes see Fig. I.4. (A) Tanks seen from the front side. (B) Tanks seen from the back side with all ends of the drainage tubes and the hooks to attach them to the bulkhead of the ship.

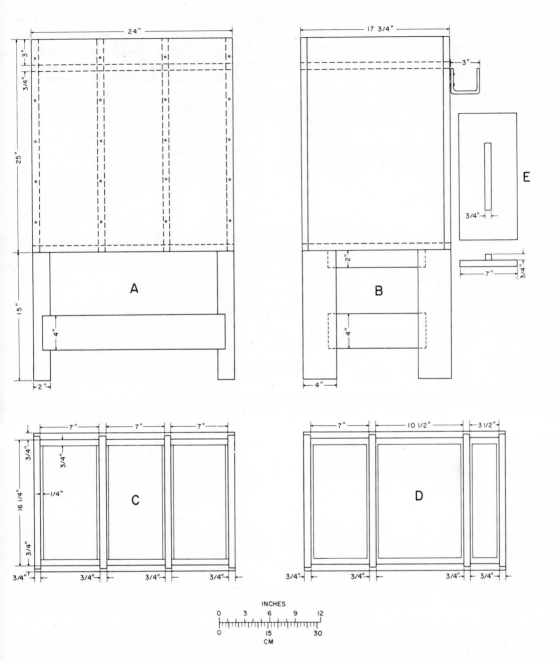

Fig. I.4. Drawings of the developing tanks. All measurements are in inches. (A) Tank seen from the front side with three tanks, all of equal size. (B) View from the side. The hoses are not presented. (C) View from above. From left to right, developing tank (10 gallons), stop bath (10 gallons), and fixer bath (10 gallons). (D) View from above with from left to right, the hypo-remover bath (10 gallons), washing tank (15 gallons), and Photo-flo bath (5 gallons). (E) Tank cover seen from above and from the side.

lath is attached as a hand grip. The inside and outside of the tanks are coated with two layers of an epoxy paint from Napko (Epoxycote "F," No. 5623, color shell gray) (Napko, 1966).

To the back of each unit two ⊔-shaped steel pieces are attached. They serve to fix the units to the bulkhead.

At the bottom of each tank a hole is drilled in which a rubber stopper fits with a glass tube through its center. To the lower part of each glass tube a plastic hose is connected for drainage. The end of the hose hangs on a nail at the back side when not in use. A small rubber stopper closes the hose. A similar hole is drilled in the washing tank just underneath the frame on which the tank cover rests. Through the bottom pipe water flows into the washing tank, while the outflow goes through the upper hole. In order to secure a tight closure of the rubber stoppers in the holes drilled in the tanks, the contacts are covered with a thick film of pipe fitting cement (Permatex: "form a gasket" from Permatex Co., Inc., 1966a, b, 1967). The epoxy paint slowly dissolves in some of the baths, but a total cruise time of 3 months/year is not enough to justify repainting.

APPENDIX TWO

COMPARISON OF SOME DIFFERENT TYPES OF X-RAY FILM

A small number of available X-ray films was tried out in order to make a comparison between the films. All films were exposed under the same conditions and the same sample slices were used. The developing-stop bath-fixer-washing series was the same during the complete experiment.

As standard for comparison Kodak Industrial X-ray film, type AA, tinted estar safety base, double coated, was used. The reason for this is that practically all radiographic work carried out by the present author is done with this type of film.

The focal distance was 29 in. (73.6 cm), using the Picker "Hotshot" X-ray apparatus, set at 40 kV. The sample slices used were No. X 108 (Fig. 3.27D), X 26 (Fig. 3.27E), X 8, X 71, X 149, X 152 (Fig. 3.27F), and a slice of concrete. Four exposures (30, 45, 60, 75 sec) at 5 mA were made and the 45 sec exposure was selected as standard. With all film types tried, exposures were made until results comparable to the Kodak AA standard were obtained.

The films experimented with are the following:

1. Kodak Blue Sensitive Medical X-ray film, single coated, tinted safety base.
2. Kodak Industrial X-ray film, type T, double coated, tinted safety base.

3. Kodak Industrial X-ray film, type R, double coated, tinted safety base.
4. Kodak Industrial X-ray film, type R, single coated, tinted safety base.
5. Kodak Industrial X-ray film, type M, double coated, tinted safety base.
6. Kodak Industrial X-ray film, type KK, double coated, tinted safety base.
7. DuPont de Nemours Cronex 506 Industrial X-ray film.
8. General Aniline & Film Corp., Ansco high-speed X-ray film, double coated, tinted base, class B, speed 80-100, catalog No. 1401-252.
9. General Aniline & Film Corp., Ansco Industrial X-ray film Superay 'A,' class II X-ray film, double coated, tinted base, fine grain.
10. General Aniline & Film Corp., Ansco Industrial X-ray film Superay 'B,' class I X-ray film, double coated, tinted base, extra fine grain, high contrast.
11. General Aniline & Film Corp., Ansco Industrial X-ray film Superay 'C,' class III X-ray film, double coated, tinted base.

The results of these experiments are presented in Table AII.1. They have to be considered as "rules of thumb" since no densitometer readings have been carried out.

Table AII.1 Comparison of Some Different X-Ray Films

Type Film	kV	mA x sec	Remarks
Kodak AA	40	225	Used as comparison standard.
Kodak M	40	1050	Finer grain than AA can be observed with some difficulty using an eyeglass.
Kodak KK	40	30	Very grainy. Result not as bright as AA, details less clear. White parts on AA are gray on KK.
Kodak T	40	600	Image slightly sharper than AA. No difference observable in gray shades.
Kodak Blue Sensitive	40	65	Very grainy. Image sharpness in between AA and KK. White of AA is gray but less than in KK.
Kodak R double coated	40	18,000	Image is excellent, but the exposure time (10 mA and 30 min) rather long. Ideal for sec. enlargements.
Kodak R single coated	–	–	Exposure took too long. Should be twice as much as R double coated.
Cronex 506	40	600	Result similar to Kodak T, only base is slightly yellower, which gives a less sharp impression.
Ansco	40	75	Very grainy. No yellow base. Image comparable to Blue Sensitive from Kodak.
Ansco A	40	150	Very little difference from Ansco.
Ansco B	40	1950	Excellent result. No difference observable from Kodak M.
Ansco C	40	85	Result between Kodak Blue Sensitive and KK. Slightly more yellow than Blue Sensitive.

APPENDIX THREE

FORMULAS AND DIRECTIONS FOR MAKING PROCESSING LIQUIDS FOR X-RAY FILMS

DEVELOPERS

It is essential to dissolve the constituents in the order given in the formula so that undesirable reactions are avoided. Each constituent should also be dissolved completely before the next one is added.

When sodium metabisulfite appears in the formula it should be added with the sulfite. Developers containing sodium hydroxide (caustic soda) or potassium hydroxide (caustic potash) should be prepared in water at a temperature of about 32°C (90°F), whereas for other developers the temperature of the water can be raised to about 50°C (122°F).

Filtering is not necessary if clear water and clean chemicals are used. However, the solution should be filtered before use or storage if there is any sediment or suspension in it.

Many formulas give only the anhydrous or the crystalline form of certain chemicals. If the alternative forms of chemicals are present, one has to adjust the formulas as follows (Kodak Ltd., 1964 f):

Crystalline sodium sulfite in place of anhydrous: multiply the weight by 2.
Crystalline sodium carbonate in place of anhydrous: multiply the weight by 2¾.

Anhydrous hypo in place of crystalline: multiply the weight by ⅝.

The mixed solutions should be stored in tightly stoppered bottles with a minimum of air above the liquid. Glass-stoppered bottles are not desirable because the alkali is likely to make the stopper stick.

Developers that are rather susceptible to aerial oxidation are often divided into two or three solutions in which the developing agent is kept separate from the alkali in order to reduce oxidation.

If the concentrated solutions are stored at low temperatures, it might be possible that some ingredients crystallize out. These crystals must not be discarded or filtered-out but should be redissolved before use by warming the solution.

In the Kodak formulas some ingredients are mentioned under their sales name (Kodak Ltd., 1964f); for example, "Elon" is a specially purified form of monomethyl para-aminophenol sulfate. This developing agent is also known under the names "Metol," "Genol," etc.

Kodak developers

Beside the previously mentioned (Fig. 3.20) Kodak DX-80 developer and the DX-80R replenisher, one can use D-19b high-contrast developer and the replenisher D-19bR (Tables AIII.1, AIII.2).
For developing films under tropical conditions the D-19b developer with sodium sulfate added should be used (Table AIII.3).

Gevaert developers

The Gevaert Company manufactures several developers. Its developer G. 230 insures a very favorable contrast for relation. It also contains

Table AIII.1   D-19b, High-Contrast
X-Ray Film Developer

"Elon"	2.2 g
Sodium sulfite (anhydrous)	72 g
Hydroquinone	8.8 g
Sodium carbonate (anhydrous)	48 g
Potassium bromide	4 g
Water to make	1000 cc

Kodak Ltd. (1965c).
Note. This solution can be used in undiluted form. The minimum developing
time at 20°C (68°F) is 5 min.

Table AIII.2   D-19b R, Replenisher
for D-19b Developer

"Elon"	4 g
Sodium sulfite (anhydrous)	72 g
Hydroquinone	16 g
Sodium carbonate (anhydrous)	48 g
Sodium hydroxide (caustic soda)	7.5 g
Water to make	1000 cc

Kodak Ltd. (1965c).
Note. The D-19b developer should be
maintained at a constant level in the tank
by frequent addition of this replenisher
solution. The total volume of the replenisher used should not be greater than the
original volume of the developer.

Table AIII.3   Sulphated D-19b High-
Contrast X-Ray Film Developer for
Processing at Temperatures between
24 and 32°C (75 and 90°F) (For
development times see Table III.6.

"Elon"	2.2 g
Sodium sulfite (anhydrous)	72 g
Hydroquinone	8.8 g
Sodium carbonate (anhydrous)	48 g
Potassium bromide	4 g
Water to make	1000 cc
Sodium sulfate (anhydrous)	45 g

After Kodak Ltd. (1965b,c).
Note: The replenisher given in Table AIII.2 can
be used here.

a bactericide. The pH is about 10.1. This developer should not be used for dish development
but only in tanks. In combination with its replenisher it can be stored for a long time. The
ideal development temperature is 20°C (68°F),
but it can be used between 18 and 25°C.

G. 150c is available in a very concentrated
form. In less dilute form the developer acts as
its own replenisher.

As a normal developer Gevaert recommends
GP. 209 and the replenisher GP. 209R (Table
AIII.4). Developing times are given in Table
AIII.5.

Ilford Developers

The Ilford Company offers two X-ray developers with replenishers. They both contain
"Phenidone," which is an Ilford developing
agent. Phen-X is a Phenidone-hydroquinone
developer, supplied in powder form. The replenisher is also sold as a powder. A total volume
of replenisher of at least twice that of the
original developer may be added before the
solution should be discarded (Ilford, not dated,
1961b). In Table AIII.6 the development times
are presented at 20°C (68°F).

Phenisol X-ray developer and replenisher are
both liquids. The developer is also based on
Phenidone-hydroquinone (Ilford, not dated,
1961c). A temperature-time chart is given in
Ilford (not dated); in Table AIII.6 only the
recommended development times at 20°C are
given. The temperature range for the Ilford
products varies between 13 and 24°C (55 to
75°F).

STOP BATH

Normal stop baths contain (glacial) acetic
acid, in which the films should be kept for 30 to
60 sec under moderate agitation. Five gallons
(18 liters) of these types of stop bath will treat
about 100 films of 14 x 17 in. or equivalent at
temperatures between 65 and 70°F.

Kodak Formulae

There is some difference between the Eastman Kodak and Kodak Limited formulae. In
Tables AIII.7 to AIII.10 some formulae are
given for normal, hardening, and tropical stop
baths.

Gevaert Formulae

Gevaert Company prescribes an acid bath of
30 cc glacial acetic acid per liter of water. The
film should be rinsed in this bath for about 60
sec (Gevaert, not dated, 1964).

Table AIII.4   Formulas for the Preparation of
Gevaert GP. 209 Developer and GP. 209R Replenisher

	GP. 209	GP. 209 R
Water	800 cc	800 cc
"Metol"	4 g	6 g
Sodium sulfite (crystalline)	130 g	140 g
Hydroquinone	10 g	20 g
Sodium carbonate (crystalline)	120 g	160 g
Potassium bromide	5 g	—
Sodium hydroxide (caustic soda)	—	10 g
Water to be added up to	1000 cc	1000 cc

After Gevaert (1964).
Note. The developer must be topped up regularly with replenisher (about 400 cc/m² film). When about 4 liters of
replenisher has been added to each original liter of developer
one can develop another ¼ m² of film per liter liquid before
the tank has to be renewed completely. If the developer is not
used regularly, it has to be renewed earlier since part of its
activity is lost by oxidation.

Table AIII.5.   Developing Times in Minutes for Gevaert
Structurix D_2, D_4, D_7, and D_{10} X-Ray Films

Temperature	18°C	20°C	22°C	24°C	26°C	28°C	30°C
A	6	5	4	3½	3	2½	2
B	7	6	5	4½	4	3½	3

After Gevaert (1964).
A: developers G. 230 and G. 150c.
B: developer GP. 209.

FIXING LIQUIDS AND FORMULAE

Beside a number of fixers in powder or concentrated liquid form, the manufacturers provide some formulae for making one's solutions.

Kodak Fixing Products

Kodak "Unifix" powder is a well-known
product from the Kodak fixers. It is an acid
hardening fixer which renders good service for
the fixation of X-ray films. There are three basic
solutions of which we need dilution C (see Fig.
3.20). Its pH is about 4.44 (Kodak Ltd., 1963c).
In Tables AII.11 and AIII.12 the Kodak
formulae F-5 and F-5a are given.

Gevaert Fixing Products

Fixing bath G. 305 is the classical bath based
on sodium thiosulfate (hypo), which gives clear
radiographs. The rapid-fixing bath G. 334c is
based on ammonium thiosulfate, which works
twice as fast as normal baths. "Tannofix" is a
hardening-fixing bath, sold in powder form.
When the temperature is above 25°C it is
recommended to add 25 cc "Aditan" per liter
G. 305 or G. 334c to transfer the baths into
hardening baths.

Under the numbers GP. 350, GP. 605, and
GP. 308 two fixing baths and a hardening solution are given in formula form (Tables AIII.13
to AIII.15).

Ilford Fixing Products

The Ilford IF-22 X-ray acid hardening fixing
salt is a single powder potassium alum acid
hardening fixer (Ilford, not dated, 1962a). Ilford
IF-9 chrome alum quick fixing salt is a fast-
working fixer in which the films must stay at
least 10 min to get results from the hardening
part of it. This fixer should be used only when a
renewal of the bath can be carried out every 2 or
3 days (Table AIII.16). Ilford Hypam fixer and
hardener are supplied in concentrated liquid
form. It contains ammonium thiosulfate as
fixing agent (Ilford, not dated, 1962b).

Table AIII.6. Development Times for Ilford Products at 20°C (Where a range of times is shown, the shorter one indicates the standard time.)

Films	Phen-X	Phenisol
Industrial A	5-8 min	4-8 min
Industrial B	5-8 min	4-8 min
Industrial C	5-15 min	4-12 min
Industrial CX	5-15 min	4-12 min
Industrial F	5-15 min	4-12 min
Industrial G	5-8 min	4-8 min

After Ilford (not dated, 1961b,c).

Table AIII.7. Kodak SB-1 Stop Bath

Glacial acetic acid	17 cc
Water	1000 cc

Kodak Ltd. (1965d).
Rinse films for 5 sec at least. Working life: 25 sheets of 8 x 10 in./liter.

Table AIII.8. Kodak SB-1a Stop Bath for Photomechanical Materials

Glacial acetic acid	50 cc
Water	1000 cc

Kodak Ltd. (1965d).
Note. Eastman Kodak (Kodak, 1957) prescribes 35 cc glacial acetic acid or 125 cc (28%) acetic acid per liter.

Table AIII.9. Kodak SB-3 Hardening Stop Bath

Potassium chrome alum	30 g
Water	1000 cc

Kodak Ltd. (1965d).
This is for use in hot weather. Agitate the film for a few seconds immediately after immersion. Maximum hardening takes 3 to 5 min in a fresh bath.

Table AIII.10. Kodak SB-4, Tropical Hardening Stop Bath

Potassium chrome alum	30 g
Sodium sulphate (crystalline)	140 g
Water	1000 cc

Kodak Ltd. (1965d).
Use for temperature between 24 and 32°C (75 to 90°F). Agitate for 30 to 45 sec after immersion. Rinse for 3 min. Working life: equivalent of 20 films of 8 x 10 in.

Note. The freshly made bath has a violet-blue color. A partly used bath will deteriorate on standing for a few days as the color changes to yellow-green.

Table AIII.11. Kodak F-5 Acid Hardening-Fixing Bath (also for use under tropical conditions)

Sodium thiosulfate (hypo) (crystalline)	240 g
Sodium sulfite (anhydrous)	15 g
Glacial acetic acid	17 cc
Boric acid (crystalline)	75 g
Potassium alum	15 g
Water	1000 cc

Kodak Ltd. (1965d).
In freshly prepared baths a fixing time of 10 min is correct. Prolonged immersion at high temperatures is harmful.
Note. Kodak "Unifix" powder is available as an alternative.

Table AIII.12. Kodak F-5a Hardener Stock Solution (for adding to a plain hypo solution to produce an acid hardening-fixing bath).

Water at about 50°C (122°F)	600	cc
Sodium sulfite (anhydrous)	75	g
Acetic acid (glacial)	88	cc
Boric acid (crystalline)	37.5	g
Potassium alum	75	g
Water to make	1000	cc

Kodak Ltd. (1965d).
Note. Add 1 part of the cool hardener stock solution to 4 parts of 30% hypo solution, while stirring the hypo rapidly. As an alternative, "Kodak Liquid Hardener" can be used.

Table AIII.13. Gevaert Acid-Fixing Bath GP. 350

Water (40°C)	800 cc
Sodium thiosulfate (hypo) (crystalline)	200 g
Sodium sulfite (crystalline) (anhydrous 12 g)	25 g
Potassium metabisulfite	12 g
Water to add up to	1000 cc

Gevaert (1964).

Table AIII.14. Gevaert Hardening Solution GP. 605

Water (40°C)	800 cc
Sodium sulfite (crystalline) (anhydrous 35 g)	75 g
Glacial acetic acid	55 cc
Potassium alum	75 g
Water to add up to	1000 cc

Gevaert, (1964).
Note. 200 cc of this solution should be added per liter GP. 350 solution in order to transfer it into a hardening-fixing bath.

Table AIII.15. Gevaert Hardening-Fixing Bath GP. 308

Water	800 cc
Sodium thiosulfate (hypo) (crystalline)	300 g
Potassium metabisulfite	12 g
Glacial acetic acid	12 cc
Borax	20 g
Potassium alum	15 g
Water to add up to	1000 cc

Gevaert (1964).

Table AIII.16. Ilford IF-9 Chrome Alum Quick-Fixing Salt

Water, 38°C (100°F)	750 cc
Chrome alum	12.5 g
Sodium metabisulfite	12.5 g
Sodium sulfite (anhydrous)	6.25 g
Sodium thiosulfate (hypo)	400 g
Water to add up to	1000 cc

Ilford (1956).

APPENDIX FOUR

SOME CONVERSION DATA FOR DIFFERENT UNITS

All nonmetric units are U. S. units, unless indicated. Data are obtained from various sources such as advertisement cards from many companies. They are completed from *The Handbook of Chemistry and Physics* [1960, Hodgman, Weast and Selby (eds.)]; *Handbook of Chemistry and Physics* [1965, Weast, Selby, and Hodgman (eds.)]; *Handbook of Chemistry and Physics* [1966, Weast and Selby (eds.)], and *American Institute of Physics Handbook* [1957, Gray (ed.)].

I. LENGTHS

1 km (kilometer)	1000 m	0.621372 mile (statute)
		0.53961 mile (nautical)
		1093.6133 yd
1 m (meter)	100 cm	0.54681 fathoms
	1000 mm	1.093611 yd
		1.093614 yd (British)
		39.3700 in.
		39.3701 in. (British)
1 cm (centimeter)	10 mm	0.3937 in.
1 mm (millimeter)	10^{-1} cm	3.937×10^{-2} in.
1 μ (micron)	10^{-4} cm	3.937×10^{-5} in.
1 mμ (millimicron)	10^{-7} cm	3.937×10^{-8} in.
1 Å (Ångstrøm)	10^{-8} cm	3.937×10^{-9} in.
1 $\mu\mu$ (millionth micron)	10^{-10} cm	3.937×10^{-11} in.
1 mile (nautical)	1.151 miles (statute)	1.852 km
		1852 m
1 mile (statute)	0.86836 mile (nautical)	1.60935 km
	1012.6859 fm.	
	5280 ft	
	63,360 in.	
1 fm. (fathom)	6 ft	1.828804 m
	72 in.	
1 yd (yard)	3 ft	91.440183 cm
	36 in.	

392

1 ft (foot)	⅓ yd	0.3048006 m
	12 in.	30.48006 cm
1 in. (inch)	1⁄12 ft	2.540005 cm
	1000 mils	
1 in. (British)		2.53998 cm
1 mil	0.001 in.	0.02540005 cm
		25.40005 μ
1⁄64 in.	0.015625 in.	0.0396875 cm
1⁄32 in.	0.03125 in.	0.0793745 cm
1⁄16 in.	0.0625 in.	0.15875 cm
⅛ in.	0.125 in.	0.3175 cm
¼ in.	0.25 in.	0.635001 cm
½ in.	0.5 in.	1.2700025 cm

II. SURFACE

1 m² (square meter)	10^{-6} km²	0.0002471044 acre
	10,000 cm²	1.195985 yd²
		10.76387 ft²
		1550 sq. in.
1 cm² (square centimeter)	10^{-4} m²	0.0010764 ft²
		0.1550 sq. in.
1 acre	0.0015625 mile²	0.4046873 hectare
	4840 sq. yd²	4046.873 m²
	4.3560×10^4 ft²	
1 yd² (square yard)	2.06612×10^{-4} acre	0.83613 m²
	9 ft²	8361.31 cm²
	1296 in.²	
1 ft² (square foot)	0.111111 (= ⅑) yd²	0.09290341 m²
	144 in.²	929.0341 cm²
1 in.² (square inch)	0.000771605 yd²	6.4516258×10^{-4} m²
	1⁄1296 yd²	6.4516258 cm²
	0.0069444 ft²	
	1⁄144 ft²	

III. VOLUME

1 l (liter)	10 dl (deciliter)	61.025 in.³
	100 cl (centiliter)	33.8147 fl. oz.
	1000 ml	2.1134 liquid pt.
	1000.027 cm³	1.8162 dry pt.
		1.05671 liquid qt.
		0.908102 dry qt.
		0.26417762 gal.
		0.21998 gal. (British)
1 ml (cc) (milliliter)	0.001 l	0.0338147 fl. oz.
	1.000027 cm³	
1 gal (gallon)	0.13368 ft³	3.7853 l
	0.83268 gal (British)	3785.4 cm³
	4 liquid qt.	
	8 liquid pt.	
	128 fl. oz.	
	231.0 in.³	

1 gal (gallon) (British)	1.20094 gal 4 liquid qt. (British) 8 liquid pt. (British) 160 fl. oz. (British) 277.3 in.³	4.54596 l 4546.1 cm³
1 in.³ [cubic (cu.) inch]	0.004329 gal 0.017316 liquid qt. 0.014881 dry qt. 0.5541 fl. oz.	0.0163868 l 16.387162 cm³
1 fl. oz. [fluid (fl.) ounce]	0.0078125 gal ¹⁄₁₂₈ gal 0.03125 liquid qt. 0.0625 (= ¹⁄₁₆) liquid pt. 1.80469 in.³	0.0295729 l 29.5729 ml 29.5737 cm³
1 liquid pt. (liquid pint)	0.125 gal 0.5 liquid qt 16 fl. oz. 28.875 in.³	0.473167 l 473.167 ml 473.179 cm³
1 liquid qt. (liquid quart)	0.25 gal 2 liquid pt. 32 fl. oz. 57.749 in.³	0.946333 l 946.333 ml 946.358 cm³

IV. WEIGHT   (av. = avoirdupois)

1 ton	2000 lb av.	
1 kg (kilogram)	1000 gr 100,000 cg (centigram) 10⁶ mg (milligram)	2.2046223 lb av. 35.273957 oz. av. 564.38 dr av.
1 gr (gram)	0.001 kg 1000 mg	0.00220462 lb av. 0.0352739 oz av. 0.564383 dr. av. 15.4324 gr.
1 lb av. (pound)	16 ounces av. 256 dr. av. 7000 gr.	0.4535924 kg 453.5924 g
1 oz (av. ounce)	0.0625 (= ¹⁄₁₆) lb av. 16 dr. av. 437.5 gr.	28.349527 gr 28349.527 mg
1 dr. (av. dram)	0.00390625 lb av. ¹⁄₂₅₆ lb av. 0.0625 (= ¹⁄₁₆) oz av. 27.34375 gr.	.1.771845 gr 1771.845 mg
1 (grain)	1.42857 × 10⁻⁴ lb av. ¹⁄₇₀₀₀ lb av. 2.2857 × 10⁻³ oz av. 0.016667 dr. av.	0.064798918 gr 64.798918 mg

Note: The Troy and Apothecaries' systems for pounds, ounces, and drams are not given.

V. PRESSURE

1 kg/m²	1.4223 × 10⁻³ lb/in.² 0.20482 lb/ft² 9.6784 × 10⁻⁵ atm	0.073556 mm Hg at 0°C 0.0028959 in. Hg at 32°F 0.1 g/cm²

1 kg/cm²	14.223 lb/in.²	735.56 mm Hg at 0°C
	2048.2 lb/ft²	28.959 in. Hg at 32°F
	0.96784 atm	1000 g/cm²
1 g/cm²	0.014223 lb/in.²	0.73556 mm Hg at 0°C
	2.0482 lb/ft²	0.028959 in. Hg at 32°F
	9.6784×10^{-4} atm	10 kg/m²
1 lb/ft²	6.9445×10^{-3} lb/in.²	0.35913 mm Hg at 0°C
	4.788×10^{-4} bar	0.48824 g/cm²
	4.7254×10^{-4} atm	
1 lb/in.²	5×10^{-4} ton/in.²	51.715 mm Hg at 0°C
	0.068947 bar	2.036 in. Hg at 32°F
	0.068046 atm	70.307 g/cm²
1 ton/ft²	13.899 lb/in²	9764.8 kg/m²
	0.94509 atm	
1 ton/in.²	2000 lb/in.²	140.62 kg/cm²
1 cm Hg (mercury) at 0°C	0.013158 atm	135.95 kg/m²
	0.19337 lb/in.²	13.595 g/cm²
	27.845 lb/ft²	
1 inch Hg (mercury) at 32°F	0.033421 atm	345.31 kg/m²
	0.49116 lb/in.²	34.531 g/cm²
	70.727 lb/ft²	
1 atm (atmosphere)	7.348×10^{-3} tons/in.²	760 mm Hg at 0°C
	1.0581 tons/ft²	29.921 in. Hg at 32°F
	14.696 lb/in.²	10,332 kg/m²
	2116.2 lb/ft²	1033.2 g/cm²
	1.0133 bars	
1 bar	0.98692 atm	10,197.1 kg/m²
	14.504 lb/in.²	1019.71 g/cm²

VI. VISCOSITY

1 poise = 100 cp (centipoise)

$$= \text{cgs unit of absolute viscosity} = \frac{g}{\sec \times cm} = \frac{dyn \times \sec}{cm^2}$$

1 stoke = 100 centistoke

$$= \text{cgs unit of kinematic viscosity} = \frac{g}{\sec \times cm \times \text{density in °F}}$$

1 centipoise = 1 centistoke × density (at given temperature)

$$= 2.4190883 \, \frac{lb}{ft \times h}$$

VII. TEMPERATURE IN DEGREES CENTIGRADE AND FAHRENHEIT

Temp. in C° = ⁵⁄₉ (F° − 32)		Temp. in F° = ⁹⁄₅ (C° + 32)	
C°	F°	C°	F°
−20	−4	−15.5	4
−19	−2.2	−15	5
−18	−0.4	−14.4	6
−17.7	0	−14	6.8
−17	1.4	−13.3	8
−16.6	2	−13	8.6
−16	3.2	−12.2	10

C°	F°	C°	F°
−12	10.4	18	64.4
−11.1	12	18.8	66
−11	12.2	19	66.2
−10	14	20	68
−9	15.8	21	69.8
−8.8	16	21.1	70
−8	17.6	22	71.6
−7.7	18	22.2	72
−7	19.4	23	73.4
−6.6	20	23.3	74
−6	21.2	24	75.2
−5.5	22	24.4	76
−5	23	25	77
−4.4	24	25.5	78
−4	24.8	26	78.8
−3.3	26	26.6	80
−3	26.6	27	80.6
−2.2	28	27.7	82
−2	28.4	28	82.4
−1.1	30	28.8	84
−1	30.2	29	84.2
0	32	30	86
1	33.8	31	87.8
1.1	34	31.1	88
2	35.6	32	89.6
2.2	36	32.2	90
3	37.4	33	91.4
3.3	38	33.3	92
4	39.2	34	93.2
4.4	40	34.4	94
5	41	35	95
5.5	42	35.5	96
6	42.8	36	96.8
6.6	44	36.6	98
7	44.6	37	98.6
7.7	46	37.7	100
8	46.4	38	100.4
8.8	48	38.8	102
9	48.2	39	102.2
10	50	40	104
11	51.8	43.3	110
11.1	52	45	113
12	53.6	48.8	120
12.2	54	50	122
13	55.4	54.4	130
13.3	56	55	131
14	57.2	60	140
14.4	58	65	149
15	59	65.5	150
15.5	60	70	158
16	60.8	71.1	160
16.6	62	75	167
17	62.6	76.6	170
17.7	64	80	176

C°	F°	C°	F°
82.2	180	180	356
85	185	182.2	360
87.7	190	187.8	370
90	194	190	374
93.3	200	193.3	380
95	203	198.9	390
98.8	210	200	392
100	212	204.4	400
104.4	220	232.2	450
110	230	250	482
114.4	240	260	500
120	248	287.8	550
121.1	250	300	572
126.7	260	315.6	600
130	266	343.3	650
132.2	270	350	662
137.8	280	371.1	700
140	284	398.8	750
143.3	290	400	752
148.9	300	426.7	800
150	302	450	842
154.4	310	454.4	850
160	320	482.2	900
165.8	330	500	932
170	338	510	950
171.1	340	537.8	1000
176.7	350		

APPENDIX FIVE

ADDRESSES OF MANUFACTURERS

AKEMI
  Akemi "HS"
  E. G. Laschet
  Au Convent 9-11, Eupen, Belgium

  "Akemi" Chemisch-techn. Spezialfabrik
  Erik Höntsch
  Hintere Marktstrasse 9
  Nürnberg, Germany

ALLIED CHEMICAL
  Allied Chemical Corporation
  Plastics Division
  2801 Lynwood Road
  Lynwood, California 90262, U.S.A.

ALPINE
  Alpine Geophysical Associates, Inc.
  Oak Street
  Norwood, New Jersey 07648, U.S.A.

AMERICAN UNDERSEA
  American Undersea
  101 Townsend Street
  San Francisco, California 94107, U.S.A.

BAKELITE
  BXL Plastics Materials Group Ltd.
  12-18 Grosvenor Gardens
  London, S.W. 1, Great Britain

  *Argentina***
    Compania Sudamerican a de Industria y
    Comercio *S.A.* "INSUD"
    25, De Mayo 460
    Buenos Aires
  Australia
    O. H. O'Brien Pty. Ltd.
    166/176, Parramatta Road
    Box 24, Burwood, N.S.W.

Austria
  Cisar & Mayr
  Anton Baumgartnerstrasse 2
  Vienna 23, Erlaa
Belgium
  Anciens Establissements Sharland &
  Russel, s.p.r.l.
  267, Rue Royale
  Brussels 3
*Brazil***
  "BRASIMET" Comercio e Industria A/S
  Caixa Postal 2787
  Sao Paulo
Denmark
  Søren Bording
  8, Aurevang
  Copenhagen, Hel.
Finland
  Hanstrom Trading
  Mantytie 6
  Helsinki
France
  Comptoir Commercial d'Importation
  42, rue Etienne Marcel
  Paris 2e
East Germany
  Anglo Austrian Trading Co. Ltd.
  1-11, Hay Hill
  London, W. 1., Great Britain
Hong Kong
  A. R. Burkill & Sons (Hongkong) Ltd.
  109 Edinburgh House
  Queen's Road Central
  P.O. Box 603

Hungary
    Orientex (Agencies) Ltd.
    Craven House
    121 Kingsway
    London, W.C. 2., Great Britain
India
    Hylam Ltd.
    Secunderabad
    Hyderabad
Luxembourg   (see Belgium)
The Netherlands
    A. Helffer N. V.
    350 Wm. de Zwijgerlaan
    Amsterdam-West
New Zealand
    Joseph Nathan & Co. Ltd.
    13 Grey Street
    Wellington, C. 1.
Norway
    A/S Elektrisk Isolasjon
    Skippergt 32
    Oslo
Pakistan
    Aminsons & Co.
    Rehman Court
    Advani Street
    Plaza Square, Karachi-3
Paraguay   (see Argentina)
Portugal, North
    Augusto Guimaraes & Irmao
    Rua de Camoes 353
    Porto
Portugal, South
    Santos Mendonca Lda.
    Rua da Boavista 83
    Lisbon
Sweden
    A/B Ewebe
    Strandvagen 5
    Vastervik
Switzerland
    Wasem & Olbrecht
    Papiermuhle-Bern
*Uruguay***
    Rodermonte Ltda.
    P.O. Box 1575
    Montevideo

Yugoslavia
    Orientex Agencies Ltd.
    Craven House
    121 Kingsway
    London, W.C. 2, Great Britain

BALTEAUX
    Balteaux Electric Corp.
    New Meadow
    Stamford, Connecticut 06902, U.S.A.
BARECO
    Bareco Wax Company
    Barnsdal, Oklahoma, U.S.A.
BENTHOS
    Benthos, Inc.
    Main Street
    North Falmouth, Massachusetts 02556, U.S.A.
BORDEN
    The Borden Chemical Company
    350 Madison Avenue
    New York, New York 10017, U.S.A.
    Argentina
        Cia. Casco S. A. I. C.
        Suipacha 268
        Buenos Aires
    Australia
        Borden Chemical Company (Australia) Pty. Ltd.
        P.O. Box 57
        Gransville, N.S.W.
    Brazil
        Alba, S. A.
        Caixa Postal 438
        Sao Paulo
    Canada
        The Borden Chemical Company (Canada) Ltd.
        Box 610
        West Hill, Ontario
    Colombia
        Cia. Quimica Borden S. A.
        Apartado Aereo 5039
        Cali
    England
        Leicester, Lovell & Co. Ltd.
        North Baddesley,
        Southampton

**Copies of all correspondence with the marked addresses should be sent to:
    "EXSUD" South American Minerals & Products Co. Ltd.
    "Trident House"
    55/6 Aldgate High Street
    London, E.C. 3, Great Britain

France
Borden Chemical Company (France)
S. A.
13, Rue Sedaine
Paris 11^e

Mexico
Casco Quimica de Mexico S. A.
Atenas 40-401
Mexico 6, D.F.

Philippines
Borden Chemical Company (Philippines)
Inc.
P.O. Box 3006
Manila

BORGHGRAET
Chemical-Products R. Borghgraet
34 Rue du Chatelain
Bruxelles 5, Belgium

BUEHLER
Buehler Ltd.
2120 Greenwood St.
P.O. Box 830
Evanston, Illinois 60204, U.S.A.

Argentina
Sirex
Libertad 836
Buenos Aires

Australia
A. A. Guthrie Pty. Ltd.
16-18 Meeks Road
Marrickville, N.S.W.

Austria
M. R. Drott K. G.
Johannesgasse 18
Vienna 1

Belgium
S. A. Analis
14, Rue Dewez
Namur

Bolivia
Gamma Limitada
Casilla No. 1775
La Paz

Brazil
Equipamentos Industriais Eisa Ltd.
Rua Marconi 23-8
Caixa Postal 4584
Sao Paulo

Canada
Carveth Metallurgical Ltd.
901 Yonge Street
Toronto, Ontario

Chile
Forestier, Weinreich & Cia. Ltda.
Esmeralda 1069
Casilla 191-V
Valparaiso

Colombia
Milciades Sanchez
Apartado Aereo No. 4675
Av. Jimenez No. 7-25
Bogota

Denmark
Buch & Holm A/S
34-36 Solvgade
Copenhagen, K.

Ecuador
Josueth Gonzales
Apartado 297
Quito

England
Shandon Scientific Co. Ltd.
65 Pound Lane, Willesden
London, N.W. 10

Finland
Havulinna Oy
Vuorikatu 16-P.O. Box 468
Helsinki

France
Osi—Omnium Scientifique et Industriel
de France
141, Rue de Javel
Paris XVe

Germany
Muller Kg
Postfach 248
Weidenau/Sieg

Greece
Michel D. Coussios
Parnithos Str. 24
Psychico, Athens

Hong Kong
Schmidt & Co. (Hong Kong) Ltd.
155 Prince's Bldg.
10 Charter Rd. 15th Fl.
Hong Kong

India
Adair, Dutt & Co. (India) Pvt. Ltd.
5, Dalhousie Square, East
Calcutta 1

Indonesia
N. V. Bah Bolon Trading Co.
Djalan Merdeka 29
Bandung

Israel
    L. Kardos
    3, Frug Street
    Tel-Aviv

Italy
    D. Gagliani
    Piazza Luigi di Savoia, 24
    Milano

Japan
    Kasai Trading Co. Ltd.
    No. 45, Ikegami-Tokumochi-Cho
    Ota-Ku, Tokyo

Korea
    Asia Science & Company
    International P.O. Box 1250
    28, 1-KA Choongmoo-Ko, Choong-Ku
    Seoul

Mexico
    Hoffmann-Pinther & Boxworth S. A.
    Apartado 101-Bis
    Mexico 1, D.F.

The Netherlands
    N. V. Vereenigde Ingenieursbureaux
    Viba
    Koningin Emmakade 199
    The Hague

New Zealand
    Watson Victor Ltd.
    P.O. Box 1180
    16, The Terrace
    Wellington

Nicaragua
    Roberto Teran G.
    Apartado Postal 689
    Managua, D.N.

Norway
    Nerliens Kemisk-Tekniske Akt.
    Tollbodgt, 32
    Oslo

Pakistan
    Akhtar Trading Corp.
    P.O. Box 4852
    Karachi 2

Panama
    Med Lab S.A.
    Apartado No. 2014
    Colon

Peru
    American Hospital Supply Corp.
    Apartado 4436
    Lima

Portugal
    Federico Bonet Lda.
    Rua do Acucar, 82
    Lisbon

Portuguese East Africa
    Arthur Ballossini
    P.O. Box 1288
    Lourenco Marques, Mozambique

Portuguese West Africa
    Equipamentos Tecnicos Lda.
    Caixa Postal 6319
    Luanda, Angola

Puerto Rico
    American Hospital Supply
    International Division
    No. 6-1213 N.E. St.
    Puerto Nuevo, San Juan

South Africa
    Taeuber & Corssen Pty. Ltd.
    Wrench Road, Isando, Transvaal
    P.O. Box 1366
    Johannesburg

Spain
    P.A.C.I.S.A.
    P.O. Box 7023
    Gral. P. Rivera, 35
    Madrid 5

Sweden
    Aktiebolaget Elur
    Norrlandsgatan 16
    Stockholm

Switzerland
    Strubin & Co.
    Gerbergasse 25
    Basle

Taiwan
    San Kwang Instruments Co.
    No. 20 Yungsui Road
    Taipei

Thailand
    Multi Trades Co. Ltd.
    59/1 Mansion 8, Rajadamnern Ave.
    Bangkok

Turkey
    Alemdar Ticaret Ve Sanayi A. S.
    Veli Alemdar Han
    Galata, Istanbul

Venezuela
    Ferrum Ca
    Apartado 4596
    Caracas

Yugoslavia
Jugomontana
Obilicev Venac No. 4/IV
Belgrad

CEMENTEX
Cementex Company, Inc.
336 Canal Street
New York, New York 10013, U.S.A.

CETA-BEVER
Ceta-Bever
Grote Houtweg 152
Beverwijk, The Netherlands

CHEMICAL PRODUCTS
Chemical Products Corporation
King Philip Road
East Providence, Rhode Island 02914,
U.S.A.

CHEM SEAL
Chem Seal Corporation of America
12910 Panama Street
Los Angeles, California 90066, U.S.A.

CIBA
CIBA Aktiengesellschaft
Kunststoffabteilung (Plastics Dept.)
141 Klybeckstrasse
Basel, Switzerland

Argentina
Productos Químicos CIBA Sociedad
Anónima
Casilla de correo 1660
Libertad 1056
Buenos Aires

Australia
CIBA Company Pty. Ltd.
P.O. Box 22, East Brunswick
22-24 Ryan Street
Melbourne N. 11, Victoria

Austria
CIBA Gesellschaft m. b. H.
Neustiftgasse 11, Wien VII

Belgium
Société Anonyme CIBA
25, rue Léopold Courouble
Bruxelles 111

Bolivia
Graner, Miranda y Cia. Ltda.
Casilla 997, Ed. Saenz, 6° pisa
La Paz

Brazil
Produtos Químicos CIBA S.A.
C. P. 3678, Avenida Adolfo Pinheiro
3414
São Paulo

Canada
CIBA Company Limited
200 Metropolitan Blvd.
Dorval P. Q.

Chile
Productos Químicos CIBA
Casilla 2864, Santo Domingo 1142
Santiago de Chile

Colombia
CIBA Colombiana S.A.
Apartado aéreo 1944, Calle 43, N° 43-44
Medellin

Cuba
Materias primas Químicas S.A.
Industria 8, Bajos
Habana

Denmark
CIBA Farver & Farmaceutika A/S
Amaliegade 14, København K

Ecuador
Juan Strebel
Calle Tarqui 361, P. O. Box 1099
Quito

Egypt
Ets H. Ghosn Henri et Hubert
15, Midan Saad Zaghlout
Alexandrie

Finland
Aktiebolaget Ekströms Maskinaffär
Postbox 310, Alexandersgatan 7 A
Helsingfors

France
Société des Produits Chimiques de
l'Allier, PROCHAL
18 bis, rue d'Anjou
Paris VIII*e*

Germany, West
CIBA Aktiengesellschaft
Wehr/Baden

Great Britain
CIBA (A.R.L.) Limited
Duxford/Cambridge

Greece
Jean Mountrihas
Place Métropole 11 & Ypatias,
Athènes (K)

Guatemala
Juan Waelti
9 A, Calle N° 4-11
Zona 1, Guatemala

Hong Kong
CIBA (China) Ltd.
G. P. O. Box 285
French Bank Building, 5 Queen's Road

Hungary
Jankó Dénes, Dipl. Ing. chem.
Mérleg-utca 2
Budapest V

India
CIBA of India Limited
P.O.B. 479
Royal Insurance Building, 14 Jamshedji
Tata Road
Churchgate Reclamation
Bombay 1

Indonesia
L. E. Tels & Co's Handelsmaatschappij
N. V.
Posttrommel 743
Gedong Pandjang 2
Djakarta-Kota

Israel
M. D. Lewenstein Ltd.
P.O. Box No. 1296
10 Hagdud Haivri Street
Tel-Aviv

Italy
CIBA Industria Chimica
25, Viale Premuda
Milano

Japan
Nagase & Co. Ltd.
Itachibori Minamidori 1-chome
Osaka

Kenya
Geroge Farkas
P.O. Box 6057
Nairobi

Lebanon
Omnicolor S. C.
B. P. 444, Rue Riad Solh, Imm. Stephan
Beyrouth

Malaya
Diethelm & Co. Ltd.
P.O. Box 191, 139-149 B Market Street
Singapore 1

Mexico
CIBA de Mexico, S. A.
Apartado 10262
Calzada de Tlalpam 1779
Mexico 21, D. F.

The Netherlands
CIBA N. V.
Postbus 241
Van der Duyn van Maasdamlaan 2
Arnhem

New Zealand
CIBA Company Pty. Ltd.
C.P.O. Box 2103
A.M.P. Chambers, 189 Featherston
Street
Wellington C. 1

Norway
O. Pers-Pleym & Co.
Oscarsgatan 12
Oslo-N.V.

Pakistan
CIBA (Pakistan) Limited
P.O. Box No. 166
Spencer's Building, McLeod Road
Karachi

Peru
La Química Suiza S.A.
Casilla 1837, Avenida Uruguay 172
Lima

Portugal
Produtos CIBA Ltda.
Rua de Gonçalo Cristóvão 277
Porto

South Africa
Carst & Walker Pty. Ltd.
Liberal House, 56 Marshall Street
Johannesburg

Spain
CIBA Sociedad Anónima de Productos
Químicas
Balmes 117
Barcelona

Sweden
CIBA Produkter Aktiebolag
Abel Beckers gatan
Norrköping

Thailand
Diethelm & Co. Ltd.
New Road
Bangkok

United States
CIBA Products Company
556 Morris Avenue
Summit, New Jersey 07901

Uruguay
CIBA Uruguana Sociedad Anónima
Maldonado 1220
Montevideo

Venezuela
  CHEMOTEX, Dr. Cornelio Barany
  Av. Francisco de Miranda, Aptdo. 3993
  Caracas

Yugoslavia
  Intertrade
  Cankarjeva 1/11
  Ljubljana

CLAUDIUS ASH
  Cladius Ash, Sons and Co., Ltd.
  26-40 Broadwick St.
  London W 1, England

CM²
  CM², Inc.
  Constitution Drive
  Bohannon Industrial Park
  Menlo Park, California 94025, U.S.A.

  The Tsurumi Precision Instruments Co. Ltd.
  No. 1506, Tsurumi-Cho, Tsurumi-Ku
  Yokohama, Japan

COMEX
  Compagnie Maritime d'Expertises
  131 Av. Joseph-Vidal
  13 Marseille (8^e), France

CONRAD
  Wolfgang Conrad
  Pulverweg 19
  3392 Claustal-Zellerfeld, Germany

COURTRIGHT
  Hugh Courtright & Co.
  1209 West 74th Street
  Chicago, Illinois 60836, U.S.A.

Australia
  H. B. Selby & Co., Pty., Ltd.
  393 Swanston St.
  Melbourne C. 1

Belgium
  S. A. Analis
  14-37 Rue Dewez
  Namur

Canada
  Micro Metallurgical Ltd.
  Thornhill, Ontario

England
  Cutrock Eng. Co. Ltd.
  35 Ballards Lane
  London N. 3

Finland
  Havulinna Oy
  P.O. Box 468
  Helsinki

France
  Touzart & Matignon
  3 Rue Amyot
  Paris 5^e

Germany
  Fema Salzgitter Rudolf Stratmann
  Postfach 27, 3327 Salzgitter-Bad

India
  Geologists Syndicate Private Ltd.
  137 Canning St.
  Calcutta 1

Japan
  Japan Import & Export Co., Ltd.
  275-1 Miyashita-Yugawara Machi
  Kanagawa Prefecture

New Zealand   (see Australia)

Norway
  A/S Christian Falchenberg
  Olav Tryggvasons Gate 1
  Trondheim

Portugal
  Federico Bonet Lda.
  Rua do Acucar 82
  Lisboa

South Africa
  Taeuber & Corssen Pty., Ltd.
  P.O. Box 1366
  Johannesburg

Sweden
  Kebo Aktiebolog
  Fack
  Stockholm 6

CUTROCK
  Cutrock Engineering Co. Ltd.
  35 Ballards Lane
  London N. 3, England

CYANAMID
  American Cyanamid Company
  Industrial Chemicals Division
  Berdan Avenue
  Wayne, New Jersey 07470, U.S.A.

Australia
  Cyanamid Australia Pty. Ltd.
  CRA Building
  95 Collins Street
  Melbourne C. 1

Canada
  Cyanamid of Canada Ltd.
  635 Dorchester Boulevard West
  Montreal 2, Quebec

England
  Cyanamid of Great Britain Ltd.
  Bush House, Aldwych
  London W.C. 2

Hong Kong
Cyanamid (Far East) Ltd.
G.P.O. Box 14217

Japan
Cyanamid (Japan) Ltd.
C.P.O. Box No. 1687
Tokyo

Mexico
Cyanamid de Mexico S.A. de C. V.
Calzada de Tlalpan, No. 3092
Mexico 22, D.F.

Peru
Cyanamid Peruana, S. A.
Casilla 4393
Lima

South Africa
South African Cyanamid (Pty.) Ltd.
P.O. Box 7552
Johannesburg

Zambia
Dr. D. M. I. Dobrovic
P.O. Box 2111
Kitwe

DESMEDT
Etablissements Carlos Desmedt
Département "Carsolith"
1306 Chaussée de Gand
Bruxelles-Berchem, Belgium

DIAMANT BOART
Diamant Boart S. A.
74 Avenue de Pont de Luttre
Brussel 19, Belgium

A. Bruyaux N. V.
Willem Fenegastraat 33-35
Amsterdam, The Netherlands

DONKA
Donka Apparatenfabriek N. V.
Crogtdijk 79
Breda, The Netherlands

DOW CHEMICAL
The Dow Chemical Company
Plastics Department
Midland, Michigan 48641, U.S.A.

DOW CORNING
Dow Corning Corporation
Midland, Michigan 48640, U.S.A.

DUPONT DE NEMOURS
E. I. DuPont de Nemours & Company, Inc.
Dyes and Chemicals Division, Organic
Chemicals Dept. (for dyes)
Petroleum Chemicals Division, Organic
Chemicals Dept. (for Ortholeum 162)

Plastics Dept. (for Lucite)
Nemours Building
Wilmington, Delaware 19898, U.S.A.

DYNAMIT NOBEL
Dynamit Nobel Aktiengesellschaft
Postfach 114-117
521 Troisdorf (Bez. Köln), Germany

Angola
OPELANA, LDA
Rua Duarte Pacheco Pereira N° 16/3°
C.P. 6329
Luanda

Argentina
L. Mayrhofer & Cia. S. R. Ltda.
Av. Pueyrredon 930
Buenos Aires

Australia
Henry H. York & Co. Pty.
47-67 Wilson Street, Botany-Sydney

Austria
VENDITOR Kunststoffe und Chemikalien
Verkaufsgesellschaft m.b.H.
Schwarzenbergplatz 7
Wien III

Belgium
G. Arion
67 rue de la Loi
Bruxelles 4

Brazil
Sociedade Geco Limitada
Rua Senador Pompeu 154
Caixa Postal 2332
Rio de Janeiro

Chile
"BAYER" Quimicas Unidas S. A.
Casilla 9429
Carlos Fernández 260
Santiago de Chile

Columbia
Representaciones Juan Beyer
Apartado aéreo 33-31
Medellin

Denmark
Wilh. F. Hoffmann & Co.
Lundtoftevej 1 B
Lyngby

Ecuador
Juan H. Krüger S.A.C.
Casilla 2440
Quito

Finland
Oy Dynos Ab

Georgsbatan 11.A, Postfach 89
Helsinki

France
Produits Chimiques de la Seine
13 Rue Paul Valéry
Paris 16°

Great Britain
TROVIPLAST Limited
Greener House, 66-68 Haymarket
London S.W. 1

Greece
Costa Flores
Rue Colocotroni 9
Athen 125

Hong Kong
Holland-China Trading Co. Ltd.
Alexandra House, Room 301/310,
P.O.B. 67

India
Chika Limited
Mehta Chambers, 13 Mathew Road
Bombay 4

Iran
Heinrich Overath
Avenue Hoghoughi 114, Postfach 1798
Teheran

Ireland
H. E. Clissmann
44 Dartmouth Square
Dublin 6

Israel
James Pels Ltd.
P.O. Box 2597
Tel-Aviv

Italy
Th. Mohwinckel
Via G. Mercalli 9
Mailand

Morocco
Société Chimique et Industrielles Maro-
caine S.C. 1.M.
283 Boulevard Ibn Tachfine
B. P. 339
Casablanca

The Netherlands
Chemimpo, N. V. Chamicalien Im and
Export Mij.
Sarphatikade 13, Postbus 1506
Amsterdam

New Zealand
Henry H. York & Co. Pty. Ltd.
4-13 Chuznee Street, P.O. Box 6434
Wellington C 2

Norway
PLASTKOMPANIET A/S
Fr. Stangsgatan 2
Oslo 2

Pakistan
Hakim Brothers
Near Sind Madressah, Frere Road
Karachi 2

Peru
Union Metalurgica S.A.
Apartado 1952, Calle Nunez 260-0F407,
Lima

Spain
Lemmel S.A.
Ava. del Gmo. Franco 309
Barcelona

South Africa
Internatio-Rotterdam (S.A.) (Pty.) Ltd.
Milldor House, 2nd Floor
Cor. Main & Delvers Streets
Johannesburg

Sweden
A. B. Dalma
Sveavägen 21, Postbox 3119
Stockholm 3

Switzerland
Scheller A. G.
Hottingerstrasse 21
CH-8021 Zürich

Turkey
Richard Ehrngruber
Persembe Pazari Caddesi No. 20
Yogurtcu Han 6
Istanbul-Karaköy

United States
Rubber Corporation of America
New South Road
Hicksville, New York 11801

Yugoslavia
Commerce
P. O. Box 48-1
Titova Cesta St. 3
Ljubljana

E G & G
E G & G International, Inc.
95 Brookline Avenue
Boston, Massachusetts 02215, U.S.A.

EKCO
Ekco Electronics Ltd.
Ekco Works, Southend-on-sea
Essex, England

Argentina
Tecnica Industrial, M.Y.B.
C. Pellegrini 651
Buenos Aires

Australia
Australian Electrical Industries Pty. Ltd.
P. O. Box 110
Crows Nest, N.S.W.

Austria
Unilabor M.B.H.
Vienna 1 X
Schliessfach 33
Rummelhardgasse 6/3

Canada
Instronics Ltd.
P. O. Box 100
Stittsville, Ontario

Ceylon
Laboratory Equipment Co.
P. O. Box 1040
Colombo

Chile
Jaime Colomer
Casilla 3941
Santiago

Denmark
Semler & Matthiassen
Aebeløgade 1
Copenhagen ø

East Africa
International Aeradio (East Africa) Ltd.
Nairobi Airport
P. O. Box 19012
Embakasi, Kenya

Finland
Oy Scienta Ab.
Ramsaynranta 3
Helsinki 33

France
Ets. Jouan
113 Bvd. St. Germain
Paris 6e

Greece
K. Karayannis
Klafthmonos Square
Athens

Hong Kong
Harvey Main & Co. Ltd.
P. O. Box 268

India
Greaves Cotton & Co. Ltd.
P.O. Box 13
H.P.O. Bangalore 1

Iran
Etabs. F. Hariri & Co.
88 Takht Jamshid Ave.
Kakh
Teheran

Iraq
Forster & Sabbagh Co.
436/1 Rashid Street
Baghdad

Ireland
Kelly & Shiel Ltd.
United Works
Distillery Road
Dublin 3

Israel
Electronic Equipment
P.O. Box 4277
Tel-Aviv

Italy
Ether Italiana S.p.a.
Via A. Magni N. 2
Iverigo (Como)

Japan
Japan Trading Co. Ltd.
P.O. Box 369
Kobe

Libya
Messrs. Hetcolite
26 Bumadian Street
Benghazi

Malaysia
Scientific Supply Co. (Malaysia) Ltd.
187 Pudu Road
Kuala Lumpur

Mexico
Harry Mazal S. A.
Roma 19
Mexico 6, D. F.

The Netherlands
Direct Sales—Resident Technical Sales Eng.
Mr. J. A. G. Lemmens
Moleneindplein 18
Vught

New Guinea (see Australia)

New Zealand
Pye Ltd
P.O. Box 2839
Auckland

Norway
Lehmkuhl Elektronikk
A/S Lehmkuhl
Hovfaret 11
Skøyen

Pakistan
Lindeteves (Pakistan) Ltd.
P.O. Box 4844
Karachi

Papua   (see Australia)

Portugal
Soc. Com. Crocker, Delaforce & Co.
Ltd.
P.O. Box 2738
Lisbon

Rhodesia
Protea Medical Services (Pvt) Ltd.
28 Jameson Avenue
Salisbury, S. Rhodesia

South Africa and S.W. Africa
Protea Physical and Nuclear Instrumen-
tation (Pty) Ltd.
P.O. Box 7793
Johannesburg

Spain
Tecnicas Nucleares S. A.
Serrano 46
Madrid 1

Thailand
Sino British (Siam) Ltd.
287 Surawong Road
Bangkok

United States
Amtradair Incorporated
30-95 32nd Street at 31st Avenue
Long Island City, New York 11102

FABELTA
Union Chimique Société Anonyme
Division Fabelta
Matières Plastiques
190 Rue de Bruxelles
Tubize, Belgium

FIL
N. V. Fabriek van Industrieële Lakpro-
ducten
Woudenbergseweg 19A
Zeist, The Netherlands

Pieter Schoen & Zoon N. V.
Oostzijde 39
Zaandam, The Netherlands

FLÜGGER
J. D. Flügger
Postfach 1744
2000 Hamburg 11, Germany

FRENCKEN
Frencken's Fabrieken N. V.
Chemische-Technische Produkten
Postbus 20
Weert, The Netherlands

FULLER
W. P. Fuller & Co.
450 East Grand Avenue
South San Francisco, California 94080,
U.S.A.

FURANE
Furane Plastics Incorporated
4516 Brazil Street
Los Angeles, California 90039, U.S.A.

GENERAL ELECTRIC COMPANY
Insulating Materials Department
305 Eastern Avenue
Chelsea, Massachusetts 02150, U.S.A.

International General Electric Company
(for requests from outside the U.S.A.)
159 Madison Avenue
New York, New York 10016, U.S.A.

X-Ray Department
4855 Electric Avenue
Milwaukee, Wisconsin 53201, U.S.A.

GENERAL MILLS
General Mills, Inc.
Chemical Division
So. Kensington Rd.
Kankakee, Illinois 60901, U.S.A.

Great Britain
Cray Valley Products, Ltd.
St. Mary Cray
Orpington, Kent

Mexico
General Mills de Mexico
Apartado Postal 85
Tlalnepantla

West Germany
Schering A. G.
Waldstrasse 14, 4619 Bergkamen
Westfalen

GEVAERT-AGFA
Gevaert-Agfa N. V.
Mortsel, Belgium

G. M. MFG. & INSTR. CORP.
G. M. Mfg. Instrument Corp.
2417 Third Avenue
New York, New York 10451, U.S.A.

Kahl Scientific Instr. Corp.
P.O. Box 1166
El Cajon, California 92022, U.S.A.

Germany
Willy Guenther K. G.
8500 Nürnberg 9
Humboldtstrasse 39

Israel
L. Kardos
P.O.B. 11033, 3 Frug Street
Tel-Aviv

Turkey
Geotek Company
Sosyal Kurumlar Hanikat-Daire 407
Findikli, Istanbul

GUSTAV RUTH
Gustav Ruth Temperol-Werke
Walddörferstrasse 136-142
2 Hamburg-Wandsbek
Germany

H.B.N.P.C.
Houillères du Basin du Nord et du Pas-de-
Calais
7, Rue du Général Foy
Paris, 8*e*, France

HENKEL
Henkel International G.m.g.H.
Postfach 4320
4000 Düsseldorf 1, Germany

Angola
Soc. Luso-Alema Ltda.
Caixa Postal 1222
Luanda

Austria
Persil Gesellschaft Mbh.
Wien III/40
Dietrichgasse 4

Belgium
S. A. Persil
66, Avenue du Port
Brussel

Bolivia
Herbol S. A.
Ayacucho 208
La Paz

Brazil
Henkel do Brasil S. A.
Caixa Postal 1183
Sao Paulo 1

Chile
L. Goth
Casilla 906
Santiago de Chile

Costa Rica
Hermann Schmidt & Cie. Ltda.
Apartado 2766
San Jose

Cyprus
G. P. Michaelides & Sons
P.O. Box 37
Nicosia

Denmark
Skandinavisk Henkel AB
Vesterbrogade 2D
København V

Dominican Republic
L. A. Quimica
Apartado 962
Santo Domingo

England
Cox Engineering Co. Ltd.
14 Park Lane
Sheffield 10

Finland
Suomen Henkel AB
Kansakoulukato 10
Helsinki

France
Unichima S. A.
Boite Postale 93-12
67, Boulevard Soult
Paris 12*e*

Greece
Eteka S. A.
18, Rue Omirou
Athen

Guatemala
Agro Comercial S. A.
Apartado Postal No. 193
Guatemala

Honduras
Ralph Fiechter & Cia.
Apartado 93
San Pedro Sula

Hong Kong
H. D. Isler
P. B. 14429
Hong Kong

Iran
Dr. S. Abdollah Rezai & Gebr. Kg.
Postfach 289
Teheran

Ireland
Southern Engineering Co. Ltd.
Parnell Place
Cork

Italy
Societa Italiana Persil S.p.A.
Lomazzo (Como)

Jamaica
National Trading Co.
133 Orange Street
Kingston

Japan
Nippon Henkel Chemical Co. Ltd.
82, Higashi Tarumi
Kobe

Kenya
Omniafrica Trading Co. Ltd.
P.O. Box 1266
Nairobi

Kuwait
Bahman Trading Co.
P.O. Box 327
Kuwait

Lebanon
Gabriel Chahine & Fils
Souk el-Kezaz
Beirut

Malaya
Lim Teck Kee Co. Ltd.
P.O. Box 328
Singapore

Morocco
S.C.I.M.
Boite Postale 339
Casablanca

The Netherlands
Chemphar
Stadhouderskade 1 A
Amsterdam

Nicaragua
Agencias Industriales Cia Ltd.
Apartado 1489
Managua

Norway
Getränke-und Nahrungsmittelindustrie
Trygve Andersen A/S
Karl 12tes Gt. 11
Oslo

Panama
Representaciones Panamericanas S. A.
Apartado 6772
Panama

Peru
A. Camarero Cebrecos S. A.
Apartado 2970
Lima

Philippines
Botica de Santa Cruz
P.O. Box 278
Manila

Portugal
Bouhon & Ca Lda.
Rua Julio Dinis, 891-1° DtO
Porto

South Africa
Southern Chemical Manufacturers (Pty)
Ltd.
P.O.B. 1620
311, South Coast Road
Durban

Spain
Henkel Iberica S. A.
Corcega 480 al 492
Barcelona–13

Sweden
Skandivaniska Henkel Ab
Stortoget 29
Malmö

Switzerland
Henkel & Cie. Ag.
Abt. P3-Industriereiniger
Labor-fac
Hardstr. 45

Trinidad
J. N. Harriman & Co. Ltd.
P.O.B. 232
Port-of-Spain

Turkey
Kimyadermen
P. K. Sisli 240
Istanbul

HERCULES
Hercules Incorporated
International Department
Wilmington, Delaware 19899, U.S.A.
(Contact a chemical supply house for small
quantities)

HICOL
Hicol
Postbus 1151
Rotterdam, The Netherlands

HOECHST
Farbwerke Hoechst AG.
Postfach 70
6230 Frankfurt (M)
Germany

HUDSON CO.
H. D. Hudson Manufacturing Company
Sprayer and Duster Dept.
589 East Illinois Street
Chicago, Illinois 60611, U.S.A.

HÜLS
Chemische Werke Hüls Aktiengesellschaft
4370 Marl
Kreis Recklinghausen, Germany

HYDRO PRODUCTS
Hydro Products
11803 Sorrento Valley Road
San Diego, California 92112, U.S.A.

Argentina
Coasin S. A.
Virrey Del Pino 4071
Buenos Aires

Australia
N.I.C. Instrument Company Pty., Ltd.
1-5 Sydney Street
P.O. Box 139
Marrickville, N.S.W.

Austria (see Germany)

Belgium
Regulation-Mesure
22 Rue Saint-Hubert
Brussels 15

Brazil
Ambriex
Avenida Pacaembu, 811
Sao Paulo

Chile
Coasin Chile Ltda.
Cosilla 14588 — Correo 15
Santiago

Denmark (see Norway)

France
Compagnie Maritime d'Expertises
131 Avenue Joseph Vidal
Marseille 6

Germany
C. T. (Hamburg) GmbH
2 Hamburg 1
Beim Strohhause 20

Greece
Marios Dalleggio Representations
2 Alopekis Street
Athens 139

India
Toshniwal Bros. Private Ltd.
198 Jamshedji Tata Road
Bombay

Israel
L. Kardos
3 Frug Street
P.O. Box 11033
Tel-Aviv

Italy
Aeromaritime
00144 Roma Via Del
Caucaso 49
Rome

Japan
Hakuto Co., Ltd.
Foreign Division
P.O. Box 25
Tokyo Central

Korea
Asia Science and Company
International P.O. Box 1250
28, 1-KA Choongmoo-Ru
Choong-ku, Seoul

The Netherlands
Ingenieursbureau Eurotechniek N. V.
Groothandelsgebouw
Conradstraat 38
Rotterdam 4

New Guinea (see Australia)

New Zealand
Pye Ltd.
P.O. Box 3374
Wellington, C. 3

Nigeria
Delta Systems, Ltd.
P.O. Box 3374
Lagos

Norway
Jan Staubo A/S
Pilestredet 7
Oslo 1

Pakistan
Jaffrey Enterprises
35 Farid Chambers
Victoria Road
Karachi 3

Papua (see Australia)

Portugal
Concessus, S.A.R.L.
Apartado 1455
Rua Dom Francisco
Manuel Del Melo, 9-9A
Lisbon 1

South Africa
Beckman Instruments, Pty.
Temple House
Corner Suitenkant & Roeland Streets
Cape Town

Sweden   (see Norway)

Switzerland   (see Germany)

Taiwan
Emerson Co., Ltd.
#10, Alley 1, Lane 70
Ha-Mi Street
P.O. Box 22551
Taiwan

Thailand
Universal Exploration Co., Ltd.
35/11 Phyathai Road
Bangkok

Turkey
Geotek
Sosyal Kurumlar Hani Kat 4
Daire 407
Findikli, Istanbul

United Kingdom
Underwater and Marine Equipment, Ltd.
Rounton Road
Crookham, Hampshire

Uruguay
Coasin Uruguaya S. A.
Cerrito 617 − 4° Piso
Montevideo

Venezuela
Trabajos Submarinos
Apartado del Este 5264
Caracas

HYDROWERKSTÄTTEN
Hydrowerkstätten G.m.b.H.
Uhlenkrog 38
Kiel-Hassee, Germany

ILFORD
Ilford Limited
Ilford, Essex
England

IMPERIAL CHEMICAL
Imperial Chemical Industries Ltd.
Plastics Division
Bessemer Road, Welwyn Garden City
Hertfordshire, England

KAHL   (see G. M. Mfg. & Instr. Corp.)

KALLE
Kalle Aktiengesellschaft
Postfach 9165
6202 Wiesbaden-Biebrich
Germany

KLAIBER
Maschinenfabrik Fritz Klaiber
Karl Marxstrasse 30
Schwenningen/Neckar, Germany

KODAK, KODAK LTD.
Eastman Kodak Company
343 State Street
Rochester, New York 14650, U.S.A.

Kodak Limited
Kingsway, London, W. C. 2, England

Kodak-Pathé S. A.
37-39 Avenue Montaigne, Paris 8e, France

Kodak A. G.
Hedelfinger Strasse
Postfach 369, 7000 Stuttgart-Wangen,
Germany

Aden
A. Besse & Co. (Aden) Ltd.
Crater

Afghanistan
Rustomji A. Dubash
Deh-Afghanan
Kabul

Algeria
Kodak-Pathé S. A.
24, Rue de Tanger
Algiers

Andorra
Valentin Claverol
Andorra-la-Vella

Angola
Lello & Cia., Ltda.
Luanda

Argentina
Kodak Argentina, Ltd.
Alsina 951
Buenos Aires R 71

Aruba, Netherlands Antilles
Neme Import Export Company, Inc.
Oranjestad

Australia
Kodak (Australasia) Pty. Ltd.
Southampton Crescent
Abbotsford No. 9, Victoria

Austria
Kodak Ges.m.b.H.
Jacquingasse 29
Vienna III

Bahamas, B. W. I.
Moseleys, Ltd.
Boyle Building
Nassau

Bahrein Islands
Ashraf Brothers
Manama

Barbados, B.W.I.
J. N. Harriman & Co., Ltd.
Lower Bay Street
Bridgetown

Belgium
Kodak S. A.
21 Avenue de la Toison D'Or
Brussels

Bermuda
Stuart's
Reid Street
Hamilton

Bolivia
Casa Kavlin, Enrique Kavlin y Cía.
Calle Potosí 259-63
La Paz

Brazil
Kodak Brasileira
Comércio e Industria Ltda.
Av. Brigadeiro Luiz Antônio, 453
Sâo Paulo

British Honduras
James Brodie & Co., Ltd.
Albert Street
Belize

British North Borneo
The Borneo Company Limited
Jesselton

Brunei
The Borneo Company Limited
Brunei Town

Burma
T. N. Ahuja & Company
86 Phayre Street
Rangoon

Burundi
Hatton and Cookson, S.A.R.L.
Usumbura

Cambodia
Comptoir General du Cambodge
116-B Vithei Oknha In
Phnom-Penh

Canada
Canadian Kodak Company Limited
Toronto 15, Ontario

Canary Islands
Foto Suministros
29 de Abril, No. 32
Las Palmas

Ceylon
Millers Ltd.
746 Galle Road
Colombo

Chile
Kodak Chilena, Ltd.
Alonso Ovalle 1180-88
Santiago

Colombia
Kodak Colombiana, Ltd.
Carrera 13, No. 18-66
Bogotá

Congo Republic
W. D. Wooden and Company
Elizabethville, Katanga

Costa Rica
Importadora Fotografica S. A.
San José

Curacao, Netherlands Antilles
El Louvre, S. A.
Willemstad

Cyprus
A. Y. Tilbian & Sons, Ltd.
29, Onasagoras Street
Nicosia

Denmark
Kodak Aktieselskab
Roskildevej 16
Glostrup

Dominican Republic
R. Esteva & Cía., C. por A.
Calle El Conde No. 59
Santo Domingo

Ecuador
Cía General de Commercio y Mandato
9 de Octubre 739
Guayaquil

Egypt
Kodak (Egypt) S. A.
Sharia Adly 20
Cairo

El Salvador
Distribuidora Fotográfica, S. A.
Edificio Ruben Dario
2a Calle Poniente
San Salvador

Eritrea
A. Besse & Co., (Ethiopia) Ltd.
Asmara

Ethiopia
A. Besse & Co., (Ethiopia) Ltd.
Addis Ababa

Finland
Oy Valovarjo, A.b.
Iso Roobertinkatu 8.A. 1,
8 Stora Robertsgatan
Helsinki

French Somaliland
A. Besse & Cie. (Mer Rouge) S.A.R.L.
Djibouti

Gambia
United Africa Co. of Gambia, Ltd.
Bathurst

Ghana
The United Africa Company of Ghana
Ltd.
Accra

Greece
Kodak (Near East) Inc.
15 Valaoritous St.
Athens

Guadeloupe
George Legros, Optique de Paris
14 Rue du Docteur Cabre
Basse Terre

Guam
The Town House
Jones & Guerrero Co.
Agana

Guatemala
Biener, Tabush & Cía. Ltda.
6a Avenida No. 10-71
Guatemala

Guiana
Bookers Drug Stores
Georgetown, Demerara

Haiti
Don Mohr Sales Corporation, S. A.
26-28 Rue Roux
Port-au-Prince

Hashemite Kingdom of the Jordan
Mr. Roupin Ketchijian
The King's Store
King Feisal Street
Amman

Hawaii
Eastman Kodak Co., Hawaiian Sales
Division
1065 Kapiolani Blvd.
Honolulu

Honduras
Rivera y Companía, Calle La Merced,
Tegucigalpa
V. Flores & Cía., Callejón J. C. del Valle
San Pedro Sula

Hong King
Kodak (Far East) Limited
Shell House

Iceland
Hans Petersen Ltd.
4 Bankastraeti
Reykjavik

India
Kodak House, Dr. Dadabhai Naoroji
Road
Bombay No. 1

Iran
Hasso Company, Inc.
210-12 Avenue Shah
Teheran

Iraq
Messrs. N. A. Hasso & Co. W.L.L.
Rashid Street
Baghdad

Irish Republic
Kodak Limited
Kodak House, Rathmines
Dublin

Israel
Delta Trading Co., Ltd.
Citrus House
Tel-Aviv

Italy
Kodak S.p.A.
Via Vittor Pisani 16
Milan

Jamaica
Stanley Motta Ltd.
109 Harbour Street
Kingston

Japan
Nagase & Co., Ltd.
Nishi Ginza Bldg.
No. 5 Ginza Nishi 4 Chome, Chuo-Ki
Tokyo

Kenya, East Africa
Kodak (East Africa) Ltd.
Zebra House, Government Road
Nairobi

Korea
Han Chang Industrial Co., Ltd.
No. 17 3-St Namdaimoon-Ro
Seoul

Kuwait
Ashraf & Co.
Share-Jadid
Al-Kuwait

Lebanon
Kodak (Near East) Inc.
Rizk-Helou Bldg.
Rond Point Kantari
Beirut

Liberia
Liberia Trading Corp.
Water St.
Monrovia

Libya
Mr. Jack Di M. Abravanel
Photo Aula
Giaddat Istiklal 52-4
Tripoli

Madagascar
M. F. Richard (Sté Opticam)
15, Rue Amiral Pierre
Tananarive

Malaya
Kodak (Malaya) Ltd.
54 Pudu Road
Kuala Lumpur

Malta
P. Cutajar & Co., Ltd.
12 St. Paul Street
Valletta

Martinique
Pierre Milon
16 Rue Schoelcher
Port-de-France

Mexico
Kodak Mexicana, Ltd.
Londres 16
México D. F.

Morocco
Kodak-Pathé S.A.
66-8 Avenue Moulay Hassan ler
Casablanca

Mozambique
A. W. Bayly & Cia., Ltda.
101 Avenida da Republica
Lourenco Marques

The Netherlands
Kodak N. V.
Anna Paulownastraat 76
The Hague

New Caledonia
Ventrillon & Cie
26 Rue de L'Alma
Noumea

New Zealand
Kodak New Zealand, Ltd.
18 Victoria Street
Wellington

Nicaragua
Roberto Teran G.
Edificio Teran
Managua, D. N.

Nigeria
Photographic Department
Kingsway Chemists of Nigeria Ltd.
Lagos

Norway
J. L. Nerlien A/S
Nedre Slottsgate 13
Oslo

Okinawa
American Photo Service, Ltd.
Naha

Pakistan (East)
Kodak Limited
328 Station Road
Chittagong

Pakistan (West)
Kodak Limited
Karachi Co. Bldg.
Wallace Street
Karachi 2

Panama
Kodak Panamá, Ltd.
98 Avenida Central
Panamá

Paraguay
Manuel Barrios
Calle Estados Unidos 129
Asunción

Peru
Kodak Peruana, Ltd.
Pasaje Nueva Rosita 160
Lima

Philippines
Kodak Philippines, Ltd.
2247 Pasong Tamo, Makati, Rizal
Manila

Portugal
Kodak Portuguesa, Ltd.
Rua Garrett 33
Lisbon

Puerto Rico
Kodak Puerto Rico. Ltd.
Ponce de Leon 305
San Juan

Qatar
Ali Bin Ali
Doha

Sarawak
The Borneo Company Limited
Kuching
Sarawak

Saudi Arabia
Sham Sudoin Ashraf
Al Khobar

Senegal
Kodak-Pathé S. A.
35 bis Rue Thiong Dakar

Sierra Leone
Kingsway Stores Limited
Freetown

Singapore
Kodak (Malaya) Ltd.
305 Alexandra Road
Singapore 3

Somaliland
A. Besse & Co. (Aden) Ltd.
Mogadiscio

South Africa
Kodak (South Africa) (Pty.) Ltd.
45 Shortmarket St.
Cape Town

South Vietnam
Vietnam Photographic Supply Co.
57 Rue Tu Do
Saigon

Southern Rhodesia
Kodak (Central Africa) Ltd.
Jameson Avenue
Salisbury

Spain
Kodak S. A.
Irún 15
Madrid 8

St. Pierre and Miquelon Islands
Leon Briand-Sasco
Rue du General Leclerc

Surinam
A. Van der Voet, N. V. Handelmaatschappij
36 Waterkant
Paramaribo

Sweden
Hasselblads Fotografiska A/B
Ostra Hamngatan 41-3
Goteborg

Switzerland
Kodak S. A.
50, Avenue de Bellerive
Lausanne

Syria
Kodak (Near East) Inc.
Kotob Bldg., Fardoss Street
Damascus

Tahiti
Sté Tahitienne de Distribution Photo-
graphique
Papeete

Tanganyika, East Africa
Kodak (East Africa) Ltd.
Dar-es-Salaam

Thailand
Borneo Company, Ltd.
Bangkok

Trinidad
J. N. Harriman & Co., Ltd.
61 Marine Square
Port-of-Spain

Tunisia
Sté Tunisienne Ciné Photo
6, rue d'Angleterre
Tunis

Turkey
Burla Biraderler Ve Ssi
Hezaren Cadessi 61-3
Istanbul

Uganda, East Africa
Kodak (East Africa) Ltd.
Kampala

Uruguay
Kodak Uruguay, Ltd.
Colonia 1222
Montevideo

Venezuela
C. Hellmund W. & Cia., Sucs. C.A.
Avenida Principal de Bello Monte
Caracas

Zanzibar
C. P. Jani & Sons

KRYLON
Krylon Inc. (see Borden Company)
P.O. Box 390
Norristown, Pennsylvania 19404, U.S.A.

KUNDE & CIE
Kunde & Cie
Remscheid, Germany

G.J.M. Verlaan N. V.
Postbus 91
Nijmegen, The Netherlands

LEUTERT
  Friedrich Leutert
  3141 Erbstorf bei Lüneburg, Germany

MATRA
  Matra-Werke G.m.b.H.
  Postfach 9195, Dieselstrasse 30-40
  6 Frankfurt/M. 9, Germany

  Miller Holding Mij., N. V.
  Technische Afdeling
  Droogbak 1-3
  Amsterdam-C, The Netherlands

MILLIGAN
  Milligan Electronics Ltd.
  30/32 Havelock Walk
  Forest Hill, London SE 23, England

MIMOSA
  Mimosa Film Produkte
  Kiel, Germany

MMM
  Minnesota Mining and Manufacturing
  Company
  General Offices
  P.O. Box 3800
  St. Paul, Minnesota 55119, U.S.A.

  Argentina
    Fabrica Argentina de Materiales Adhe-
    sivos
    "Fadma"S.A.C.I.
    Tucuman 117, Ier. piso
    Buenos Aires

  Australia
    Minnesota Mining and Manufacturing
    (Australia) Pty. Ltd.
    2 Wentworth Avenue
    Sydney, N.S.W.

  Austria
    Minnesota "3M" technische Vertrieb-
    sgesellschaft mbH
    P.O. Box 99
    Vienna 64

  Belgium
    Minnesota Mining and Manufacturing
    (Belgium) S. A.
    36 rue Demasthene
    Anderlecht, Brussels 7

  Brazil
    Minnesota Manufactureira e Mercantil,
    Ltda.
    Caixa Postal 22-060
    Sao Paulo, Estado de Sao Paulo

Canada
  Minnesota Mining & Manufacturing of
  Canada, Limited
  P.O. Box 2757
  London, Ontario

Colombia
  Minnesota Manufacturera y Mercantil,
  S.A.
  Apartado Aereo 11091
  Bogota

Denmark
  Minnesota Mining and Manufacturing
  A/S
  Sjaellandsbroen 2
  Copenhagen, SV

England
  Minnesota Mining & Manufacturing Co.
  Ltd.
  3M House, Wigmore Street
  London, W. 1

France
  Minnesota de France, S. A.
  135 Boulevard Serurier
  Paris, 19me

Germany
  Minnesota Mining and Manufacturing
  Company mbH
  P.O. Box 5629
  Düsseldorf

Hong Kong
  Minnesota (3M) Far East, Ltd.
  Luk Hoi Tong Building, 7th floor
  Queen's Road Central

Italy
  3M Minnesota Italia, S.P.A.
  Via G. Gozzi 1
  Milano

Japan
  Sumitomo 3 M Ltd.
  Central P.O. Box 490
  Tokyo

Mexico
  Minnesota Manufacturera de Mexico,
  S.A. de C. V.
  Apartado Postal 7533
  Mexico, D. F.

The Netherlands
  Minnesota (Nederland) N. V.
  Rooseveltstraat 55
  Leiden

Norway
Minnesota Mining & Manufacturing A/S
Økern Torvvei 9
Økern

Puerto Rico
Minnesota Mining and Manufacturing
Co. (P.R.) Inc.
270 Canals Street
Stop 20
Santurce

Republic of South Africa
Minnesota Mining & Manufacturing
Company (South Africa) (Pty.) Ltd.
P.O. Box 10465
Johannesburg

Spain
Minnesota De Espana, S. A.
Espronceda 36
Madrid 3

Sweden
Minnesota Mining & Manufacturing A/B
P.O. Box 6071
Stockholm

Switzerland
Minnesota Mining Products, AG
P.O. Box 232
Zürich 1

MÜLLER
Georg Müller, Kugellagerfabrik KG
Äussere Bayreuther Strasse 230
85 Nürnberg 30, Germany

NAPKO
Napko Corporation
5300 Sunrise Street
Houston, Texas 77021, U.S.A.

Mexico
Pinturas Limsa
685 Apartado
Monterrey, N. L.

NIELSEN
Nielsen Hardware Corporation
P.O. Box 568
770 Wethersfield Avenue
Hartford, Connecticut 06101, U.S.A.

Canada
Metal & Wood Fastening Devices Inc.
6302 Papineau Avenue
Montreal 35, Quebec

Denmark
T. Praestmark
Vester Farimagsgade 1
Copenhagen V

Great Britain
Hairlock Limited
Magna Works
Kathie Road, Bedford

The Netherlands
Uni-office N. V.
P.O. Box 1122
Rotterdam

Sweden
BPG Industrier AB
Margartelundsvagen 17
Stockholm-Bromma

NOURY & VAN DER LANDE
Noury & Van der Lande N. V.
P.O. Box 10, Bergschild 39
Deventer, The Netherlands

NUODEX
Nuodex Division, Tenneco Chemicals, Inc.
1075 Magnolia Avenue
Elizabeth, New Jersey, 07201, U.S.A.

OBSERVATOR
N. V. Observator
Observatorhuis, Vaste Land 30
Rotterdam-1, The Netherlands

PERMATEX
Permatex Company, Inc.
Flagler Court Building
P.O. Box 1350
West Palm Beach, Florida 33402, U.S.A.

PERFECTA
Chemische Fabriek "Perfecta Goes N. V."
Naereboutstraat 25
Goes, The Netherlands

PHILIPS
Philips Bedrijfsapparatuur Nederland N. V.
Groep Meet-en Regeltechniek
Eindhoven, The Netherlands

PICKER
Picker X-ray Corporation
595 Miner
Cleveland, Ohio 44124, U.S.A.

PITTSBURGH PLATE GLASS CO.
Pittsburgh Plate Glass Company
Coatings & Resins Division
One Gateway Center
Pittsburgh, Pennsylvania 15222, U.S.A.

Canada
Canadian Pittsburg Industries Ltd.
3730 Lakeshore Blvd. West
Toronto 14, Ontario

Italy

    Bombrini Parodi Delfino
    Via Lombardia 31
    Roma

RAM

    Ram Chemicals, Inc.
    P.O. Box 192
    210 East Alondra Blvd.
    Gardena, California 90247, U.S.A.

REEK

    N. V. Handelmij. Jan Reek
    Oranje Straat 49
    Zaandam, The Netherlands

REICHHOLD

    Reichhold Chemicals, Inc.
    RC 1 Building
    White Plains, New York 10602, U.S.A.

Argentina

    Plastica Bernabo Industrial y Commercial S. A.
    Terrada 658-664
    Buenos Aires

Australia

    A. C. Hatrick Pty. Ltd.
    P.O. Box 59
    Botany, N.S.W.

Austria

    Reichhold Chemie AG
    Breitenleerstrasse 97/99
    Vienna-Kagran XXII

Belgium

    Comechin, S. A.
    108-110, rue des Palais
    Brussels 1

Brazil

    Reichhold Quimica S.A.
    Caixa Postal 570
    Rio de Janeiro

Canada

    Reichhold Chemicals (Can.) Ltd.
    Executive Offices
    1919 Wilson Avenue
    Weston (Toronto), Ontario

Chile

    COMEX S.A.
    Casilla 9306
    Santiago

Colombia

    Quimica Proco S.A.
    Apartado Aereo 919
    Medellin

Costa Rica

    Rene Sonderegger & Cia., Ltda.
    Apartado 839
    San Jose

Denmark

    Afridana A/S
    Dronning Olgasvej 6
    Copenhagen, F.

Ecuador

    Industrias "Akios"
    Calle Gorivar 250
    (P.O. Box 219)
    Quito

England

    James Beadel & Co., Ltd.
    Frodsham House, Edwards Lane
    Speke, Liverpool 24

Finland

    Lejos Oy
    P.O.Box 78
    Helsinki

France

    Reichhold-Beckacite S. A.
    119/151 Route de Carrières
    Bezons (S. & O.)

Germany

    Reichhold Chemie A G
    Iversstrasse 57
    Hamburg-Wandsbek 1

Guatemala

    M. A. Nicol Representaciones, Lt.
    8a Avenida 10-24, Zona 1
    Guatemala City, C.A.

Hong Kong

    Reichhold Chemicals (Hong Kong) Ltd.
    23 Hing Yip Street
    Kun Tong, Kowloon

India

    Reichhold Chemicals, India Ltd.
    Post Bag No. 3
    Kannabiran Koil St.
    Madras 60

Israel

    Jack Iscovitch
    P.O. Box 2264
    Tel-Aviv

Italy

    Resia S.p.A.
    Via Sannitica
    Casoria (Naples)

Japan
  Japan Reichhold Chemicals, Inc.
  3, Tori-Sanchome Nihonbashl
  Chuo-ku, Tokyo

Korea
  Amspac Corporation
  Taeil Building—Room 502
  40, Bukchang-Dong
  Chung-ku, Seoul

Mexico
  Reichhold Chemicals de Mexico, S.A.
  Calle Norte 45, # 731
  Col. Industrial Vallejo
  México 15, D. F.

The Netherlands
  Verkoopkantoor Reichhold Holland
  Koningsweg 11
  s-Hertogenbosch

New Zealand
  A. C. Hatrick (N.Z.) Ltd.
  66 Main Road
  Tawa Flat, Wellington

Norway
  Erichsen & Langseth
  Klingenberggaten 4
  Oslo

Peru
  A. C. Mulanovich
  Casilla 2206
  Lima

Philippine Islands
  Resins Incorporated
  P.O. Box 2534
  Manila

Portugal
  Dansep Sepulchre Limitada
  Calcada de Santos No. 19
  Lisbon

Puerto Rico
  C. Withington Co.
  RFD # 2
  Buzon 505-A
  Rio Piedras

Southern Rhodesia
  Lewis & Everitt (Rhodesia) (Pvt.)
  P. Bag 21
  Kopje, Salisbury

Republic of South Africa
  Lewis & Everitt (Pty.) Ltd.
  P.O. Box 785
  Durban

Spain
  Enrique Arp
  Apartado 450
  Ronda Universidad 12
  Barcelona

Sweden
  Lenbro AB
  Ostra Hamngatan 15
  Gothenburg

Switzerland
  Theodor Baumann & Co.
  Sophienstrasse 2
  Zürich 32

Taiwan
  Yah Sheng Chong
  Yung Kee Co., Ltd.
  198 Nan King East Road
  Section 2
  Taipei

Trinidad, W. I.
  Gordon Grant & Co., Ltd.
  P.O. Box 286
  Port-of-Spain

Turkey
  Oscar J. Fuchs
  Boite Postale Galata 332
  Istanbul

Uruguay
  Compania Bao Soc. Anon.
  Real. 4338
  Montevideo

Venezuela
  Resimon C.A.
  Apartado 245
  Valencia

REXTANT
  Rextant Inc.
  P.O. Box 891
  La Jolla, California 92037, U.S.A.

RHODIUS
  Gebrüder Rhodius
  Burgbrohl, Bez. Koblenz, Germany

The Netherlands
  Machinefabriek en Handelmaatschappij
  Van Voorden, N.V.
  Parallelweg 3
  Etten-Leur (N.B.)

ROHM & HAAS
  Rohm & Haas Company
  Washington Square
  Philadelphia, Pennsylvania 19106, U.S.A.

RÖHM & HAAS
  Röhm & Haas GmbH, Chemische Fabrik
  Mainzer Strasse 42
  61 Darmstadt, Germany

  or contact: Carl Roth
  Herrenstrasse 26-28
  7500 Karlsruhe, Germany

SCADO
  Scado-Archer-Daniels N.V.
  Postbus 15
  Zwolle, The Netherlands

  America, Latin   (see United States)

  Australia
    Jordan Chemical Works (A'sia) Pty. Ltd.
    10-14 Young Street
    Sydney, N.S.W.

  Austria
    Heinrich Kail
    Neulinggasse 12
    Vienna III

  Belgium
    B.N.L. Chemicals s.p.r.l.
    70 Avenue de la Couronne
    Brussels-5

  The Canaries
    Henri Moreau
    Apartado 132
    Santa Cruz de Tenerife

  Congo
    Collchimie S. A.
    111, Chaussée de Charleroi
    Brussels-6, Belgium

  Denmark
    Scancolor A/S
    38, Hagens Alle
    Hellerup

  England
    Rex Campbell & Co. Ltd.
    7, Idol Lane, Eastcheap
    London, E. C. 3

  Finland
    Marner Oy
    Frederikinkatu 45a
    Helsinki

  France
    Société Graham
    146, Av. des Champs-Elysées
    Paris-VIII

  Germany, East   (see Poland)

Greece
    Daffa Brothers
    7, P.P. Germanou Str. & St. Theodores'
    Square
    Athens

Hong Kong
    China United Chemical Corporation
    Limited
    P.O. Box 2045
    Hong Kong

Iceland
    Efnavörur h.f.
    Skúlagata 42
    Reykjavik

India
    Indian Vegetable Products Ltd.
    Forbes Building
    Home Street
    Bombay-1

Iran
    A. Niknafs & Sons
    General Importers
    Avenue Nasserkhosrow
    P.O. Box 1216
    Teheran

Ireland
    R. S. O'Connor
    Maynooth Road
    Celbridge co Kildare

Israel
    N. Jarden
    24, Geulah Street
    Tel-Aviv

Italy
    A.L.C.R.E.A., S.p.A.
    Via Rubens 19
    Milan

Lebanon
    Nametallah Saadé & Co.
    P.O.B. 2093
    Beirut

Luxembourgh (see Belgium)

Malaysia
    Lee Bin & Co.
    No. 7 Road 25
    Petaling Jaya, Selangor

Morocco
    Comptoir Marocain Chimique Comachi
    75-77, Bd de la Résistance
    Boite Postale 4.606
    Casablanca

New Zealand
The Fletcher Chemical Company Limited
P.O. Box 22-210
16, Beach Road
Otahuhu/Auckland

Norway
Christen Høeg A/S
Brogt. 11
Oslo-1

Pakistan
Mohammed Siddiq Abdul Majid
South Napier Road
Karachi-2

The Philippines
Advance Resins Corporation
P.O. Box 2081
Manila

Poland
Stemmler-Imex N. V.
P.O. Box 168
Hilversum, The Netherlands

Portugal
José Lino da Mota
Rua da Fabrica 46-2°
Porto

Republic of South Africa, Cape Province
Scado (Pty.) Ltd.
P.O. Box 7
Athlone Street
Tiervlei, C.P.

Republic of South Africa, rest
Africa General Corporation (Pty.) Ltd.
P.O. Box 6553
Johannesburg

Russia   (see Poland)

South Vietnam
Orient-Chimie
103/15, Rue Phát-Diêm
P.O. Box 27
Saigon

Southern Rhodesia
Dowson & Dobson Ltd.
P.O. Box 2312
Salisbury

Spain
"INDARCO"
Juan Cohrs Longsdon
Apartado de Correos 1506
Barcelona

Sweden
Elof Hansson

1, Langgatan 19
Göteborg-SV

Switzerland
Albert Isliker & Co.
Löwenstrasse 35a
Zürich

Syria
A, Mouallem
East-West Trading
P.O. Box 680
Aleppo

Taiwan
Lee Lai Enterprises Limited
P.O. Box 3101
Taipei

Turkey
Importeks
P.O. Box 16
Galata, Istanbul

United States
Archer-Daniels-Midland Company
Overseas Division
733, Marquette Avenue
Minneapolis, Minnesota 55440, U.S.A.

Yugoslavia
Intertrade
Podjetje za mednarodno trgovino
International Trade Corporation
Ljubljana, Titova 1

SCOTT BADER
Scott Pader & Co. Ltd.
Wollaston, Wellingborough, Northamptonshire,
Great Britain

Argentina
Ulivi Bianchi y Cia. S.A.
Venezuela 1578
Buenos Aires

Australia
Monsanto Chemicals (Australia) Ltd.
P.O. Box 62
West Footscray Wiz, Victoria

Austria
Ing. Egon Wildschek und Cie., Chemische und Lakfabrik
Postfach 26
Wien-Atzgersdorf

Brazil
Alba S.A.
Caixa Postal 438
Sao Paulo

German Federal Republic (see The Netherlands)

Israel

Makhteshim Chemical Works Ltd.
P.O. Box 60
Beer-Sheva

Italy

Savid S.p.A.
Cernobbio
Como

The Netherlands

Kunstharsfabriek Synthese N. V.
Industrieweg 6
Katwijk aan Zee

New Zealand (see Australia)

Scandinavia

A B Syntes
No 1, Sweden

South Africa

The Natal Chemical Syndicate Ltd.
P.O. Box 46
New Germany, Natal

Switzerland

Firma Dr. Walter Mäder
Lack- und Farbenfabrik AG.
Killwangen-8956

(The interested reader should write to Wollaston for addresses of agents.)

SEPPIC

Société d' Exploitation de Produits Pour les
Industries Chimiques
70 Avenue des Champs-Elysées
Paris 8e, France

SHELL

Shell Chemical Company
A Division of Shell Oil Company
Plastics and Resins Division
110 West 51st Street
New York, New York 10020, U.S.A.

(The company suggests contacting the Shell Chemical Company in the capital city of each country or writing to the given address.)

SIKKENS

Sikkens Verkoop Nederland N.V.
Rijksstraatweg 31
Sassenheim, The Netherlands

Belgium

Sikkens Groep België N.V.
Koekelbergaan 56, Berchem/Brussel

France

Franco Hollandaise de Peinture S.A.

29 Rue d'Astorg
Paris 8^e

Germany

Sikkens Lackfabriken G.m.b.H.
Loewenbergerhaf 4, Emmerich

Italy

Sikkens S.p.A.
Strada Nazionale del Sempione
Dormelletto (Novara)

SPRINGSTOFFENFABRIEKEN

Koninklijke Nederlandsche Springstoffenfabrieken N.V.
Kruitpad 16
Muiden, The Netherlands

STOCKHAUSEN

Chemische Fabrik Stockhausen & Cie.
Postfach 570
415 Krefeld, Germany

Austria

Bayer-Pharma BmbH
Biberstr. 15
Wien 1

Belgium

N.V. Desinfecta S.A. Croes & Co.
De Vrièrestraat 27
Antwerpen

Denmark

A/S Dansk Ilt- & Brintfabrik
29, Scandiagade
Kopenhagen

Finland

Karske Oy
Nokiantie 2-4
Helsinki 51

France

F.U.S.A.
20, rue de Villiers
Levallois-Hauts-de-Seine (92)

Italy

PINE Sas
Via Sangro 9
Milano

The Netherlands

Bayer (Nederland) N.V.
Velperweg 28
Arnhem

Sweden

A. B. Tegma
Kopparbergsgatan 29
Malmö SÖ

Switzerland
Opopharma A. G.
Kirchgasse 42
Zürich

United States
Stiefel Lab., Inc.
Oak Hill, New York 12460, U.S.A.

SUUNTO OY
Suunto Oy
Itämerenkatu 52
Helsinki, Finland

SYNRES
NV Chemische Industrie Synres
Hook of Holland, The Netherlands

SYNTHESE
Kunstharsfabriek Synthese N.V.
Industrieweg 6
Katwijk aan Zee, The Netherlands

UNION CARBIDE
Union Carbide International Company
Division of Union Carbide Corporation
270 Park Avenue
New York, New York 10017, U.S.A.

Argentina
Union Carbide Inter-America Inc.
Avenida Belgrano 367
Buenos Aires

Australia
Union Carbide Australia Ltd.
Chemicals Division
G.P.O. Box 5322
Rhodes, Sydney, N.S.W.

Austria
Gebrüder Schoeller
Renngasse 3
P.O. Box 568
Vienna 1

Belgium
Union Carbide Europa S.A.
Benelux Branch
66 rue Royale
Brussels

Brazil
Union Carbide do Brasil S.A.
Rua Formosa No. 367, 30°
Sao Paulo

Chile
Carlos Braun y Cia. Ltda.
Huerfanos No. 1189, 6° Piso, Oficina
No. 5
Santiago

Colombia
Union Carbide Colombia S.A.
Carrera 10 No. 16-18-Zer Piso
Apartedo Aero 4704
Bogota

Denmark
Union Carbide Europa S.A.
40 rue du Rhône
Geneva, Switzerland

Ecuador
Industrios Akios
Calle Gorivar 250, P.O. Box 219
Quito

Finland
Oy Falcken & Co.
Fredriksgatan 48
Helsinki

France
Union Carbide Europa S.A.
22 Avenue de la Grande Armée
Paris 17

Germany
Brenntag G.m.b.H.
Wilhelmstrasse 7
Muelheim/Ruhr

Greece
Union Carbide Middle East Limited
P.O. Box 740
Athens

Hong Kong
Union Carbide Asia Ltd.
Shell House

India
Union Carbide India Limited
Industrial Products Division
Faraday House, P17, Mission Row Extension
P.O. Box 2170
Calcutta 1

Italy
S.p.A. Celene
Corso Italia 13
Milan

Japan
Nagase & Company, Ltd.
Itachibori Minamidori 1-Chome
Osaka

Kenya
Union Carbide Africa Ltd.
P.O. Box 4765
Nairobi

Korea
    Bando Trading Co. Ltd.
    U1-Chi Building (The Han-II Bank)
    P.O. Box 1899
    Seoul

Malaysia
    National Carbon (Eastern) Ltd.
    P.O. Box 4042
    Singapore 21

Mexico
    National Carbon-Eveready S.A.
    Presidente Mazarik 8
    Mexico 5, D. F.

Middle East
    Union Carbide Middle East Limited
    P.O. Box 740
    Athens, Greece

The Netherlands
    Heybroek & Co's
    Handel Maatschappij N.V.
    Postbox 555
    Amsterdam-C

New Zealand
    Union Carbide New Zealand, Proprietary Limited
    P.O. Box 2542
    Wellington C 3

Norway
    C. H. Norsted A/S
    Revierstredet 9
    Oslo

Pakistan
    National Carbon Co. (Pakistan) Ltd.
    P.O. Box 4785
    Karachi

Panama
    Union Carbide Inter-America Inc.
    P.O. Box 7216, 7504
    Via Espana 120
    Panama City, R.P.

Peru
    Representaciones Rodval S.A.
    P.O. Box 3709, 377 Abancay Ave.
    Lima

Philippines
    Union Carbide Philippines Inc.
    P.O. Box 677
    Manila

Portugal
    Ahlers Lindley, Lda.
    Apartado 2884
    Lisbon

Puerto Rico
    Peerless Chemicals (P.R.) Inc.
    Calle Amor No. 3, Box 951
    Ponce

South Africa
    Union Carbide South Africa (Pty.) Ltd.
    P.O. Box 8194
    Johannesburg

Spain
    Quimidroga, S.A.
    Tuset 26
    Barcelona-6

Sweden
    Schweiziska Union Carbide Europa S.A.
    Drottninggatan 49
    Stockholm C

Switzerland
    Union Carbide Europa S.A.
    40 rue du Rhône
    Geneva

Taiwan
    Formosa Chemical Import Company
    No. 46 Nan Yang Street, P.O. Box 1332
    Taipei

Uruguay
    Representaciones Nuevo-Mundo Ltda.
    Ituzaingo 1482
    Montevideo

Venezuela
    Union Carbide de Venezuela, C.A.
    Apartado 5363
    Chacao Caracas

U.S. PLYWOOD
    United States Plywood Corporation
    Lawrence Ottinger Research Center
    P.O. Box 126
    Brewster, New York 10509, U.S.A.

WACKER
    Wacker-Chemie G.m.b.H.
    Prinzregentenstrasse 22
    8 München 22, Germany

WALLACE & TIERNAN
    Wallace & Tiernan Inc.
    Lucidol Division
    1740 Military Road
    Buffalo, New York 14240, U.S.A.

WARD
    Ward's Natural Science Establishment, Inc.
    P.O. Box 1712
    Rochester, New York 14603, U.S.A.

WATSON, MANASTY & CO.
  Watson, Manasty & Co, Ltd.
  Orleans House, Manor Road
  Teddington, Middlesex, Great Britain

WIRTZ
  Jean Wirtz
  Charlottenstrasse 73
  4 Düsseldorf 1, Germany

BIBLIOGRAPHY

Adams, G. D., 1963a. Units and measurements of radiation. In G. L. Clark (Ed.), *Encyclopedia of X-rays and gamma rays*, Reinhold, New York, pp. 1100–1101.

_____, 1963b. Chemical dosimetry. In G. L. Clark (Ed.), *Encyclopedia of X-rays and gamma rays*, Reinhold, New York, p. 290.

_____, 1963c. Detectors of radiation. In G. L. Clark (Ed.), *Encyclopedia of X-rays and gamma rays*, Reinhold, New York, pp. 225–228.

Agfa-Gevaert, 1966. *Structurix Industrial X-ray films,* Agfa-Gevaert N. V., 12 pp.

Ahrens, W. and H. Weyland, 1928. Die Herstellung von Dünnschliffen aus lockerem Material für petrografische Untersuchungen, *Centralbl. Min., Geol. und Pal.,* **Abt. A**, 370–376.

Aime, G., 1845. *Recherches de physique générale sur la Méditerranée. Exploration Scient. de l'Algérie,* Imprimerie Royale, Paris, 211 pp.

Alexander, F. E. S. and R. M. Jackson, 1954. Examination of soil microorganisms in their natural environment, *Nature,* **174,** 750–751.

_____ and R. M. Jackson, 1955. Preparation of sections for study of soil microorganisms, *Soil Zool. Easter School Agricul. Sci., Nottingham Univ.,* **2,** 433–440.

Allison, H. B., 1962. *The plastisol concept in new product development,* Soc. Plastics Industry, Inc., Sect. 5-B, 4 pp.

Alpine, not dated a. 244 Heavy duty bottom grab, Alpine Geoph. Assoc., Inc., *Catalog of Oceanographic Instruments,* 32.

_____, not dated b. *Catalog of Oceanographic Instruments.* Alpine Geoph. Assoc., Inc., 66 pp.

Altemüller, H. J., 1956. Neue Möglichkeiten zur Herstellung von Bodendünnschliffen. *Z. Pflanzenern., Düngung u. Bodenk.,* **72,** 56–62.

_____, 1957. *Bodentypen aus Löss im Raume Braunschweig und ihre Veränderungen unter dem Einfluss des Ackerbaues.* Inaugural-diss., Rheinischen Friedrich-Wilhelms-Univ., Bonn, 250 pp.

_____, 1962. Verbesserung der Einbettungs-und Schleiftechnik bei der Herstellung von Bodendünnschliffen mit Vestopal. *Z. Pflanzenern., Düngung u. Bodenk.,* **99,** 164–177.

Amstutz, G. C., 1960. The preparation and use of polished thin sections. *Am. Min.,* **45**(9, 10), 1114–1116.

Anderson, T. F., 1951. Techniques for the preservation of three-dimensional structures in preparing specimens for electron microscopy. *Trans. New York Acad. Sci.,* **13,** 130–134.

Andresen, A., S. Sollie, and A. F. Richards, 1965. N.G.I. gas-operated sea-floor sampler. *Proc. 6th Intern. Conf. Soil Mech.,* 3 pp.

Andrews, H. N., Jr. 1961. *Studies in paleobotany.* Wiley, New York, 487 pp.

Andrews, K. W., and W. Johnson, 1963. Applications of microradiography of iron and steel. In G. L. Clark (Ed.), *Encyclopedia of X-rays and gamma rays,* Reinhold, New York, pp. 618–621.

Appel, J. E., 1933. A film method for the study of textures. *Econ. Geol.,* **28,** 383–388.

Arbeidsinspectie, 1964. *Aanwijzingen voor de verwerking van onverzadigde polyesters.* De Hoofdinspecteur-Directeur van de Arbeid, Hoofd v/h 4^e district der Arbeidsinspectie, special circular, 5 pp.

Armstrong, G. C., 1967. An instrument for measuring planes and vectors in space, *J. Sediment. Petrol.*, **37** (4), 1241–1243.

Arthur, M. A., 1949. Preserving wet, fragile, geologic specimens, methods of, *J. Sediment. Petrol.*, **19**, 131–134.

Auerbach, M., 1934. Ein quantitativer Bodengreifer. *Beitr. Naturkundi. Forsch. Südwestdeutschld.*, **Bd XII**, 17–22.

Bailey, E. H., and W. P. Irwin, 1959. K-feldspar content of Jurassic and Cretaceous graywackes of northern Coast Ranges and Sacramento Valley, California, *Bull. Amer. Assoc. Petrol. Geologists*, **43**(12), 2797–2809.

———— and R. E. Stevens, 1960. Selective staining of K-feldspar and Plagioclase on rock slabs and thin sections, *Am. Min.*, **45**, 1020–1025.

Bakelite, 1958a. *Catalysts Q.17446/1 and Q.17447 for Bakelite Polyester Resins,* BXL Plastics Materials Group Ltd. Mem. 45520/1875, 3 pp.

————, 1958b. *Accelerator Q.17448 for Bakelite Polyester Resins,* BXL Plastics Materials Group Ltd. Mem. 25525/2279, '3 pp.

————, 1962. *Bakelite Resin SR. 19038,* BXL Plastics Materials Group Ltd., Tech. Mem. P. 122, 3 pp.

————, 1963. *Bakelite Resin DSR-19098,* BXL Plastics Materials Group Ltd., Tech. Mem. P. 126, 3 pp.

Balteau, 1964. *X informations.* Balteau, Industrial X-ray Division No. 2, 4 pp.

————, not dated. *Baltospot 140, portable industrial X-ray unit.* Balteau Electric Corp., descriptive bull., 404–708, 4 pp.

Barclay, A. E., 1947. Microarteriography, *Brit. J. Radiol.*, **20**, 394.

Barkow, D., 1957. Warum und wie Hautschutz für Lackierer? *Industrie-Lackier-Betrieb*, B. 25, 315–316.

————, 1959, Hautschutz und Hautpflege im Betrieb, *Berufsdermatosen*, **VII**(2), 1–12.

Barret, R. E. and I. Fatt, 1953. An improved method of mounting core specimens in Lucite, *J. Petrol. Techn.*, **V**(1), 14.

Basumallick, S., 1966. Size differentiation in a cross-stratified unit. *Sedimentology*, **6** (1), 35–68.

Beales, F. W., 1960. Limestone peels. *J. Alberta Soc. Petrol. Geologists*, **8**, 132–135.

Begemann, H. K. S. Ph., 1961. A new method for taking samples of great lengths, *Proc. 5th Intern. Conf. Soil Mech. Fdtion. Enging.*, 2/2, 1–4.

Bell, J. F., 1939. Notes on the use of methyl methacrylate "Lucite," *Econ. Geol.*, 804–811.

Benthos, not dated. *Boomerang sediment corer.* Benthos, data sheet No. 27, 4 pp.

Berger, K. C. and R. J. Muckenhirn, 1945. Soil profiles of natural appearance mounted with vinylite resin, *Soil Sci. Soc. Amer.*, **10**, 368–370.

Berndtsson, B. and L. Turunen, 1954. Wirkung verschiedener Zusätze auf ungesättigte Polyesterharze, *Kunststoffe*, **44**, 430–436.

Berry, F. G., 1960. A new instrument for determining planar orientation, *Geol. Mag.*, **97**, 309–312.

Bezrukov, P. L. and V. P. Petelin, 1962. Experience in working with impact core-samplers. Translated from Trudy Instituta Okeanologiyi, *Akad. Nauk SSSR,* **5** (1951). 14–16. Directorate of Scien. Inform. Serv. DRB Canada, T 370 R, 3 pp.

Billiton, 1963a. *Lamellon, onverzadigde Polyester harsen,* N.V. Billiton Chemie, overzichtsbrochure nr. 5, 4 pp.

————, 1963b. *Lamellon, hulpmiddelen,* N. V. Billiton Chemie, bijlage behorende bij overzichtsbrochure nr. 5, 2 pp.

————, 1963c. *Lamellon 230,* N. V. Billiton Chemie, 2 pp.

Bissel, H. J., 1957. Combined preferential staining and cellulose peel technique. *J. Sediment. Petrol.*, **27**, 417–420.

Bohatirchuk, F., 1942. Die Fragen der Mikrolöortgenographie, *Fortschr. Geb. Röntgenstrahlen,* **65**, 253.

————, 1961. Medicobiologic research by microradiography. *Encyclopedia of Microscopy,* pp. 591–627.

————, 1963. Magnified X-ray images in medicine and biology (micro-macroradiography and historadiography). In G. L. Clark (Ed.), *Encyclopedia of X-rays and gamma rays,* Reinhold, New York, pp. 567–571.

Borchert, H., 1961a. Einfluss der Bodenerosien auf die Bodenstruktur und Methoden zu ihrer Kennzeichnung, *Geol. Jahr.*, **178**, 439–502.

————, 1961b. Ein methodischer Beitrag zur Entnahme von Bodenproben in ungestörter Lagerung. *Z. Pflanzenern., Düngung u. Bodenk.*, 210–214.

————, 1962. Die Herstellung von Bodendünnschliffen mit dem Festlegungsmittel Plexigum M7466, *Z. Pflanzenern., Düngung u. Bodenk.*, **99**, 159–165.

————, 1963. Bodengefüge-Untersuchungen mit Hilfe von Bodenschliffen, *Die Umschau in Wissenschaft u. Technik*, **14**, 443–446.

————, 1964. Eine Methode zur Untersuchung des Bodengefüges meliorationsbedürftiger Standorte, *Mitt. Deutschen Bodenk. Gesell.*, **2**, 225–229.

Borden, 1961. *Elmer's wood-gluing handbook,* The Borden Chemical Co., a division of the Borden Co., 11 pp.

————, 1965. The Borden Chemical Co., a division of the Borden Co., 25 pp.

Borghgraet, not dated. *Concent. R. B. S. 25.* Chemical-Products R. Borghgraet, bull., 4 pp.

Bouma, A. H., 1962. *Sedimentology of some Flysch deposits, a graphic approach to facies interpretation,* Elsevier, Amsterdam, 168 pp.

_____, 1963. Facies model of salt marsh deposits, *Sedimentology*, 2, 122–129.

_____, 1964a. Self-locking compass, *Marine Geol.*, I, 181–186.

_____, 1964b. Sampling and treatment of unconsolidated sediments for study of internal structures. *J. Sediment. Petrol.*, 34, 349–354.

_____, 1964c. Turbidites. In A. H. Bouma and A. Brouwer (Eds.), *Turbidites*, Elsevier, Amsterdam, pp. 247–256.

_____, 1964d. Ancient and recent turbidites, *Geol. en Mijnbouw*, 43, 375–379.

_____, 1964e. Notes on X-ray interpretation of marine sediments, *Marine Geol.*, II, 278–309.

_____, 1965. Sedimentary characteristics of samples collected from some submarine canyons, *Marine Geol.*, III, 291–320.

_____, and J. A. K. Boerma, 1968. Vertical disturbances in piston cores, *Marine Geol.*, 6(3), 231–241.

_____ and N. F. Marshall, 1964. A method for obtaining and analyzing undisturbed oceanic sediment samples, *Marine Geol.*, II, 81–99.

_____, E. Mutti, and M. H. Maarschalkerweerd, 1965. Tilt-compensating compass. *Geol. Mijnbouw*, 406–408.

_____ and F. P. Shepard, 1964. Large rectangular cores from submarine canyons and fan valleys, *Bull. Am. Assoc. Petrol. Geologists*, 48, 225–231.

Bouma, J., 1965. The preservation of soil properties. Dept. Reg. Soil Science, Agricul. State Univ., Wageningen, The Netherlands (unpub. report), 16 pp.

Bourbeau, G. A. and K. C. Berger, 1947/48. Thin sections of soils and friable materials prepared by impregnation with the plastic "Castolite." *Soil Sci. Soc. Amer.*, 12, 409–412.

Bowen, V. T. and P. L. Sachs, 1964. The free corer, *Oceanus*, 11(2), 2–6.

Bowman, D. W., 1963. Generation of X-rays. In G. L. Clark (Ed.), *Encyclopedia of X-rays and gamma rays*, Reinhold, New York, pp. 416–421.

Brewer, R., 1956. A petrographic study of two soils in relation to their origin and classification, *J. Soil Sci.*, 7, 268–279.

Brison, R. J., 1951. A method for the preparation of polished thin sections of mineral grains, *Am. Min.*, 36, 731–736.

Brown, W. E. and H. W. Patnode, 1953. Plastic lithification of sands in situ, *Bull. Am. Assoc. Petrol. Geologists*, 37, 152–157.

Brownell, C. G., 1951. Making copies of radiographs, *Med. Radiogr. Photogr.*, 27, 114–121.

Bucher, W. H., 1944. The stereographic projection, a handy tool for the practical geologist, *J. Geol.*, 52, 191–212.

Buehler, E. J., 1948. The use of peels in carbonate petrology. *J. Sediment. Petrol.*, 18, 71–73.

Buehler, 1957–1967. *The AB Metal Digest, News of interest for Metallography and Geology*, Buehler Ltd., a bulletin published 4-6 times a year.

_____, 1962. *The metal analyst, a review of selected apparatus and equipment for metallurgical and mineralogical testing*, Buehler Ltd., 296 pp.

_____, 1966. *The Mineralogy-Petrography Catalog, a review of selected apparatus for mineralogy and petrography*, Buehler Ltd., 48 pp.

Buffington, E. C. and C. J. Shipek, 1963. Three dimensions on the sea floor, *Sea frontiers*, 9(2), 78–84.

Buol, S. W. and D. M. Fadness, 1961. New method of impregnating fragile materials for thin sectioning, *Soil Sci. Soc. Amer.*, 25, 253.

Burachek, A. G., 1933, On the method of measurement of orientation of pebbles and cross-bedding. *Repts. All-Russian Mineral. Soc.*, 62, 432–434 (Russian).

Burges, A. and D. P. Nicholas, 1961. Use of soil sections in studying the amount of fungal hyphae in soil, *Soil Sci. U.S.A.*, 92, 25–29.

Burns, R. E., 1963. A note on some possible misinformation from cores obtained by piston-type coring devices, *J. Sediment. Petrol.*, 33, 950–952.

_____, 1966. Free-fall behavior of small, light-weight gravity corers, *Marine Geol.*, 4, 1–9.

Butler, A. J., 1935. Use of cellulose films in Paleontology, *Nature*, 135, 150.

Byrne, J. V. and L. D. Kulm, 1962. An inexpensive lightweight piston corer, *Limnol. and Oceanogr.*, 7(1), 106–108.

Calvert, S. E. and J. J. Veevers, 1962. Minor structures of unconsolidated marine sediments revealed by X-radiography, *Sedimentology*, 1, 296–301.

Carruthers, J. N., 1958. *A leaning-tube current indicator ("Pisa")*, Bull. Inst. Oceanogr. Monaco, 1126, 34 pp.

Carsola, A. J., 1954. Microrelief on Arctic Sea floor, *Bull. Am. Assoc. Petrol. Geologists*, 38, 1587–1601.

Carson, H. L., 1964. Some fundamental aspects of industrial radiography, *Third Intern. Symp. on Industrial Radiography*, Gevaert Photo-producten N. V., Cl-9.

Catt, J. A., and P. C. Robinson, 1961. The preparation of thin sections of clays, *Geol. Mag.*, 98, 511–515.

Cavanaugh, R. J. and C. F. Knutson, 1960. Laboratory technique for plastic saturation of porous rocks, *Bull. Am. Assoc. Petrol. Geologists*, 44, 628–640.

Cementex, 1964. *Directions for Cementex #600, #660 and #600-B (Pre-vulcanized Latex base)*, Cementex Co., Inc., 2 pp.

Ceta Bever, not dated a. *Cetaflex*. N. V. Ceta Bever, Beverwijk, techn. gegevens C 1 (in Dutch), 3 pp.

_____, not dated b. *Fineren en verlijmen van kunststofplaten in de koude pers*, N. V. Ceta Bever, Beverwijk, techn. gegevens C 2, 1 p.

_____, not dated c. *Weerstandverlijmingen voor kantenfineer*. N. V. Ceta Bever, Beverwijk, techn. gegevens C 6, 1 p.

————, not dated d. *Snelfix*, N. V. Ceta Bever, Beverwijk, techn. gegevens S 1 (in Dutch), 1 p.

Charlesby, A., 1963. Plastics and polymers: effects of X- or gamma radiations. In G. L. Clark (Ed.), *Encyclopedia of X-rays and gamma rays*, Reinhold, New York, pp. 768–772.

Chayes, F., 1952. Notes on the staining of potash feldspar with sodium cobaltinitrite in thin section, *Am. Min.*, **37**, 337–340.

Chem. Prod., not dated a. *Chem-o-sol for general purpose dipping*, Chemical Products Corp., Bull. 142, 5 pp.

————, not dated b. *Dip coating with chem-o-sol*, Chemical Products Corp., Instr. Bull. No. 1, 2 pp.

Chem Seal, 1961. *CS 3501, CS 3502, CS 3503 Polyurethane potting and encapsulating compounds*, Chem Seal Corp. of America, data sheet, 4 pp.

————, 1963. *CS 3501, clear polyurethane molding compound*, Chem Seal Corp. of America, data sheet, 6 pp.

————, not dated. *Degassing instructions for polyurethanes*, Chem Seal Corp. of America, data sheet, 2 pp.

Chilingar, G. V., H. J. Bissell, and R. W. Fairbridge, (Eds.), 1967. *Carbonate rocks, physical and chemical aspects*. Devel. in Sedimentology, 9 B, Elsevier, Amsterdam, 413 pp.

Chmelik, F. B., 1967. Electro-osmotic core cutting, *Marine Geol.*, **5**, 321–325.

————, A. H. Bouma and W. R. Bryant, 1968. Influence of sampling on geological interpretation, Trans. Gulf Coast Assoc. Geol. Soc., **XVIII**, 256–263.

CIBA, 1953a, *Araldit-Gieszharz F*, CIBA Akt., Basel, Kunststoffabt., Techn. Bull., 6 pp.

————, 1953b. *Araldite Casting Resin F*, CIBA Ltd., Plastics Dept., Tech. Bull. 245, 7 pp.

————, 1963. *Die Araldit-Produkte der CIBA*, CIBA Ltd., Bull. 30584, 630.425, 29 pp.

————, 1964a. *Araldit*, CIBA Ltd., Bull. 0676-2/600.728, 55 pp.

————, 1964b. *Araldit Epoxyd-Gieszharz-Systeme*, CIBA Ltd., Plastics Dept., Tech. Bull., 4 pp.

————, not dated a. *Araldit F (CY 205), Härter HT 901, HT 902 oder HT 903*, CIBA Ltd., Bull. 32 588, 641.109, 6 pp.

————, not dated b. *Araldit F (CY 205), Härter HY 951*, CIBA Ltd., Tech. Bull. 30423, 621. 120, 2 pp.

————, not dated c. *Araldite 6010*, CIBA Prod. Co., N.J., Tech. Bull. 20820/7, 2 pp.

————, not dated d. *Araldit D (CY 230)*, CIBA Akt., Tech. Broch. 30792, 10 pp.

————, not dated e. *Araldit DY 021 (frühere Bezeichnung: RD-1)*, CIBA Akt., Tech. Broch. 621.119, 4 pp.

————, not dated f. *Araldite RD-1*, CIBA Prod. Co., N.J., Tech. Bull. 111 27/3, 2 pp.

————, not dated g. *Araldite 6005*, CIBA Prod. Co., N.J., Tech. Bull. 20820/7, 2 pp.

————, not dated h. *Araldit E (CY 232), Härter HY 951, HY 943, HY 938, HY 956, oder Lancast A.* CIBA Akt., Tech. Broch. 640.526, 6 pp.

————, not dated i. *Araldit D (CY 230), Härter HY 951*, CIBA Akt., Tech. Broch. 640.929, 8 pp.

————, not dated j. *Araldit D (CY 230), Härter HY 956*, CIBA Akt., Tech. Broch. 640.521, 15 pp.

————, not dated k. *Araldite 502*, CIBA Prod. Co., N.J., Tech. Bull. 30905/3/2, 2 pp.

————, not dated l. *Araldite 506*, CIBA Prod. Co., N.J., Tech. Bull. 10210/2/2/1.5, 2 pp.

————, not dated m. *Araldite Hardener 906 (Revision 1)*, CIBA Prod. Co., N.J., Tech. Bull. 41012/2, 2 pp.

————, not dated n. *Araldite Hardener 956*, CIBA Prod. Co., N.J., Tech. Bull. 20312/7, 2 pp.

————, not dated o. *Lancast A*, CIBA Prod. Co., N.J., Tech. Bull. 11121/3/1.5, 3 pp.

————, not dated p. *Arigal C*, CIBA Ltd., Tech. Bull. 611.001, 8952, 21 pp.

Cintel, 1961. *Cintel electronic controlled enlarger, types 75/54*, Rank Cintel Ltd. (now Milligan Electronics Ltd.), 20 pp.

————, not dated. *Cintel electronic controlled printers and copiers, types 30411/2*, Rank Cintel Ltd. (now Milligan Electronics Ltd.), 20 pp.

Clark, G. L. (Ed.), 1963. *Encyclopedia of X-rays and gamma rays*, Reinhold, New York, 1149 pp.

Clark, W., 1946. *Photography by infrared, its principles and applications, 2nd ed.*, Wiley, New York, 472 pp.

CM², not dated. *CM² Oceanographic and Limnological instrumentation catalog*, 48 pp.

Cochran, J. B., 1959. *Instruction manual for oceanographic observations*, Hydrographic Office Pub. No. 607, Washington 25, D.C., 210 pp.

Cochran, W. G., 1965. *Sampling techniques, 2nd ed.*, Wiley, New York, 413 pp.

Coffee, C. E., 1968. A new technique in sand coring, *Undersea Tech.*, **9**(3), 35–37.

Coleman, J. M., 1963a. Preparation of thin sections by plastic impregnation, Coastal Studies Inst., Louisiana State Univ., written comm., 2 pp.

COMEX, not dated. *Emploi du vibro-carottier C.O.M.E.X.*, Compagnie Maritime d'Expertises, Inform. Bull., 3 pp.

Conil, R. et M. Lys, 1964. *Matériaux pour l'étude micropaléontologique du Dinantien de la Belgique et de la France (Avesnois)*, Mém. Inst. Géol., Univ. Louvain, Tome XXIII, 1ᵉ partie: Alques et Foraminifères, 296 pp.

Conix, A., 1950. De Kinetica van de polymerisatiereacties, *Plastica*, **3**, 253–256; 289–293.

Conkin, J. E., 1956. Plastic spray in laboratory and field, *J. Sediment. Petrol.*, **26**, 68.

Conrad, not dated a. *Universal-Trennmaschinen*, Wolfgang Conrad; Druckschrift/o, 2 pp.

————, not dated b. *Gesteinssägen mit Diamant Trennscheiben*, Wolfgang Conrad, Druckschrift 13/62, 2 pp.

————, not dated *c. Gesteinssägen Zusatzeinrich-tungen*, Wolfgang Conrad, Druckschrift 14/62, 2 pp.

Corney, G. M., 1963. X-ray film. In G. L. Clark (Ed.), *Encyclopedia of X-rays and gamma rays*, Reinhold, New York, pp. 375–381.

Cosslett, V. E. and W. C. Nixon, 1955. Projection microradiography, *Brit. J. Radiol.*, **28**, 532.

Courtright, not dated a. *No. 70C Lakeside brand, thermoplastic transparent cement in bar form*, Hugh Courtright & Co., tech. broch., 6 pp.

————, not dated b. *No. 30 L Lakeside brand, temporary cement in bar form*, Hugh Courtright & Co., tech. broch., 6 pp.

Curray, J. R., 1955. Instructions for artificial lithification of sand, Scripps Inst. Oceanogr., unpubl. report, 3 pp.

Cutrock, not dated a. *Core and rock cutting machine, model VCC 500*, Cutrock Engineering Co., data sheet, 2 pp.

————, not dated b. *Refractory brick cutting machine, model BCM 200*, Cutrock Engineering Co., data sheet, 2 pp.

————, not dated c. *Auto feed rock cutting machines, models ARC 100 P and ARC 100 H*, Cutrock Engineering Co., data sheet, 2 pp.

————, not dated d. *Overarm cut off machine, model OAC 200*, Cutrock Engineering Co., data sheet, 2 pp.

————, not dated e. *Mark II Unicutta, thin section cutting and grinding machine*, Cutrock Engineering Co., data sheet, 2 pp.

————, not dated f. *Core drilling and cutting machine, model CD 200*, Cutrock Engineering Co., data sheet, 2 pp.

————, not dated g. *Geological laboratory and field equipment*, Cutrock Engineering Co., data sheet, 32 pp.

————, not dated h. *Core and cube cutting machine, model GSP 250*, Cutrock Engineering Co., data sheet, 2 pp.

————, not dated i. *Thin section cutting and grinding machine, model GH 3*, Cutrock Engineering Co., data sheet, 2 pp.

————, not dated j. *The Croft parallel grinder*, Cutrock Engineering Co., data sheet, 1 p.

————, not dated k. *High-speed diamond wheel grinding machine, model LSC 100*, Cutrock Engineering Co., data sheet, 1 p.

————, not dated l. *Cutrock single and double spindle lapping machines*, Cutrock Engineering Co., data sheet, 1 p.

Cyanamid, 1961. *AM-9 chemical grout*. American Cyanamid Co., AM-9 tech. data, A-2907-2000-6/61, 60 pp.

Cywinski, J. W., 1960. *The role of organic peroxides in curing polyester resins and their influence on the physical properties of reinforced plastics*, Novadel Ltd., Lucidol Prod. Div., London.

Dahl, O. and J. E. Fjeldstad, 1949. A new repeating current meter, *P. V. Int. Assn. phys. Oceanogr.*, **4**, 131–132. Gen. Assem., Oslo 1948.

Dalrymple, J. B., 1957. Preparation of thin sections of soils, *J. Soil Sci.*, **8**, 161–165.

Daniels, F., 1963. Dosimetry: thermoluminescence. In G. L. Clark (Ed.), *Encyclopedia of X-rays and gamma rays*, Reinhold, New York, pp. 316–317.

Darrah, W. C., 1936. The peel method in paleobotany, *Harvard Univ. Botanical Leaflets*, **4**, 69–93.

Day, P. R., 1949. Experiments in the use of the microscope for the study of soil structure, *Soil Sci. Soc. Amer., Proc.*, **13**, 43–50.

De Boer, J., 1963. *The Geology of the Vicentinian Alps* (NE Italy), Thesis, Univ. Utrecht, The Netherlands, 178 pp.

Debyser, J., 1957. Note sur un procédé de préparation des plaques minces dans les sédiments fins actuel. *Rev. Inst. France Petrole.*, **12**, 489–492.

Deevey, E. S., Jr., 1965. Sampling lake sediments by use of the Livingstone sampler. In B. Kummel and D. Raup (Eds.), *Handbook of paleontological techniques*, Freeman and Co., San Francisco, pp. 521–529.

De Ridder, N. A., 1960. Minor structures of some Holocene and Pleistocene sediments from the southwestern part of The Netherlands, *Geol. en Mijnbouw*, **39**, 679–685.

Diamant Boart, 1961. *Outils diamantés et machines pour laboratoires*, Diamant Boart S. A., instr. leaflet 5/5.61/1F, 16 pp.

————, 1964. *Gediamanteerde gereedschappen voor het bewerken van steen, marmer en graniet*, Diamant Boart S. A., instr. leaflet 9/4.62/3 F1, 71 pp.

Dill, R. F. and G. Shumway, 1954. Geologic use of self-contained diving apparatus, *Bull. Am. Assoc. Petrol. Geologists*, **38**, 148–157.

Doeglas, D. J., J. Ch. L. Favejee, D. J. G. Nota, and L. Van der Plas, 1965. On the identification of feldspars in soils, *Mededelingen Landbouwhogeschool, Wageningen, Nederland*, **65**-9, 1–14.

Dollar, A. T. J., 1942. Laminar molds in cellulose acetate for the study of rock and mineral structures, *Geol. Mag.*, **79**, 253–255.

Dollé, P., 1959. Note sur les méthodes de travail employées au laboratoire du service géologique des Houillères du Bassin du Nord et du Pas-de-Calais, *Rev. de l'Industrie Minérale*, **41**(2), 1–10.

Dow Chemical, not dated a. *Dow liquid epoxy resins*, The Dow Chemical Co., Plastics Dept., 31 pp.

————, not dated b. *Dow liquid epoxy resins D.E.R. 331, D.E.R. 332 and D.E.R. 334*, The Dow Chemical Co., Plastics Dept., 27 pp.

Dow Corning, 1963a. *Silastic RTV room temperature vulcanizing silicone rubber*, Dow Corning Corp., Bull. 08-034, 12 pp.

————, 1963b. *Silastic RTV selector for mold making*, Dow Corning Corp., Bull. 08-039, 1 p.

————, 1963c. *Making and using release agents for molds of Silastic RTV,* Dow Corning Corp., Bull. 08-041, 1 p.

————, 1963d. *Silastic 732 RTV Silicone rubber for repairing flexible molds,* Dow Corning Corp., Bull. 08-063, 2 pp.

————, 1963e. *Silastic 502 RTV Silicone rubber,* Dow Corning Corp., Bull. 08-064, 4 pp.

————, 1963f. *Silastic 588 RTV Silicone rubber,* Dow Corning Corp., Bull. 08-065, 3 pp.

————, 1963g. *Silastic 589 RTV Silicone rubber,* Dow Corning Corp., Bull. 08-066, 3 pp.

————, 1964a. *Silastic 1202 RTV primer,* Dow Corning Corp., Bull. 02-015, 2 pp.

————, 1964b. *Silastic RTV mold release,* Dow Corning Corp., Bull. 08-107, 2 pp.

————, 1966a. *Silastic A RTV Mold Making Rubber,* Dow Corning Corp., Bull. 08-240, 3 pp.

————, 1966b. *Silastic B RTV Mold Making Rubber,* Dow Corning Corp., Bull. 08-241, 3 pp.

————, 1966c. *Silastic C RTV Mold Making Rubber,* Dow Corning Corp., Bull. 08-242, 3 pp.

————, 1966d. *Silastic D RTV Mold Making Rubber,* Dow Corning Corp., Bull. 08-243, 3 pp.

Drake, J., 1962. *Salt for short gel times at low temperatures,* Am. Cyanamid Co., Soils Engin. Res., Princeton, 4 pp.

Dunn, J. A., 1937. Polished and thin-section technique in the laboratory of the Geological Survey of India; *Geol. Surv. India, Records,* 72, 207–226.

DuPont, 1955. *"Lucite" acrylic resin. Embedment of specimens,* DuPont de Nemours & Co., Polychemicals Dept., Tech. Bull. No. X-28c, 10 pp.

————, 1959. *Ortholeum 162 lubricant assistant,* DuPont Petroleum Chem. Division, Bull. A-10008, 4 pp.

————, 1966. *Lucite acrylic resins. Embedment and casting techniques,* DuPont de Nemours & Co., Plastics Dept., Tech. Bull. A-51886, 8 pp.

————, not dated. *"Lucite" acrylic resin. Molding of monomer-polymer doughs,* DuPont de Nemours & Co., Polychemicals Dept., Tech. Bull. No. X-27c, 5 pp.

Dynamit Nobel, not dated. *Lack collodiumwollen,* Dynamit Nobel Aktiengesellschaft, Tech. data, 1 p.

Easton, MW. H., 1942. An improved technique for photographing peel sections of corals, *J. Paleontol.,* 16, 261–263.

Ecko, not dated. *Ecko radiation monitor type N 596 B,* Tech. instructions (issue 2), Ecko Electronics Ltd., 15 pp.

Eering, J. Th., 1965. Enkele aspecten rondom de röntgenfotografie, Lecture Röntgen Technische Dienst, written by A. H. Bouma, 15 pp.

Ekman, V. W., 1901. On a new current-meter invented by Prof. Fridtjof Nansen, *Nyt Mag. Naturzid., Christiania,* 39, 163–187.

————, 1905. *Kurze Beschreibung eines Propellstrommessers,* Publ. Circ. Cons. int. Explor. Mer., 24, 4 pp.

Emery, G. R. and D. E. Broussard, 1954. A modified Kullenberg piston corer, *J. Sediment. Petrol.,* 24, 207–211.

Emery, K. O. and R. S. Dietz, 1941. Gravity coring instrument and mechanics of sediment coring, *Bull. Geol. Soc. Am.,* 52, 1685–1714.

———— and J. Hülsemann, 1964. Shortening of sediment cores collected in open barrel gravity corers, *Sedimentology,* 3, 144–154.

———— and R. E. Stevenson, 1950. Laminated beach sand, *J. Sediment. Petrol.,* 20, 220–223.

Emiliani, C., 1951. Notes on thin sectioning of smaller foraminifera, *J. Paleontol.,* 25, 531–532.

Engström, A., 1949. Microradiography, *Acta Radiol.,* 31, 503.

———— and Lindström, B., 1951. The properties of fine-grained photographic emulsions used for microradiography, *Acta Radiol.,* 35, 33.

Evamy, B. D., 1963. The application of a chemical staining technique to a study of dedolomitization, *Sedimentology,* 2, 164–170.

Ewing, M., A. C. Vine, and J. L. Worzel, 1946. Photography of the ocean bottom, *J. Opt. Soc. Amer.,* 36, 307–321.

Exley, C. S., 1956. A method of impregnating friable rocks for cutting of thin sections, *Mineral. Mag.,* 31, 347–349.

Fabelta, 1963. Caractéristiques moyennes des qualités d'acétate Tubize, Fabelta, Tubize, Belgique, letter July 12, 1963, 1 p.

Fairbridge, R. W., 1966. *The Encyclopedia of Oceanography,* Reinhold, New York, 1056 pp.

Farkas, A. and E. Passaglia, 1950. The decomposition of cyclohexylhydro-peroxide and the peroxide-catalyzed polymerization of styrene, *J. Am. Chem. Soc.,* 72, 3333–3337.

Farrow, G. E., 1966. Bathymetric zonation of Jurassic trace fossils from the coast of Yorkshire, England, *Paleogeography, Polaeoclimatology, Palaeoecology,* 2, 103–151.

Fay, R. O., 1961. *Blastoid studies,* The University of Kansas, Paleontological contributions: Echinodermata article 3, The University of Kansas Publ., 147 pp.

Felsher, M., 1964. A core-aligner designed to recover oriented recent marine cores, *Limnol. Oceanogr.,* 9 (4), 603–605.

Fenton, M. A., 1935. Nitrocellulose sections of fossils and rocks, *Amer. Midland Naturalist,* 16, 410–412.

Fleming, C. S. and S. E. Calvert, not dated. *The piston immobilizer, a further modification to the piston corer.* Rep. Scripps Inst. of Ocean., Univ of California, San Diego, 12 pp.

Flügger, 1964a. Impredur Luftlack 784 für die Konservierung geologischer Objekte. J. D. Flügger, Lack- und Farbenfabrik, Hamburg, written comm., August 4, 1964, 2 pp.

————, 1964b. *Kulturschichten auf Lack gezogen,* Flügger Rundbrief, 9. Jahrgang, Nr. 2-3. J. D. Flügger, Lack- und Farbenfabrik, Hamburg, paper by W. Hähnel, pp. 379–381.

Folk, R. L., 1959. Practical petrographic classification of limestones, *Bull. Am. Assoc. Petrol. Geologists,* **43**, 1–38.

Foucar, K., 1938. Die Bergung eines Flieszerdeprofils aus dem Saale-Weichsel-Interglazial bei Halle a.d. Saale mittels des Lackfilmverfahrens, *Z. Geschiebeforschg. Flachlandsgeol.,* **14**, 104–126.

Fowler, J. W. and J. Shirley, 1947. A method of making thin sections from friable materials and its use in the examination of shales from the Coal Measures, *Geol. Mag.,* **84**, 354–359.

Franzisket, L., 1962a. Präparation des Torfprofils eines lebenden Hochmoores, *Der Präparator, Z. Museum Technik,* **2**, 47–53.

———, 1962b. Die Darstellung von Pfanzengesellschaften aus den zugehörigen Bodenprofielen im Landesmuseum für Naturkunde in Munster, *Museumkunde,* 53–61.

Frei, E., 1947. Mikromorphologische Bodenuntersuchungen anhand von Dünnschliffen, *Mitt. Lebensmittelunters. und Hygiene,* **38**, 138–144.

Frencken, 1962. *Het insluiten van voorwerpen in heldere kunsthars,* Frencken's Fabrieken, Tech. Broch. No. 60, 7 pp.

Friedman, G. M., 1959. Identification of carbonate minerals by staining methods, *J. Sediment. Petrol.,* **29**, 87–97.

Fuchs, A. W. and G. M. Corney, 1965. Radiographic recording media and screens. In *The Science of Ionizing Radiation,* Lewis E. Etter (Ed.), Charles C. Thomas, London, 173–184.

Füchtbauer, H. und H. E. Reineck, 1963. Porosität und Verdichtung rezenter, mariner Sedimente, *Sedimentology,* **2**, 294–306.

Fuller, 1957. *Marine finishes 6605 Glas-skin plastic resin and 6606 catalyst,* Fuller & Co., tech. data No. 153, 3 pp.

Furane, 1962. *Epocast 530-2, sand binder for flooring and grouting,* Furane Plastics Inc., Tech. Bull. 62-17 (Mod. 1), 3 pp.

Gabriel, A. and E. P. Cox, 1929. A staining method for the quantitative determination of certain rock minerals, *Am. Min.,* **14**, 290–292.

Gadgil, P. D., 1962. Soil sections of grassland. *Prel. Proc. Coll. on soil fauna and soil micro flora and their relationships,* Oosterbeek, pp. 244–249.

Gendron, N. J., 1958. *Mounting of metallographic samples in clear, cold setting plastic,* General Electric Co., Materials and Processes Lab., DF 58 SL 306, 11 pp.

———, 1959. Mounting of geological specimens in clear cold-setting plastic, *Econ. Geol.,* **54**, 308–310.

General Electric, 1958a. *1202 insulating varnish,* Product data-401, General Electric Co., Insulating materials, October 15, 1958, 2 pp.

———, 1958b. *G-E 1557 Glyptal Clear Lacquer,* Product data, General Electric Co., Insulating Materials, March 27, 1958, 2 pp.

———, 1963. *1276, 1285, 1286 Glyptal Cements,* General Electric, Insulating Materials Dept., Product data 601, 2 pp.

General Electric, X-ray Dept., 1961. *Mobile "90-II" X-ray unit,* General Electric, X-ray Dept., Product Data Sheet A 0024, 2 pp.

General Mills, 1961. *Genamid 250 liquid epoxy coreactant,* General Mills, Chemical Division, data sheet, 2 pp.

———, 1966. *Plus value chemicals from General Mills,* General Mills, Chemical Division, data sheet G-30, 8 pp.

Germann, K., 1965. Die Techniek des Folienabzuges und ihre Ergänzung durch Anfarbemethoden, *N. Jb. Geol. Paläont. Abh.,* **121**(3), 293–306.

Gevaert, 1964. *Industriële röntgenfilms,* Gevaert Photo-producten N. V., 29 pp.

———, not dated. *Industrial radiography,* Gevaert Photo-producten N. V., 100 pp.

Geyger, E., 1962. Zur Methodik der mikromorphometrischen Boden Untersuchung, *Z. Pflanzenern., Düngung u. Bodenk.,* **99**, 118–129.

Gianturco, C., 1963. Low exposure radiography. In G. L. Clark (Ed.), *Encyclopedia of X-rays and gamma rays,* Reinhold, New York, pp. 900–902.

Gibson, H. L., 1951. Copying radiographs with miniature Kodachrome film, *Med. Radiogr. Photogr.,* **27**, 125–128.

———, 1962. Copying radiographs with the new Kodak colour films, *Med. Radiogr. Photogr.,* **38**, 117.

Ginsburg, R. N., H. A. Bernard, R. A. Moody, and E. E. Daigle, 1966. The Shell method of impregnating cores of unconsolidated sediments, *J. Sediment. Petrol.,* **36**, 1118–1125.

———, and R. M. Lloyd, 1956. A manual piston coring device for use in shallow water, *J. Sediment. Petrol.,* **26**, 64–66.

Glashoff, H., 1935. Ein einfaches Gerät zur Messung von Schrägschüttungen, *Mitt. geol. Staatsinst. Hamburg,* No. 24, 46–47.

G. M. Mfg., 1967. *The 1967 Model shallow water current-indicating bottle (Ref NIO/4972) for use in all waters at depths not exceeding about 60 fathoms (120 meters),* G. M. Mfg. & Instrument Corp., data sheet with calibrating curves, 4 pp.

———, not dated. *Carruthers Current Indicating Bottles (PISA) No. 231 WA 605,* G. M. Mfg. & Instrument Corp., data sheet, 2 pp.

G.M.M.C., not dated a. *No. 214 WA 150 Emery dredge,* G. M. Manufacturing Co., Bull. WAP-0463, 2 pp.

———, not dated b. *No. 425 Heavy duty bottom sampler,* G. M. Manufacturing Co., Bull. I-J, 2 pp.

———, not dated c. *No. 214 WA 130 Dietz-LaFond bottom sampler, No. 214 WA 110 mud snapper, and Nos. 214 WA 200-220-230-240 orange peel bucket,* G. M. Manufacturing Co., Bull. WAP-1064, 2 pp.

———, not dated d. *Instrument sales program,* G. M. Manufacturing & Instrument Corp., Bull., 4 pp.

———, not dated e. *The 1967 model shallow water current-indicating bottle (Ref. NIO/4972) for use in all waters at depths not exceeding about 60 fathoms (120 meters)*, G. M. Manufacturing & Instrument Corp., 5 pp.

Goby, P., 1913. New application of the X-rays: microradiography, *Royal Micr. Soc. J.*, 4, 373–375.

Goemann, H. B. 1937. Ein Härtungsverfahren mit Dioxan und Parraffin, *Senckenbergiana*, 19, 77–80.

———, 1940. Die Konservierung angetriebener meerischer Holz- und Torffunde mit I. G.-Wachs, *Senckenbergiana*, 22, 5/6, 9.

Gray, D. E.(Coordinating Ed.). 1957. *American Institute of Physics Handbook*, McGraw-Hill, New York, 1532 pp.

Grosskopf, W., 1937. Der Bodenfilm, *Weltforstwirtsch.*, 4, 331–332.

Gumenskii, B. M. and N. S. Komarov, 1961. *Soil drilling by vibration*, Transl. from Russian ed. 1959 by Consultants Bureau, New York, 80 pp.

Gwinner, M. P., 1963. Lackfilme als Hilfsmittel bei der Herstellung von Gesteinsschnittbildern, *Photographische Korrespondenz, Int. Z. wissenschaft angewandte Photographie und gesamte Reproduktionstechnik*, Bd. 99, Nr. 11, 188–190.

Haarlov, N. and T. Weiss-Fogh, 1963. A microscopical technique for studying the undisturbed structure of soils, *Z. Pflanzenern., Düngung u. Bodenk.*, 99, 140–144.

Hageman, B. P., 1963. A new apparatus for taking cores in unconsolidated sediments, *Verh. Koninkl. Ned. Geol. Mijnb. Gen., Geol. Ser.*, 21–2, 221–223.

Hagn, H., 1953. Polestar, ein Hilfsmittel für die Herstellung orientierter Dünnschliffen kleiner Objekte, *Mikrokosmos*, 43, 68–72.

Hähnel, W., 1961. Die Lackfilmmethode zur Konservierung. *Der Präparator. Z. Museumtechnik*, 4, 243–263.

———, 1962. The Lacquer-film Method of Conserving Geological Objects, *Curator*, 5(4), 353–368.

Halmshaw, R., 1966. The image on a radiograph and its interpretation. *Fourth Intern. Symp. Industrial Radiography*, Gevaert-Agfa N. V., H 1-8.

Hamblin, W. K., 1962a. X-ray radiography in the study of structures in homogeneous sediments, *J. Sediment. Petrol.*, 32, 201–210.

———, 1962b. Staining and etching techniques for studying obscure structures in clastic rocks, *J. Sediment Petrol.*, 32, 530–533.

———, 1963. Radiography of rock structures. In G. L. Clark (Ed.), *Encyclopedia of X-rays and gamma rays*, Reinhold, New York, pp. 940–941.

———, 1965. *Internal structures of "homogeneous" sandstones*, State Geol. Survey of Kansas, Bull. 175, part 1, 37 pp.

——— and J. Van Sant, 1963. Radiography in paleontology. In G. L. Clark (Ed.), *Encyclopedia of X-Rays and gamma rays*, Reinhold, New York, pp. 684–686.

Hamilton, E. L., 1960. Ocean basin ages and amounts of original sediments, *J. Sediment. Petrol.*, 30, 370–379.

———, 1963. Sea floor relief, *Trans. Am. geophys. Un.* 44, 493–494.

Hanna, G. D., 1927. Synthetic resin as a mounting medium, *Science, N. S.*, 55, 575–576.

Hanna, M. H., 1927. A simple thin-section lap, *J. Paleontol.*, 1, 219–220.

Häntzschel, W., 1938. Die Anwendung des Lackfilms in der Meeresgeologie, *Natur u. Volk*, 68, 117–119.

Harbaugh, J. W., 1953. Scotch tape aids thin-section studies, *Bull. Am. Assoc. Petrol. Geologists*, 37, p. 452.

Harper, H. J. A., 1932. A study of methods for preparation of permanent soil profiles, *Okla. Agr. Exp. Sta. Bull.*, 201.

Harper, H. J. and G. W. A. Volk, 1936. A method for the microscopic examination of the natural structure and pore space in soils, *Soil. Sci. Soc. Amer., Proc.* 1, 39–42.

Harrison, C. G. A., J. C. Belshé, A. S. Dunlap, J. D. Mudie, and A. I. Rees, 1967. A photographic compass inclinometer for the orientation of deep-sea sediment samples, *J. Ocean Technology*, I(2), 37–39.

Harrison, J. B., O. L. Mageli, and S. D. Stengel, 1962. Polyester polymerisation with mixed catalyst systems, *Modern Plastics*, 39, 9 pp.

Hayes, J. R. and M. A. Klugman, 1959. Feldspar staining methods, *J. Sediment. Petrol.*, 29, 227–232.

H.B.N.P.C., 1959. *Polyesters norsodyne, norsomix et norsopore*, Houillères du Bassin du Nord et du Pas-de-Calais, 64 pp.

———, 1962. *Polyesters Norsodyne et Norsomix*, Houillères du Bassin du Nord et du Pas-de-Calais, 61 pp.

———, not dated. *Inclusions d'objects divers et de pièces anatomiques dans les Norsodyne*. Houillères du Bassin du Nord et du Pas-de-Calais, 6 pp.

Heezen, B. C., 1954. Methods of exploring the ocean floor: a discussion. *Oceanographic Instrumentation*, Nat. Acad. Science, Nat. Res. Coun. Publ. No. 309, 200–205.

——— and C. Hollister, 1964. Deep-sea current evidence from abyssal sediments, *Marine Geol.*, I, 141–174.

——— and G. L. Johnson III. 1962. A peel technique for unconsolidated sediments, *J. Sediment. Petrol.*, 32, 609–613.

Henkel, 1961a. *P3-asepto*, Henkel-Werke, Bull. 104, 3 pp.

———, 1961b. *P3-zinnfest*, Henkel-Werke, Bull. 130, 3 pp.

Hepple, S. and A. Burges, 1956. Sectioning of soil, *Nature*, 177, 1186.

Hercules, not dated a. *Hercules Cumene Hydroperoxide*, Hercules Inc., Synthetics Dept., product data no. 260-1, 1 p.

———, not dated b. *Hercules hydroperoxides*, Hercules Inc., Synthetics Dept., technical data, Bull. no. S-143, 10 pp.

Herrnbrodt, A., 1954. Eine neue Lackfilmmethode: Das Capaplex-Verfahren, *Bonner Jahrb. d. rhein-Landesmuseums in Bonn,* Heft 154, Kevelaer, 182–184.

Hickam, R., 1956. The preparation of thin sections of well cuttings, *J. Sediment. Petrol.,* **26,** 165.

Hicol, not dated. *RBS-25 concentraat,* "Hicol," Laboratorium chemicalien, chemische producten, industrie chemicalien, specialiteiten, bulletin, 4 pp.

Hill M. N. (Ed.), 1962–63. *The Sea: Ideas and Observations on Progress in the Study of the Sea,* Interscience, New York, Vol. I, 844 pp; Vol. II, 554 pp; Vol. III, 963 pp.

Hodgman, C. D., R. C. Weast, and S. M. Selby, (Eds.-in-chief), 1960. *Handbook of Chemistry and Physics, 41st ed.,* Chemical Rubber, Cleveland, Ohio, 3472 pp.

Hoechst, 1959. *Mowilith,* Farbwerke Hoechst AG. (in German), 160 pp.

———, 1960a. *Mowilith,* Farbwerke Hoechst AG. (in English), 135 pp.

———, 1960b. *Mowilith,* Farbwerke Hoechst AG. (in French), 154 pp.

———, 1964. *Mowilith 35/73,* Farbwerke Hoechst AG., 3 pp.

———, 1965a. *Synthetic Resins and Raw Materials for the Paint and Varnish Industry: Tables.* Farbwerke Hoechst AG., 40 pp.

———, 1965b. *Mowilith 35/73,* Farbwerke Hoechst AG. (in English), 2 pp.

Holme, N. A., 1964. Methods of sampling the benthos. In F. S. Russell (Ed.), *Advances in marine biology, Vol. 2.* Academic Press, London, 171–260.

Hook, G., 1952. Hautschutz bei stark verschmutzenden Arbeiten, *Haut- und Geschlechtskrankheiten,* **XII**(5), 211–214.

Hopkins, T. L., 1964. A survey of marine bottom samplers. In Mary Sears (Ed.), *Progress in oceanography, Vol. 2,* Pergamon-MacMillan, 213–255.

Hospers, J., 1950. Over het conserveren van bodemprofielen, *Geol. en Mijnbouw,* **12,** jrg., 155–157.

Howard, J. D. and V. J. Henry, Jr., 1966. Sampling device for semiconsolidated and unconsolidated sediments, *J. Sediment. Petrol.,* **36,** 818–820.

Hudson, 1961. *Hudson sprayers and dusters,* Booklet No. 566 of the H. D. Hudson Manufacturing Company, 12 pp.

Hüls, 1961. *Vestopal,* Chemische Werke Hüls Aktiengesellschaft, 52 pp.

Hulshof, H. J., 1955. *Handleiding voor het vervaardigen van lakfilms en bodemprofielen,* Tuinbouwgids, p. 281.

Hvorslev, M. J., 1949. *Subsurface exploration and sampling of soils for civil engineering purposes,* Vicksburg, Miss., (U.S. Army, Corps of Engineers) Waterways Experiment Station, 521 pp.

——— and T. B. Goode, 1960. *Core drilling in frozen ground,* U.S. Army Engineer Waterway Exp. St.,

CE, Vicksburg, Miss., Tech. Rep. No. 3-534, 36 pp.

———and H. C. Stetson, 1946. Free-fall coring tube; a new type gravity bottom sampler, *Bull. Geol. Soc. Am.,* **57,** 935–950.

Hydro Products, not dated a. *Bouma Box Sampler, model 610,* Hydro Products, Bull. 610, 1 p.

———, not dated b. *Shipek sediment sampler, model 860,* Hydro Products, Broch., 2 pp.

———, not dated c. *The Shipek sediment sampler,* Hydro Products, Spec. Bull., 3 pp.

———, not dated d. *Underwater instruments and accessories,* Hydro Products, Spec. Bull., 63 pp.

Ilford, 1956. *Ilford IF-9 Chrome Alum quick fixing salt,* Ilford Ltd., Techn. Inf. Sheet D 352, 1 p.

———, 1958a. *Ilford Industrial B X-ray film,* Ilford Ltd., Techn. Inf. Sheet A 762, 1 p.

———, 1958b. *Ilford Industrial F X-ray film,* Ilford Ltd., Techn. Inf. Sheet A 765, 1 p.

———, 1958c. *Ilford Industrial A X-ray film,* Ilford Ltd., Techn. Inf. Sheet A 761, 1 p.

———, 1959. *Ilford Industrial CX X-ray film,* Ilford Ltd., Techn. Inf. Sheet A 766, 1 p.

———, 1961a. *Ilford Industrial G X-ray film,* Ilford Ltd., Techn. Inf. Sheet A 71.5, 8 pp.

———, 1961b. *Ilford Phen-X X-ray developer,* Ilford Ltd., Techn. Inf. Sheet D 26.5, 2 pp.

———, 1961c. *Ilford Phenisol X-ray developer,* Ilford Ltd., Techn. Inf. Sheet D 26.6, 2 pp.

———, 1962a. *Ilford IF-22 X-ray acid hardening fixing salt,* Ilford Ltd., Techn. Inf. Sheet D 30.3, 1 p.

———, 1962b. *Ilford Hypam fixer and hardener,* Ilford Ltd., Techn. Inf. Sheet D 35.2, 2 pp.

———, not dated. *X-ray films, screens and chemicals for industrial radiography,* Ilford Ltd., 27 pp.

Ingerson, E. and J. L. Ramisch, 1954. Studies of unconsolidated sediments, *Tschermaks Min. und Petrogr. Mitt.,* **4,** 117–124.

Jackson, E. D. and D. C. Ross, 1956. A technique for modal analyses of medium- and coarse-grained (3-10 mm) rocks, *Am. Min.,* **41,** 648–651.

Jager, A., 1959. *Handleiding voor het conserveren van bodemprofielen,* Wageningen, Landbouwhogeschool, Afd. Regionale Bodemkunde, 11 pp.

——— en A.C.F.M. Schellekens, 1963. Handleiding voor het conserveren van zware en/of natte bodemprofielen (Manual for conservation of heavy and wet soil profiles), *Boor en Spade* (Auger and Spade), **XIII,** 61–66.

——— en W. J. M. Van der Voort, 1965. *Het nemen en conserveren van bodemmonolieten van zandgronden en gerijpte kleigronden boven en beneden de grondwaterspiegel,* Soil Survey Papers, No. 2, Soil Survey Institute, Wageningen, The Netherlands, 13 pp.

Jenny, H. and K. Grossenbacker, 1963. Root Soil Boundary zones as seen in the electron microscope, *Soil Sci. Soc. Amer., Proc.,* **27,** 273–275.

Jessen, W., 1938a. Versteifung von Lackprofilen bei schwer ablösbaren Gesteinen, *Z. Deut. Geol. Ges.,* **90,** 51–54.

————, 1938b. Lackprofile in der erdgeschichtlichen Heimat-Schausammlung, *Natur u. Volk*, **68**, 120–123.

Johnson, C. R. and O. M. Kiel, 1961. A simple drilling tool for orienting cores at shallow depth, *J. Sediment. Petrol.*, **31**, 609–610.

Jonasson, A. and E. Olausson, 1966. New devices for sediment sampling, *Marine Geol.*, **4**, 365–372.

Jongbloed, W. L., 1965. Recent microscopisch onderzoek, *TNO Nieuws (The Netherlands)*, **20**, 227–233.

Jongerius, A., 1957. *Morfologische onderzoekingen over de bodemstructuur*, Diss. Wageningen. Verslagen Landbouwk. Onderzoek No. 63.12. Bodemkundige studies 2. Den Haag, 93 pp.

———— en G. Heintzberger, 1962. *De vervaardiging van mammoet-slijpplaten*, Stichting voor Bodemkartering, Wageningen, no. 2816, 17 pp.

————, 1963. *The preparation of mammoth sized thin sections*, Soil Survey papers nr. 1. Neth. Soil Survey Instit. Wageningen, 37 pp.

Kahl, not dated. *Instrumentation for Science and Industry*, Kahl Scientific Instr. Corp., Different Bull. combined, incl. from G. M. Mfg. & Instr. Corp., 107 pp.

Kahlsico, not dated. *No. 507/SM mud snapper*, Kahl Scientific Instr. Corp., Bull. H-1-63, 2 pp.

Kalle, not dated a. *Glutofix 600*, Kalle Aktiengesellschaft, Bull. 4948-6.200.511 V, 2 pp.

————, not dated b. *Tylose C 10, C 30, C 300, C 600, C 6000, C 10,000 und C 30 p, C 1000 p, C 10,000 p*, Kalle Aktiengesellschaft, Bull. 4753-6.130.012 XIX, 12 pp.

————, not dated c. *Tylose MH 20, MH 50, MH 200, MH 300, MH 1000, MH 2000, MH 4000 und MH 50 p, MH 200p, MH 300 p, MH 1000 p, MH 2000 p, MH 4000 p*, Kalle Aktiengesellschaft, Bull. 4753-6.130.012 XIX, 12 pp.

————, not dated d. *Tylose C (Carboxymethylcellulose)*, Kalle Aktiengesellschaft, Bull. 4955-6.200.812 1/engl., 24 pp.

————, not dated e. *Tylose MH, MB and MH-K grades (methylcellulose)*, Kalle Aktiengesellschaft, Inf. Bull. 4869-6.200.505 1/engl., 24 pp.

————, not dated f. *Tylose H 20 p, H 4000 p*, Kalle Aktiengesellschaft, Inf. Bull. 67e (4804-6.200.202 VI), 2 pp.

Kallstenius, T., 1958. Mechanical disturbances in clay samples taken with piston samplers, *Roy. Swed. Geotech. Inst. Proc.*, **6**, 75 pp.

Karol, R. H., 1961. *Erosion of stabilized soils*, American Cyanamid Co., Soils Eng. Res. Center Princeton, 5 pp.

————, 1962. *Gel-time data for high temperatures*, American Cyanamid Co., Soils Eng. Res. Princeton, 4 pp.

————, 1963a. *Procedure for checking gel times*, American Cyanamid Co., Soils Eng. Res. Center Princeton, 3 pp.

————, 1963b. *Gel charts for high temperatures*, American Cyanamid Co., Soils Eng. Res. Center Princeton, 4 pp.

————, 1964. *Optimum catalyst ratios for AM-9 Chemical Grout*, American Cyanamid Co., Eng. Chem. Res. Center Princeton, 2 pp.

————, 1965a. *Permanence of AM-9 Gel*, American Cyanamid Co., Eng. Chem. Res. Center Princeton, 3 pp.

————, 1965b. *Aging tests AM-9 Chemical Grout*, American Cyanamid Co., Eng. Chem., Soils Eng. Res. Center Princeton, 4 pp.

————, 1966a. *Stability of AM-9 Chemical Grout Gels*, American Cyanamid Co., Eng. Chem. Res. Center Princeton, 5 pp.

————, 1966b. *Chemical Grout Field Manual*, American Cyanamid Co., Eng. Chem. Res. Center, 3rd Ed., 97 pp.

Katz, A. and G. M. Friedman, 1965. The preparation of stained acetate peels for the study of carbonate rocks, *J. Sediment. Petrol.*, **35**, 248–249.

Kaufman, W. J. and D. P. Jackson, 1962. Plastisol molding, *Modern Plastics Encyclopedia*, 6 pp.

Kawai, K. and M. Oyama, 1962. A method of preparing thin sections of soils with the polyester resin "Polylite," *Soil Sci. Plant Nutrit., Japan*, 8(4), 18–22.

Keith, M. L., 1939a. Petrology of the alkaline intrusive at Blue Mountain, Ontario, *Bull. Geol. Soc. Am.*, **50**, 1795–1826.

————, 1939b. Selective staining to facilitate Rosiwal analysis, *Am. Min.*, **24**, 561–565.

Keller, G. H., A. F. Richards, and J. H. Recknagel, 1961. Prevention of water loss through CAB plastic sediment core liners, *Deep-Sea Res.*, **8**, 148–151.

Kennedy, G. C., 1945. The preparation of polished thin sections, *Econ. Geol.*, **40**, 355–359.

Kermabon, A., 1964. *A theoretical and experimental study of the kinematics of a coring tube accelerated in water by hydrostatic pressure*, NATO Saclant ASW Res. Center, Tech. Rep. No. 21, 46 pp.

————, P. Blavier, V. Cortis, and H. Delauze, 1966. The "Sphincter" corer: a wide-diameter corer with watertight core-catcher, *Marine Geol.*, **4**, 149–162.

————, and V. Cortis, in press. A new "Sphincter" corer with recoilless piston, *Marine Geol.*

Keyes, M. G., 1925. Making thin sections of rocks, *Am. J. Science*, **10**, 538–551.

Kistler, S. S., 1932. Coherent expanded aerogels, *J. Phys. Chem.*, **36**, 52–64.

Kjellman, W., T. Kallstenius, and O. Wager, 1950. *Soil sampler with metal foils, device for taking undisturbed samples of very great length*, Proc. Royal Swedish Geotech. Inst., No. 1, 76 pp.

Klaiber, 1964a. *Horizontal- und Vertikal- Flächen-Schleifmaschine, Modell HV 11*, Klaiber Maschinenfabrik, Broch. 1/64-3000, 2 pp.

_____, 1964b. *Horizontal- and vertical- surface grinding machine, model HV 11,* Klaiber machine manufactory, Broch. 2 pp.

_____, 1964c. *Machine à rectifier plane horizontale et verticale, modèle HV 11,* Klaiber machinerie, Broch. 4/64-2000, 2 pp.

_____, 1966. *Combined precision surface grinding machine HV 11a,* Klaiber machine manufactory, Broch., 4 pp.

Klingebiel, A., A. Rechiniac, et M. Vigneaux, 1967. Étude radiographique de la structure des sédiments meubles, *Marine Geol.,* 5, 71–76.

Kodak, 1957. *Radiography in modern industry,* Eastman Kodak Co., X-ray Division, 136 pp.

_____, 1961. *Kodak industrial X-ray films,* Eastman Kodak Co., X-ray Sales Div., Data Sheet F 3-401, 6 pp.

_____, 1963. *Kodak advanced data book, infrared and ultraviolet photography,* Eastman Kodak Co., Rochester, N.Y., 48 pp.

_____, 1966. *Radiography in modern industry, Supplement No. 3,* Eastman Kodak Co., Radiography Markets Div., Data Sheet M3-17, 3 pp.

Kodak Ltd. 1959. *Penumbral unsharpness,* Kodak Ltd., London, Kodak Data Sheet XR-2, 4 pp.

_____, 1961. *Faults in processing X-ray films,* Kodak Ltd., London, Kodak Data Sheet XR-7, 8 pp.

_____, 1962a. *Kodak "Crystallex" X-ray film,* Kodak Ltd., London, Kodak Data Sheet FM-28, 2 pp.

_____, 1962b. *Processing X-ray films,* Kodak Ltd., London, Kodak Data Sheet XR-6, 7 pp.

_____, 1962c. *Reducer, intensifier, and bleach formulae,* Kodak Ltd., London, Kodak Data Sheet FY-5, 7 pp.

_____, 1962d. *Professional Darkroom Design,* Kodak Ltd., London, 63 pp.

_____, 1962e. *Contact microradiography,* Kodak Ltd., London, Kodak Data Sheet IN-12, 2 pp.

_____, 1963a. *Lead intensifying screens,* Kodak Ltd., London, Kodak Data Sheet XR-4, 2 pp.

_____, 1963b. *Darkroom design,* Kodak Ltd., London, Kodak Data Sheet PR-7, 8 pp.

_____, 1963c. *Kodak "Unifix" powder,* Kodak Ltd., London, 8 pp.

_____, 1963d. *Kodak "Industrex" X-ray film, type S,* Kodak Ltd., London, Kodak Data Sheet FM-26, 2 pp.

_____, 1964a. *Gamma-Radiography,* Kodak Ltd., London, Kodak Data Sheet IN-16, 16 pp.

_____, 1964b. *"Kodirex" X-ray film,* Kodak Ltd., London, Kodak Data Sheet FM-17, 2 pp.

_____, 1964c. *Kodak "Industrex" X-ray film, type D,* Kodak Ltd., London, Kodak Data Sheet FM-27, 2 pp.

_____, 1964d. *Kodak "Microtex" X-ray film,* Kodak Ltd., London, Kodak Data Sheet FM-25, 2 pp.

_____, 1964e. *The selection of materials for the construction of photographic processing apparatus,* Kodak Ltd., London, Kodak Data Sheet PR-6, 8 pp.

_____, 1964f. *Formulary: index and general information,* Kodak Ltd., London, Kodak Data Sheet FY-1, 8 pp.

_____, 1964g. *Localisation in radiography,* Kodak Ltd., London, Kodak Data Sheet XR-3, 5 pp.

_____, 1965a. *The storage of X-ray films and radiographs,* Kodak Ltd., London, Kodak Data Sheet XR-5, 2 pp.

_____, 1965b. *Tropical processing of "Kodak" X-ray films,* Kodak Ltd., London, Kodak Data Sheet XR-8, 2 pp.

_____, 1965c. *Developer formulae,* Kodak Ltd., London, Kodak Data Sheet FY-2, 10 pp.

_____, 1965d. *Stop, hardening, and fixing bath formulae,* Kodak Ltd., London, Kodak Data Sheet FY-4, 5 pp.

_____, 1965e. *Copying radiographs and other transparencies,* Kodak Ltd., London, Kodak Data Sheet GN-2, 9 pp.

_____, 1967. *"Kodak" X-ray film speeds,* Kodak Ltd., London, Kodak Data Sheet XR-1, 2 pp.

Koehler, 1951. Klinischer Untersuchungs- und Erfahrungsbericht über Medizinal- Praecutan, *Therapie der Gegenwart,* 2, 1–7.

Kögler, F. C., 1963. Das Kastenlot, *Meyniana,* 13, 1–7.

Krebs, A. T., 1963. Dosimeters: recent developments. In G. L. Clark (Ed.), *Encyclopedia of X-rays and gamma rays,* Reinhold, New York, pp. 274–276.

Krieger, P. and P. H. Bird, 1932. Mounting polished sections in bakelite, *Econ. Geol.,* 27, 675–678.

Krumbein, W. C., 1935. Thin section mechanical analysis of indurated sediments, *J. Geol.,* 43, 482–496.

_____, and F. J. Pettijohn, 1938. *Manual of sedimentary petrography,* Appleton-Century-Crofts, New York, 549 pp.

Kruseman, G. P., 1962. *Étude Paléomagnetique et sédimentologique du Bassin Permien de Lodève, Hérault, France.* Thesis, Univ. of Utrecht, The Netherlands, 66 pp.

Krylon, 1956. *Krylon clear coating technical characteristics,* Krylon, Inc., 2 pp.

_____, not dated. *Krylon aerosol products for industry,* Form No. K-2273, Krylon, Inc., 16 pp.

Kubiena, W. L., 1937. Verfahren zur Herstellung von Dünnschliffen von Böden in ungestörter Lagerung, *Zeiss-Nachr.,* 2(3), 81-91.

_____, 1938. *Micropedology,* Collegiate Press Inc., Ames, Iowa, 243 pp.

_____, 1942. Die Dünnschlifftechnik in der Bodenuntersuchung, *Forschungsdienst.* Sonderheft, 16, 91-97.

Kudinov, E. I., 1957. Vibro-piston core sampler, *Inst. of Oceanology, USSR Acad. Sciences,* 25, 143-152 (in Russian).

Kuenen, Ph. H., 1961. Some arched and spiral structures in sediments, *Geol. en Mÿnbouw,* 40, 71-74.

Kullenberg, B., 1944. "Kolvlodet." In B. Kullenberg and E. Fromm (Eds.), Nya Försök att Upphämta

Långa Sedimentprofiler från Havsbottnen (New Experiments to obtain long sediment cores from the ocean bottom), *Geol. För. Stockh. Förh.,* **66,** 501–504.

———, 1947. The piston core sampler, *Svenska hydrogr.-biol. Komm. Skr.,* 1(2), 46.

———, 1955. A new core sampler. *Oceanog. Inst. Göteborg Meddelanden,* No. 26, 17.

Kullmann, A., 1953. Die Anfertigung und Bedeutung von Bodenlackfilmen, *Die Deutsche Landwiztschaft, Berlin,* 4, 203–206.

———, 1954. Zur Anfertigung und Bedeutung von Bodenlackfilmen, *Die Deutsche Landwiztschaft, Berlin,* 11, 589–592.

Kunde & Cie., 1964. *Catalog Qualitäts-Werkzeuge für Gartenbau, Obstbau, Weinbau und Baumschulen; für Töpfer, Ofensetzer, Fliesenleger und Bauhandwerk,* Kunde & Cie., 26 pp.

Kuron, H. und E. Homrighausen, 1959. Die Herstellung von Bodendünnschliffen mit dem Festlegungsmittel Plexigum 7466, *Mitt. Inst. Bodenkunde und Bodenerhaltung der Justus Liebig-Universität,* Giessen, I, 2 pp.

Kurotori, T. and H. Matsumoto, 1958. Microscopical study of forest soil I. On the method of making thin sections of soils for microscopic observation and the application of this technique to the study of forest soils, *Forest Soils Japan,* Rep. 9, 1–12.

LaFond, E. C. and R. S. Dietz, 1948. New snapper-type sea floor sediment sampler. *J. Sediment. Petrol.,* 18, 34–37.

Lamar, J. E., 1950. Acid etching in the study of limestones and dolomites, *Illinois St. Geol. Surv., Circ.,* 156, 47 pp.

Lamarque, P., 1936. Historadiographie, *Radiology,* 27, 563.

Lammers, J., 1965. Hand-drilling tools for geological investigation, *Geol. en Mijnbouw,* 44, 94–95.

Lane, D. W., 1962. Improved acetate peel technique, *J. Sediment. Petrol.,* 32, 870.

Laughton, A. S., 1963. Microtopography. In M. N. Hill (Ed.), *The Sea: Ideas and Observations on Progress in the Study of the Seas,* 3, 437–472.

Lees, A., 1958. Etching technique for use on thin sections of limestones, *J. Sediment. Petrol.,* 28, 200–202.

———, 1962. Sedimentology finds a new use for a standard industrial measuring projector, *Res. Develop. Industry,* 11, 37–38.

Leggette, M., 1928. The preparation of thin sections of friable rocks, *J. Geol.,* 36, 549–557.

Lehmann, W. M., 1932. Stereo-Röntgenaufnahmen als Hilfsmittel bei der Untersuchung von Versteinerungen, *Natur und Museum,* 62, 323–330.

Leitz, 1961. ARISTOPHOT, Universelle Photoeinrichtung für Mikroaufnahmen und Makroaufnahmen im Durchlicht und Auflicht, Liste 54-8a, Ernst Leitz GMBH, 12 pp.

———, 1963. ARISTOPHOT, Universelle Photoeinrichtung für Mikro- und Makroaufnahmen in Durchlicht und Auflicht, Liste 54-8b, Ernst Leitz GMBH, 12 pp.

Leutert, 1954. *Operating instructions for deviation recorder "Docenette MK."* Friedrich Leutert K. G., Tech. Inf., 13 pp.

———, 1964. *Special Leutert deep-sea clinometer.* Friedrich Leutert Deviation Instruments, Tech. Descr., 1 p.

Livingstone, D. A., 1955. A lightweight piston sampler for lake deposits, *Ecology,* 36, 137–139.

Lockwood, W. N., 1950. Impregnating sandstone specimens with thermo-setting plastics for study of oil-bearing formations. *Bull. Am. Assoc. Petrol. Geologists,* 34, 2061–2067.

Lyford, W. H., 1940. Preservation of soil profiles by Voigt's method, *Soil Sci. Soc. Amer. Proc.,* 4, 355–357.

Maarse, H. and J. H. J. Terwindt, 1964. A new method of making lacquer peel sections, *Marine Geol.,* I, 98–105.

———, and R. S. Piekhaar, 1962. Toepassing van polyesterhars (Vestynox) voor het maken van lakprofielen. Rep. Rijkswaterstaat, Deltadienst Waterloopkundige Afdeling, Bureau Hellevoetsluis, May 1962, 35 pp.

Mackenzie, A. F. and J. E. Dawson, 1961. The preparation and study of thin sections of organic soil materials, *J. Soil Sci., G. B.,* 12, 142–144.

Mackereth, F. J. H., 1958. A portable core sampler for lake deposits, *Limnol. and Oceanogr.,* 3, 181–191.

Mageli, O. L., S. D. Stengel, and D. F. Doehnert, 1959. Correlation of peroxide half-life with polymerization. *Modern Plastics,* 36(7), 135–140.

Maltha, P., 1957. Die Aushärtung von Polyesterharz-Katalysator-Gemischen und einige der diese Aushärtung beeinflussende Faktoren, *Fette, Seifen, Anstrichmittel,* 59, 163–169.

Matra, not dated a. *Genauigkeits-Flachschleifmaschinen, Modell MF 60/40, 80/40, 100/50,* MATRA-Werke GMBH, Tech. paper 0114/5000/856, 2 pp.

———, not dated b. *Betriebsanleitung Hydraulische Flachschleifmaschine Modell MF 80/40,* MATRA-Werke GMBH, Tech. paper, 23 pp.

———, not dated c. *Genauigkeits-Flachschleifmaschine. Modell MF 60/40,* MATRA-Werke GMBH, Tech. paper 072/3000/265, 20 pp.

McClure, G. M. and G. D. Converse, 1940. A method for taking and mounting monolithic soil profile samples, *Soil Sci. Soc. Amer. Proc.,* 14, 120–121.

McCrone, A. W., 1963. Quick preparation of peel-prints for sedimentary petrography, *J. Sediment. Petrol.,* 33, 228–230.

McKee, E. D., 1957a. Flume experiments on the production of stratification and cross-stratification, *J. Sediment. Petrol.,* 27, 129–134.

———, 1957b. Primary structures in some recent sediments, *Bull. Am. Assoc. Petrol. Geologists,* 41, 1704–1747.

_____, 1959. *Procedure for Latex peels.* Intern. Rep. U.S. Dept. Interior, Geol. Surv., Denver (Colorado), 2 pp.

_____, M. A. Reynolds, and C. H. Baker, Jr., 1962a. Laboratory studies on deformation in unconsolidated sediment, *U.S. Geol. Surv., Prof. Paper 450-D,* 151–155.

_____, M. A. Reynolds, and C. H. Baker, Jr., 1962b. Experiments on intraformational recumbent folds in crossbedded sand, *U.S. Geol. Surv., Prof. Paper 450-D,* 155–160.

_____, and T. S. Sterrett, 1961. Laboratory experiments on form and structure of longshore bars and beaches. *Geometry of sandstone bodies,* (*Am. Assoc. Petrol. Geologists*), 13–28.

McMillan, N. J. and J. Mitchell, 1953. A microscopic study of platy and concretionary structures in certain Saskatchewan soils, *Can J. of Agric. Sci.,* 33, 178–183.

McMullen, R. M. and J. R. L. Allen, 1964. Preservation of sedimentary structures in wet unconsolidated sands using polyester resins, *Marine Geol.,* I, 88–97.

Mégnien, C., 1957. Differentiation calcite-dolomite et anhydrite-gypse par colorations sélectives sur échantillons macroscopiques, *Bull. Geol. Soc. France,* Sér. 6, (7), 27–30.

Menard, H. W., 1952. Deep ripple marks in the sea, *J. Sediment. Petrol.,* 22, 3–9.

Menzies, R. J., L. Smith, and K. O. Emery, 1963. A combined underwater camera and bottom grab: a new tool for investigation of deep-sea benthos, *Int. Revue ges. Hydrobiol.,* 48(4), 529–545.

Meyer, C., 1946. Notes on cutting and polishing thin sections, *Econ. Geol.,* 41, 166–172.

Meyer, K. O., 1960. Lackfilme für die Geologische Heimat-Schausammlung in diorama-ähnlicher Aufstellung, *Verh. Vereins. Naturw. Heimatforsch, Hamburg,* Bd. 34, 48–50.

Miller, T. H. and R. M. Jeffords, 1962. Some properties of acetate films used in peels, *J. Paleontol.,* 36, 1382–1383.

Milner, H. B., 1962. *Sedimentary petrography, Vol. I: Methods in sedimentary petrography,* George Allen & Unwin, London, 643 pp.

Mitchel, G.A.G., 1963. Microradiography in biology. In G. L. Clark (Ed.), *Encyclopedia of X-rays and gamma rays,* Reinhold, New York, pp. 607–618.

Mitchell, J. K., 1956. The fabric of natural clay and its relation to engineering properties, *35th Ann. Highway Res. Board Proc.,* 35, 693–713.

MMM, not dated a. *Scotchcast electric resins, product and engineering guide,* Intern. Div., 3M Co., IDE-SSC (73.25)JR, 8 pp.

_____, not dated b. *Scotchcast electrical resins, insulate, seal, protect,* Intern. Div., 3M Co., IDE-RCAT (10.25)JR, 29 pp.

_____, not dated c. *Scotchcast resin, no. 3,* Electrical Prod. Div., 3M Co., ESRL(3510)S, 9 pp.

Moore, D. G., 1961. The free corer, sediment sampling without wire and winch, *J. Sediment. Petrol.,* 31, 627–630.

Moore, J. R., III, and S. H. Garraway, 1963. Use of polymethyl methacrylate in preparing thin sections of recent sediments, *Nature,* 200(4901), 62–63.

Müller, 1959, *Müller Präzis-Schleifmaschine MPS 3H,* Georg Müller Kugellagerfabrik, prospect 533/3/864/Pf., 16 pp.

Müller-Beck, H. und A. Haas, 1961. Ein neues Verfahren zur Konservierung von Feuchthölzern, *Der Präparator, Z. Museumtechnik,* 7, 157–168.

Nansen, F., 1906. *Methods for measuring direction and velocity of current in the sea,* Publ. Circ. Cons. into Explor. Mer. 34, 42 pp.

Napko, 1966. *Epoxycote "F."* Napko Corp., Indust. Coatings Div., data sheet SA-195 5M, 4 pp.

Neschen, 1964. *"Filmomatt" im Dienste der Wissenschaft,* Die Esche, Freunde des Hauses Neschen, Bückeburg, no. 1, p. 3.

Neuenhaus, H., 1940. Glutofix, ein neues Klebe- und Härtung- mittel zum Konservieren von Fossielen und Sedimenten, *Paleont. Zeitschr.,* 22(2), 134–138.

Nielsen, not dated a. *A wide range of catches, handles and grapples,* Nielsen Hardware Corp., catalog sheet, 1 p.

_____, not dated b. *Compression spring catches,* Nielsen Hardware Corp., Form 1020, 4 pp.

_____, not dated c. *Alternate strikes for use with various Neilsen catches,* Nielsen Hardware Corp., Form 500R, 1 p.

Nikiforoff, C. C., R. P. Humbert, and J. G. Cady, 1948. The hardpan in certain soils of the coastal plain, *Soil Sci. U.S.A.,* 65, 135–153.

Noller, D. C., S. D. Stengel, and O. L. Mageli, 1961. *Peroxide curing of unsaturated polyesters; I: Effect of some accelerators and inhibitors,* Soc. Plastics Indust., Inc., sect. 5-C, 10 pp.

Northrop, J., 1951. Ocean-bottom photographs of the neritic and bathyal environment south of Cape Cod, Massachusetts, *Bull. Geol. Soc. Am.,* 62, 1381–1384.

Noury & Van der Lande, 1956. *Methyl ethyl ketone peroxide butanox M-50, Iso-butanox M-50, butanox L, butanox M-105, butanox HBO-50 and butanox 28.* Noury & Van der Lande, Tech. Broch. 1/50 E 0956, 9 pp.

_____, 1961a. *Accelerators for the polymerisation of unsaturated polyester resins,* Noury & Van der Lande, Tech. Broch. 2/10 E, 4 pp.

_____, 1961b. *Release agents in the application of polyester resins,* Noury & Van der Lande, Tech. Broch. 2/57/E, 1 p.

_____, 1961c. *Noury U.V. absorbers,* Noury & Van der Lande, Tech. Broch. 2/71E, 5 pp.

_____, 1961d. *Noury thixotropy powder, an auxiliary material for use with unsaturated polyester resins,* Noury & Van der Lande, Tech. Broch. 2/60E, 2 pp.

———, 1962. *Versnellers voor de polymerisatie van onverzadigde polyesterharsen,* Noury & Van der Lande, Tech. Broch. 2/10E, 5 pp.

———, 1964a. *Organic peroxides for polyester resins,* Noury & Van der Lande, Tech. Broch. 5/3E, 13 pp.

———, 1964b. *Cyclohexanone peroxide cyclonox, cyclonox B-50, cyclonox B-60, cyclonox LE-50, cyclonox LTM-50 and cyclonox G-20,* Noury & Van der Lande, Tech. Broch. 1/30E, 8 pp.

———, not dated a. *The role of peroxides in radical polymerisations,* Noury & Van der Lande N.V., 7 pp.

———, not dated b. *Accelerators for use with organic peroxide catalysts in the polymerisation of unsaturated resins,* Noury & Van der Lande N. V., Tech. Broch. 2/10E, 1246, 9 pp.

———, not dated c. *Lucidol B-50,* Noury & Van der Lande, Tech. Broch. 1/11E1146, 4 pp.

———, not dated d. *Butanox M-50, iso-butanox M-50, iso-butanox M-60 methylethylketonperoxyde,* Noury & Van der Lande, Tech. Broch. 1/40N, 8 pp.

———, not dated e. *Cyclonox, cyclonox B-50, cyclonox B-60, cyclonox LE-50, cyclonox LTM-50, cyclonox G-20 cyclohexanon peroxyde,* Noury & Van der Lande, Tech. Broch. 1/30N, 9 pp.

———, not dated f. *Trigonox K-70,* Noury & Van der Lande, Tech. Broch. 1/61E 0756, 3 pp.

———, not dated g. *Inhibitor NLC-1 and NLC-10,* Noury & Van der Lande, Tech. Broch. 2/90E, 2 pp.

Nuodex, 1960. *Nuodex M.E.K. Peroxide,* Nuodex Prod. Div., Tech. data sheet #6, 2 pp.

———, 1961a. *Nuodex B.P. Paste,* Nuodex Prod. Div., Tech. data sheet #2, 2 pp.

———, 1961b. *Nuodex cobalt accelerator,* Nuodex Prod. Div., Tech. data sheet #8, 2 pp.

———, 1962. *Nuodex polycure M.E.K. Peroxide,* Nuodex Prod. Div., Tech. data sheet #6A, 2 pp.

———, not dated. *Storage and handling precautions for organic peroxides and organic peroxide compounds at warehouses,* Nuodex Prod. Div., Nuodex organic peroxides: Bull. #8, 4 pp.

Nuss, W. F. and R. L. Whiting, 1947. Technique for reproducing rock pore space, *Bull. Am. Assoc. Petrol. Geologists,* 31, 2044-2049.

Oele, E., 1964. *Sedimentological aspects of four lower-Paleozoic formations in the northern part of the province of Léon (Spain).* Ph. D. thesis, Univ. of Leiden, The Netherlands, 99 pp.

Ong Sing Poen, 1963. Projection microscopy. In G. L. Clark (Ed.), *Encyclopedia of X-rays and gamma rays,* Reinhold, New York, pp. 846-851.

Oomkens, E. and J. H. J. Terwindt, 1960. Inshore estuarine sediments in the Haringvliet (Netherlands), *Geol. en Mijnbouw,* 22, 701-710.

Orviku, K., 1940. Geoloogiliste Jäädvustamine Lakkfilmmeetodi Abil. (Die Lackfilmmethode von E. Voigt), *Eesti Loodus,* 3, 1-5.

Osmond, D. A. and I. Stephan, 1957. The micropedology of some red soils from Cyprus. *J. Soil Sci.,* 8, 19-27.

Overlau, P., 1963. Particularités sédimentaires du Calcaire de Basècles, *Bull. Soc. belge de Géol., de Paléontol. et d'Hydrol.,* LXXII, 261-271.

———, 1965a. Acetate peels; method Kaisin-Overlau, Written comm., April 1965, 7 pp.

———, 1965b. The photographic printing of acetate peels and thin sections, Written comm., April 1965, 2 pp.

———, 1966. *La sédimentation Viséenne dans l'ouest du Hainaut Belge,* Thèse l-Univ. Catholique de Louvain, 130 pp.

Pantin, H. M., 1960. Dye-staining technique for examination of sedimentary microstructures in cores, *J. Sediment. Petrol.,* 30, 314-316.

Parizak, R. R., 1967. Strike-dip indicator, *J. Sediment. Petrol.,* 37, 1249-1250.

Perfecta, 1964. Bison Kit. Chemische Fabriek "Perfecta Goes N.V." Written Comm., June 5, 1964, 2 pp.

Permatex, 1966a. *Permatex adhesives and sealants selector guide,* Permatex Co., Inc., form 556-5-66, 2 pp.

———, 1966b. *Permatex,* Permatex Co., Inc., form 555-5-66-2, 8 pp.

———, 1967. *Permatex Service Repair Center.* Permatex Co., Inc., form 680-3-67, 4 pp.

Petterson, H. and B. Kullenberg, 1940. A vacuum core-sampler for deep sea sediments. *Nature,* 145 (3369), 306.

Philips, not dated. *Industriële röntgenapparatuur catalogus,* Philips Bedrijfsapparatuur Nederland N.V., Groep Meet-en Regeltechniek, 21 pp.

Philips Electronic Instruments, 1963a. *X-ray, methods.* In G. L. Clark (Ed.), *Encyclopedia of X-rays and gamma rays,* Reinhold, New York, pp. 593-594.

———, 1963b. Detectors of X-radiation (Norelco). In G. L. Clark (Ed.), *Encyclopedia of X-rays and gamma rays,* Reinhold, New York, pp. 228-230.

Picker, not dated. *Picker "Hotshot" 110 kV portable X-ray unit,* Picker X-ray Corp., PD sheet 1248, 4 pp.

Pincus, I. and N. J. Gendron, 1959. *The vacuum impregnation of brush carbons and metal samples with epoxy resin for the purpose of polishing for microscopic examination,* General Electric Co., Materials and Processes Lab, 7 pp.

———, and N. J. Gendron, 1960. Microscopy and structures of brush carbons. In *Proc. Fourth Conf. on Carbon,* Pergamon Press, New York, pp. 687-701.

Pittsburgh, 1958. *Selectron 5003, general purpose molding and casting polyester resin,* Pittsburgh Plate Glass Co., Paint and brush Div., Tech. data sheet, 4 pp.

Plafker, G., 1956. A technique for modal analyses of some fine- and medium-grained (0.1-5mm) rocks. *Am. Min.,* 41, 652-655.

Plankeel, F. H. en J. P. Van der Sluis, 1960. Een registrerende kompas/hellingmeter, *Geol. en Mijnbouw*, **39**, 326–329.

Potsaid, M. S., 1963. Solid "Phantom" dosimeters. In G. L. Clark (Ed.), *Encyclopedia of X-rays and gamma rays*, Reinhold, New York, pp. 279–289.

Potter, P. E. and F. J. Pettijohn, 1963. *Paleo-currents and basin analysis*, Springer Verlag, Berlin, 296 pp.

Pozaryska, K., 1958. Museum Geiseltalskie, *Kosmos "A" Zeszyt* 1, Rok VII, 74–78.

Pratje, O., 1933. Gewinnung und Untersuchung der Meeresgrundproben, In Abderhalden (Ed.), *Handbuch biol. Arbeitsmeth.*, Abt. 9, Tl. **6** (2), 377–542.

———, 1952. Die Erfahrungen bei der Gewinnung von rezenten, marinen Sedimenten in den letzten 25 Jahren, *Mitt. geogr. Ges. Hamb.*, Bd. L, 118–197.

Pryor, W. A., 1958. Dip direction indicator, *J. Sediment. Petrol.*, **28**, 230.

Ram, 1958. *Garalease 915*, Ram Chemicals, prod. data, 1 p.

———, 1959. *Plastilease 512B, a Garan product*, Ram Chemicals, prod. data, 1 p.

———, 1960. *Clear adhesive A-33*, Ram Chemicals, prod. data 61560, 1 p.

———, 1961. *Garox QZA, a stable and unique benzoyl peroxide catalyst paste*, Ram Chemicals, prod. data 121161, 1 p.

———, 1962a. *Pigment dispersions 44-00 series*, Ram Chemicals, prod. data 11962-11762, 3 pp.

———, 1962b. *Ram mold release agents*, Ram Chemicals, prod. data, 5 pp.

———, 1963a. *Ram organic peroxide pastes*, Ram Chemicals, prod. data 61363, 3 pp.

———, 1963b. *Ram mold release agents*, Ram Chemicals, prod. data, 5 pp.

———, 1965. *Ram clear adhesive A-33*, Ram Chemicals, prod. data 91665, 2 pp.

———, not dated a. *Garox MEK, methyl ethyl ketone peroxide*, Ram Chemicals, prod. data, 1 p.

———, not dated b. *Silastic RTV. The Dow Corning silicone rubber that cures at room temperature*, Ram Chemicals, form 08-010, 8 pp.

———, not dated c. *Silastic RTV room temperature vulcanizing silicone rubber*, Ram Chemicals, Bull. 08-043, 4 pp.

———, not dated d. *Release agents*, Ram Chemicals, 4 pp.

Redlich, G. C. 1940. Een nieuwe methodiek voor de beoordeling van den grond in zijn natuurlijke ligging met behulp van slijpplaatjes, *De Ingenieur*, **55**(6), 19–23.

Reed, F. S. and J. L. Mergner, 1953. Preparation of thin sections, *Am. Min.*, **38**, 1184–1203.

Reeder, S. W. and A. L. McAllister, 1957. A staining method for the quantitative determination of feldspars in rocks and sands from soils, *Can. J. Soil Sci.*, **37**, 57–59.

Reichhold, 1965. *32-032 Polylite (HU-332)*, Reichhold Chemicals, Inc., Specialty Chem. Div., Tech. Bull., 1 p.

Reineck, H. E., 1957. Stechkasten und Deckweisz, Hilfsmittel des Meeresgeologen, *Natur Volk*, **87**, 132–134.

———, 1958a. Über Gefüge von orientierten Grundproben aus der Nordsee, *Senckenberg Lethea*, **39**, 25–41.

———, 1958b. Über das Härten und Schleifen von Lockersedimenten, *Senckenberg Lethea*, **39**(1/2), 49–56.

———, 1960a. *Über die Entstehung von Linsen- und Flaserschichten*, Abh. dtsch. Akad. Wiss. Berlin, Klasse III, Heft 1, "Festschrift zum 70. Geburtstag von Ernst Kraus," 369–374.

———, 1960b. Über Zeitlücken in Rezenten Flachsee-Sedimenten, *Geol. Rundschau*, 49/1, 149–161.

———, 1961a. Die Herstellung von Meeresboden-Präparaten in Senckenberg-Institut Wilhelmshaven, *Museums Kunde*, **2**, 87–89.

———, 1961b. Versteinerte Nordsee, *Natur Volk*, **91** (5), 151–162.

———, 1962. Reliefgüsse ungestörter Sandproben, *Z. Pflanzenern. Düngung u. Bodenk.*, **99**, 151–153.

———, 1963a. Der Kastengreifer. *Natur Mus.*, **93** (3), 102–108.

———, 1963b. Sedimentgefüge im Bereich der südlichen Nordsee, *Abh. d. Senckenborgischen Natur f. Gesell.*, **505**, 138 pp.

———, 1963c. Naszhärtung von ungestörten Bodenproben im Format 5 x 5 cm für projizierbare Dickschliffe, *Senckenberg Lethea*, **44** (4), 357–362.

———, 1967. Ein Kolbenlot mit Plastik-Rohren, *Senckenberg Lethea*, **48** (3/4), 285–289.

Reish, D. J. and K. E. Green, 1958. Description of a portable piston corer for use in shallow water, *J. Sediment Petrol.*, **28**, 227–229.

Rhoads, D. C. and D. J. Stanley, 1966. Transmitted infrared radiation: a simple method for studying sedimentary structures, *J. Sediment. Petrol.*, **36**, 1144–1149.

Richards, A. F., 1961. *Investigations of deep-sea sediment cores. I. Shear strength, bearing capacity, and consolidation*, U.S. Navy Hydrographic Office, Techn. Rept. 63, 70 pp.

——— and G. H. Keller, 1961. A plastic-barrel sediment corer, *Deep-Sea Res.*, **8**, 306–312.

———, and H. W. Parker, in press. Surface coring for shear strength measurements. In *Civil engineering in the ocean*, Amer. Soc. Civil Engineering, New York.

Richard, J., 1908. *L'Océanographie*, Vuibert et Nony, Paris, 398 pp.

Richardson, L. M. and R. E. Deane, 1961. Thin sections of unconsolidated material, *Proc. Geol. Assoc. Canada*, **13**, 135–136.

Richter, K., 1961. Die Geologische Geländeaufnahme. In A. Bentz (Herausg.), *Lehrbuch der angewandten Geologie, Bd. 1. Algemeine Methoden,* Ferdinand Enke Verlag, Stuttgart, 1–160.

Rioult, M. et R. Riby, 1963. Examen radiographique de quelques minerais de fer de l'Ordovicien normand. Importance des rayons X en sédimentologie, *Bull. Soc. Géol. France,* Sér. 7, T. V, 59–61.

Ritch, H. B. and P. H. Cardwell, 1951. New technique for application of plastics in consolidating formations, *World Oil,* **133**(7), 117–184.

Ritz, V. H., 1963. Dosimeters, silver-activated glass block: low energy X-ray intensity measurements. In G. L. Clark (Ed.), *Encyclopedia of X-rays and gamma rays,* Reinhold, New York, pp. 277–279.

Roberts, J. E., 1963. Dosimetry, radiation and units. In G. L. Clark (Ed.), *Encyclopedia of X-rays and gamma rays,* Reinhold, New York, pp. 313–316.

Rodgers, J., 1940. Distinction between calcite and dolomite on polished surfaces, *Am. J. Sci.,* **238,** 788–798.

Rogers, T. H., 1963. X-ray tubes. In G. L. Clark (Ed.), *Encyclopedia of X-rays and gamma rays,* Reinhold, New York, pp. 1090–1097.

Rohm & Haas, 1962. *Paraplex P-43 and Paraplex P-43 HV,* Rohm & Haas Co., plastics dept., Paraplex techn. data, Bull. 442, 8 pp.

———, not dated. *Plexiglas, acrylic plastic molding powders,* Rohm & Haas Co., data sheet PL-474, 4 pp.

Röhm & Haas, not dated. *Plexigum und pleximon zur Herstellung von Schliffen und Schnitten,* Röhm & Haas GmbH, 5 pp.

Rolfe, W.D.I., 1965. Uses of infrared rays. In B. Kummel and D. Raup (Eds.), *Handbook of paleontological techniques,* W. H. Freeman and Co., San Francisco, pp. 345–350.

Römer, A., 1934. Bohrer zur Entnahme von Bodenprofilen, *Bodenkundl. Forschung,* 4, 84.

———, 1951. Die Herstellung von transportabelen Bodenprofilen, *Landw. Forsch.,* 3, 212–216.

Rosenblum, S., 1956. Improved techniques for staining potash feldspars, *Am. Min.,* 41, 662–664.

Rosfelder, A. M., 1966a. A tubular spring valve used as a tight and thin-walled core-retainer, *J. Sediment. Petrol.,* 36, 973–976.

———,1966b. Hydrostatic actuation of deep-sea instruments, *J. Ocean Technology,* 1(1), 53–63.

———, 1966c. Subsea coring for geological and geotechnical surveys, *Proc. OECON* (Offshore Exploration Conf.), 709–734.

———, 1967. Obtaining located samples from sandy and rocky formations in deep water, *Proc. World Dredging Conf.,* 1967, 487–516.

——— and N. F. Marshall, 1966. Oriented marine cores: a description of new locking compasses and triggering mechanisms, Sears Foundation, *J. Marine Res.,* 24(3), 353–364.

———, and N. F. Marshall, 1967. Obtaining large, undisturbed, and oriented samples in deep water. In A. F. Richards (Ed.), *Marine Geotechnique,* Univ. of Illinois Press, Urbana, pp. 243–263.

Ross, C. S., 1924. A method of preparing thin-sections of friable rock, *Amer. J. Science,* Sec. 5, 483–485.

———, 1926. Methods of preparation of sedimentary materials for study, *Econ. Geol.,* 21, 454–468.

Ross, D. A. and W. R. Riedel, 1967. Comparison of upper parts of some piston cores with simultaneously collected open-barrel cores, *Deep-Sea Res.,* 14, 285–294.

Rotter, W., 1941. Herstellung von Bodendünnschliffen nach dem Resinolverfahren, *Bodenk. Pflanzenern,* 25, (70), 251–256.

Rusnak, G. A. and S. J. Luft, 1963. *A suggested outline for the megascopic description of marine sedimentary cores,* Rep. Inst. Marine Sci., Univ. of Miami, 23 pp.

Sachs, P. L. and S. O. Raymond, 1965. A new unattached sediment sampler, *J. Marine Res.,* 23(1), 44–53.

Sanders, J. E., 1960. The Kudinov vibro-piston core sampler: Russian solution to underwater sand-coring problem, *Intern. Geol. Review,* 2, 174–178.

———, 1966. *Summary of research on the Atlantic shelf and adjoining coastal areas carried out at Hudson Laboratories in 1965–1966,* Hudson Laboratories of Columbia Univ., Informal Documentation No. 123, 26 pp.

Sawdon, W., 1948. Plastic tubes preserve core information, *Petrol. Engineer,* **XX** (3), 139–140.

Sayles, R. W., 1921. Microscopic sections of till and stratified clay, *Bull. Geol. Soc. Am.,* 32, 59–62.

Scado, 1965. *Algemene handleiding Lamellon harsen,* Scado-Archer-Daniels N.V., Bull., 7 pp.

———, not dated a. *Lamellon 230,* Scado-Archer-Daniels N.V., Bull., 2 pp.

———, not dated b. *Scadopol S 810, S 811, S 812, S 813,* Scado-Archer-Daniels N.V., Bull., 2 pp.

Schaffer, R. J. and P. Hirst, 1930. The preparation of thin sections of friable and weathered materials by impregnation with synthetic resin, *Proc. Geol. Assoc. London,* 41, 32–43.

Schlossmacher, K., 1919. Ein Verfahren zur Herrichtung von schiefrigen und lockeren Gesteinen zum Dünnschleifen, *Centralblatt Min., Geol. u. Paläont.,* 190–192.

Schurmann, J. J. and M. A. J. Goedewaagen, 1955. A new method for the simultaneous preservation of profiles and root systems, *Plant and Soil,*6 (4), 373–381.

Schwarz, A., 1929. Ein Verfahren zum Härten nicht verfestiger Sedimente, *Natur und Museum,* 59, 204–207.

Schwarzacher, W. and K. Hunkins, 1961. Dredged gravels from the Central Arctic Ocean, *Geol. of the Arctic,* 666–677.

Scott Bader, 1965. *Polyester Handbook,* Scott Bader & Co., 120 pp.

Seidel, G., 1951. Über die Gewinnung unzerlegter Schlitzproben aus Steinkohlenflözen Unterlage *Bergfreiheit,* **5,** 1-2.

Seppic, not dated. *Ortholeum 162, agent de démoulage,* S.E.P.P.I.C., Bull. 61-25/66-91, 2 pp.

Shannon, J. P., Jr., and C. W. Lord, 1967. Preservation of unconsolidated sediment cores in plastic, *J. Sediment. Petrol.,* **37,** 1200-1203.

Shell, 1962a. *Epon resins,* Shell Chemical Co., Plastics and Resins Div., Tech. Bull. SC, 62-131, 5 pp.

———, 1962b. *Epon resin 828,* Shell Chemical Co., Plastics and Resins Div., Data sheet SC, 60-146, 2 pp.

Shepard, F. P., 1963. *Submarine geology, 2nd ed.,* Harper & Row, New York, 527 pp.

———, J. R. Curray, D. L. Inman, E. A. Murray, E. L. Winterer, and R. F. Dill, 1964. Submarine geology by diving saucer, *Science,* **145**(3636), 1042-1046.

Shipek, C. J., 1960. Photographic study of some deep-sea floor environments in the Eastern Pacific, *Bull. Geol. Soc. Am.,* **71,** 1067-1074.

———, 1962. Photographic survey of sea floor on southwest slope of Eniwetok Atoll, *Bull. Geol. Soc. Am.,* **73,** 805-812.

———, 1966. *Photo analysis of sea floor microrelief,* U.S. Navy Electronics Lab., San Diego, Calif., Rep. 1374, 70 pp.

Short, M. N., 1931. Microscopic determination of the ore minerals, *U.S. Geol. Surv. Bull. 825,* 7-10.

Showalter, W. E., 1950. Mounting core analysis specimens in thermoplastic, *J. Petrol. Tech.,* **II** (12), 8-9.

Shrock, R. R., 1940. Lucite as an aid in studying hard parts of living and fossil animals, *J. Paleontol.,* **14,** 86-88.

Sikkens, 1965. *Metakote,* Sikkens Groep N.V., Techn. Broch., 5 pp.

Silverman, M. and R. C. Whaley, 1952. Adaptation of the piston coring device to shallow water sampling, *J. Sediment. Petrol.,* **22,** 11-16.

Simon, J., 1966. *Infrared radiation,* D. Van Nostrand Co., Princeton, N. J., 119 pp.

Sinha, E. and L. Strauss, 1967. *A selected bibliography of oceanography books published between 1959 and 1966,* Oceanic Library and Information Center, a division of the Oceanic Res. Inst., La Jolla, Calif., 13 pp.

Slager, S., 1964. A study of the distribution of biopores in some sandy soils in the Netherlands. In A. Jongerius (Ed.), *Soil micromorphology,* Elsevier, Amsterdam, pp. 421-427.

———, 1966. *Morphological studies of some cultivated soils,* Agricultural Res. Rep. 670, Centre for Agricultural publ. a. document., Wageningen, The Netherlands, 111 pp.

Smith, A. J., 1959. Description of the Mackereth portable core sampler, *J. Sediment. Petrol.,* **29,** 246-250.

Smith, H. W. and C. P. Moodie, 1947. Collection and preservation of soil-profiles, *Soil Sci.,* **64,** 61-69.

Smith, L. O., 1964. *Deep sea grab-photography for oceanographic sampling,* Undersea Techn., Febr., 4 pp.

Snodgrass, W. L., 1960. Glue peels of wet sediment cores, Rep. Scripps Inst. of Oceanogr., La Jolla, California (unpubl.), 4 pp.

Stanley, D. J. and L. R. Blanchard, 1967. Scanning of long unsplit cores by X-radiography, *Deep-Sea Res.,* **14,** 379-380.

——— and D. C. Rhoads, 1967. Dune sands examined by infrared photography, *Bull. Am. Assoc. Petrol. Geologists,* **51,** 424-430.

Steinlein, H., 1950. Eine einfache graphische Methode zur indirekten Bestimmung des Streichens und Fallens und der Schrägschichtungs-Richtung. *Neues Jb. Min., Geol. Paläontol.,* **91,** 149-160.

Sternberg, R. M. and H. F. Belding, 1942. Dry-peel technique, *J. Paleontol,* **16,** 135-136.

Sternberg, R. W. and J. S. Creager, 1965. An instrument system to measure boundary-layer conditions at the sea floor, *Marine Geol.,* **3,** 475-482.

St. John, E. G. and D. R. Craig, 1957. Log Etronography. *Am. J. Roentgenology, Radium Therapy and Nuclear Med.,* **78,** 124-133.

Stockhausen, 1962a. *Waarom huidbescherming?* Chemische Fabrik Stockhausen & Cie., Bull., 5 pp.

———, 1962b. *Arretil-T,* Chemische Fabrik Stockhausen & Cie., Bull. 10.62, 2 pp.

———, 1962c. *Kosmosan,* Chemische Fabrik Stockhausen & Cie., Bull., 2 pp.

———, 1962d. *Stokolan,* Chemische Fabrik Stockhausen & Cie., Bull. 10.62, 2 pp.

———, 1962e. *Stoko-Emulsion,* Chemische Fabrik Stockhausen & Cie., Bull. 10.62, 2 pp.

———, 1962f. *Industrie-Praecutan,* Chemische Fabrik Stockhausen & Cie., Bull. 6/62, 2 pp.

———, 1963a. *Merkblatt für die Verarbeitung von Mehrcomponenten-Harzen,* Chemische Fabrik Stockhausen & Cie., Bull., 8 pp.

———, 1963b. *Warum Hautschutz?* Chemische Fabrik Stockhausen & Cie., Bull., 6 pp.

———, 1963c. *Spezieller Hautschutz,* Chemische Fabrik Stockhausen & Cie., Bull., 6 pp.

———, 1963d. *Es-Te Lackentferner,* Chemische Fabrik Stockhausen & Cie., Bull. J.W.5.63, 2 pp.

———, 1964a. *Stockhausen-Hautschutz zur Verhütung von beruflichen Schäden und Erkrankungen der Haut,* Chemische Fabrik Stockhausen & Cie., Bull., 4 pp.

———, 1964b. *Stockhausen-huidbescherming ter voorkoming van aandoeningen en beroepsziekten van de huid,* Chemische Fabrik Stockhausen & Cie., Bull., 4 pp.

Stürmer, W., 1965. Röntgenaufnahmen von einigen Fossilien aus dem Geologischen Institut der Universität Erlangen-Nürnberg, *Geol. Blätter, N. O. Bayern*, **15** (4), 217–223.

Swarbrick, E. E., 1964. A peel technique for the study of sedimentary structures, *Sedimentology*, **3**, 75–78.

Swedish Committee on Piston Sampling, 1961. *Standard piston sampling*, Swedish Geotechnical Institute Proc. No. 19, 45 pp.

Synres, 1963. *Synolite 711*, N. V. Chemische Industrie Synres: 4 pp.

———, 1967. *Synolite 336*, N. V. Chemische Industrie Synres, 4 pp.

Synthese, 1962. *Werken met polyester harsen*. Kunstharsfabriek Synthese N. V., afd. Plastic-chemie, Broch. 10-62-231-SYR-N-3, 21 pp.

Tanis, K., 1952. Het conserveren van bodemprofielen, *Maandblad Landbouwvoorlichtingsdienst*, **9**, 449–454.

———, 1954a. Nieuwe handleiding voor het conserveren van bodemprofielen, *Landbouwvoorlichting*, **77**, 170–173.

———, 1954b. Nieuwe handleiding voor het conserveren van bodemprofielen, *Boor en Spade*, **VII**, 178–180.

Taylor, J. C. M., 1960. Impregnation of rocks for sectioning, *Geol. Mag.*, **97**, 261.

Ten Haaf, E., 1959. *Graded beds of the Northern Apennines*, Thesis Univ. Groningen, The Netherlands, 102 pp.

——— and H. Wensink, 1962. Geological solar compasses, *Geol. Rundschau*, **52**, 541–548.

Terry, R. D., 1961. *Oceanography, its tools, methods, resources and applications*, **Vol. I.**, Autonetics, a division of North American Aviation, Inc. Rep., 347 pp.

———, 1966. *The deep submersible*, Western Periodicals Co., North Hollywood, California, 456 pp.

Thiessen, R., G. E. Sprunk, and H. J. O'Dunnel, 1938. Preparation of thin sections of coal, *Circular U.S. Bureau of Mines*, No. 7021, 1–18.

Thissen, J., 1959. Une méthode pour la fabrication des lames minces. *Naturalistes Bèlges*, **40**(6), 208.

Thorpe, S. H. and D. W. Davison, 1944. Improved methods of printing from radiograph negatives, *Engineering*, **157**, 241–242.

Tourtelot, H. A., 1961. Thin sections of clay and shale, *J. Sediment. Petrol.*, **31**, 131–132.

Troedsson, G., 1938. Voigts Lackfilmmetod, *Geol. Fören. Stockh. Förh.*, **60**(4), 646–647.

Tsuda, S., 1963. Gamma dosimetry: a comparison of all known methods. In G. L. Clark (ed.), *Encyclopedia of X-rays and gamma rays*, Reinhold, New York, pp. 291-304.

Turnbull, W. J., 1953. Field investigations, technique of field observations including compaction control, soil stabilization. *Proc. 3rd Intern. Conf. Soil Mech. Fndtn. Engring., Switzerland*, **II**, 319–333.

Tüxen, R., 1957. Ein vereinfachtes Verfahren zur Herstellung von Lackabzügen von Bodenprofilen, *Mitt. d. floristisch-soziol. Arbeitsgem., N.F., Stolzenau, Weser*, Heft 6/7, 336–339.

Udinzew, G. B., A. P. Lisizyn, and L. P. Konstantinow, 1956. Konstrukzija porschewoi trubki s awtomatitsches koi stabilisazijei porschnja, *TIO AN SSSR*, **19**, 232–257.

Union Carbide, 1962. *Carbowax, polyethylene glycols for pharmaceuticals and cosmetics*, Union Carbide Corp., Tech. Bull., 40 pp.

U.S. Plywood, not dated. *Weldwood Plastic Resin Glue*, United States Plywood Corp., Tech. data sheet no. 100, 6 pp.

Vallentyne, J. R., 1955. A modification of the Livingstone piston sampler for lake deposits, *Ecology*, **36**(1), 139–141.

Van den Bussche, H. K. J. and J. J. H. C. Houbolt, 1964. A corer for sampling shallow-marine sands, *Sedimentology*, **3**, 155–159.

Van der Plas, L. and S. Slager, 1964. A method to study the distribution of biopores in soils. In A. Jongerius (Ed.), *Soil micromorphology*, Elsevier, Amsterdam, pp. 411–419.

Van der Sluis, J. P. and A. I. Schaafsma, 1963. A drilling rig for coring unconsolidated sediments, *Verh. Koninkl. Ned. Geol. Mijnb. Gen., Geol. Ser.*, 21-2, 225–227.

Van Dorn, W. G., 1953. *The marine release delay timer*, Oceanographic Equipment Report No. 2, Scripps Inst. Oceanogr., Ref. 53-23, 10 pp.

Van Hinte, J. E., 1963. Zur Stratigraphie und Mikropaläontologie der Oberkreide und des Eozäns des Krappfeldes (Kärnten), *Jb. Geol. Bundesanstalt*, Sonderband 8, 147 pp.

Van Luyt, E., 1964. *Epoxyharsen*. De Veiligheid–40, U.D.C. 678.5, spec. circ. Arbeidsinspectie, 15–18.

Van Olphen, H., 1965. *An Introduction to Clay Colloid Chemistry*, Interscience, New York, 301 pp.

Van Straaten, L. M. J. U., 1951. Texture and genesis of Dutch Wadden Sea sediments, *Proc. 3rd Intern. Congr. Sedimentology, Groningen-Wageningen, Netherlands*, 5–12 July 1951, 225–244.

———, 1954. Composition and structure of Recent marine sediments in the Netherlands, *Leidse Geol. Med.*, **XIX**, 1–110.

———, 1956. Composition of shell beds formed in tidal flat environment in the Netherlands and in the bay of Arcachon (France), *Geol. en Mijnbouw*, 18, 209–226.

———, 1957a. Recent sandstones of the coasts of the Netherlands and of the Rhône delta, *Geol. en Mijnbouw*, 19, 196–213.

———, 1957b. The Holocene deposits. In The excavation at Velsen, *Verhand. Kon. Nederl. Geol. Mijnb. Gen., Geol. Ser.*, XVII, 2, 158–183.

———, 1959. Minor structures of some recent littoral and neritic sediments, *Geol. en Mynbouw (N.W.*

Ser), **21** (Symposium: sedimentology of recent and old sediments), 197–216.

———, 1965. Coastal barrier deposits in South- and North-Holland, in particular in the areas around Scheveningen and IJmuiden, *Med. Geol. Stichting*, Nw. Serie no. 17, 41–75.

Van Veen, J., 1936. *Onderzoekingen in de Hoofden, in verband met de gesteldheid der Nederlandsche Kust*, Diss. University Leiden.

Vargas, M., 1957. Techniques of field measurement and sampling, *Proc. 4th Intern. Conf. Soil Mech. Fndtn. Engring.*, Div. 2, 436–440.

Varney, F. M. and L. E. Redvine, 1937. A hydraulic coring instrument for submarine geologic investigations, *Nat. Res. Council, Rep. Comm. Sedim.*, 107–113.

Vernet, J. P. et A. Gautier, 1962. La technique des coupes minces appliqué à l'étude de l'halloysite au microscope électronique, *C. R. Acad. Sci. Franc.*, **254** (14), 2608–2610.

Voigt, E., 1933. Die Übertragung fossiler Wirbeltierleichen auf Zellulose-Filme, eine neue Bergungsmethode für Wirbeltiere aus der Braunkohle, *Paläont. Zeitschr.*, **15**, 72–78.

———, 1934. Die Fische aus der mitteleozänen Braunkohle des Geiseltales mit besonderer Berücksichtigung der erhaltenen Weichteile, *Nov. Act. Leopoldina*, N. F. 2, 21–146.

———, 1935a. Die Erhaltung von Epithelzellen mit Zellkernen, von Chromatophoren und Corium in fossiler Froschhaut aus der mitteleozänen Braunkohle des Geiseltales, *Nov. Act. Leopoldina*, N.F. 3, 339–360.

———, 1935b. Die Bedeutung der Lackfilmmethode für die vorgeschichtliche Forschung, *Nachrichtenblatt Deutsche Vorzeit*, **2**, 2–4.

———, 1936a. Weichteile an Säugetieren aus der eozänen Braunkohle des Geiseltales, *Nov. Act. Leopoldina*, N. F. 4, 301–310.

———, 1936b. Über das Haarkleid einiger Säugetiere aus der mitteleozänen Braunkohle des Geiseltales, *Nov. Act. Leopoldina*, N.F. 4, 317–334.

———, 1936c. Ein neues Verfahren zur Konservierung von Bodenprofilen, *Z. Pflanzenern., Düngung u. Bodenk.*, **45**, 111–115.

———, 1936d. Die Lackfilmmethode, *Die Umschau*, 1–3.

———, 1936e. Die Lackfilmmethode, ihre Bedeutung und Anwendung in der Paläontologie, Sedimentpetrographie und Bodenkunde, *Z. deutsch. geol. Ges.*, **88**, 272–292.

———, 1937a. Weichteile an Fischen, Amphibien und Reptilien aus der eozänen Braunkohle des Geiseltales, *Nov. Act. Leopoldina*, N.F. 5, 115–142.

———, 1937b. Paläohistologische Untersuchungen an Bernsteineinschlüssen, *Paleont. Zeitschr.*, **19**, 35–46.

———, 1937c. Ein Fischskelett aus dem unteroligozänen Grünsand von Palmnicken im Samland, *Z. deutsch. geol. Ges.*, **89**, 72–76.

———, 1938a. Ein fossiler Saitenwurm (Gordius tenufibrosus n. sp.) aus der eozänen Braunkohle des Geiseltales, *Nov. Act. Leopoldina*, N.F. 5, 351–360.

———, 1938b. Die Konservierung geologischer und bodenkundlicher Profile mit Hilfe der Lackfilmmethode, *Veröff. Reichsst. Unterrichtsfilm Hochschulfilm.*, C. 236, 3–11.

———, 1938c. Weichteile an fossilen Insekten aus der eozänen Braunkohle des Geiseltales bei Halle a.d.S., *Nov. Act. Leopoldina*, N. F. 6, 3–38.

———, 1938d. Die mikroscopische Spurenuntersuchung im durchfallenden Licht mit Hilfe der Lackfilmmethode, *Kriminalistik, Monatshefte ges. kriminal. Wissenschaft und Praxis*, **12**, 265–269.

———, 1939. Fossil red blood corpuscles found in a Lizard from the Middle Eocene Lignite of the Geiseltal near Halle, *Research and Progress*, **5**, 53–56.

———, 1947. Der Block-Lackfilm, *Z. deuts. Geol. Ges.*, **99**, 124–131.

———, 1949a. Fortschritte der Lackfilmmethode, *Technik und Wissenschaft*, **1**, 22–23.

———, 1949b. Die Anwendung der Lackfilm-Methode bei der Bergung geologischer und Bodenkundlicher Profile, *Mitt. Geol. Staatsinstitut Hamburg*, **19**, 111–129.

Volk, G. W. and H. J. Harper, 1939. A revised method for the microscopic examination of natural structures and pore space in soils, *Soil Sci. USA*, **48**, 141–149.

Wacker, not dated a. *Vinnapas*, Wacker-Chemie GMBH, nr. 1767.639.C.K. (in English), 11 pp.

———, not dated b. *Vinnapas*, Wacker-Chemie GMBH München, nr. 1767.C.K. 6112 (in German), 11 pp.

———, not dated c. *Vinnapas*, Wacker-Chemie GMBH, nr. 1767.C.K. 628 (in French), 11 pp.

Walker, B., 1967. A diver-operated pneumatic core sampler. *Limnol. and Oceanogr.*, **12**, 144–146.

Wallace & Tiernan, 1961. *Lucidol cobalt naphthenate*, Lucidol division, Wallace & Tiernan, Inc., data sheet 40.10, 5 pp.

———, 1962a. *Luperco ABB*, Lucidol division, Wallace & Tiernan, Inc., data sheet 1.102, 3 pp.

———, 1962b. *Luperco ATC*, Lucidol division, Wallace & Tiernan, Inc., data sheet 1.107, 2 pp.

———, 1962c. *Lupersol DDM*, Lucidol division, Wallace & Tiernan, Inc., data sheet 4.301, 4 pp.

———, 1962d. *Lupersol Delta*, Lucidol division, Wallace & Tiernan, Inc., data sheet 4.302, 4 pp.

———, 1962e. *Lupersol Delta-X*, Lucidol division, Wallace & Tiernan, Inc., data sheet 4.303, 5 pp.

———, 1962f. *Luperco JDB-50-T*, Lucidol division, Wallace & Tiernan, Inc., data sheet 4.202, 3 pp.

———, not dated. *Organic peroxides, their safe handling and use*, Lucidol division, Wallace & Tiernan, Inc., data sheet 30.40, 15 pp.

Walton, J., 1928. A method of preparing sections of fossil plants contained in coal balls or in other types of petrifaction, *Nature,* **112,** 571.

Ward, not dated a. *Embedding in bio-plastic,* Ward's Natural Science Establishment, Inc.: Ward's curriculum aid number 3, 2nd Ed., 6 pp.

――――, not dated b. *Ward's popular bio-plastic kit H 39, BPM 22,* Ward's Natural Science Establishment, Inc., 1 p.

Warne, S. St.J., 1962. A quick field or laboratory staining scheme for the differentiation of the major carbonate minerals, *J. Sediment. Petrol.,* **33,** 29–38.

Warrikhoff, H. F. H., 1963. Dosimetry: energy-independent roentgen elements. In G. L. Clark (Ed.), *Encyclopedia of X-rays and gamma rays,* Reinhold, New York, pp. 290–291.

Watson, Manasty & Co., 1958. *Shadomaster, model VMP measuring projector,* Watson, Manasty and Co., Ltd., 2 pp.

Weast, R. C. and S. M. Selby, (Eds.-in-Chief), 1966. *Handbook of Chemistry and Physics, 47th ed.,* Chemical Rubber, Cleveland, Ohio, 1860 pp.

――――, S. M. Selby and C. D. Hodgman, (Eds.-in-Chief), 1965. *Handbook of Chemistry and Physics, 46th ed.,* Chemical Rubber, Cleveland, Ohio, 1715 pp.

Weatherhead, A. V., 1940. A new method for the preparation of thin sections of clay, *Min. Mag.,* **25,** 529–533.

――――, 1947. *Petrographic microtechnique,* A. Barron Ltd., London, 98 pp.

Weiss, M. P. and C. E. Norman, 1960. The American Upper Ordovician Standard. IV. Classification of the limestones of the type Cincinnatian, *J. Sediment. Petrol.,* **30,** 283–296.

Wells, C. B., 1962. Resin impregnation of soil samples, *Nature, G. B.,* **193**(4817), 804.

Werner, F., 1966. Herstellung von ungestörten Dünnschliffen aus wassergesättigten, pelitischen Lockersedimenten mittels Gefriertrocknung, *Meyniana,* **16,** 107–112.

Werner, J., 1962. Über die Herstellung fluoreszierender Bodenanschliffen, *Z. Pflanzenern., Düngung u. Bodenk.,* **99,** 144–150.

West, I. M., 1965. A new method of displaying microstructures in porous limestone, *J. Sediment. Petrol.,* **35,** 250–251.

Whitehouse, U. G. and R. S. McCarter, 1957. *Diagenetic modification of clay mineral types in artificial sea water,* Texas A&M Research Found., Project 24, Tech. Rpt., 57 pp.

Williams, R. C., 1953. A method of freeze drying for electron microscopy, *Exptl. Cell. Research,* **4,** 178–201.

Wirtz, not dated a. *Jean Wirtz, Spezialhaus für Laboratorien und Versuchsanstalten,* Jean Wirtz, different data sheets, 48 pp.

――――, not dated b. *Universal Diamond Separating Machine Models DIMA 10, 20 and 30,* Jean Wirtz, Special House for Laboratories and Research Institutes; Tech. leaflet, 1 p.

――――, not dated c. *Universal-Diamant-Trennmaschine Modellen DIMA 10, 20 und 30,* Jean Wirtz, Spezialhaus für Laboratorien und Versuchsanstalten; Tech. Mitteil., 1 p.

――――, not dated d. *Original Wirtz-Grinding and Polishing automatic device DGBM. Original Wirtz-Schleif- und Polierautomatik DBGM,* Jean Wirtz, Special House for Laboratories and Research Institutes; Tech. leaflet, 2 pp.

――――, not dated e. *Automatic Original Wirtz Vibratory Grinder and Polisher VIBROPOL,* Jean Wirtz, Special House for Laboratories and Research Institutes; Tech. leaflet, 1 p.

――――, not dated f. *Automatisches Original-Wirtz-Vibrations-Schleif- und Poliergerät VIBROPOL,* Jean Wirtz, Spezialhaus für Laboratorien und Versuchsanstalten; Tech. Mitteil., 1 p.

Wit, K. E., 1960. *Een apparaat voor het steken van ongeroerde monsters in diepe boorgaten,* Report Inst. Cultuurtechniek en Waterhuishouding, Nr. 10, 27 pp.

――――, 1961. *Technische gegevens betreffende steekapparatuur,* Nota no. 72, Inst. Cultuurtechniek en Waterhuishouding, 12 pp.

――――, 1962. An apparatus for coring undisturbed samples in deep boreholes, *Soil Sci.,* **94** (2), 65–70.

Wolf, K. H., A. J. Easton, and S. Warne, 1967. Techniques of examining and analyzing carbonate skeletons, minerals and rocks. In G. V. Chilingar, H. J. Bissell, and R. W. Fairbridge (Eds.), *Carbonate rocks, physical and chemical aspects. Developments in Sedimentology, 9B,* Elsevier, Amsterdam, pp. 253–341.

―――― and S. St. J. Warne, 1960. Remarks on the application of Friedman's staining methods, *J. Sediment. Petrol.,***30,** 496–497.

Wright, H. E., E. J. Cushing, and D. A. Livingstone, 1965. Coring devices for lake sediments, In B. Kummel and D. Raup (Eds.), *Handbook of paleontological techniques,* Freeman and Co., San Francisco, pp. 494–520.

Wyckoff, R. W. G., 1946. Frozen-dried preparations for the electron microscope, *Science,* **104,** 36–37.

Zenkevitch, N. L. and V. P. Petelin, 1956. Photography of the sea bottom, *Priroda,* **45,** 95–99 (in Russian).

Zenkovitch, V. P., 1962. Some new exploration results about sand shores development during the sea transgression, *De Ingenieur,* **17** (Bouw en Waterkunde 9), 113–121.

Züllig, H., 1956. Sedimente als Ausdruck des Zustandes eines Gewässers, *Schweizerische Zeitschrift für Hydrologie,* **XVIII** (1), 5–143.

AUTHOR INDEX

Includes names of institutions and manufacturers.

SUBJECT INDEX